Y0-BTM-397

Manufacturing Engineers' Manual

Manufacturing Engineers' Manual

American Machinist Reference Book Sheets

Edited by RUPERT LE GRAND

Formerly Senior Editor, *American Machinist*

McGRAW-HILL BOOK COMPANY

New York St. Louis San Francisco Düsseldorf Johannesburg Kuala Lumpur London Mexico Montreal New Delhi Panama Rio de Janeiro Singapore Sydney Toronto

MANUFACTURING ENGINEERS' MANUAL

 Library of Congress Catalog Card Number 73-167557

07-037066-4

1234567890 HDBP 754321

Contents

Preface

Compilation of this book is based on fortunate circumstances. For nearly five decades the *American Machinist* magazine has taken the leadership in publishing handbook-type information of practical value and *current* interest to its subscribers. Much of this effort is concentrated in a department known as Reference Book Sheets.

Although many readers take special care to tear out and file the RBS as they appear, many others would like to have them republished at intervals in a convenient book form. This was done for many years in the form of paperbacks, issued every two years, and usually handed only to new subscribers as a premium.

Since discontinuance of the paperbacks several years ago, a large reservoir of first-class untapped material has come into existence. By selecting only the best of these RBS, all of proved readership and value, it is now possible to give you a hard-cover book of greater breadth than the paperbacks, and one that is available to all, both subscribers to *American Machinist* and other persons.

The twenty-four chapters in this book are arranged in a manner similar to that used in the "American Machinist Handbook." Ten chapters cover metal-cutting subjects, three are on metal forming, two talk about welding processes, one discusses heat treatment. Other chapters are compiled on metal finishing, plating, jig and fixture details, fasteners, and mathematics.

In a compilation such as this, no attempt can be made to cover all subjects. However this material is distinctive in that much of it is original, truly useful, and has appeared nowhere else, except in *American Machinist*. In other instances, the data, tables, sketches, and how-to-do-it instructions are derived from metalworking companies, engineering societies, trade associations, and materials suppliers. The effort to obtain duplicates independently and the expense thereof would be astounding.

Rupert Le Grand

Chapter 1
Broaching

Broach geometry

Source: 'Machining Data Handbook', published by Metcut Research Associates, Inc, Cincinnati, Ohio

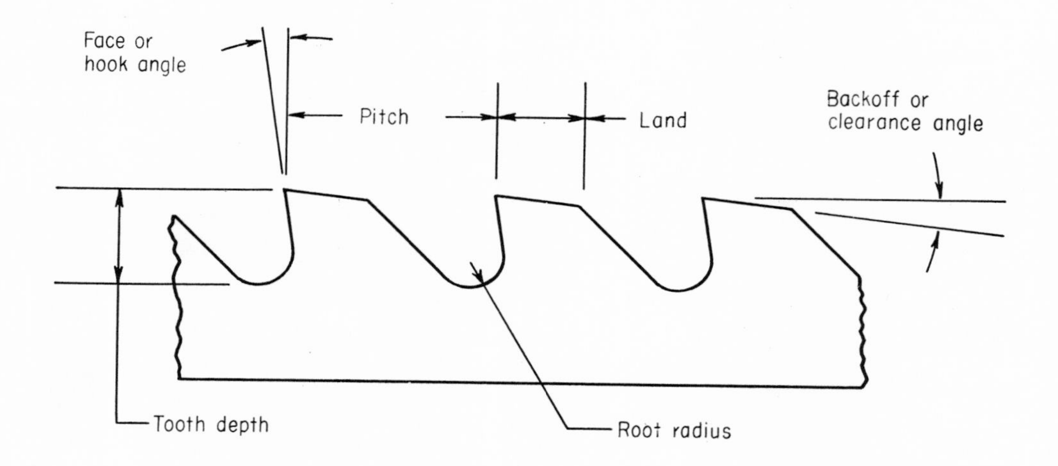

Material	Hardness BHN	Hook angle degrees	Clearance angle degrees
Free machining plain carbon steels	100-375	15 to 20	2 to 3
Plain carbon steels	85-375	15 to 20	1 to 2
Free machining alloy steels	150-375	15 to 20	2 to 3
Alloy steels	125-375	8 to 15	1 to 3
Nitriding steels	200-350	8 to 15	1 to 2
Armor plate	250-320	8 to 12	1 to 2
Ultra-high strength steels	175-375	8 to 15	1 to 3
Tool steels	100-375	8 to 12	1 to 2
Cast steels	120-350	8 to 15	1 to 3
Gray irons	110-320	6 to 8	2 to 3
Ductile irons	140-400	8 to 15	2 to 3
Malleable irons	110-280	8 to 15	2 to 3
Free machining stainless steels	135-425	8 to 12	1 to 2
Ferritic stainless steels	135-185	12 to 15	2 to 3
Austenitic stainless steels	135-275	12 to 15	½ to 2
Martensitic stainless steels	140-425	8 to 12	1 to 2
Precipitation hardening stainless steels	150-440	10 to 15	2
Titanium alloys	110-440	8 to 10	2 to 4
High temperature alloys	140-400	15 to 20	2 to 3
Nickel alloys	80-360	12 to 18	½ to 2
Aluminum alloys	30-150 500 kg	10 to 15	1 to 3
Magnesium alloys	40-90 500 kg	10 to 15	1 to 3
Copper alloys 314, 332, 340, 342, 353, 360, 370, 377, 385, 485, 544 & 356	$20R_B$-$100R_B$	-5 to 5	1 to 2
Copper alloys 226, 230, 240, 260, 268, 280, 335, 365-368, 443-445, 464-467, 651, 655, 675, 687, 770 & 796	$20R_B$-$100R_B$	0 to 10	1 to 2
Copper alloys 102, 110, 122, 170, 172, 175, 210, 220, 502, 510, 521, 524, 614, 706, 715, 745, 752, 754 & 757	$20R_B$-$100R_B$	10 to 15	2 to 3
Zinc alloys	80-100	10 to 15	2 to 5

Broaching of various materials

Source: 'Machining Data Handbook,' published by Metcut Research Associates, Inc., Cincinnati, Ohio

Material	Hardness BHN	Condition	High-speed tool steel		
			Speed fpm	Chip load in./tooth	Tool material
Free machining plain carbon steels Resulphurized B1111 1119 B1112 1212 B1113 1213 1113 1213+Te	100 to 175	Hot rolled or annealed	35	0.004	M2
	150 to 200	Cold drawn	30	0.004	M2
Resulphurized 1108 1120 1109 1126 1115 1144 1117 1211 1118	100 to 150	Hot rolled or annealed	30	0.004	M2
	150 to 200	Cold drawn	30	0.004	M2
Resulphurized 1132 1141 1137 1145 1138 1146 1139 1151 1140	175 to 225	Hot rolled, normalized, annealed or cold drawn	30	0.003	M2
	275 to 325	Quenched and tempered	20	0.003	M2
	325 to 375	Quenched and tempered	15	0.002	T5, T15
Leaded 10L18 12L13 10L20 12L14	100 to 150	Hot rolled, normalized, annealed or cold drawn	30	0.004	M2
	150 to 200	Hot rolled, normalized, annealed or cold drawn	30	0.004	M2
	200 to 250	Hot rolled, normalized, annealed or cold drawn	25	0.003	M2
Plain carbon steels 1006 1016 1022 1008 1017 1023 1009 1018 1024 1010 1019 1025 1012 1020 1026 1015 1021	85 to 125	Hot rolled, normalized, annealed or cold drawn	30	0.004	M2

Material	Hardness BHN	Condition	High-speed tool steel		
			Speed fpm	Chip load in./tooth	Tool material
Plain carbon steels (continued) 1006 1016 1022 1008 1017 1023 1009 1018 1024 1010 1019 1025 1012 1020 1026 1015 1021	125 to 175	Hot rolled, normalized, annealed or cold drawn	30	0.004	M2
	175 to 225	Hot rolled, normalized, annealed or cold drawn	20	0.004	M2
	225 to 275	Annealed or cold drawn	20	0.003	M2
1027 1052 1030 1055 1033 1060 1034 1062 1035 1064 1036 1065 1037 1066 1038 1070 1039 1074 1040 1078 1041 1080 1042 1084 1043 1085 1045 1086 1046 1090 1049 1095 1050	125 to 175	Hot rolled, normalized, annealed or cold drawn	25	0.003	M2
	175 to 225	Hot rolled, normalized, annealed or cold drawn	20	0.003	M2
	225 to 275	Hot rolled, normalized, annealed, cold drawn or quenched and tempered	20	0.003	M2
	275 to 325	Hot rolled, normalized, annealed or quenched and tempered	15	0.003	T5, T15
	325 to 375	Quenched and tempered	10	0.002	T5, T15

[Continued]

Broaching of various materials [Continued]

Material	Hardness BHN	Condition	High-speed tool steel Speed fpm	Chip load in./tooth	Tool material
Free machining alloy steels **Resulphurized** 3140 4150 4140 8640	150 to 200	Hot rolled, normalized, annealed or cold drawn	30	0.004	M2
	200 to 250	Hot rolled, normalized, annealed or cold drawn	25	0.003	M2
	275 to 325	Quenched and tempered	20	0.003	T5, T15
	325 to 375	Quenched and tempered	20	0.002	T5, T15
Leaded 41L30 51L32 41L40 86L20 41L47 86L40 41L50 52L100 43L47	150 to 200	Hot rolled, normalized, annealed or cold drawn	30	0.004	M2
	200 to 250	Hot rolled, normalized, annealed or cold drawn	25	0.003	M2
	275 to 325	Quenched and tempered	20	0.003	T5, T15
	325 to 375	Quenched and tempered	20	0.002	T5, T15
Alloy steels	125 to 175	Hot rolled, annealed or cold drawn	25	0.004	M2
1320 4128 6120 2317 4317 6317 2512 4320 6325 2515 4608 6415 2517 4615 8115 3115 4617 8615 3120 4620 8617 3125 4621 8620 3310 4720 8622 3316 4815 8625 4012 4817 8627 4017 4820 8720 4023 5015 8822 4024 5020 9310 4027 5024 9315 4028 5120 94B15 4118 6118 94B17 4125	175 to 225	Hot rolled, annealed or cold drawn	20	0.004	M2
	225 to 275	Hot rolled, normalized, annealed or cold drawn	20	0.003	M2
	275 to 325	Hot rolled, normalized, cold drawn or quenched and tempered	15	0.003	T5, T15
	325 to 375	Normalized or quenched and tempered	10	0.002	T5, T15

Material	Hardness BHN	Condition	High-speed tool steel Speed fpm	Chip load in./tooth	Tool material
Alloy steels (Continued) 1330 4337 6260 1332 4340 6270 1345 50B44 6382 2330 5046 6440 2335 50B46 6475 2340 50B50 81B45 2345 50B60 8630 3130 5075 8637 3135 5080 8640 3140 5130 8642 3141 5132 8645 3145 5135 86B45 3150 5140 8650 4030 5145 8655 4032 5147 8660 4037 5150 8740 4042 5155 8742 4047 5160 9255 4063 51B60 9260 4130 50100 9262 4135 51100 94B30 4137 52100 94B40 4140 6145 9445 4142 6150 9840 4145 6180 9845 4147 6240 9850 4150 6250	175 to 225	Hot rolled, normalized, annealed or cold drawn	20	0.004	M2
	225 to 275	Normalized, cold drawn or quenched and tempered	20	0.003	M2
	275 to 325	Normalized or quenched and tempered	15	0.003	T5, T15
	325 to 375	Normalized or quenched and tempered	10	0.002	T5, T15
Nitriding Steels **Nitralloy 125H** **Nitralloy 135G** **Nitralloy 135M** **Nitralloy N** **Nitralloy 230**	200 to 250	Annealed	20	0.003	M2
	300 to 350	Normalized or quenched and tempered	10	0.002	T5, T15
Armor plate MIL-A-1260 (ORD)	250 to 320	Quenched and tempered	10	0.002	T5, T15
Ultra-high Strength steels D6AC H11 MX-2 H13 4340	200 to 250	Annealed	20	0.003	M2
	250 to 300	Normalized	15	0.002	T5, T15
Maraging steels 18% Ni, Grade 200 18% Ni, Grade 250 18% Ni, Grade 300	275 to 325	Annealed	10	0.002	T5, T15
Maraging steels 25% Ni	175 to 225	Annealed	15	0.003	M2
Nickel alloy steels HP9-4-25 HP9-4-45	325 to 375	Annealed	10	0.002	T5, T15

Broaching malleable iron castings

Malleable iron can be broached as well or better than steels of the same hardness, but only if the setup takes into account the structure of the malleable matrix

By George Kramer,
Central Foundry Division
General Motors Corp, Saginaw, Mich

Because malleable irons are so easily cold worked, the very first roughing tooth making contact with the casting must be set to take a definite under-the-skin cut. The first five or six roughing teeth, therefore, should have a step of 0.008 to 0.010 in. The following roughing teeth can be set for a depth of 0.003 to 0.005 in. and the finishing teeth, at a step of 0.0015 to 0.002 in. minimum per tooth. Only the last two teeth will be set for the finished size of the surface to be broached.

Pearlitic and ferritic malleable iron can be machined more easily than steel of comparable hardness because small nodules of carbon present in the matrix lower the plasticity of these materials. These properties in turn will result in smaller chips, lower tool forces and greater tool life. Also speeds and chip per tooth can be increased.

Two important conditions

In setting up to broach the material—malleable is actually several ferrous materials in the hardness range of 116 to 269 Brinell—two metallurgical conditions must be considered, the condition of the skin and that of the subsurface. Other variables in the operation are the configuration of the machined area, amount of stock removal, and the ductility and hardness of the grade of malleable to be broached. As the table shows, depth of cut recommendations are the same for all four grades of malleable covered, but cutting speeds can be increased in inverse relation to surface hardness. There are other important tooling factors to consider, and experience by customers of Central Foundry Division of General Motors Corporation indicates that the following conditions give the best results in broaching malleable iron castings that are intended for automotive use:

1. Face angles:

Face angles, commonly called hook angles, will vary with the types of malleable to be cut. The face angle decreases with increase in material hardness. Face angles recommended for ferritic malleable are 10-15° and for pearlitic malleable 5-7° for the harder range and 8-10° for the softer range.

2. Depth of cut:

The general recommendations referred to previously hold for all situations involving malleable castings. It is important for the first roughing teeth to cut under the concentration of ferrite (decarburization) which occurs to a greater or lesser degree near the casting surface. If the first cut is not sufficient, glazing the surface will result.

3. Shear angle:

When surface broaching, an angle of 15° is recommended.

4. Chip breakers:

Staggered chip breakers on alternate teeth are recommended.

5. Power:

Horsepower must be great enough to overcome any variations (within specified tolerances) inherent in the material and castings to be broached.

6. Fixtures:

Fixtures must hold tight enough to prevent any movement of the casting.

7. Cutting teeth:

Roughing teeth do not usually have the same form as the finishing teeth. The finishing teeth must be the exact shape the tool is to generate. Roughing teeth are designed to remove surplus stock most economically and prepare the surface and size for the finishing teeth of the broaching tool.

Drive-line yokes are critical parts, because of the torsional loading

8. Coolant:

Use soluble oil in 1:10 ratio.

The good machinability of malleable iron castings, combined with the tooling recommendations listed here, can pay important dividends in reducing machining cycles. In one recent application of pearlitic malleable, internal gears were broached at 32 to 35 sfpm with a 0.002 to 0.003 in. chip per tooth. When the part was forged steel, chip per tooth was only 0.00075 to 0.0015 in. and broach speed was 17 sfpm. Another big factor in pearlitic malleable iron's favor was greater tool life—1600 pieces between tool resharpening vs. 1000 for the forged blank—a 60% savings in tooling costs.

Recommended speeds and feeds for broaching malleable iron

ASTM material	Brinell hardness	Rough cut per tooth	Finish cut per tooth min.	Sfpm
80002	241-269	0.003-0.005	0.0015-0.002	25-35
60003	197-241	0.003-0.005	0.0015-0.002	30-35
48004	173-207	0.003-0.005	0.0015-0.002	40
32510	156 max.	0.003-0.005	0.0015-0.002	40-45

How to specify broach tooling

A satisfactory proposal or quotation for broach tooling can be prepared only if the buyer supplies certain data. Here are 32 points about several kinds of broaches

Every order for a broach tool is treated as a special request in that each tool must be designed for a specific job, and every design is different. You were made acquainted with the number of broach types in the Special Report No. 638 (AM—Dec.15,'69, p85), but there are more kinds in use. And every day inquiries are made on tooling for jobs that have never been broached before. Because of this situation, it is wise to base a request for a quotation on an orderly sequence of points, covering both the general data required, and the specific information needed in connection with a specific type of broach. Certain information is required to make an intelligent proposal or quotation for broach tooling. Here are suggestions for several types:

General data

1. Supply a print of part, giving dimensions with tolerances. This is necessary to insure that the proper broaching machine has been selected for the job and the broach designed for maximum life.
2. Indicate prior machining operations on the part and the surfaces to be used for location, so that a determination can be made as to the proper sequence of operations.
3. Material specifications should be given. The machining characteristics of various materials makes it imperative that the best broach-tooth geometry be utilized.
4. Prior heat-treatment of part. Specify hardness so that the proper broach material can be selected to produce quality parts at low cost.
5. Give subsequent heat-treatment of the part after broaching, which may affect the accuracy of the broached surface; part dimensions after shrinkage; other factors.
6. State quality of surface finish required, in specific values if possible. While extremely high finishes are possible, depending on the material and hardness, it is usually more economical to accept surface finishes consistent with the functional requirements.
7. Give the production quantity required and production rate desired. Broach-tooling engineers require this information to assist them in the design of the fixture.
8. Make and model of broaching machine to be used should be listed, or maximum and minimum length of stroke; power capacity in pounds of pull or push; details of broach support, and special fixtures or guides that may affect the broach design.

Internal broaching

In addition to general requirements:

9. Shank type required, including length to the first tooth. Specify type and size of puller (key, automatic, or special). In case of special pullers, submit a print of shank. State whether or not rear support is used.
10. Dimensions of the hole before broaching, including the length to be broached. If direction of pull is not optional, specify the locating surfaces.
11. Dimensions of the hole after broaching, including limits.
12. Previous machining. State whether the hole is drilled, reamed, hot-or-cold-punched, forged or cored (give shape of the cored hole).
13. Recesses or counterbores. If either are required, state whether machined before or after broaching. If after, a rebroach may be necessary to remove the burrs.
14. Holes broached at one time. State whether one or more holes or parts are to be broached at once.

Spline broaching

In addition to general requirements:

15. Stock allowance for grinding, if the hole is to be ground afterward.
16. Radius or chamfer, if required on the corner of splines, ID or OD.
17. Is the broach to size the ID in addition to cutting the splines?
18. Whether close concentricity of splines is required in relation to the ID of the hole or external surfaces. If the latter, it may be necessary to machine those critical outside surfaces after the broaching operation.
19. Spline fit, whether controlled by ID, OD or sides of spline in part, and whether an exceptionally close fit is required.

Involute spline broaching

In addition to general requirements and spline requirements:

20. All gear elements for involute form broaches.
21. Lead or helix angle, if any.

Keyway broaching

In addition to general requirements:

22. The guide that is to be used with the broach, including dimensions.
23. Any important relationship of the keyway to any other surface dimensions.

Surface broaching

In addition to general requirements:

24. Specify surfaces to be broached, the maximum stock removal required per surface, and finish required.
25. Method of locating and holding the part to be broached.
26. Type of clamping desired, whether manual or power-operated.
27. Type of part-loading—manual or automatic.

Horizontal broaching machines

28. Size of machine platen.
29. Size of hole in machine platen and reducing bushings if any.

Vertical broaching machines

30. Specify whether machine has fixed or shuttle-type action; if shuttle type, specify the shuttle stroke.
31. Machine throat dimensions (distance from face of machine ram to edge of work table).
32. Prints of existing mainholder, if available.■

By Camille J. Morawski, manager
Broach Design & Mfg. Engineering
Detroit Broach & Machine Co.
Detroit, Mich

Chapter 2

Drilling

Spade drills

When holes are large, spade drills compete with twist drills, but their geometry must be correct and they must be used in rigid machines

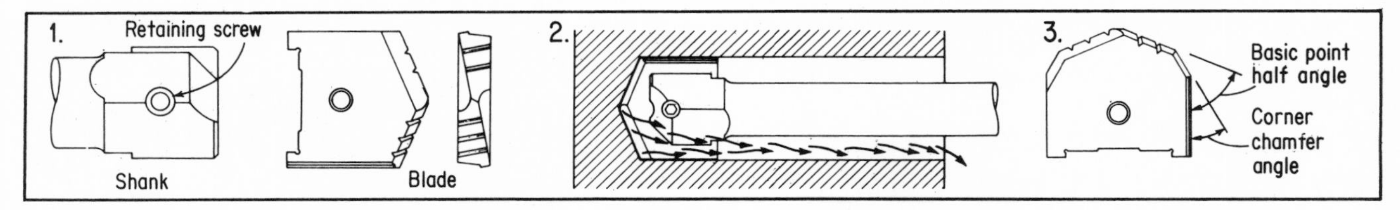

1. Spade-drill blade, holder (shank) and retaining screw

2. Restrictions in chip-escape paths require well-broken chips

3. Double-angle point lengthens the cutting edges near the OD and improves resistance to abrasive wear

Spade drills are generally available in a wide range of sizes above 1 inch. They are readily adaptable to many types of machines and most provide through-the-drill cutting fluid feed capabilities.

The basic spade drill assembly consists of at least three parts. These are cutting blade, blade holder or shank, and a retaining device—usually a screw. As is evident in Fig. 1, the pointed end of the blade resembles the cutting lips, clearance surfaces, and chisel edge of a twist-drill point.

If the flat spade-drill blade were merely sharpened on the end, the cutting action would not be satisfactory. The chisel edge would be too long, thus increasing the required thrust force, and the cutting lips would have negative rake, resulting in a high torque. These conditions are overcome by incorporating a combined rake face —web thinning grind on the leading sides of the blade.

The point geometry of a spade drill blade reflects its special operating condition. These are evident in Fig. 2. Arrows indicate how the chips must travel in order to escape from the hole. Chip-splitting grooves are staggered on the two lips so that any fin left by the groove is removed by the following cutting edge.

The rake surfaces on general-purpose spade drill blades are necessarily a compromise. Optimum cutting performance may require some modification of these surfaces to accommodate the work material and the operating conditions.

Commonly, the clearance surfaces on the blade end are ground flat to provide a clearance angle in the 6 to 10-deg range. If suitable equipment is available, spade-drill blade clearance surfaces can be produced on a twist-drill sharpening machine.

Such sharpening often results in more stable entry of the drill into the work-piece. Sometimes a secondary clearance surface is helpful.

Point angles are usually 118 to 135 deg. Most standard spade drill blades are furnished with a 135 deg point angle. Long points with small point angles are indicated for abrasive work materials because the outside corner wear is spread over a longer section of cutting edge. However, the length of the blade limits the amount of point-angle reduction permissible while maintaining adequate chip-clearance space and resharpening life. In such cases, a double-angle point may be used. (Fig. 3)

The actual web thickness used depends upon the application. For general-purpose usage on sizes up to 4¾ in. this formula is suggested:

$$W = 0.04D + 0.035$$

Where:

W = Center web thickness of blade, in.
D = blade diameter, in.

For sizes between 1¾ in. and 4¾ in. a web thickness of 5% of the blade diameter is a reasonably close approximation of the values given by the formula. Spade drills larger than about 4 in. commonly use blades of 11/16 in. thickness and above 4¾ in. diameter a web thickness of about 0.235 in. is suggested.

The grinding angle for the web-thinning cut should be adjusted to obtain both the desired web thickness and to produce a rake face which just extends to the outside diameter of the blade.

Excessive depth of this cut at the outside diameter must be avoided to allow for some regrinding while preserving some full-diameter margin where the rake cut runs out. The thinning cut is usually angled slightly to the cutting edge so that the rake surface width is reduced at the center.

An excessive rake-surface width at the center may cause an undercut behind the chisel edge, thus weakening the point. The web thickness behind the chisel edge should never be reduced by undercutting more than 5%—usually not more than a few thousandths of an inch. Sharpening a spade drill in the manner described requires that grinding angles be readjusted for each size. Estimates of torque and thrust requirements can be obtained from formulas developed for twist drills.

The torque, thrust, and power requirements of larger spade drills can be quite impressive. Consider a 4-in. spade drill operating at 60 sfm (57 rpm) at a feed of 0.020 ipr in mild steel. Assuming that there are no chip-ejection problems, the thrust force can be expected to be in the order of 8000 lb, the torque about 1200 in. lb, and the power at the cut about 11 hp. These high values point up the need for rigid machines with good tooling. ■

Source: "Metal Cuttings," Vol. 18, No. 1
Published by National Twist Drill & Tool Co.
Rochester, Michigan

Speeds and feeds for spade drills

To find the optimum speeds and feeds for spade drilling, you must consider the workpiece material, the diameter of the drill, and how deep a hole you are going to drill

The speeds and feeds given in the tables are conservative and will serve as starting points under most conditions. The tables are used in this way:

1. From Table 1 select the proper speed or speed range and the feed symbol or symbols for the material to be drilled.

2. By use of Table 2, convert the feed symbol to a feed rate. Note that the feed rate varies with drill size. [Continued]

Table 1. Suggested speeds and feeds for spade drilling

MATERIAL	SPEED SFM	FEED SYMBOL
Allegheny Metal	50-70	J
Aluminum	300-450	L-N
Z 011-T-3	300-450	N
Aluminum Bronze	60-70	J
Armor Plate*	10-12	E-G
Brass (with carbide)	200-300	C-E
Brass (HSS)	40-60	M
Admiralty or Naval	30	J
Bearing	90-100	L-M
Forgings	70-80	K-L
Free Machining	100-110	M-N
Leaded	200-300	N
Naval, leaded	75	L-M
Red	55-65	K-L
Red, leaded	130-180	M
Yellow	50-60	M
Bronze	80	M
Architectural	100	M
Arsenical	100	L
High Tensile	70-150	J
Leaded	95	L
Manganese	60-70	K-L
Phosphor	60-70	H-J
Phosphor, FM	100-110	M
Phosphor, leaded	80-90	M
Carbon	60-70	G
Cast Copper	50-60	F
Cast Iron		
Chilled*	30-50	F-G
Hard	45-50	G
Medium	70-110	J
Soft	120-150	M
Cast Steel	40-50	F-G
Cast Steel, .35%C	50-70	H
Copper	70	G
Castings	50-60	F
Leaded	85	J
Phosphorized	25-30	D-E
Rolled, 1/4H	60	F-G
Cupronickel	30-40	H
Duralumin	110-150	K-L
Everdur 1010, 1015	30	G
Everdur 1012	50-65	F
Everdur, leaded	100-125	J
Forgings, Steel	40-60	J
Gilding Metal	25-40	D-F
Gun Metal, cast	40-65	H
Inconel	60	G
Ingot Iron	45-55	H
Hastelloy A	50-70	H-K
Hastelloy B, C	45-60	H
Hastelloy D	45-50	G
Magnesium	250-450	J-L
Magnet Steel, Hard*	15	F

MATERIAL	SPEED SFM	FEED SYMBOL
Magnet Steel, Soft	35-40	G
Malleable Iron		
Ferritic	80-150	K
Pearlitic	70-80	F
Manganese Bronze	35-45	D-E
Manganese Copper, 30%	20-25	F
Manganese Steel (7-13%)*	15-20	E
Manganese Steel, OH	25	F
Molybdenum Steel		
Forgings	45-55	G
Monel	60-65	K
H, Cast	60-70	J
K	60-65	K
KR	60-70	L
R	65-75	M
S, Cast	65	J
Muntz Metal	50-60	G
Nickel Aluminum Bronze	35	E
Nickel		
Cast	60	H
Cold Drawn	50-60	H
D	60	H
Hot Rolled	30-40	F
Pure	60-75	E-G
Silver	30-35	G
Silver, leaded	50-60	G-H
Z	55	H
Nikrome	60	E
Nitralloy 135	50-60	F
Permalloy, 77% N	50	G
Silicon-Aluminum Bronze	80	H
Silicon Brass	40	F
Stainless Steels		
16-18	40-50	G
201	40-50	G
202	40-50	G
301	40-50	G
302	40-50	G
303	60-70	H
303Se	60-70	H
303Ma	75-100	H
304	40-50	F
304-L	40-50	H
305	40-50	H
309	35-45	F
309S	35-45	G
309SCb	35-45	F
310	35-45	G
314	35-45	F
316	35-45	G
316F	75-100	H
316FSe	75-100	H
316-L	45-55	H
317	35-45	F

MATERIAL	SPEED SFM	FEED SYMBOL
Stainless (cont.)		
318	35-45	F
321	40-50	G
347	40-50	F
348	40-50	G
403	40-50	E-F
405	40-50	E-F
410	40-50	F
414	40-50	G
416	80-100	H
416 Duplex	90-100	J
416F	95-120	K
420	40-50	G
420F	50-80	J
420FSe	50-60	G
430	40-50	F
430F	80-100	H
430FSe	80-100	H
431	45-55	G
440A	40-50	F
440B	40-50	F
440C	40-50	F
440F	50-60	G
440FSe	40-50	E-F
442	40-50	F
443	40-60	G
446	40-50	H
Steels		
1018	80	G
1020	75	G
1030	80	G
1035	80	G
1040	76	G
1045	72	G
1095 annealed	50	D
B1112	120	H
B1113	145	J
1117	115	H
1137	90	G
1141	90	G
1144	95	H
1212	120	H
1213	145	J
12L14	195	K
1330 annealed	75	F
1340 annealed	68	E
2317	72	F
2330 annealed	76	F
2340 annealed	68	E
3115	76	F
3120	80	F
3140 annealed	76	F
3145 annealed	76	F

(Continued)

Courtesy of the National Twist Drill & Tool Co., Rochester, Mich.

Speeds and feeds for spade drills [Continued]

Table 1. Suggested speeds and feeds for spade drilling (continued)

MATERIAL	SPEED SFM	FEED SYMBOL	MATERIAL	SPEED SFM	FEED SYMBOL	MATERIAL	SPEED SFM	FEED SYMBOL
Steels (cont.)			Steel (cont.)			Steel (cont.)		
3150 annealed	72	G	5130	65	E	8720	76	F
3135 annealed	85	G	5135 annealed	85	G	8740 annealed	80	F
4017	85	F	5140 annealed	80	G	8745 annealed	76	G
4023	85	F	5150 annealed	72	F	8750 annealed	72	F
4032	90	H	5613 annealed	60	E	E9310	62	E
4042 annealed	80	G	6120	65	F	9440 annealed	72	F
4130 annealed	85	G	6150 annealed	62	E	9445 annealed	60	E
4140 annealed	80	G	8615 annealed	85	F	E52100 annealed	47	C
4145 annealed	76	H	8617	76	F	High-Speed	35	E
4147 annealed	72	G	8620	76	G	Tool, LC, Lo Chrom	36	F
4320 annealed	72	G	8630 annealed	85	F	Tool, HC, Hi Chrom	32	D
4340 annealed	68	E	8635 annealed	85	F	Titanium Alloys*	10-25	J
4615	76	F	8640 annealed	80	G	Titanium Alloy Sheet*	50-60	L
4620	76	G	8642 annealed	80	G	Tri-ten, Man-ten, T-1	30-50	E
4640 annealed	76	F	8645 annealed	76	G	Zinc Alloys	200-250	J
5120	85	G	8650 annealed	72	F			

*Although standard blades with special clipped corners can be used, special premium HSS blades or carbide tipped blades are better for these materials.

3. Compute the starting speed and feed by applying hole-depth corrections from Table 3. The recommended coolant usage is also given in Table 3.

If drilling goes smoothly, gradually increase the feed approximately 50%. If the drill still functions smoothly, increase the speed to the high side of the range given in Table 1. Continue raising first the feed in 25% to 50% steps and then the speed in 10% steps, until optimum conditions and production are achieved. ■

Table 2. Feeds for spade drilling

Feed Symbol	Diameter Range (Inches): 1-1½	1½-3	3-5	5-8	8-13	13-15
	Feed (Inches per Revolution)					
C	.004	.006	.010	.015	.020	.025
D	.005	.007	.012	.018	.024	.030
E	.006	.008	.014	.022	.030	.040
F	.007	.010	.016	.025	.035	.050
G	.008	.012	.018	.030	.042	.057
H	.010	.014	.022	.035	.050	.065
J	.012	.016	.027	.042	.060	.080
K	.015	.020	.030	.048	.070	.095
L	.018	.026	.042	.065	.090	.120
M	.020	.028	.046	.070	.100	.140
N	.025	.035	.055	.085	.120	.160

Table 3.
Hole-depth adjustment factors and coolants

Speeds and feeds should be reduced when drilling holes over three diameters deep

Depth in Diameters	Reduce Speed	Reduce Feed	Coolant: Horizontal Hole: Flood	Thru Holder	Air Assist	Vertical Hole: Flood	Thru Holder	Air Assist
3	—5%	0	Yes	Yes		Yes	Yes +	Yes
4	—5%	0	Yes	Yes +	Yes	No	Yes +	Yes
5	—5%	—5%	No	Yes +	Need	No	Yes +	Need
6	—10%	—5%	No	Yes +	Need	No	Yes +	Must
8	—15%	—5%	No	Yes +	Need	No	Yes +	Vital
10	—20%	—10%	No	Yes +	Must	No	Yes +	Vital
12	—25%	—10%	No	Yes +	Must			
15	—30%	—20%	No	Yes +	Vital			
20	—30%	—20%	No	Yes +	Vital			
30	—40%	—25%	No	Yes +	Vital			

Point lengths for drills

There are times during setup of drilling units (such as on numerically controlled machines) when the point length of a drill is required in order to process the workpiece effectively. Cases also arise where the designer or the machinist require such information. However, the desired point length of the drill must be calculated, because tabular data are not available.

Tables 1, 2, and 3 give the required point-length information to the fourth decimal place for letter, number, and fractional-size drills ground with an included angle of 118°. Naturally, different point angles are required for drilling different materials. However, the following tables will be useful for the majority of cases.

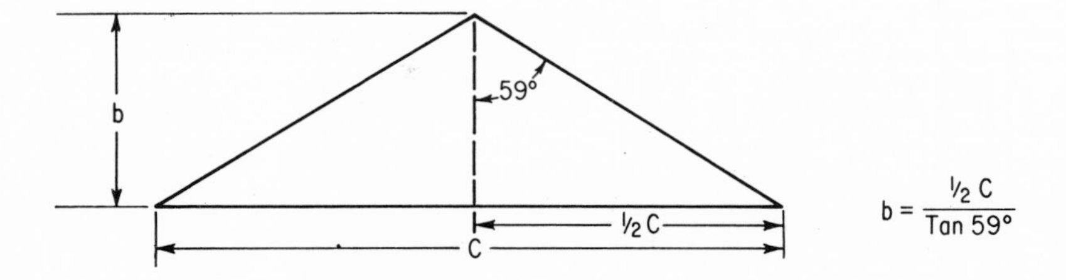

1. Point lengths for letter drills

Drill size (c)	Decimal equiv.	Point length (b)
A	0.2340	0.0703
B	0.2380	0.0715
C	0.2420	0.0727
D	0.2460	0.0739
E	0.2500	0.0751
F	0.2570	0.0772
G	0.2610	0.0784
H	0.2660	0.0799
I	0.2720	0.0817
J	0.2770	0.0832
K	0.2810	0.0844
L	0.2900	0.0871
M	0.2950	0.0887
N	0.3020	0.0908
O	0.3160	0.0950
P	0.3230	0.0971
Q	0.3320	0.0998
R	0.3390	0.1019
S	0.3480	0.1046
T	0.3580	0.1076
U	0.3680	0.1106
V	0.3770	0.1133
W	0.3860	0.1160
X	0.3970	0.1193
Y	0.4040	0.1214
Z	0.4130	0.1241

2. Point lengths for number drills

Drill size (c)	Decimal equiv.	Point length (b)	Drill size (c)	Decimal equiv.	Point length (b)
80	0.0135	0.0040	40	0.0980	0.0295
79	0.0145	0.0044	39	0.0995	0.0299
78	0.0160	0.0048	38	0.1015	0.0305
77	0.0180	0.0054	37	0.1040	0.0313
76	0.0200	0.0060	36	0.1065	0.0320
75	0.0210	0.0063	35	0.1100	0.0331
74	0.0225	0.0068	34	0.1110	0.0334
73	0.0240	0.0072	33	0.1130	0.0340
72	0.0250	0.0075	32	0.1160	0.0349
71	0.0260	0.0078	31	0.1200	0.0361
70	0.0280	0.0084	30	0.1285	0.0386
69	0.0292	0.0088	29	0.1360	0.0409
68	0.0310	0.0093	28	0.1405	0.0422
67	0.0320	0.0096	27	0.1440	0.0433
66	0.0330	0.0099	26	0.1470	0.0442
65	0.0350	0.0105	25	0.1495	0.0449
64	0.0360	0.0108	24	0.1520	0.0457
63	0.0370	0.0111	23	0.1540	0.0463
62	0.0380	0.0114	22	0.1570	0.0472
61	0.0390	0.0117	21	0.1590	0.0478
60	0.0400	0.0120	20	0.1610	0.0484
59	0.0410	0.0123	19	0.1660	0.0499
58	0.0420	0.0126	18	0.1695	0.0509
57	0.0430	0.0129	17	0.1730	0.0520
56	0.0465	0.0140	16	0.1770	0.0532
55	0.0520	0.0156	15	0.1800	0.0541
54	0.0550	0.0165	14	0.1820	0.0547
53	0.0595	0.0179	13	0.1850	0.0556
52	0.0635	0.0191	12	0.1890	0.0568
51	0.0670	0.0202	11	0.1910	0.0574
50	0.0700	0.0211	10	0.1935	0.0582
49	0.0730	0.0220	9	0.1960	0.0589
48	0.0760	0.0229	8	0.1990	0.0598
47	0.0785	0.0236	7	0.2010	0.0604
46	0.0810	0.0244	6	0.2040	0.0613
45	0.0820	0.0247	5	0.2055	0.0618
44	0.0860	0.0259	4	0.2090	0.0628
43	0.0890	0.0268	3	0.2130	0.0640
42	0.0935	0.0281	2	0.2210	0.0664
41	0.0960	0.0289	1	0.2280	0.0685

Point lengths for drills [Continued]

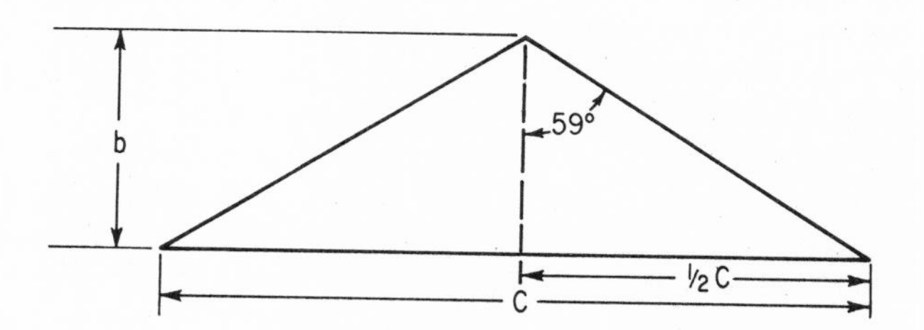

$$b = \frac{\frac{1}{2}C}{\tan 59°}$$

3. Point lengths for fractional-size drills

Drill size (c)	Decimal equiv.	Point length (b)
1/64	0.0156	0.0047
1/32	0.0312	0.0094
3/64	0.0469	0.0141
1/16	0.0625	0.0188
5/64	0.0781	0.0235
3/32	0.0938	0.0282
7/64	0.1094	0.0329
1/8	0.1250	0.0376
9/64	0.1406	0.0423
5/32	0.1562	0.0470
11/64	0.1719	0.0517
3/16	0.1875	0.0564
13/64	0.2031	0.0610
7/32	0.2188	0.0658
15/64	0.2344	0.0704
1/4	0.2500	0.0751
17/64	0.2656	0.0798
9/32	0.2812	0.0845
19/64	0.2969	0.0892
5/16	0.3125	0.0939
21/64	0.3281	0.0986
11/32	0.3438	0.1033
23/64	0.3594	0.1081
3/8	0.3750	0.1127
25/64	0.3906	0.1174
13/32	0.4062	0.1221
27/64	0.4219	0.1268
7/16	0.4375	0.1315
29/64	0.4531	0.1361
15/32	0.4688	0.1409
31/64	0.4844	0.1456
1/2	0.5000	0.1502
33/64	0.5156	0.1549
17/32	0.5312	0.1596
35/64	0.3286	0.1643
9/16	0.5625	0.1690
37/64	0.5781	0.1737
19/32	0.5938	0.1784
39/64	0.6094	0.1831
5/8	0.6250	0.1878
41/64	0.6406	0.1925
21/32	0.6562	0.1972
43/64	0.6719	0.2019
11/16	0.6875	0.2066
45/64	0.7031	0.2113
23/32	0.7188	0.2160
47/64	0.7344	0.2207
3/4	0.7500	0.2253
49/64	0.7656	0.2300
25/32	0.7812	0.2347
51/64	0.7969	0.2394
13/16	0.8125	0.2441
53/64	0.8281	0.2488
27/32	0.8438	0.2535
55/64	0.8594	0.2582
7/8	0.8750	0.2629
57/64	0.8906	0.2676
29/32	0.9062	0.2723
59/64	0.9219	0.2770
15/16	0.9375	0.2817
61/64	0.9531	0.2864
31/32	0.9688	0.2911
63/64	0.9844	0.2958
1	1.0000	0.3005

TAP-DRILL SELECTION..I

WARREN G OGDEN, JR, methods engineer, Instrumentation Laboratory
Department of Aeronautical Engineering, MASSACHUSETTS INSTITUTE OF TECHNOLOGY

Few people in engineering and mechanical fields realize how much depends on selecting the maximum allowable tap-drill size for preparing a hole for tapping. This is true, in particular, when tapping some of the materials currently used in miniature components of military goods. Pure beryllium, hafnium, the high-temperature alloys give trouble in tapping, when the percentage of thread engagement is a bare minimum, say 50 to 55%. Tapping difficulties also arise with stringy metals like copper, drawn aluminum, Monel metal, nickel silver, and soft stainless steel, which, because of their malleable nature, have a tendency to flow toward the crest of the minor diameter. Thus, the effective thread engagement is increased.

Most tap-drill selection tables have been worked up for a 75% engagement of thread. In use for many years, and based on experience with a free-machining material like brass, these tables have little relationship to the wide range of metals machined now, and the characteristics of those metals.

The man doing experimental work or small-lot production on pure beryllium parts, for example, won't get very far with a 75% depth of thread. He will break taps. Actually, he can get satisfactory strength in the fastening with a 50% thread depth in most instances, and particularly when the hole depth is more than 1½ times the diameter. For the vast majority of tapped-hole requirements, a minor diameter that provides a 55 to 65% thread height is considered adequate. The 55% minimum is safe, and more than 60% is not really needed. Why?

Thread strength tests clearly show that any increase in thread height over 60% for the tapped member does not increase the static strength of a threaded fastening. In general, the bolt will break at 55% of engagement.

On the other hand, the minor diameter of the internal thread has an important relationship to the torque required to drive the tap. As a general rule, torque is doubled when the percentage of thread height is increased from 60 to 72%, and tripled when thread height is increased to 80%. Thus, the chances of tap breakage are vastly increased without securing any benefit in respect to strength.

Drills Cut Oversize. Under most conditions, drills cut slightly larger than their own size. Some light on this subject, as far as drilling of cast iron and steel is concerned, was developed through extensive tests reported by the Metal Cutting Tool Institute in "Drilled Holes for Tapping—1954," and further reported in Reference Book Sheets under the title "Tolerances for Tap-Drilled Holes" (*AM*—Apr 26 '54, p143). In general, it may be said that the oversize amounts to a reduction of thread height by 6 or 7%. For example: in tap-drilling for a No. 4-40 screw, a No. 42 drill (0.0935 in.) would presumably be selected to obtain a thread engagement of 57%. Actually the mean probable oversize, according to these tests, would be 0.0020 in., reducing the engagement to 51%. For a No. 43 drill (0.0890) the theoretical engagement is 71% and the probable engagement is 65%.

How to Use the Thread-Engagement Tables

For modern conditions the tables of thread engagement for machine-screw sizes (see pp. 14-18) are based on engagements from 50% to 83 1/3% in increments of 5%. The columns for 83 1/3% engagement are included solely for reference when boring holes in a screw-cutting engine lathe in preparation for chasing close-tolerance threads with a single-point tool.

EXAMPLE: Again take the No. 4-40 screw. Assume you want a theoretical engagement of 70%. There are four figures in the block under 70% engagement and opposite 4-40. The top figure, 0.0893, is the theoretical hole size. The next two figures show the size of the nearest metric drill in millimeters (2.25 mm) and in parentheses the inch equivalent (0.0886). The bottom figure gives the number of the American Standard drill, in this case No. 43 (0.0890). Allowing for the drill cutting oversize, the probable thread height will be around 65% in this case.

In some cases the blocks contain only one number, the theoretical hole size, because there is no equivalent standard drill, either inch or metric. Obviously, another height of thread must be selected. In other cases, only an inch or a metric drill is available.

The great advantage of thread-engagement tables worked out in this fashion is the facility with which the shop man or engineer can select a tap drill that provides only the degree of engagement he needs. When he runs into trouble he can easily adjust to suit conditions without hunting through drill tables and making extensive calculations.

Tables on thread engagement for taps from ¼-20 to 1¼-12 (coarse and fine series) will be published in an early number. Likewise, a table of suggested percentages of thread for various materials will appear in the same number.

TAP-DRILL SELECTION . . II

Percentage of Thread for Machine-Screw Sizes

	83⅓%	75%	70%	65%	60%	55%	50%
0—80	.0465 (.0465) 1.18mm 56(.0465)	.0479 (.0477) 1.21mm 3/64(.0469)	.0486 (.0485) 1.23mm	.0494 (.0492) 1.25mm	.0502 (.0500) 1.27mm	.0510 (.0508) 1.29mm	.0519 (.0516) 1.31mm 55(.0520)
1—64	.0561 (.0564) 1.43mm 54(.0550)	.0578 (.0579) 1.47mm	.0588 (.0587) 1.49mm	.0599 (.0591) 1.50mm 53(.0595)	.0609 (.0611) 1.55mm	.0619	.0629 (.0630) 1.60mm 1/16(.0625)
1—72	.0580 (.0579) 1.47mm	.0595 (.0591) 1.50mm 53(.0595)	.0604	.0613 (.0611) 1.55mm	.0622	.0631 (.0630) 1.60mm 1/16(.0625)	.0640 52(.0635)
2—56	.0667 (.0670) 1.70mm	.0686 (.0689) 1.75mm 51(.0670)	.0698 50(.0700)	.0710 (.0709) 1.80mm	.0721 (.0729) 1.85mm	.0732 49(.0730)	.0744 (.0748) 1.90mm
2—64	.0691 (.0689) 1.75mm	.0708 (.0709) 1.80mm 50(.0700)	.0718	.0729 (.0729) 1.85mm	.0739 49(.0730)	.0749 (.0748) 1.90mm	.0759 48(.0760)
3—48	.0764 (.0768) 1.95mm 48(.0760)	.0788 (.0787) 2.00mm 47(.0785)	.0801 (.0807) 2.05mm	.0815 46(.0810)	.0828 (.0827) 2.10mm 45(.0820)	.0841 (.0846) 2.15mm	.0855 44(.0860)
3—56	.0797 (.0787) 2.00mm	.0816 (.0807) 2.05mm 46(.0810)	.0828 (.0827) 2.10mm 45(.0820)	.0840 (.0846) 2.15mm	.0851	.0862 44(.0860)	.0874 (.0866) 2.20mm
4—40	.0849 (.0846) 2.15mm 45(.0820)	.0877 (.0866) 2.20mm 44(.0860)	.0893 (.0886) 2.25mm 43(.0890)	.0909 (.0906) 2.30mm	.0926 (.0925) 2.35 mm	.0942 (.0945) 2.40mm 3/32(.0938)	.0958 (.0965) 2.45mm 41(.0960)
4—48	.0894 43(.0890)	.0918 (.0906) 2.30mm	.0931 (.0925) 2.35mm 42(.0935)	.0945 (.0945) 2.40mm 3/32(.0938)	.0958 41(.0960)	.0971 (.0965) 2.45mm	.0985 (.0984) 2.50mm 40(.0980)
5—40	.0979 40(.0980)	.1007 (.1004) 2.55mm 39(.0995)	.1023 (.1024) 2.60mm 38(.1015)	.1039 37(.1040)	.1056 (.1044) 2.65mm	.1072 (.1063) 2.70mm 36(.1065)	.1088 (.1083) 2.75mm 35(.1100)

TAP-DRILL SELECTION . . III

Percentage of Thread for Machine-Screw Sizes

	$83\frac{1}{3}\%$	75%	70%	65%	60%	55%	50%
5 – 44	.1004 (.1004) 2.55mm 39(.0995)	.1029 (.1024) 2.60mm 38(.1015)	.1044 (.1044) 2.65mm 37(.1040)	.1059 (.1063) 2.70mm	.1073 36(.1065)	.1087 (.1083) 2.75mm	.1102 (.1102) 2.80mm 35(.1100)
6 – 32	.1042 (.1044) 2.65mm 37(.1040)	.1076 36(.1065)	.1096 (.1082) 2.75mm 7/64(.1094)	.1117 (.1122) 2.85mm 34(.1110)	.1137 33(.1130)	.1157 (.1142) 2.90mm	.1177 32(.1160)
6 – 40	.1109 (.1102) 2.80mm 34(.1110)	.1137 (.1122) 2.85mm 33(.1130)	.1153 (.1142) 2.90mm	.1169 (.1160) 2.95mm 32(.1160)	.1185 (.1181) 3.00mm	.1201 31(.1200)	.1218 (.1220) 3.10mm
8 – 32	.1302 (.1299) 3.30mm 30(.1285)	.1336 (.1339) 3.40mm	.1356	.1377 (.1378) 3.50mm 29(.1360)	.1397	.1417 (.1417) 3.60mm 28(.1405)	.1437 27(.1440)
8 – 36	.1339 (.1339) 3.40mm	.1370 29(.1360)	.1388 (.1378) 3.50mm	.1406 28(.1405)	.1424 (.1417) 3.60mm	.1442 27(.1440)	.1460 (.1457) 3.70mm 26(.1470)
10 – 24	.1449 27(.1440)	.1495 (.1496) 3.80mm 25(.1495)	.1522 24(.1520)	.1549 23(.1540)	.1576 (.1575) 4.00mm 22(.1570)	.1602 21(.1590)	.1629 (.1614) 4.10mm 20(.1610)
10 – 32	.1562 5/32(.1562)	.1596 21(.1590)	.1616 (.1614) 4.10mm 20(.1610)	.1637	.1657 (.1654) 4.20mm	.1677 (.1673) 4.25mm 19(.1660)	.1697 18(.1695)
12 – 24	.1709 18(.1695)	.1755 17(.1730)	.1782 16(.1770)	.1809 (.1811) 4.60mm 15(.1800)	.1836 14(.1820)	.1862 (.1850) 4.70mm 13(.1850)	.1889 (.1890) 4.80mm 3/16(.1875)
12 – 28	.1773 (.1772) 4.50mm 16(.1770)	.1813 (.1811) 4.60mm 15(.1800)	.1836 14(.1820)	.1859 (.1850) 4.70mm 13(.1850)	.1882 (.1870) 4.75mm 3/16(.1875)	.1905 (.1890) 4.80mm 12(.1890)	.1928 (.1929) 4.90mm 11(.1910)

TAP-DRILL SELECTION . . IV

WARREN G OGDEN, JR, methods engineer, Instrumentation Laboratory
Department of Aeronautical Engineering, MASSACHUSETTS INSTITUTE OF TECHNOLOGY

Suggested Percentages of Thread for Various Materials

FERROUS MATERIALS

Steels:

Material	%
A 5150	60
B 1111	70
B 1112	70
B 1113	75
C 1010	60
C 1020	65
C 1025	65
C 1040	65
C 1095	60
E 4340	65
E 4640	65
E 52100	60
High chrome-high carbon	50
High speed	55
Manganese steel	55
Tool (low tungsten, chrome and carbon)	60
Cast	65

Stainless Steels:

Material	%
Type 302	50
303	60
309	50
310	50
316	50
345	50
410	55
416	60
430	55
446	55

Irons:

Material	%
Cast Iron (soft)	75
" " (medium)	70
Cast Iron (hard)	65
Malleable	75
Wrought	60
Ni-Resist	60
Stainless iron	65

SPECIAL MATERIALS:

Material	%
Fansteel	50
Hevimet alloy	50
Mallory 1000 metal	50

NON-FERROUS MATERIALS

Aluminum

Thread lengths should be increased from 10% to 40% as compared to length of engagement in steel

Material	%
Wrought Alloys	65
Cast alloys	75
Ampco Metal	55
Beryllium copper	60
Beryllium metal (pure)	50

The use of serial taps is recommended even with 50% thread height, to prevent this very brittle metal from crumbling on the crest of thread.

Material	%
Columbium	50
Copper and copper-zinc alloys (hard-tempered)	
Electrolytic tough pitch copper	55
Oxygen-free copper	55
Yellow brass	65
Muntz metal	65
Free-cutting yellow brass	75
Copper (rolled)	60
Copper (cast)	65
Copper-silicon alloys:	
Everdur No. 1010	55
" " 1012	65
" " 1014	65
" " 1015	55
Duronze I	50
" II	55
" III	65
Herculoy	55
Gold	65
Hafnium	55
Magnesium	75
Manganese bronze	55
Nickel-base alloys, wrought and cast:	
R monel	60
Monel (cast)	55
KR monel	55
Hastelloy	55
Illium R	55
Inconel	50
Molybdenum	55
Nickel (commercial, pure)	50
Nickel Silvers:	
10% to 20% nickel silver	50
10% extruded nickel silver	70
13% " " "	60
Silver	60
Tantalum	50
Thorium	55
Titanium (use CCl_4 or lithopone paste as a lubricant)	60
Uranium	50
Zirconium	60

Tables of tap drills for machine-screw sizes were published in earlier RBS

TAP-DRILL SELECTION . . V

Coarse- and Fine-Thread Series

	83⅓%	75%	70%	65%	60%	55%	50%
1/4-20	.1959 9(.1960)	.2012 7(.2010)	.2046 (.2047) 5.20 mm 6(.2040)	.2078 (2067) 5.25 mm 5(.2055)	.2111 4(.2090)	.2143 3(.2130)	.2175 (.2165) 5.50 mm
1/4-28	.2113 4(.2090)	.2153 3(.2130)	.2176 (.2165) 5.50 mm	.2199 7/32(.2188)	.2222 2(.2210)	.2245 (.2244) 5.70 mm	.2268 (.2264) 5.75 mm 1(.2280)
5/16-18	.2524 (.2520) 6.40 mm 1/4(.2500)	.2584 F(.2570)	.2620 G(.2610)	.2656 (.2657) 6.75 mm 17/64(.2656)	.2692 (.2677) 6.80 mm H(.2660)	.2728 (.2717) 6.90 mm I(.2720)	.2764 (.2756) 7.00 mm J(.2770)
5/16-24	.2674 (.2677) 6.80 mm H(.2660)	.2720 (.2717) 6.90 mm I(.2720)	.2747 (.2756) 7.00 mm	.2774 J(.2770)	.2801 (.2795) 7.10 mm	.2827 (.2835) 7.20 mm K(.2810)	.2854 (.2854) 7.25 mm
3/8-16	.3073 (.3071) 7.80 mm	.3142 (.3150) 8.00 mm 5/16(.3125)	.3182 (.3189) 8.10 mm O(.3160)	.3223 (.3228) 8.20 mm	.3263 (.3268) 8.30 mm P(.3230)	.3303 (.3307) 8.40 mm 21/64(.3281)	.3344 (.3346) 8.50 mm Q(.3320)
3/8-24	.3299 21/64(.3281)	.3345 (.3346) 8.50 mm Q(.3320)	.3372 (.3386) 8.60 mm	.3399 R(.3390)	.3426 (.3425) 8.70 mm	.3452 (.3445) 8.75 mm 11/32(.3438)	.3479 (.3465) 8.80 mm S(.3480)
7/16-14	.3602 23/64(.3594)	.3680 U(.3680)	.3726 (.3701) 9.40 mm	.3772 V(.3770)	.3819 (.3819) 9.70 mm	.3865 W(.3860)	.3911 25/64(.3906)
7/16-20	.3834 (.3819) 9.70 mm	.3888 W(.3860)	.3921 25/64(.3906)	.3953 (.3937) 10.00 mm	.3986 X(.3970)	.4018	.4050 Y(.4040)
1/2-13	.4167 (.4134) 10.50 mm Z(.4130)	.4251 27/64(.4219)	.4301	.4351 (.4331) 11.00 mm	.4401 7/16(.4375)	.4450	.4500 (.4528) 11.50 mm 29/64(.4531)
1/2-20	.4459	.4513 (.4528) 11.50 mm	.4546 29/64(.4531)	.4578	.4611	.4643	.4675 15/32(.4688)
9/16-12	.4723 (.4724) 12.00 mm	.4814	.4868 31/64(.4844)	.4922 (.4921) 12.50 mm	.4976	.5030 1/2(.5000)	.5084 (.5118) 13.00 mm
9/16-18	.5024 1/2(.5000)	.5084	.5120 (.5118) 13.00 mm	.5156 33/64(.5156)	.5192	.5228	.5264 17/32(.5312)

TAP-DRILL SELECTION . . VI

Coarse- and Fine-Thread Series

	83⅓%	75%	70%	65%	60%	55%	50%
5/8–11	.5266	.5365 (.5315) 13.50mm 17/32(.5312)	.5424	.5483 35/64(.5469)	.5542 (.5512) 14.00mm	.5601	.5660 9/16(.5625)
5/8–18	.5649 9/16(.5625)	.5709 (.5709) 14.50mm	.5745	.5787 37/64(.5781)	.5817	.5853	.5889 (.5906) 15.00mm
3/4–10	.6417 41/64(.6406)	.6526 (.6496) 16.50mm	.6591 21/32(.6562)	.6656	.6721 (.6693) 17.00mm 43/64(.6719)	.6785	.6850 (.6890) 17.50mm 11/16(.6875)
3/4–16	.6823	.6892 (.6890) 17.50mm 11/16(.6875)	.6932	.6973	.7013	.7053 45/64(.7031)	.7094 (.7087) 18.00mm
7/8–9	.7547 3/4(.7500)	.7668 49/64(.7656)	.7740 (.7677) 19.50mm	.7812 25/32(.7812)	.7884 (.7874) 20.00mm	.7956 51/64(.7969)	.8028 (.8071) 20.50mm
7/8–14	.7977 51/64(.7969)	.8055	.8101 (.8071) 20.50mm	.8147 13/16(.8125)	.8194	.8240	.8286 (.8268) 21.00mm 53/64(.8281)
1"–8	.8647 (.8661) 22.00mm 55/64(.8594)	.8783 7/8(.8750)	.8864 (.8858) 22.50mm	.8945 57/64(.8906)	.9026	.9107 (.9055) 23.00mm 29/32(.9062)	.9188 59/64(.9219)
1"–12	.9098 29/32(.9062)	.9188	.9242 (.9252) 23.50mm 59/64(.9219)	.9296	.9350	.9404 15/16(.9375)	.9459 (.9449) 24.00mm
1 1/8–7	.9704 31/32(.9688)	.9859 63/64(.9844)	.9951	1.0044 1.000	1.0137 1 1/64(1.0156)	1.0229 (1.0236) 26.00mm	1.0322 1 1/32(1.0312)
1 1/8–12	1.0348 1 1/32(1.0312)	1.0439 (1.0433) 26.50mm	1.0493 1 3/64(1.0469)	1.0547	1.0601	1.0655 (1.0630) 27.00mm 1 1/16(1.0625)	1.0709
1 1/4–7	1.0954 1 3/32(1.0938)	1.1109 1 7/64(1.1094)	1.1201 (1.1220) 28.50mm	1.1294 1 1/8(1.1250)	1.1387	1.1479 1 9/64(1.1406)	1.1572 1 5/32(1.1562)
1 1/4–12	1.1598 (1.1614) 29.50mm 1 5/32(1.1562)	1.1689	1.1743 1 11/64(1.1719)	1.1797	1.1851 (1.1811) 30.00mm	1.1905 1 3/16(1.1875)	1.1959 (1.2008) 30.50mm

Drilling malleable iron, I

Drilling speeds for malleable iron can be increased 100% if the as-cast surface is removed first. And use of a cutting fluid enables cutting speeds to be increased 25 to 50% for all four grades of malleable iron generally produced by the industry.

These conclusions are the main points in a study recently made by Metcut Research Associates, Inc., Cincinnati. This organization compiled feed and speed data for irons ranging in hardness from 116 to 285 Bhn. Table I summarizes their recommendations.

Malleable iron contains graphite nodules, or temper carbon, which aids machinability in two ways. It acts as a lubricant between the tool and the workpiece. In addition, it causes formation of a discontinuous chip.

All tests were conducted with type M-10 high-speed-steel jobbers' length drills in a Cincinnati box column drilling machine equipped with a variable

By Hans J Heine
Technical director
Malleabale Founders Society
Cleveland

Table I: Recommendations for drilling malleable irons

Material	Surface condition	Feed ipr	Cutting speed, fpm	Drill life No. of holes
Grade 32510 116 Bhn	Casting skin removed	0.005	300	510
		0.009	300	500
		0.015	250	500
	Through casting skin	0.005	150	515
		0.009	150	510
		0.015	150	600
Grade 48004 156 Bhn	Casting skin removed	0.005	250	430
		0.009	250	500
		0.015	200	375
	Through casting skin	0.005	75	520
		0.009	100	595
		0.015	125	330
Grade 60003 217 Bhn	Casting skin removed	0.005	150	490
		0.009	150	405
		0.015	125	325
	Through casting skin	0.005	100	550
		0.009	110	510
		0.015	100	450
Grade 80002 269 Bhn	Casting skin removed	0.005	125	360
		0.009	100	380
		0.015	75	325
	Through casting skin	0.005	100	270
		0.009	100	215
		0.015	75	300

Drill material: M-10 HSS
Drill diameter: 27/64 in.
Point angle: 118°
Clearance angle: 7°
Point style: plain chisel point
Depth of hole: 1 in. deep through hole
Cutting fluid: soluble oil (1:20)
Drill-life end point: 0.015 in. wearland

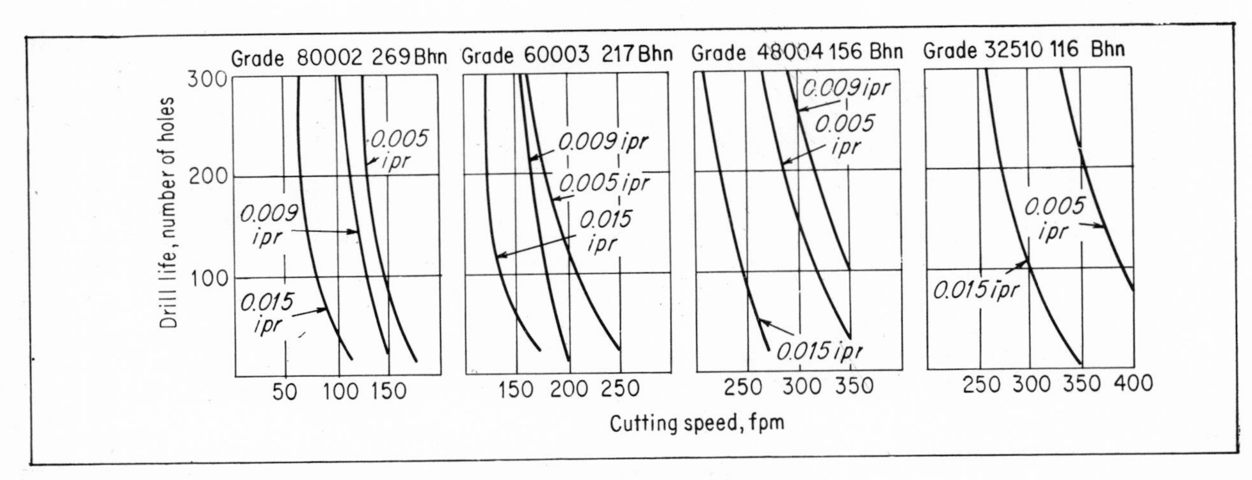

Drill life values were obtained for four grades of malleable iron

Drilling malleable iron, II

spindle speed and feed system. Tests were discontinued when a uniform wearland of 0.015 in. was observed.

Two types of plates, one machined and the other with an as-cast surface, were used for each grade of malleable iron in the test. Tests were carried out using a 1:20 soluble oil emulsion and dry.

These data show that for a drill life of approximately 500 holes, ferritic grades of malleable iron can be drilled at 300 fpm, Grade 48004 at 225 fpm, Grade 60003 at 150 fpm, and Grade 80002 at 100 fpm, when using feeds of .005 and 0.009 ipr respectively, and drilling through a machined surface. When the feed is increased to 0.015 ipr, cutting speeds must be reduced by about 25% for all four grades.

Workpiece surface conditions affected cutting speed considerably when drilling the ferritic grades. Without the use of a cutting fluid, the cutting speed can be increased 100% if the casting skin is removed, and 200% with cutting fluid. Higher feed rates, 0.010 to 0.015 ipr, provide the best drill life when drilling the lower strength grades.

When drilling pearlitic malleable irons, soluble oil permitted cutting speeds to be increased up to 50%.

Optimum cutting speeds for pearlitic malleable irons, at constant feed, decreases as the hardness increases; it being on the order of 225 fpm for Grade 48004, 150 fpm for Grade 60003, and 125 fpm for Grade 80002.

Tapping

Tap life of 500 holes was obtained with Grades 32510 and 48004 at 200 fpm while only 400 holes could be tapped in Grade 60003 at the same speed. Cutting speed had to be reduced to 60 fpm to obtain a tap life of 500 holes in the harder Grade 80002.

Table II: Tapping recommendations for malleable irons

	Hardness Bhn	Cutting speed, fpm	No. of holes tapped	Wearland on tap, in.
Grade 32510	116	200	500	0.008
Grade 48004	156	200	500	0.010
Grade 60003	217	200	400	0.020
Grade 80002	269	60	500	0.010
B-1112 screw stock	163	200	500	0.006
AISI 1018 C R steel	165	10	500	0.010*

*Highly chlorinated oil

Tap material: M-10 HSS
Tap size: ½-13 NC
Tap design: 4-flute plug

Tap-drill size: 27/64 in.
Percent thread: 75%
Cutting fluid: soluble oil

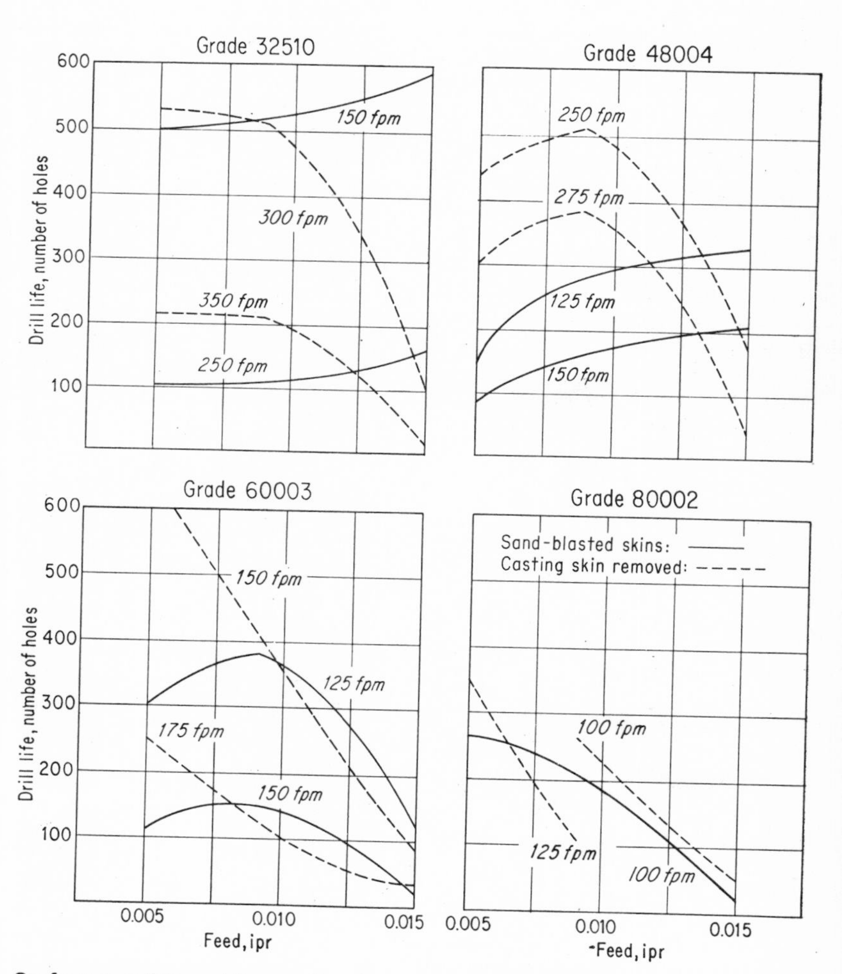

Surface condition of the casting has a profound effect upon drill life when the feed rate is varied

Drilling Superalloys—I

By Carl T Olofson, principal metallurgist, and Francis W Boulger, chief,
Metalworking Research Div, Battelle Memorial Institute, Columbus, Ohio

Drilling is the most difficult machining operation to perform on superalloys. In the first place, the thrust and torque forces are higher than those needed for drilling conventional materials. The center web of the drill does not cut but extrudes the metal in its vicinity. Consequently, the bottom of the hole can work harden sufficiently to cause early drill failure. Work hardening can be minimized by using drills with thinner webs. A constant, positive feed is necessary, and the drill must be kept sharp.

Drilling Setup

Successful drilling requires a true-running spindle with no end play. Parts must be properly supported at the point of thrust by fixturing.

Drills should be as short as possible. A drill bushing should be incorporated in the drilling setup for additional rigidity.

Proper alignment of the supported work and drill is also necessary to prevent premature drill breakage.

Drill Materials

High-speed steel drills are usually employed for drilling superalloys. The types used include M-2, M-33, and M-36. The cobalt grades generally perform better than the standard grades of high-speed steels.

Carbide drills can be useful on specific jobs. However, their high cost coupled with the high incidence of breakage usually prohibits their use. Only small amounts of lip and corner wear are permissible with carbide drills.

The high-strength superalloys often require spindle speeds lower than are available on conventional drill presses. Optimum speeds depend on the depth of the hole and on the feed rate. Speeds must be reduced for deep holes to compensate for difficulty of getting the cutting fluid to the cutting area. Speeds should be reduced if the feed is increased.

In every case, a constant, positive feed is essential. The drill must not dwell in the hole without cutting. Therefore, off-hand drilling is not recommended. The drill should be pulled out of the hole frequently to free it from chips, and to permit intermittent cooling of the drill.

Ordinarily, mineral oils with 2.5% active sulfur are superior to water-base cutting fluids. When drilling deep holes, a soluble-oil coolant or a chemically active, soluble-oil coolant may be put into the hole under pressure. The low heat conductivities of the superalloys necessitate the use of large quantities of coolant.

Drill Angles for Superalloys

Geometry	Relief Clearance	Cutting Edge Point Angle/2
DA	10-15	59-67.5
DB	10	67.5-70
DC	10	59
DD	10-12	59
DE	9	67.5
DF	12-15	
DG	6-12	65-70
DH(a)	10-12	59
DI(b)	10-12	67.5
DJ		67.5
DK		
DL		59

(a) For deep holes
(b) For sheet

Drilling Superalloys—II

Feeds and Speeds for Drilling Superalloys

Alloy	Size Drill, inch	Tool Material	Feed, ipr	Speed, fpm	Geometry	Cutting Fluids
René 41, aged		Co-HSS M-33-HSS Carbide tip	0.002-0.006 0.001 0.002-0.006	4-8 10 10-20	DA DJ DA	Sulfur-base oils or sulfo-chlorinated oils
Hastelloy X, Sol. treated		Co-HSS Co-HSS	0.003-0.007	14-21	DB	Sulfur-base oils
Waspaloy, Sol. treated	⅛-½ ⅛-½	Co-HSS Carbide tip	0.002-0.006 0.002-0.006	8-12 20-35	DA DA	
Inconel X Sol. treated " " "	1/16-¼ ¼-¾ ¾-2 ⅛-½	Co-HSS Co-HSS Carbide tip	0.0005-0.002 0.002-0.004 0.004-0.006 0.002-0.006 0.002-0.006	10-20 10-15 25-40	DG DA	Sulfurized-chlorinated, with or without kerosene
Aged	⅛-½	Co-HSS	0.002-0.006	7-12	DA	
J-1500, Sol. treated Nimonic 90 Udimet 500	⅛-½ 0.063-0.25 0.25-0.75 0.75-2	Carbide Co-HSS-M33 Co-HSS Co-HSS	0.002-0.006 0.0005-0.002 0.002-0.004 0.004-0.006	20-50 10-20	DA	
M-252	⅛-½	Co-HSS	0.002-0.006	7-12	DA	
Haynes Alloy No. 25 Sol. treated	¼	HSS-M-3 Carbide tip	0.005 0.001-0.003	30 12-20	DA DB	
J-1650, aged J-1570		T4-HSS M-35-HSS Carbide C-1	0.002-0.006 0.002-0.006	8-12 20-35	DA	
S-816, aged	⅛-½ ⅛-½	HSS Carbide C-1	0.002-0.006 0.002-0.006	10-15 25-40	DA DA	
Haynes Alloy No. 31, aged		Carbide tipped	0.001-0.003	12-18	DL	
Haynes Alloy No. 21, cast		Carbide tipped	0.001-0.003 0.0005	12-18 100	DL	
L605, annealed		HSS HSS-M33	0.003 0.002	4 30	DC DC	Sulfurized oil with or without kerosene (1:1)
N-155, aged S-590	0.75-2.0 0.25-0.75	HSS HSS	0.003-0.007 (rough) 0.002-0.004 (finish)	14-20	DB	Soluble oil or sulfur-base oils
J-1300 (no cobalt)	⅛-½ ⅛-½	HSS Carbide	0.002-0.006 0.002-0.006	8-12 20-35	DA	Soluble oil or sulfur-base oils

Drilling titanium alloys, 1

Excerpted from 'Machinability Data for Titanium Alloys,' published by the Air Force Machinability Data Center, Cincinnati, Ohio

Titanium alloys may be difficult to drill unless proper cutting conditions are used. As a drill dulls, the chips flow with increasing difficulty, the lip and margin become smeared with titanium, and the chips clog the flutes. Therefore, it is very important to inspect the drills frequently, particularly for smearing. Also, it is important to note the type of chip being produced. Feathered or discolored chips indicate dullness of the drill.

The drilling recommendations herewith are based on using a drill length no longer than required for the depth of hole to be drilled and still allow for the free flow of chips. Short-length drills with a crankshaft point will provide good results.

As indicated, M-1, M-2, or M-10 high-speed-steel drills are satisfactory for the lower hardness alloys.

Highly chlorinated oils are effective in drilling titanium. If it is necessary to use a water-base type, a chlorinated extreme-pressure type is recommended.

Recommended drilling conditions are given for commercially pure titanium, alpha and alpha-beta alloys, and beta alloys.

Nominal tool life to be expected for the recommended drilling conditions, using HSS drills, is 75 holes.

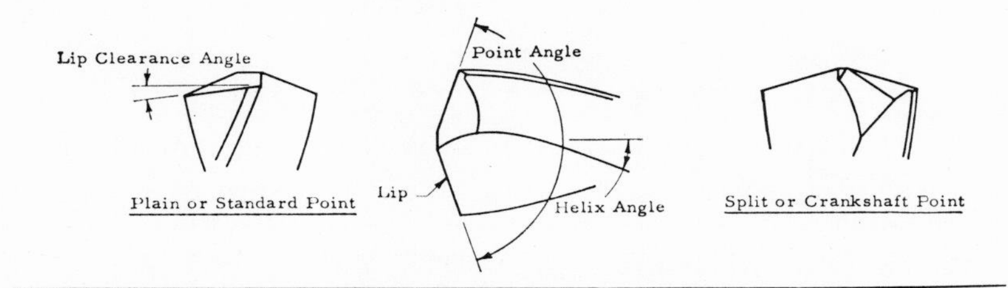

Plain or Standard Point

Split or Crankshaft Point

Drilling recommendations

Tool geometry—stub length drill, PA 118°, CL 7°, PG: crankshaft
Cutting fluid—highly chlorinated oil or sulpho-chlorinated oil

Material	Condition	BHN	Speed fpm	Feed, ipr — Hole diameter, in. ⅛	¼	½	¾	1	1½	2	3
Commercially pure titanium											
99.5	Annealed Y. S. - 20,000-30,000	110-170	100	0.0005	0.002	0.006	0.007	0.008	0.010	0.013	0.015
99.2 99.0 0.15 to .20 Pd	Annealed Y. S. - 45,000-60,000	140-200	80	0.0008	0.003	0.006	0.007	0.008	0.010	0.013	0.015
99.0 98.9	Annealed Y. S. - 60,000-85,000	200-275	50	0.002	0.005	0.006	0.007	0.008	0.010	0.013	0.015
Alpha and alpha-beta alloys											
2.5Al-16V	Solution treated Y. S. - 45,000-60,000	150,200	70	0.002	0.005	0.006	0.008	0.009	0.010	0.012	0.013
3Al-2.5V	Annealed Y. S. - 80,000-90,000	200-260	50	0.002	0.005	0.006	0.008	0.009	0.010	0.012	0.013

Drilling titanium alloys, 2

Drilling recommendations

Cutting fluid—highly chlorinated oil or sulpho-chlorinated oil
Tool geometry—stub length drill, PA 118°, CL 7°, PG: crankshaft

Material	Condition	BHN	Speed fpm	Feed, ipr Hole diameter, in. ⅛	¼	½	¾	1	1½	2	3
Alpha and alpha-beta alloys (continued)											
2Fe-2Cr-2Mo 5Al-2.5Sn 5Al-2.5Sn (low O) 7Al-2Cb-1Ta 4Al-3Mo-1V 5Al-5Sn-5Zr	Annealed Y. S. - 115,000- 125,000	300-340	40	0.002	0.005	0.006	0.007	0.008	0.010	0.011	0.012
7Al-12Zr 6Al-4V 4Al-4Mn 8Mn	Annealed Y. S. - 125,000- 135,000	310-350	30	0.002	0.005	0.006	0.007	0.008	0.009	0.010	0.011
7Al-4Mo 8Al-1Mo-1V 5Al-1.25Fe-2.75Cr 5Al-1.5Fe-1.4Cr-1.2Mo 6Al-6V-2Sn-1(Fe, Cu)	Annealed Y. S. - 140,000- 150,000	320-370	20	0.002	0.005	0.006	0.007	0.008	0.009	0.010	0.011
1Al-8V-5Fe	Annealed Y. S. - 160,000- 180,000	320-380	15	0.002	0.004	0.005	0.006	0.007	0.008	0.009	0.010
6Al-4V 4Al-4Mn 2.5Al-16V	Solution treated & aged Y. S. - 143,000- 165,000	350-400	25	0.001	0.002	0.004	0.005	0.006	0.007	0.008	0.008
2Fe-2Cr-2Mo 5Al-1.25Fe-2.75Cr 6Al-6V-2Sn-1(Fe, Cu) 7Al-4Mo 5A1-1.5Fe-1.4Cr-1.2Mo 4Al-3Mo-1V	Solution treated & aged Y. S. - 165,000- 185,000	375-420	20	0.001	0.002	0.003	0.004	0.004	0.005	0.005	0.005
1Al-8V-5Fe	Solution treated & aged Y. S. - 175,000- 245,000	375-440	15	0.0005	0.001	0.0015	0.0015	0.002	0.002	0.003	0.004
Beta alloys											
3Al-13V-11Cr	Solution treated Y. S. - 135,000- 145,000	310-350	20	0.001	0.003	0.004	0.005	0.006	0.007	0.008	0.009
3Al-13V-11Cr	Solution treated & aged Y. S. 175,000- 245,000	375-440	15	0.0005	0.001	0.0015	0.0015	0.002	0.002	0.003	0.004

Chapter 3

Gears and Splines

Extended-Center Helical Gears—I

CHARLES TIPLITZ, chief engineer CHALON, CONSULTING ENGINEERS, CEDAR GROVE, N J

Methods for calculation of pin dimensions for standard spur and helical gears have been presented in many books and magazines. However, in some applications, extended-center helical gears are in use, such as in roll mills for paint and ink, and in feeding devices for round or flat stock not of constant diameter or thickness. Inspection of these gears with pins is complicated by computations. Accordingly, a systematical method based on the traditional formula* and a step-by-step form (page 27) was developed.

Equations 1 and 2 relate cutting data to backlash of a gear set. Equations 3 and 4 convert normal angles to angles in the plane of rotation. The other equations determine the measurements across pins. Equation 2 can be simplified, first by recognizing that for standard hobs, tooth thickness at the pitch radius equals half the circular pitch, i.e., $2t = p$. Then since $p \cos \phi = p_b$, equation (2) reduces to equation (9).

The first term on the right-hand side can be reduced further by substituting $\cos \phi \tan \phi$ for $\sin \phi$:

$$\frac{p_b\,(N_a + N_b)}{2\,\pi \cos \phi}\left[\frac{\text{inv}\,\phi_r - \text{inv}\,\phi}{\tan \phi}\right]$$

Now $p_b \div \cos \phi = C_p$ (circular pitch), and $C_p\,(N_a + N_b)$ = total circumference of both gears, for standard pitch gears. Total circumference $\div\, 2\pi$ = total radii or center distance for standard pitch gears at tight mesh (zero clearance) and let this equal C_1. Then long addendum is given by equation (10).

The basic formulas are:

(1) $t_c = t + 2\,\Delta \tan \phi$

(2) $\Delta_a + \Delta_b = \dfrac{p_b - 2t \cos \phi - b \cos \phi_r}{2 \sin \phi} + \dfrac{p_b\,(N_a + N_b)\,(\text{inv}\,\phi_r - \text{inv}\,\phi)}{2\,\pi \sin \phi}$

(3) $\sin \psi_B = \sin \psi \cos \phi_n$

(4) $\tan \phi \cos \psi = \tan \phi_n$

(5) $\text{inv}\,\phi_p = \dfrac{t_c}{2r} - \dfrac{\pi}{N} + \text{inv}\,\phi + \dfrac{W}{r_B \cos \psi_B}$

(6) $r_p = \dfrac{r \cos \phi}{\cos \phi_p}$

(7) $M_{\text{even}} = 2(r_p + W)$

(8) $M_{\text{odd}} = 2(r_p \cos \dfrac{90^\circ}{N} + W)$

(9) $\Delta_a + \Delta_b = \dfrac{p_b\,(N_a + N_b)\,(\text{inv}\,\phi_r - \text{inv}\,\phi)}{2\pi \sin \phi} - \dfrac{b \cos \phi_r}{2 \sin \phi}.$

(10) $\Delta_a + \Delta_b = C_1 \left[\dfrac{\text{inv}\,\phi_r - \text{inv}\,\phi}{\tan \phi}\right] - \dfrac{b \cos \phi_r}{2 \sin \phi}.$

Where:

- t_c = actual tooth thickness at standard pitch
- t = standard thickness at standard pitch
- Δ = amount cutter is moved out from its standard position
- ϕ = pressure angle in plane of rotation at standard pitch circle
- Δ_a = amount cutter is moved out from standard position to cut gear a
- p_b = circular pitch at base circle
- b = backlash at meshing pitch circle
- ϕ_r = pressure angle in plane of rotation at meshing pitch
- ψ_B = helix angle at base circle
- ψ = helix angle of hob setting
- ϕ_n = pressure angle of hob
- ϕ_p = pressure angle of measuring roll or pin
- r = radius of standard pitch circle
- W = radius of measuring pin
- r_B = radius of base circle
- r_p = distance from gear center to measuring pin center
- M = measurement across outside of rolls or pins

The listing of formulas may not be adequate because 2 hr may be required for a single solution. Time-study has showed that substantial time was lost referring to the equations and determining successive steps. To get around this, a list of necessary steps was prepared. The list is self contained, except for mathematical tables, and thus eliminates the time each person need spend to set up his own procedure. Checking is simplified, likewise adaptability to digital computer programming.

Use of the form will reduce computing time to 30 min. or less. The right-hand column shows the number of decimal places required in each step to obtain 3-place accuracy for gears between 3 and 30 in. Other requirements can be scaled up or down.

---

*Manual of Gear Design, Sections I-III. Industrial Press, 1935-7, and Analytic Mechanics of Gears, McGraw-Hill, 1949—both by Earle Buckingham; and Involute Gears, by W. Steeds, Longmans Green & Co, 1948.

Extended-Center Helical Gears—II

Tabulation of Cutting and Pin Calculations

STEP	Description or Source	Gear *a*	Gear *b*	Dec. Places
1	N, Number of teeth	----	----	—
2	D_p, Diametral pitch			5
3	ϕ_n, Hob pressure angle (normal)			3
4	ψ, Helix angle, hob			3
5	cos Step 4 (std table)			5
6	tan Step 3 (std table)			5
7	Step 6 ÷ Step 5 = tan ϕ (equat. 4)			5
8	ϕ, pressure angle at standard pitch in rotation plane (std table)			3
9	cos Step 8 (cos ϕ)			5
10	Step 2 × Step 5 (D_p cos ψ)			5
11	Step 1 ÷ Step 10 = standard diameter = D	----	----	4
12	Step 11 ÷ 2 = standard pitch radius = R	----	----	4
13	Step 12, Col. *a* + Step 12, Col. *b* = standard gear center = C_1			4
14	Step 11 × Step 9 = base circle diameter = D_B	----	----	4
15	Step 14 ÷ 2 = base circle radius = R_B = R cos ϕ	----	----	4
16	Step 15, Col. *a* + Step 15, Col. *b* = total base circle radii			4
17	Required centers = C_2			4
18	Step 16 ÷ Step 17 = cos ϕ_r			5
19	ϕ_r, actual or mesh angle (std table)			3
20	involute Step 19, inv ϕ_r (std table)			6
21	inv Step 8, inv ϕ (std table)			6
22	Step 20 — Step 21			6
23	Step 22 ÷ Step 7			5
24	Step 23 × Step 13 = $C_1\left[\dfrac{\text{inv }\phi_r - \text{inv }\phi}{\tan\phi}\right]$			4
25	sin Step 8, sin ϕ (std table)			5
26	Step 25 × 2 = 2 sin ϕ			5
27	*b*, backlash at meshing pitch circle			4
28	Step 18 × Step 27 = *b* cos ϕ_r			4
29	Step 28 ÷ Step 26 = $\dfrac{b\cos\phi_r}{2\sin\phi}$			4
30	Step 24 — Step 29 = $\Delta_a + \Delta_b$ (equat. 10)	----	----	4
31	Δ_a & Δ_b, amount of long addendum			4
32	1 ÷ Step 2, standard addendum	----	----	4
33	Step 31 + Step 32, actual addendum	----	----	4
34	Step 33 × 2	----	----	4
35	Step 34 + Step 11, actual OD			3
36	π ÷ 2			4
37	Step 36 ÷ Step 10 = *t*, std tooth thickness at *R*			4
38	2 × Step 7 = 2 tan ϕ	----	----	5
39	Step 38 × Step 31 = 2 Δ_i tan ϕ, extra tooth thick.	----	----	4
40	Step 39 + Step 37 = t_c, actual tooth thick. (equat. 1)	----	----	4
41	Step 40 ÷ Step 11 = t_c/2R	----	----	6
42	π ÷ Step 1 = π/N			6
43	cos Step 3 (std table)			5
44	sin Step 4 (std table)			5
45	Step 43 × Step 44 = sin ψ_B (equat. 3)			5
46	ψ_B, helix angle at base, R_B (std table)			2
47	cos Step 46 (std table)			5
48	Step 14 × Step 47 = D_B cos ψ_B	----	----	5
49	1.728 ÷ Step 2 = 2W = Pin or roll diameter			4
50	Step 49 ÷ Step 48 = $\dfrac{W}{R_B\cos\psi_B}$	----	----	6
51	Step 41 + Step 21 + Step 50 — Step 42 = inv ϕ_p (equat. 5)	----	----	6
52	ϕ_p, pressure angle of pin (std table)	----	----	2
53	cos Step 52, cos ϕ_p (std table)	----	----	5
54	Step 14 ÷ Step 53 = $2r_p$ (equat. 6)	----	----	4
55	If Step 1 is even write Step 54 in Step 58 and skip to Step 59	----	----	—
56	If Step 1 is odd: 90° ÷ Step 1	----	----	2
57	cos Step 56 (std table)	----	----	5
58	Step 54 × Step 57 (If Step 1 is odd = 2 r_p cos $\dfrac{90°}{N}$	----	----	4
59	Step 58 + Step 49 = M, across roll or pins (equat. 7 or 8)	----	----	3
60	2.157 ÷ Step 2, standard cutter depth	----	----	3

NOTES:

1. Subscript *a* or *b* refers to value in column for step number for gear *a* or gear *b*, respectively.

2. Data in steps 1 to 4, 17, 27 and 31*a* or 31*b* must be supplied for each problem as base or design data.

3. (std table) refers to values obtained from standard tables for trigonometric or involute functions.

4. Dash lines in columns for Gear *a* and Gear *b* indicate that data must be entered for both gears. Where no dash lines are given, a single entry will suffice.

Form Cutters for Helical Gears...I

By OLIVER SAARI, Development Engineer, ILLINOIS TOOL WORKS

Most handbooks and other sources still give the formula for the selection of the proper spur tooth cutter for milling involute helical gear teeth as being

$$N_c = \frac{N}{\cos^3 \psi} \qquad (1)$$

where

N = actual number of teeth in gear to be cut

ψ = helix angle of gear to be cut

N_c = number of spur gear teeth for which cutter has been made.

No spur-tooth form cutter can give an exact involute shape when used to mill helical teeth—in fact, even for spur gears the form is correct for only one number of teeth in the range specified for a given cutter. But Formula (1) gives a needlessly poor approximation, particularly with small numbers of teeth and helix angles above 20°. Its continued presentation to the gear-cutting public is strange in view of the fact that thirty years ago Ernest Wildhaber published a much more exact and almost as simple formula for finding the cutter number (*AM*—Dec. 20, 1923). This formula is

$$N_c = \frac{N}{\cos^3 \psi} + (P_n D_c) \tan^2 \psi \qquad (2)$$

where the symbols are the same as before, and in addition

P_n = normal diametral pitch of gear and cutter

D_c = pitch diameter of cutter = OD of cutter − 2 × dedendum of gear

This formula gives values which are theoretically exact in both tooth-thickness and pressure angle at one point of the tooth profile, and the spur cutter itself provides a crowning or relieving effect above and below this point.

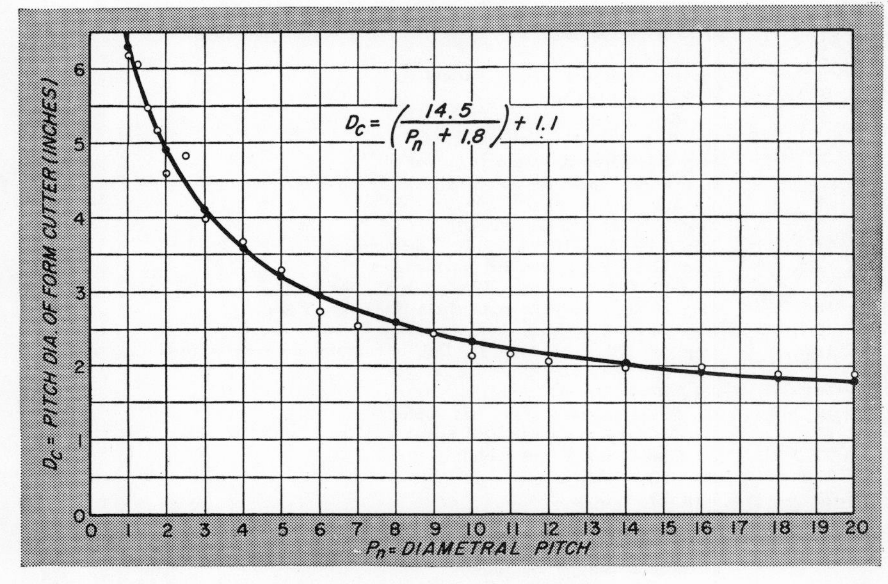

FIG. 1... Empirical relationship of gear-cutter pitch diameter to diametral pitch (Standard ITW cutters)

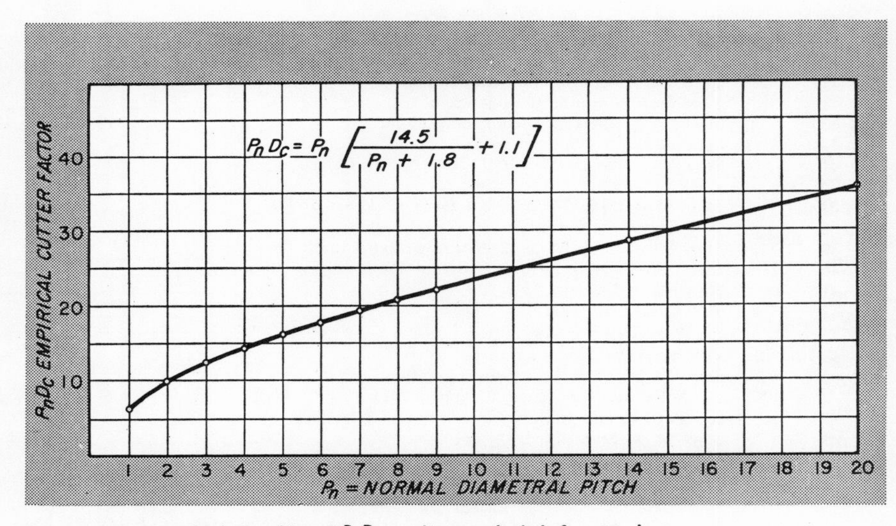

Fig. 2... Empirical relationship of $P_n D_c$ to diametral pitch for standard ITW cutters

Possibly one reason that this simple formula has not been included in the readily available

Form Cutters for Helical Gears...II

gear literature is the fact that it involves the cutter pitch diameter. This varies with the diametral pitch and with the various manufacturers. It is not generally known to the tool designer or anyone except the man in the shop who actually gets the cutter from the tool crib. Then it is too late to do anything but "cut and try."

This difficulty can be eliminated by establishing an empirical relationship between the cutter diameter and diametral pitch. Spur cutter sizes can be directly related to diametral pitches, even though such a relation may not be intentional on the part of the manufacturer, because the proportions of the tooth form and the size of hole required almost automatically dictate the "natural" size of the cutters. A curve of this relationship, together with an empirical formula, is shown in Fig. 1. Even though the individual cutters may deviate slightly from this curve, the errors caused by the deviation are far smaller than those resulting if the effects of cutter diameter are ignored altogether.

The combined relationship of the product P_nD_c, which appears in the exact equation (2), to the diametral pitch is shown in the curve and empirical formula of Fig 2. This curve and Equation (2) are sufficient for all calculations in the range of pitches from one to 20. A combined formula which may be used alternatively if Curve (2) is not available is

$$N_c = \frac{N}{\cos^3 \psi} + P_n \left(\frac{14.5}{P_n + 1.8} + 1.1 \right) \tan^2 \psi \quad (3)$$

A comparison of the values obtained by the use of Equations (1) and (3) for one particular case—10-tooth, 6-pitch gears of various helix angles — is shown by the curves of Fig 3. These curves show that the exact formula gives a better choice of cutter number for all helix angles above 17°.

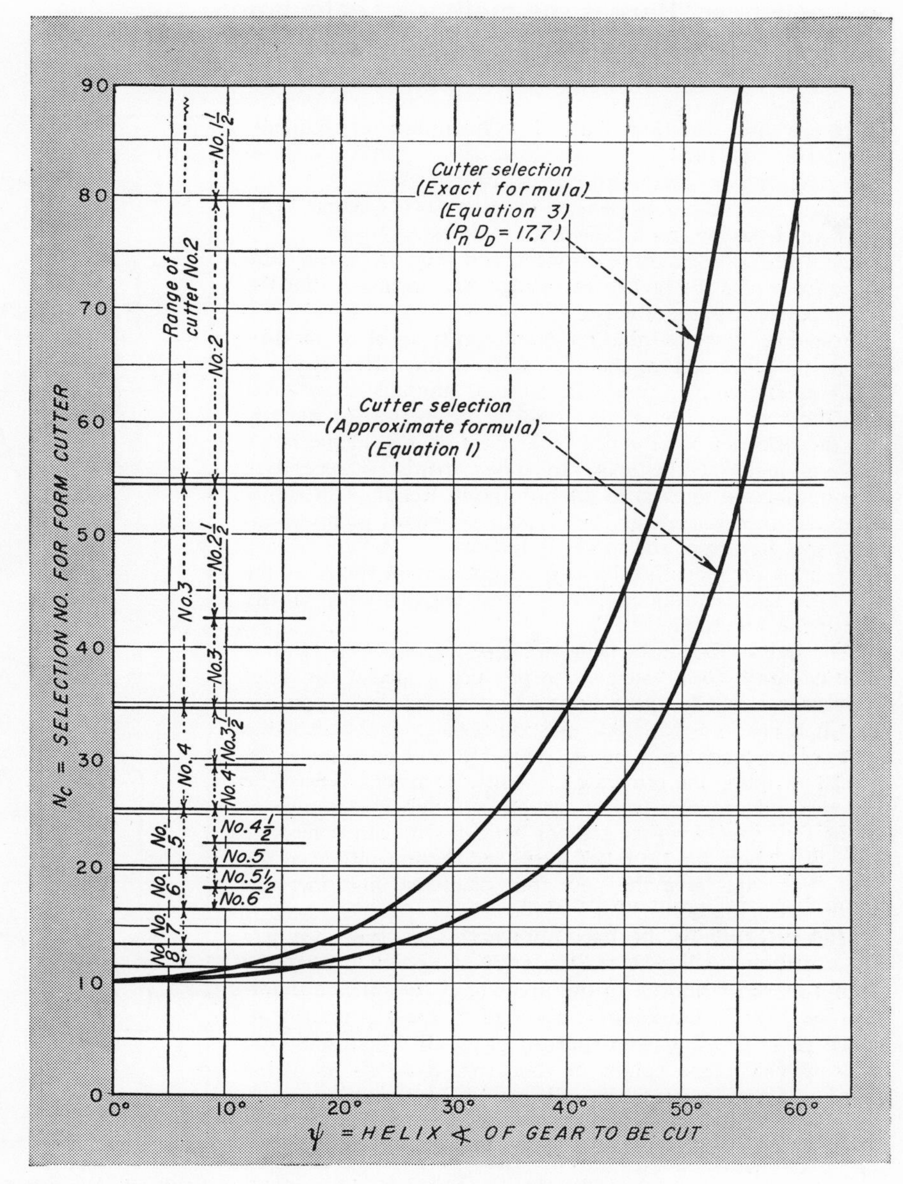

FIG. 3 . . . Curves for 6-DP gear, 10 teeth, showing the variation in form-cutter number determined by the exact and approximate selection formulas

Tip relief produced by a hob, 1

Will an existing hob produce acceptable tip relief on a new gear? Here is the method of calculation that lets you find out well in advance

There are three questions that a hob manufacturer frequently gets from the field:

1. Are my hobs producing correct tip relief?
2. Can the same hobs be used to relieve other gears?
3. Am I buying more hobs than is necessary?

These questions deserve answers, and they are given here in the form of a method of calculation and formulas that the gear producer can readily use.

From Fig. 1, we see that there are two types of tip modification: tip relief and tip chamfer. What is the difference?

Tip relief involves removing a small amount of material from the basic profile of the tooth in the region of the tip. This modification is achieved by a slight change in the hob's pressure angle at the root. In other words, a secondary pressure angle is applied to the hob tooth. Relief of this kind is applied to ease meshing, to alleviate problems of misalignment, spacing errors and tooth deflection.

Chamfer on the other hand is a protectional break at the gear OD, and is produced by a broad angular ramp at the root of the hob tooth.

Gear shops frequently have on hand various hobs with a modified tooth form intended to produce a special tip relief or chamfer. Usually these hobs are designed for a specific gear. It is possible that these existing hobs can cut acceptable tip relief on other gears of the same DP and pressure angle, thereby avoiding the purchase of additional special hobs. The necessary information can be established well in advance of cutting the gears, whether spur or helical. One can determine the radius where the tip relief starts and the magnitude of the relief. Formulas give the new root diameter, also the new OD of a topping hob.

After determining the hob tooth parts and gear specifications, calculate radius R_s to the start of tip relief, next the amount of relief normal to the involute at the outside diameter. See Fig. 2. Equations 1 through 7 apply primarily to helical gears and are basic steps to bring all dimensions into the transverse plane (plane of rotation). Symbols are listed in order of their appearance in Table 1. The formulas are:

$$\tan \phi_t = \tan \phi_n \div \cos \psi \quad \ldots\ldots 1$$

$$P_t = P_n \times \cos \psi \quad \ldots\ldots 2$$

$$D = N \div P_t \quad \ldots\ldots 3$$

$$D_b = \cos \phi_t \times D \quad \ldots\ldots 4$$

$$\tan \phi_{t2} = \tan \phi_{n2} \div \cos \psi \quad \ldots\ldots 5$$

$$D_{b2} = \cos \phi_{t2} \times D \quad \ldots\ldots 6$$

$$\tan \psi_o = (\pi \times D_o) \div L_G \quad \ldots\ldots 7$$

In equations 8 and 9, note that when $t_{GT} > t_{HS}$ the hob is held out to obtain correct gear tooth thickness and Δ_{hs} is positive. When $t_{GT} < t_{HS}$, the hob is lowered in the cut to obtain correct thickness and Δ_{hs} is negative.

Formulas that apply to Fig. 2 are:

$$\Delta h_s{}^* = \cot \phi_n (t_{Gt} - t_{HS}) \div 2 \quad \ldots\ldots 8$$

$$h_{sa} = h_s + \Delta h_s \quad \ldots\ldots 9$$

$$\tan \phi_{t2s} = h_{sa} \div (R_{b2} \times \sin \phi_{t2}) + \tan \phi_{t2} \quad \ldots\ldots 10$$

$$R_s = R_{b2} \div \cos \phi_{t2s} \quad \ldots\ldots 11$$

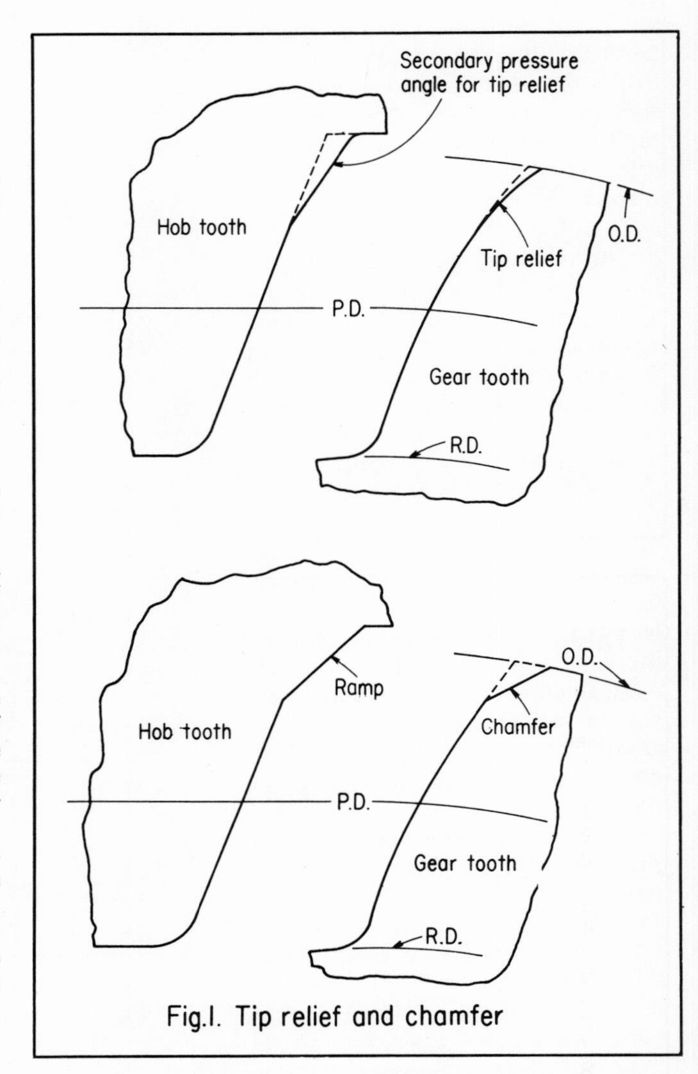

Fig.1. Tip relief and chamfer

*Δh_s can be used to determine new gear diameters thus:
new root diameter = D − 2 (hob addendum − Δh_s)
Having a topping hob:
New outside diameter = D + 2 (hob dedendum + Δh_s)

By D L Bargeron
Asst chief tool engineer
Union Twist Drill Div,
Litton Industries
Athol, Mass

Tip relief produced by a hob [Continued]

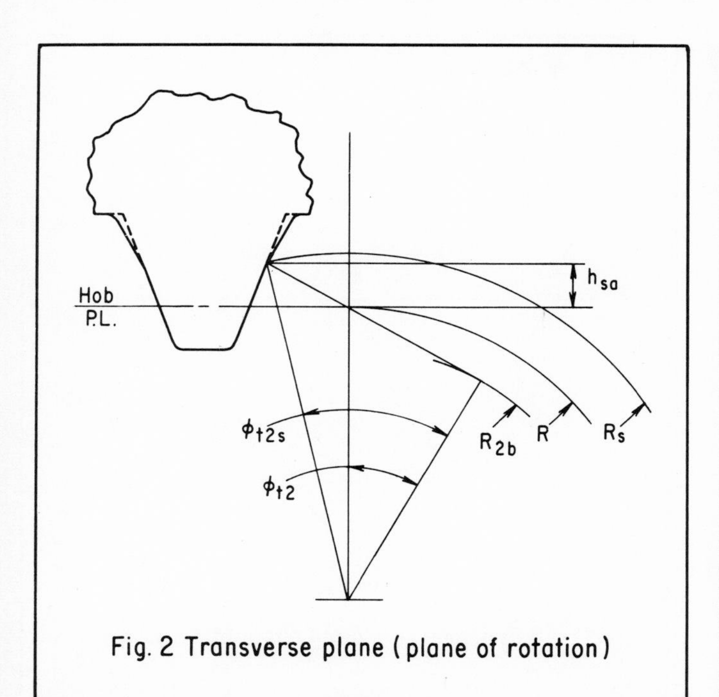

Fig. 2 Transverse plane (plane of rotation)

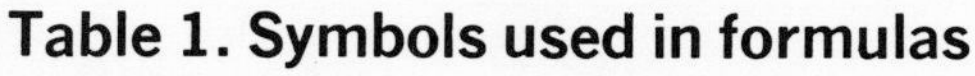

Table 1. Symbols used in formulas

Equation first used		Symbols in order of use
1	ϕ_t	= transverse pressure angle
1	ϕ_n	= normal pressure angle
1	ψ	= helix angle at P D
2	P_t	= transverse diametral pitch
2	P_n	= normal diametral pitch
3	D	= pitch diameter
3	N	= no. of teeth in gear
4	D_b	= base diameter
5	ϕ_{t2}	= secondary transverse P. A.
5	ϕ_{n2}	= sec. nor. P. A. (mod. for tip relief)
6	D_{b2}	= base dia. of secondary involute
7	ψ_o	= helix angle at D_o (outside diameter)
7	L_G	= lead of gear helix at pitch line
8	Δh_s	= increment of change in h_s due to cutting depth adjustment
8	t_{Gt}	= nor. cir. thickness of gear tooth at P. L.
8	t_{Hs}	= nor. thickness of hob space at P. L.
9	h_{sa}	= adjusted h_s
9	h_s	= depth on hob from P. L. to start of ϕ_{n2}
10	ϕ_{t2s}	= sec. trans. P. A. at start of tip relief
11	R_s	= radius to start of tip relief
12	PLD	= pitch line differential
12	R	$= D \div 2$
13	R_{b2}	$= D_{b2} \div 2$
14	ϕ_{ts}	= trans. P. A. at start of tip relief
14	R_b	$= D_b \div 2$
15	ϕ_{t2o}	= sec. trans. P. A. at O. D.
16	R_o	$= D_o \div 2$
17	ϕ_{to}	= trans. P. A. at O. D.
18	A	= arc of tip relief
19	x	≐ tip relief along O. D.
20	ϕ_{n2o}	= sec. nor. P. A. at O. D.
21	x_n	= tip relief nor. to involute in nor. plane

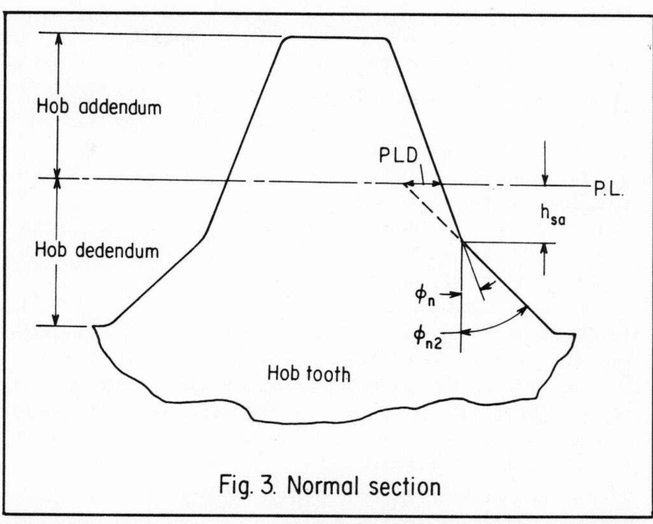

Fig. 3. Normal section

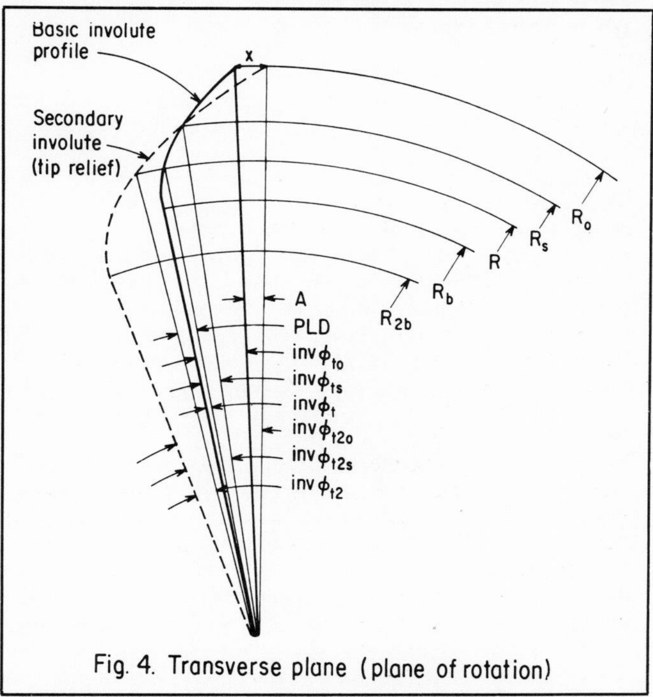

Fig. 4. Transverse plane (plane of rotation)

In equations 10 and 11, R_s is calculated under the assumption that the intersection of ϕ_n and ϕ_{n2} on the hob will contact the gear tooth at the junction of the primary and secondary involutes. This, theoretically, is not true. When calculating the start of tip relief, where the difference between ϕ_n and ϕ_{n2} is less than 5°, R_s as obtained in 10 and 11 is entirely satisfactory. Any error can be considered negligible. However, in extreme cases of size or broad chamfer angles, a more precise value of R_s is desirable. R_s must satisfy equation 15 within ± 0.0001.

$$PLD = h_{sa} (\tan \phi_{n2} - \tan \phi_n) \div (R \times \cos \psi) \quad . . 12$$

PLD, see Fig. 3, is the distance along the hob pitch line from ϕ_n to ϕ_{n2} extended, and is also the arc length along the generating pitch circle between the primary and secondary involutes, see Fig. 4. R_s is determined by the intersection of the converging involutes. Since the relationship of R_s and the involute of ϕ is transcendent, two or three trial calculations may be necessary.

Tip relief produced by a hob, 2

For equations below, see Fig. 4

$\phi_{t2s} = \cos^{-1}(R_{b2} \div R_s)$ 13

$\phi_{ts} = \cos^{-1}(R_b \div R_s)$ 14

$(\text{inv}\,\phi_{t2s} - \text{inv}\,\phi_{ts}) - (\text{inv}\,\phi_{t2} - \text{inv}\,\phi_t + \text{PLD}) = 0$ 15

Having established a satisfactory value for R_s, we can proceed to calculate the amount of tip relief.

$\cos\phi_{t20} = R_{b2} \div R_o$ 16

$\cos\phi_{to} = R_b \div R_o$ 17

$A = \text{inv}\,\phi_{t20} - \text{inv}\,\phi_{t2s} + \text{inv}\,\phi_{ts} - \text{inv}\,\phi_{to}$ 18

$x = A \times R_o$ 19

$\tan\phi_{n20} = \tan\phi_{t20} \times \cos\psi_0$ 20

$x_n = x \times \cos\psi_0 \times \cos\phi_{n20}$ 21

In equations 20 and 21, tip relief along the OD is converted to a more meaningful figure in the normal plane normal to the involute profile. Geometrically, the use of $\cos\phi_{n20}$ can be questioned in equation 21. However, it is sufficiently accurate when applied to tip relief as to preclude use of the lengthy exact calculation.

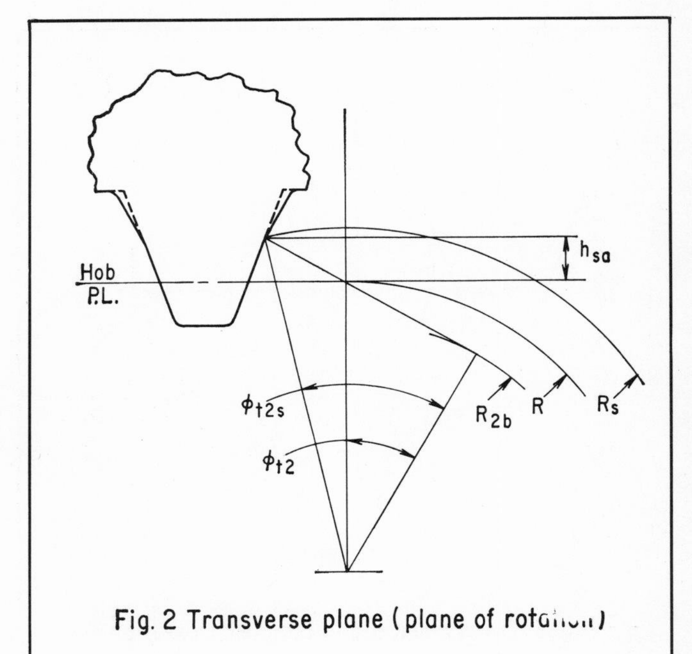

Fig. 2 Transverse plane (plane of rota...n)

How to calculate tip relief on a helical gear

Given:

$\phi_n = 20°$; $\phi_{n2} = 22°$; $P_n = 24$; $N = 10$; $\psi = 23°$;
$D_o = 0.567$; $t_{GT} = 0.0762$; $t_{HS} = 0.0654$; $h_s = 0.0087$;
$L_G = 3.350122$

Calculate:

R_s and x_n

Equation

1 $\tan\phi_t = 0.3639702 \div 0.9205049 = 0.3954027$
$= \tan 21.5739742°$ (cos = 0.9299436)

2 $P_t = 24 \times 0.9205049 = 22.0921176$

3 $D = 10 \div 22.0921176 = 0.4526501$

4 $D_b = 0.9299436 \times 0.4526501 = 0.4209390$

5 $\tan\phi_{t2} = 0.4040262 \div 0.9205049 = 0.4389180$
$= \tan 23.6975348°$ (cos = 0.9156799; sin = 0.4019084)

6 $D_{b2} = 0.9156799 \times 0.4526501 = 0.4144825$

7 $\tan\psi_o = (\pi \times 0.567) \div 3.350122 = 0.5317069$
$= \tan 27.9998883°$ (cos = 0.8829486)

8 $\Delta_{hs} = 2.747477\,(0.0762 - 0.0654) \div 2 = 0.0148363$

9 $h_{sa} = 0.0087 + 0.0148363 = 0.0235363$

10 $\tan\phi_{t2s} = 0.0235363 \div (0.2072412 \times 0.4019084) + 0.4389180 = 0.7214938$
$= \tan 35.810215°$ (cos = 0.8109595)

11 $R_s = 0.2072412 \div 0.8109595 = 0.2555506$

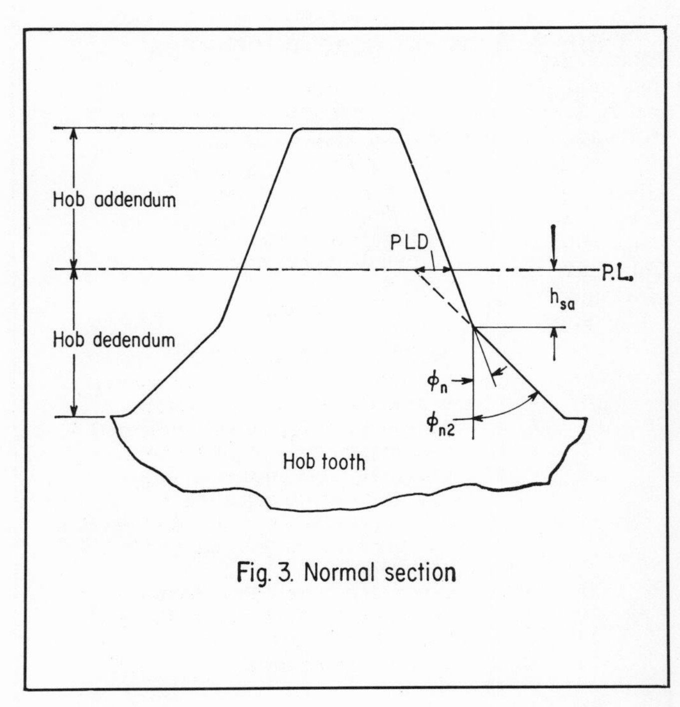

Fig. 3. Normal section

By D L Bargeron
Asst chief tool engineer
Union Twist Drill Div,
Litton Industries
Athol, Mass

Proof of R_s

12 $PLD = 0.0235363\ (0.4040262 - 0.3639702) \div (0.2263250 \times 0.9205049) = 0.0045252$

13 $\phi_{t2s} = 35.810215°$ from equation 10 (inv = 0.0964877)

14 $\cos \phi_{ts} = 0.2104695 \div 0.2555506 = 0.8235922 = \cos 34.5539798°$ (inv = 0.0855882)

15 $(0.0964877 - 0.0855882) - (0.0253180 - 0.0188657 + 0.0045252) = -0.0000780$

16 $\cos \phi_{t2o} = 0.2072412 \div 0.2835 = 0.7310095 = \cos 43.0289085°$ (inv = 0.1824627)

17 $\cos \phi_{to} = 0.2104695 \div 0.2835 = 0.7423968 = \cos 42.0640120°$ (inv = 0.1682735)

18 $A = 0.1824627 - 0.0964877 + 0.0855882 - 0.1682735 = 0.0032897$

19 $x = 0.0032897 \times 0.2835 = 0.0009326$

20 $\tan \phi_{n2o} = 0.9334588 \times 0.8829486 = 0.8241961 = \tan 39.4952132°$ (cos = .7716778)

21 $x_n = 0.0009326 \times 0.8829486 \times 0.7716778 = 0.0006353$

How to calculate tip relief on a spur gear

Spur gear calculations are essentially the same as for helicals. All normal figures are in the plane of rotation and and subscript t becomes subscript n. Equations 1, 2, 5, 7, and 20 are not required. In equation 12, eliminate $\cos \psi$ and in equation 21, eliminate $\cos \psi_o$.

Given:

$\phi_n = 20°$; $\phi_{n2} = 25°$; $P_n = 5$; $N = 80$; $D_o = 16.400$
$t_{Gt} = 0.310$; $t_{HS} = 0.3142$; $h_s = 0.100$

Calculate:

R_S and X_n

Equation

1

2

3 $D = 80 \div = 16.000$

4 $D_b = 0.9397 \times 16.000 = 15.0351$; $R_b = 7.5175$

5

6 $D_{b2} = 0.9063 \times 16.000 = 14.5009$; $R_{b2} = 7.2504$

7

8 $\Delta h_s = 2.7475\ (0.310 - 0.3142) \div 2 = -0.005$

9 $h_{sa} = 0.100 + (-0.0057) = 0.0942$

10 $\tan \phi_{n2s} = 0.0942 \div (7.2505 \times 0.4226) + 0.4663 = 0.4971 = \tan 26.43°$ (cos = 0.8955)

11 $R_s = 7.2505 \div 0.8955 = 8.0968$

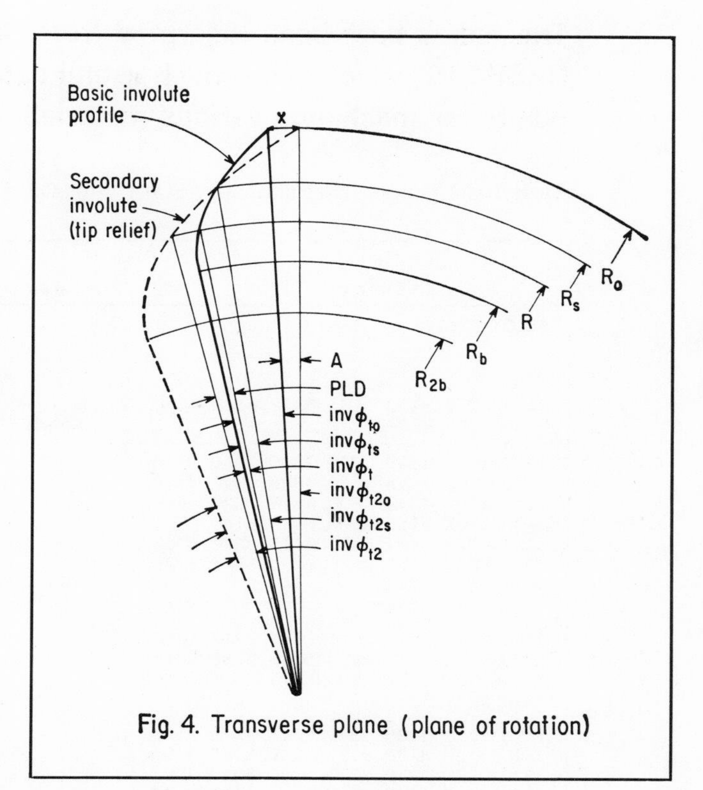

Fig. 4. Transverse plane (plane of rotation)

Proof of R_s

12 $PLD = 0.942\ (0.4663 - 0.3640) \div 8 = 0.0012$

13 $\phi_{n2s} = 26.43°$ (from eq. 10); inv = 0.0358

14 $\cos \phi_{ns} = 7.5175 \div 8.096 = 0.9285 = \cos 21.80°$; inv = 0.0195

15 $(0.0358 - 0.0195) - (0.0300 - 0.0149 + 0.0012) = 0.0000085$

16 $\cos \phi_{n2o} = 7.2505 \div 8.200 = 0.8842 = \cos 27.85°$; inv = 0.0423

17 $\cos \phi_{no} = 7.5175 \div 8.200 = 0.9168 = \cos 23.54°$; inv = 0.0248

18 $A = 0.0423 - 0.0358 + 0.0195 - 0.0248 = 0.0012$

19 $X = 0.0012 \times 8.200 = 0.0098$

20

21 $X_n = 0.0098 \times 0.8842 = 0.0087$

Involute Splines

The involute spline chamfer calculation is the same as for spur gears except that we need to solve for R_s and x only and equation 21 can be eliminated.

Hobbing speeds and feeds, 1

These data have been excerpted from Ordnance Corps ORDP40-1, which gives much useful data on starting points for machining various materials

By Herbert Gregg, chief engineer, Star Cutter Co, Farmington, Mich

MATERIAL	Condition	Brinell Hardness Number	Type of Cut	Diametral Pitch	Hob Speed fpm	Hob Feed ipr	HSS Hob Material	Cutting* Fluid
ALLOY STEELS								
2115, 2315, 3130, 3135, 3312, 4028, 4050, 5135, 5140, 8637	Annealed	150 to 210	Conv. Climb Climb Climb Climb	1-4 5-10 11-19 20-48 48 & Finer	125 145 160 175 190	.060 .050 .030 .015 .008	M2	I
1320, 1330, 1335, 2320, 2330, 2340, 3115, 3120, 3140, 3145, 3150, 4027, 4047, 4140, 4145, 4150, 4340, 4608, 4615, 4620, 4640, 5130, 5145, 5150, 5160, 6120, 6145, 6150, 6317, 6325, 6382, 6415, 8615, 8620, 8625, 8640, 8645, 8650, 8655, 8720, 8740, 8745, 9437, 9440, 9445	Annealed	150 to 230	Conv. Climb Climb Climb Climb	1-4 5-10 11-19 20-48 48 & Finer	115 135 150 160 175	.060 .050 .030 .015 .008	M2	I
2345, 2512, 2515, 2517, 3310, 3316, 4063, 4068, 4125, 4815, 4820, 5046, 5132, 50100, 51100, 52100, 6342, 6440, 6475, 9255, 9260, 9310, 9315	Annealed	160 to 240	Conv. Climb Climb Climb Climb	1-4 5-10 11-19 20-48 48 & Finer	105 120 140 145 160	.060 .050 .030 .015 .008	M2	I
1320, 1330, 1335, 1340, 2515, 2330, 2340, 2345, 3120, 3130, 3135, 3140, 3145, 3150, 3220, 3230, 3240, 3250, 3340, 4023, 4027, 4030, 4037, 4042, 4047, 4125, 4137, 4140, 4145, 4150, 4320, 4337, 4340, 4615, 4620, 4640, 4820, 5046, 5130, 5132, 5140, 5145, 5150, 6120, 6145, 6150, 6240, 6325, 6342, 6440, 8615, 8620, 8640, 8645, 8720, 8740, 8745, 8750, 9250, 9260, 9443, 9440, 9445	Norm. or Quenched and Tempered	240 to 310	Conv. Climb Climb Climb Climb	1-4 5-10 11-19 20-48 48 & Finer	85 95 110 120 130	.035 .029 .020 .010 .008	M2	I
1335, 1340, 2340, 2345, 3135, 3140, 3145, 3150, 4037, 4042, 4047, 4053, 4137, 4140, 4145, 4150, 4337, 4340, 4640, 5046, 5132, 5145, 5150, 5160, 6135, 6145, 6150, 8637, 6260, 8645, 8650, 8655, 8660, 8735, 8740, 8745, 8750, 9520, 9255, 9437, 9440, 9445,	Quenched and Tempered	310 to 370	Conv. Climb Climb Climb Climb	1-4 5-10 11-19 20-48 48 & Finer	60 65 75 80 90	.030 .025 .018 .008 .008	M3	I
STAINLESS STEELS **Martensitic** 416, 440	Cold Drawn and Annealed	140 to 160	Conv. Climb Climb Climb Climb	1-4 5-10 11-19 20-48 48 & Finer	120 140 155 165 180	.060 .050 .030 .015 .008	M2	I
Austenitic 302, 304	Cold Drawn and Annealed	160 to 220	Conv. Climb Climb Climb Climb	1-4 5-10 11-19 20-48 48 & Finer	85 95 110 120 130	.060 .050 .030 .015 .008	M2	I
Austenitic—Free Machining 303	Cold Drawn and Annealed	160 to 220	Conv. Climb Climb Climb Climb	1-4 5-10 11-19 20-48 48 & Finer	115 145 160 175 190	.060 .050 .030 .015 .008	M2	I
NITRIDING STEELS Nitralloy G Mod., Ni-Nitralloy, AMS 6480A	Norm.	260 to 320	Conv. Climb Climb Climb Climb	1-4 5-10 11-19 20-48 48 & Finer	60 75 85 90 100	.035 .029 .020 .010 .008	M2	I

***Group I—Water Base Emulsion or Water Base Chemical Solution or Sulphurized Oil**

Hobbing speeds and feeds, 2

MATERIAL	Condition	Brinell Hardness Number	Type of Cut	Diametral Pitch	Hob Speed fpm	Hob Feed ipr	HSS Hob Material	Cutting* Fluid
FREE MACHINING PLAIN CARBON STEEL								
B1111, B1112, B1113, B1115, B1116, C1211, C1212, C1213	Cold Drawn	170 to 190	Conv.	1-4	200	.060	M2	I
			Climb	5-10	225	.050		
			Climb	11-19	260	.030		
			Climb	20-48	270	.015		
			Climb	48 & Finer	300	.008		
C1108, C1109, C1110, C1117, C1118, C1119, C1120, C1125	Cold Drawn	160 to 190	Conv.	1-4	180	.060	M2	I
			Climb	5-10	210	.050		
			Climb	11-19	240	.030		
			Climb	20-48	250	.015		
			Climb	48 & Finer	280	.008		
C1132, C1137, C1138, C1140, C1141, C1144, C1145, C1146, C1151	Cold Drawn	200 to 230	Conv.	1-4	140	.060	M2	I
			Climb	5-10	160	.050		
			Climb	11-19	170	.030		
			Climb	20-48	185	.015		
			Climb	48 & Finer	200	.008		
C1137, C1140, C1144, C1145, C1146, C1151	Quenched and Tempered	250 to 300	Conv.	1-4	100	.035	M2	I
			Climb	5-10	120	.029		
			Climb	11-19	140	.020		
			Climb	20-48	145	.010		
			Climb	48 & Finer	160	.008		
PLAIN CARBON STEELS								
1010, 1015, 1020, 1025, 1030, 1035	Annealed	110 to 165	Conv.	1-4	165	.060	M2	I
			Climb	5-10	190	.050		
			Climb	11-19	210	.030		
			Climb	20-48	230	.015		
			Climb	48 & Finer	250	.008		
1040, 1050	Annealed	120 to 185	Conv.	1-4	125	.060	M2	I
			Climb	5-10	145	.050		
			Climb	11-19	160	.030		
			Climb	20-48	175	.015		
			Climb	48 & Finer	190	.008		
1095	Annealed	170 to 220	Conv.	1-4	105	.060	M2	I
			Climb	5-10	120	.050		
			Climb	11-19	140	.030		
			Climb	20-48	145	.015		
			Climb	48 & Finer	160	.008		
1025, 1030, 1035, 1040	Quenched and Tempered	210 to 250	Conv.	1-4	100	.060	M2	I
			Climb	5-10	110	.050		
			Climb	11-19	130	.030		
			Climb	20-48	135	.015		
			Climb	48 & Finer	150	.008		
FREE MACHINING ALLOY STEELS								
Resulphurized Steels								
3140, 4140, 4150, 8640	Annealed	160 to 210	Conv.	1-4	170	.060	M2	I
			Climb	5-10	190	.050		
			Climb	11-19	210	.030		
			Climb	20-48	230	.015		
			Climb	48 & Finer	250	.008		
Leaded Steels								
1018, 8620, C1213	Annealed	140 to 190	Conv.	1-4	190	.060	M2	I
			Climb	5-10	220	.050		
			Climb	11-19	250	.030		
			Climb	20-48	260	.015		
			Climb	48 & Finer	290	.008		
4147, 4340, 4140	Norm.	250 to 300	Conv.	1-4	90	.035	M2	I
			Climb	5-10	100	.029		
			Climb	11-19	120	.020		
			Climb	20-48	125	.010		
			Climb	48 & Finer	160	.008		

*Group I—Water Base Emulsion or Water Base Chemical Solution or Sulphurized Oil

Spline data, 1

By Carl A Johnson

Pitch diameter, base diameter, and number of teeth for 30° pressure angle involute splines

This table of pitch diameters, base-circle diameters, and number of teeth applies to both internal and external splines and is worked out for splines from 2½/5 to 48/96 DP, 6 to 100 teeth, up to 24 in. pitch diameter. Note that the denominators in the pitch designations have not been included in column headings.

Use this table for a quick check of spline data on drawings of splined parts and spline gages. The figures are accurate to 6 decimal places and based on the accurate value of cos 30° = 0.866,025,4.

The table values can also be used in the formulas for measurement over pins of both internal and external splines.

Pitch dia, D, in.	Base circle dia, D_b, in.	Diametral pitch, DP: 2½	3	4	5	6	8	10	12	16	20	24	32	40	48
.125	.108253														6
.145833	.126295														7
.15	.129904													6	
.166667	.144338														8
.175	.151554													7	
.1875	.162380												6		9
.2	.173205													8	
.208333	.180422														10
.21875	.189443												7		
.225	.194856													9	
.229167	.198464														11
.25	.216506											6	8	10	12
.270833	.234549														13
.275	.238157													11	
.28125	.243570												9		
.291667	.252591											7			14
.3	.259808										6			12	
.3125	.270633												10		15
.325	.281458													13	
.333333	.288675											8			16
.34375	.297696												11		
.35	.303109										7			14	
.354167	.306717														17
.375	.324760									6		9	12	15	18
.395833	.342802														19
.4	.346410										8			16	
.40625	.351823												13		
.416667	.360844											10			20
.425	.368061													17	
.4375	.378886									7			14		21
.45	.389711										9			18	
.458333	.396928											11			22
.46875	.405949												15		
.475	.411362													19	
.479167	.414971														23
.5	.433013								6	8	10	12	16	20	24
.520833	.451055														25
.525	.454633													21	
.53125	.460076												17		
.541667	.469097											13			26
.55	.476314										11			22	
.5625	.487139									9			18		27
.575	.497965													23	
.58333	.505181								7			14			28
.59375	.514203												19		
.6	.519615							6			12			24	
.604167	.523224														29
.625	.541266									10		15	20	25	30
.645833	.559308														31
.65	.562917										13			26	
.65625	.568329												21		
.666667	.577350								8			16			32
.675	.584567													27	
.6875	.595392									11			22		33
.7	.606218							7			14			28	
.708333	.613435											17			34
.71875	.622456												23		
.725	.627868													29	
.729167	.631477														35
.75	.649519						6		9	12	15	18	24	30	36
.770833	.667561														37
.775	.671170													31	
.78125	.676582												25		
.791667	.685603											19			38
.8	.692820							8			16			32	
.8125	.703646									13			26		39
.825	.714471													33	
.833333	.721688								10			20			40
.84375	.730709												27		
.85	.736122										17			34	
.854167	.739730														41
.875	.757772						7			14		21	28	35	42
.895833	.775814														43
.9	.779423							9			18			36	
.90625	.784836												29		
.916667	.793857								11			22			44
.925	.801073													37	
.9375	.811899									15			30		45
.95	.822724										19			38	
.958333	.829941											23			46
.96875	.838962												31		
.975	.844375													39	
.979167	.847983														47
1.0	.866025					6	8	10	12	16	20	24	32	40	48
1.020833	.884068														49
1.025	.887676													41	
1.03125	.893089												33		
1.041667	.902110											25			50
1.05	.909327										21			42	
1.0625	.920152									17			34		51
1.075	.930977													43	
1.0833	.938194								13			26			52
1.09375	.947215												35		
1.1	.952628							11			22			44	
1.104167	.956236														53
1.125	.974279						9			18		27	36	45	54
1.145833	.992321														55
1.15	.995929										23			46	
1.15625	1.001342												37		
1.166667	1.010363					7			14			28			56
1.175	1.017580													47	
1.1875	1.028405									19			38		57
1.2	1.039230				6			12			24			48	
1.208333	1.046447											29			58
1.21875	1.055468												39		

Spline data, 2

By Carl A Johnson

Pitch diameter, base diameter, and number of teeth for 30° pressure angle involute splines

Pitch dia, D, in.	Base circle dia, D_b, in.	Diametral pitch, DP													
		2½	3	4	5	6	8	10	12	16	20	24	32	40	48
1.225	1.060881													49	
1.229167	1.064490														59
1.25	1.082532						10		15	20	25	30	40	50	60
1.270833	1.100574														61
1.275	1.104182													51	
1.28125	1.109595												41		2
1.291667	1.118616											31			6
1.3	1.125833							13			26			52	3
1.3125	1.136658									21			42		6
1.325	1.147484													53	
1.333333	1.154701					8			16			32			64
1.34375	1.163722												43		
1.350	1.169134										27			54	
1.354167	1.172743														65
1.375	1.190785						11			22		33	44	55	66
1.395833	1.208827														67
1.4	1.212436				7			14			28			56	
1.40625	1.217848												45		
1.416667	1.226869								17			34			68
1.425	1.234086													57	
1.4375	1.244912									23			46		69
1.45	1.255737										29			58	
1.458333	1.262954											35			70
1.46875	1.271975												47		
1.475	1.277387													59	
1.479167	1.280996														71
1.5	1.299038			6		9	12	15	18	24	30	36	48	60	72
1.520833	1.317080														73
1.525	1.320689													61	
1.53125	1.326101												49		
1.541667	1.335122											37			75
1.55	1.342339										31			62	
1.5625	1.353165									25			50		74
1.575	1.363990													63	
1.583333	1.371207								19			38			76
1.59375	1.380228												51		
1.6	1.385641				8			16			32			64	
1.604167	1.389249														77
1.625	1.407291						13			26		39	52	65	78
1.645833	1.425333														79
1.65	1.428942										33			66	
1.65625	1.434355											53			
1.666667	1.443376					10			20			40			80
1.675	1.450593													67	
1.6875	1.461418									27			54		81
1.7	1.472243							17			34			68	
1.708333	1.479460											41			82
1.71875	1.488481												55		
1.725	1.493894													69	
1.729167	1.497502														83
1.75	1.515544			7			14		21	28	35	42	56	70	84
1.770833	1.533587														85
1.775	1.537195													71	
1.78125	1.542608												57		
1.791667	1.551629											43			86
1.8	1.558846				9			18			36			72	
1.8125	1.569671									29			58		87
1.825	1.580496													73	
1.833333	1.587713					11			22			44			88
1.84375	1.596734												59		

Pitch dia, D, in.	Base circle dia, D_b, in.	Diametral pitch, DP													
		2½	3	4	5	6	8	10	12	16	20	24	32	40	48
1.85	1.602147										37			74	
1.854167	1.605755														89
1.875	1.623798						15			30		45	60	75	90
1.895833	1.641840														91
1.9	1.645448							19			38			76	
1.90625	1.650861												61		
1.916667	1.659882								23			46			92
1.925	1.667099													77	
1.9375	1.677924									31			62		93
1.95	1.688750										39			78	
1.958333	1.695966											47			94
1.96875	1.704988												63		
1.975	1.710400													79	
1.979167	1.714009														95
2.0	1.732051		6	8	10	12	16	20	24	32	40	48	64	80	96
2.020833	1.750093														97
2.025	1.753701													81	
2.03125	1.759114												65		
2.041667	1.768135											49			98
2.05	1.775352										41			82	
2.0625	1.786177									33			66		99
2.075	1.797003													83	
2.083333	1.804220								25			50			100
2.09375	1.813241												67		
2.1	1.818653							21			42			84	
2.125	1.840304						17			34		51	68	85	
2.15	1.861955										43			86	
2.15625	1.867367												69		
2.166667	1.876388					13			26			52			
2.175	1.883605													87	
2.1875	1.894431									35			70		
2.2	1.905256				11			22			44			88	
2.208333	1.912473											53			
2.21875	1.921494												71		
2.225	1.926907													89	
2.25	1.948557			9			18		27	36	45	54	72	90	
2.275	1.970208													91	
2.28125	1.975620												73		
2.291667	1.984642											55			
2.3	1.991858							23			46			92	
2.3125	2.002684									37			74		
2.325	2.013509													93	
2.333333	2.020726		7			14			28			56			
2.34375	2.029747												75		
2.35	2.035160										47			94	
2.375	2.056810						19			38		57	76	95	
2.4	2.078461	6			12			24			48			96	
2.40625	2.083874												77		
2.416667	2.092895								29			58			
2.425	2.100112													97	
2.4375	2.110937									39			78		
2.45	2.121762										49			98	
2.458333	2.128979											59			
2.46875	2.138000												79		
2.475	2.143413													99	
2.5	2.165063			10		15	20	25	30	40	50	60	80	100	
2.53125	2.192127												81		
2.541667	2.201148											61			
2.55	2.208365										51				
2.5625	2.219190									41			82		

Spline data [Continued]

Pitch dia, D, in.	Base circle dia, D_b, in.	Diametral pitch, DP 2½	3	4	5	6	8	10	12	16	20	24	32
2.583333	2.237232								31			62	
2.59375	2.246253												83
2.6	2.251666				13			26			52		
2.625	2.273317						21			42		63	84
2.65	2.294967										53		
2.65625	2.300380												85
2.666667	2.309401		8			16			32			64	
2.6875	2.327443									43			86
2.7	2.338269							27			54		
2.708333	2.345485											65	
2.71875	2.354507												87
2.75	2.381570			11			22		33	44	55	66	88
2.78125	2.408633												89
2.791667	2.417654											67	
2.8	2.424871	7			14			28			56		
2.8125	2.435696									45			90
2.833333	2.453739					17			34			68	
2.84375	2.462760												91
2.85	2.468172										57		
2.875	2.489823						23			46		69	92
2.9	2.511474							29			58		
2.90625	2.516886												93
2.916667	2.525907								35			70	
2.9375	2.543950									47			94
2.95	2.554775										59		
2.958333	2.561992											71	
2.96875	2.571013												95
3.0	2.598076		9	12	15	18	24	30	36	48	60	72	96
3.03125	2.625139												97
3.041667	2.634161											73	
3.05	2.641377										61		
3.0625	2.652203									49			98
3.083333	2.670245								37			74	
3.09375	2.679266												99
3.1	2.684679							31			62		
3.125	2.706329						25			50		75	100
3.15	2.727980										63		
3.166667	2.742414					19			38			76	
3.1875	2.760456									51			
3.2	2.771281	8			16			32			64		
3.208333	2.778498											77	
3.25	2.814583			13			26		39	52	65	78	
3.291667	2.850667											79	
3.3	2.857884							33			66		
3.3125	2.868709									53			
3.333333	2.886751		10			20			40			80	
3.35	2.901185										67		
3.375	2.922836						27			54		81	
3.4	2.944486				17			34			68		
3.416667	2.958920								41			82	
3.4375	2.976962									55			
3.45	2.987788										69		
3.458333	2.995005											83	
3.5	3.031089			14		21	28	35	42	56	70	84	
3.541667	3.067173											85	
3.55	3.074390										71		
3.5625	3.085215									57			
3.583333	3.103258								43			86	
3.6	3.117691	9			18			36			72		
3.625	3.139342						29			58		87	

Pitch dia, D, in.	Base circle dia, D_b, in.	Diametral pitch, DP 2½	3	4	5	6	8	10	12	16	20	24	32
3.65	3.160993										73		
3.666667	3.175426		11			22			44			88	
3.6875	3.193469									59			
3.7	3.204294							37			74		
3.708333	3.211511											89	
3.75	3.247595			15			30		45	60	75	90	
3.791667	3.283680											91	
3.8	3.290897				19			38			76		
3.8125	3.301722									61			
3.833333	3.319764					23			46			92	
3.85	3.334198										77		
3.875	3.355848						31			62		93	
3.9	3.377499							39			78		
3.916667	3.391933								47			94	
3.9375	3.409975									63			
3.95	3.420800										79		
3.958333	3.428017											95	
4.0	3.464102	10	12	16	20	24	32	40	48	64	80	96	
4.041667	3.500186											97	
4.05	3.507403										81		
4.0625	3.518228									65			
4.083333	3.536270								49			98	
4.1	3.550704							41			82		
4.125	3.572355						33			66		99	
4.15	3.594005										83		
4.166667	3.608439					25			50			100	
4.1875	3.626481									67			
4.2	3.637307				21			42			84		
4.25	3.680608			17			34			68	85		
4.3	3.723909							43			86		
4.3125	3.734735									69			
4.333333	3.752777		13			26			52				
4.35	3.767210										87		
4.375	3.788861						35			70			
4.4	3.810512	11			22			44			88		
4.416667	3.824946								53				
4.4375	3.842988									71			
4.45	3.853813										89		
4.5	3.897114			18		27	36	45	54	72	90		
4.55	3.940416										91		
4.5625	3.951241									73			
4.583333	3.969283								55				
4.6	3.983717				23			46			92		
4.625	4.005367						37			74			
4.65	4.027018										93		
4.666667	4.041452		14			28			56				
4.6875	4.059494									75			
4.7	4.070319							47			94		
4.75	4.113621			19			38		57	76	95		
4.8	4.156922	12			24			48			96		
4.8125	4.167747									77			
4.833333	4.185789					29			58				
4.85	4.200223										97		
4.875	4.221874						39			78			
4.9	4.243524							49			98		
4.916667	4.257958								59				
4.9375	4.276000									79			
4.95	4.286826										99		
5.0	4.330127		15	20	25	30	40	50	60	80	100		
5.0625	4.384254									81			

Spline data, 3

By Carl A Johnson

Pitch diameter, base diameter, and number of teeth for 30° pressure angle involute splines

Pitch dia, D, in.	Base circle dia, D_b, in.	Diametral pitch, DP: 2½	3	4	5	6	8	10	12	16	20	24	32
5.083333	4.402296								61				
5.1	4.416730							51					
5.125	4.438380						41			82			
5.166667	4.474465					31			62				
5.1875	4.492507									83			
5.2	4.503332	13			26			52					
5.25	4.546633			21			42		63	84			
5.3	4.589935							53					
5.3125	4.600760									85			
5.333333	4.618802		16			32			64				
5.375	4.654887						43			86			
5.4	4.676537				27			54					
5.416667	4.690971								65				
5.4375	4.709013									87			
5.5	4.763140			22		33	44	55	66	88			
5.5625	4.817266									89			
5.583333	4.835308								67				
5.6	4.849742	14			28			56					
5.625	4.871393						45			90			
5.666667	4.907477		17			34			68				
5.6875	4.925519									91			
5.7	4.936345							57					
5.75	4.979646			23			46		69	92			
5.8	5.022947				29			58					
5.8125	5.033773									93			
5.833333	5.051815					35			70				
5.875	5.087899						47			94			
5.9	5.109550							59					
5.916667	5.123984								71				
5.9375	5.142026									95			
6.0	5.196152	15	18	24	30	36	48	60	72	96			
6.0625	5.250279									97			
6.083333	5.268321								73				
6.1	5.282755							61					
6.125	5.304406						49			98			
6.166667	5.340490					37			74				
6.1875	5.358532									99			
6.2	5.369357				31			62					
6.25	5.412659			25			50		75	100			
6.3	5.455960							63					
6.333333	5.484828		19			38			76				
6.375	5.520912						51						
6.4	5.542563	16			32			64					
6.416667	5.556997								77				
6.5	5.629165			26		39	52	65	78				
6.583333	5.701334								79				
6.6	5.715768				33			66					
6.625	5.737418						53						
6.666667	5.773503		20			40			80				
6.7	5.802370							67					
6.75	5.845671			27			54		81				
6.8	5.888973	17			34			68					
6.833333	5.917840					41			82				
6.875	5.953925						55						
6.9	5.975575							69					
6.916667	5.990009								83				
7.0	6.062178		21	28	35	42	56	70	84				
7.083333	6.134347								85				
7.1	6.148780							71					
7.125	6.170431						57						

Pitch dia, D, in.	Base circle dia, D_b, in.	Diametral pitch, DP: 2½	3	4	5	6	8	10	12	16	20	24	32
7.166667	6.206515					43			86				
7.2	6.235383	18			36			72					
7.25	6.278684			29			58		87				
7.3	6.321985							73					
7.333333	6.350853		22			44			88				
7.375	6.386937						59						
7.4	6.408588				37			74					
7.416667	6.423022								89				
7.5	6.495190			30		45	60	75	90				
7.583333	6.567359								91				
7.6	6.581793	19			38			76					
7.625	6.603444						61						
7.666667	6.639528		23			46			92				
7.7	6.668396							77					
7.75	6.711697			31			62		93				
7.8	6.754998				39			78					
7.833333	6.783866					47			94				
7.875	6.819950						63						
7.9	6.841601							79					
7.916667	6.856034								95				
8.0	6.928203	20	24	32	40	48	64	80	96				
8.083333	7.000372								97				
8.1	7.014806							81					
8.125	7.036456						65						
8.166667	7.072541					49			98				
8.2	7.101408				41			82					
8.25	7.144710			33			66		99				
8.3	7.188011							83					
8.333333	7.216878		25			50			100				
8.375	7.252963						67						
8.4	7.274613	21			42			84					
8.5	7.361216			34		51	68	85					
8.6	7.447818				43			86					
8.625	7.469469						69						
8.666667	7.505553		26			52							
8.7	7.534421							87					
8.75	7.577722			35			70						
8.8	7.621024	22			44			88					
8.833333	7.649891					53							
8.875	7.685975						71						
8.9	7.707626							89					
9.0	7.794229		27	36	45	54	72	90					
9.1	7.880831							91					
9.125	7.902482						73						
9.166667	7.938566					55							
9.2	7.967434	23			46			92					
9.25	8.010735			37			74						
9.3	8.054036							93					
9.333333	8.082904		28			56							
9.375	8.118988						75						
9.4	8.140639				47			94					
9.5	8.227241			38		57	76	95					
9.6	8.313844	24			48			96					
9.625	8.335494						77						
9.666667	8.371579		29			58							
9.7	8.400446							97					
9.75	8.443748			39			78						
9.8	8.487049				49			98					
9.833333	8.515916					59							
9.875	8.552001						79						

Spline data [Continued]

Pitch dia, D, in.	Base circle dia, D_b, in.	Diametral pitch, DP 2½	3	4	5	6	8	10
9.9	8.573651							99
10.0	8.660254	25	30	40	50	60	80	100
10.125	8.768507						81	
10.166667	8.804592					61		
10.2	8.833459				51			
10.25	8.876760			41			82	
10.333333	8.948929		31			62		
10.375	8.985014						83	
10.4	9.006664	26			52			
10.5	9.093267			42		63	84	
10.6	9.179869				53			
10.625	9.201520						85	
10.666667	9.237604		32			64		
10.75	9.309773			43			86	
10.8	9.353074	27			54			
10.833333	9.381942					65		
10.875	9.418026						87	
11.0	9.526279		33	44	55	66	88	
11.125	9.634533						89	
11.166667	9.670617					67		
11.2	9.699484	28			56			
11.25	9.742786			45			90	
11.333333	9.814955		34			68		
11.375	9.851039						91	
11.4	9.872690				57			
11.5	9.959292			46			92	
11.6	10.045895	29			58			
11.625	10.067545						93	
11.666667	10.103630		35			70		
11.75	10.175798			47			94	
11.8	10.219100				59			
11.833333	10.247967					71		
11.875	10.284052						95	
12.0	10.392305	30	36	48	60	72	96	
12.125	10.500558						97	
12.166667	10.536642					73		
12.2	10.565510				61			
12.25	10.608811			49			98	
12.333333	10.680980		37			74		
12.375	10.717064						99	
12.4	10.738715	31			62			
12.5	10.825317			50		75	100	
12.6	10.911920				63			
12.666667	10.969655		38			76		
12.75	11.041824			51				
12.8	11.085125	32			64			
12.833333	11.113993					77		
13.0	11.258330		39	52	65	78		
13.166667	11.402668					79		
13.2	11.431535	33			66			
13.25	11.474837			53				
13.333333	11.547005		40			80		
13.4	11.604740				67			
13.5	11.691343			54		81		
13.6	11.777945	34			68			

Pitch dia, D, in.	Base circle dia, D_b, in.	Diametral pitch, DP 2½	3	4	5	6	8	10
13.666667	11.835680		41			82		
13.75	11.907849			55				
13.8	11.951151				69			
13.833333	11.980018					83		
14.0	12.124356	35	42	56	70	84		
14.166667	12.268693					85		
14.2	12.297561				71			
14.25	12.340862			57				
14.333333	12.413031		43			86		
14.4	12.470766	36			72			
14.5	12.557368			58		87		
14.6	12.643971				73			
14.666667	12.701706		44			88		
14.75	12.773875			59				
14.8	12.817176	37			74			
14.833333	12.846043					89		
15.0	12.990381		45	60	75	90		
15.166667	13.134719					91		
15.2	13.163586	38			76			
15.25	13.206887			61				
15.333333	13.279056		46			92		
15.4	13.336791				77			
15.5	13.423394			62		93		
15.6	13.509996	39			78			
15.666667	13.567731		47			94		
15.75	13.639900			63				
15.8	13.683201				79			
15.833333	13.712069					95		
16.0	13.856406	40	48	64	80	96		
16.166667	14.000744					97		
16.2	14.029611				81			
16.25	14.072913			65				
16.333333	14.145082		49			98		
16.4	14.202817	41			82			
16.5	14.289419			66		99		
16.6	14.376022				83			
16.666667	14.433757		50			100		
16.75	14.505925			67				
16.8	14.549227	42			84			
17.0	14.722432		51	68	85			
17.2	14.895637	43			86			
17.25	14.938938			69				
17.333333	15.011107		52					
17.4	15.068842				87			
17.5	15.155444			70				
17.6	15.242047	44			88			
17.666667	15.299782		53					
17.75	15.371951			71				
17.8	15.415252				89			
18.0	15.588457	45	54	72	90			
18.2	15.761662				91			
18.25	15.804964			73				
18.333333	15.877132		55					
18.4	15.934867	46						
18.5	16.021470			74				

Pitch dia, D, in.	Base circle dia, D_b, in.	Diametral pitch, DP 2½	3	4	5	6	8	10
18.6	16.108072				93			
18.666667	16.165807		56					
18.75	16.237976			75				
18.8	16.281278	47			94			
19.0	16.454483		57	76	95			
19.2	16.627688	48			96			
19.25	16.670989			77				
19.333333	16.743158		58					
19.4	16.800893				97			
19.5	16.887495			78				
19.6	16.974098	49			98			
19.666667	17.031833		59					
19.75	17.104002			79				
19.8	17.147303				99			
20.0	17.320508	50	60	80	100			
20.25	17.537014			81				
20.333333	17.609183		61					
20.4	17.666918	51						
20.5	17.753521			82				
20.666667	17.897858		62					
20.75	17.970027			83				
20.8	18.013328	52						
21.0	18.186533		63	84				
21.2	18.359738	53						
21.25	18.403040			85				
21.333333	18.475208		64					
21.5	18.619546			86				
21.6	18.706149	54						
21.666667	18.763884		65					
21.75	18.836052			87				
22.0	19.052559	55	66	88				
22.25	19.269065			89				
22.333333	19.341234		67					
22.4	19.398969	56						
22.5	19.485571			90				
22.666667	19.629909		68					
22.75	19.702078			91				
22.8	19.745379	57						
23.0	19.918584		69	92				
23.2	20.091789	58						
22.35	20.135091			93				
23.333333	20.207259		70					
23.5	20.351597			94				
23.6	20.438199	59						
23.666667	20.495935		71					
23.75	20.568103			95				
24.0	20.784610	60	72	96				

Roll angles for involute splines

CARL A JOHNSON

The tooth profiles of precision involute splines are often inspected on involute checking machines. For this operation the roll angles E at the limiting diameters must be established. For standard splines with working depth equal to one half that of standard gear teeth the limiting diameters are $\frac{N-1}{P}$ and $\frac{N+1}{P}$

where:

N = number of teeth in 30° P.A. spline

P = numerator of diametral pitch designation, which is the number of teeth per inch of pitch diameter

$\frac{N-1}{P}$ = True Involute Form (TIF) Diameter for external splines, all fits.
Minor diameter for internal splines, side fits, flat or fillet root.

$\frac{N+1}{P}$ = True Involute Form (TIF) Diameter for internal splines, side fits, fillet root or full dedendum flat root.
Major diameter for external splines, side fits, flat or fillet root.

For the convenience of inspectors the roll angles E_{N-1} and E_{N+1} at diameters $(N - 1)$ and $(N + 1)$ are listed in the accompanying table for splines of 9 to 56 teeth.

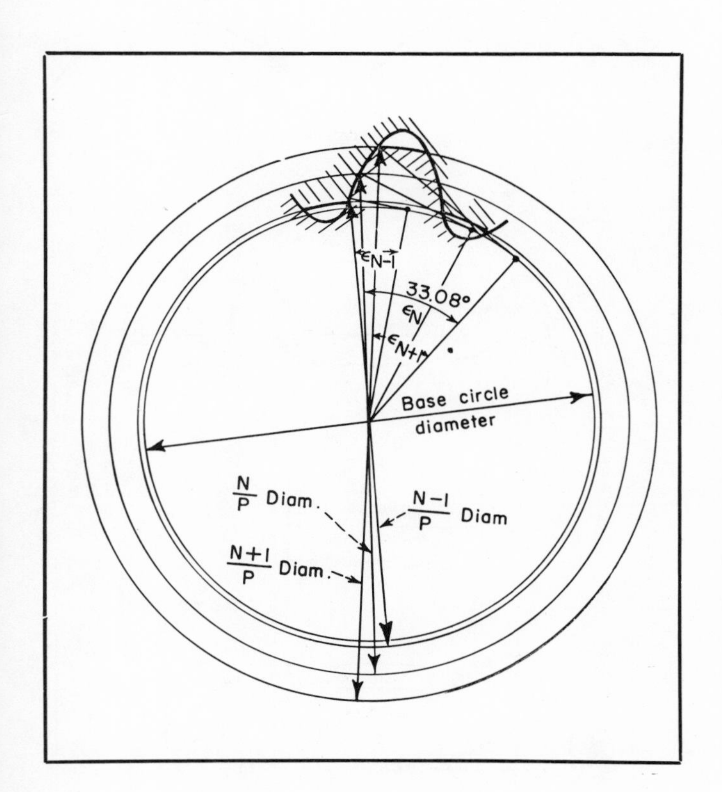

N	Roll Angle Deg. E_{N-1}	Roll Angle Deg. E_{N+1}	N	Roll Angle Deg. E_{N-1}	Roll Angle Deg. E_{N+1}
9	13.25	46.05	33	28.86	36.93
10	16.21	44.87	34	28.99	36.82
11	18.29	43.89	35	29.12	36.71
12	19.88	43.06	36	29.23	36.62
13	21.14	42.35	37	29.34	36.52
14	22.16	41.74	38	29.44	36.44
15	23.02	41.20	39	29.54	36.35
16	23.75	40.72	40	29.63	36.27
17	24.38	40.30	41	29.72	36.20
18	24.93	39.93	42	29.80	36.13
19	25.41	39.59	43	29.88	36.06
20	25.84	39.28	44	29.96	35.99
21	26.22	39.00	45	30.03	35.93
22	26.56	38.75	46	30.10	35.87
23	26.87	38.51	47	30.17	35.81
24	27.15	38.30	48	30.23	35.76
25	27.41	38.10	49	30.29	35.70
26	27.64	37.91	50	30.35	35.65
27	27.86	37.74	51	30.40	35.60
28	28.06	37.58	52	30.46	35.56
29	28.24	37.43	53	30.51	35.51
30	28.41	37.30	54	30.56	35.47
31	28.57	37.16	55	30.60	35.42
32	28.72	37.04	56	30.65	35.38

Design of Precision Sprockets . . I

ARTHUR HILL, chief designer, DACO INSTRUMENT COMPANY, BROOKLYN, N Y

Camera-type sprockets are being adapted to computers. Here are gear-tooth calculations and data for dimensioning and generating a precision sprocket, as normally used in a tracking camera or a computer for the accurate transmission of tape or film.

The sprocket will accommodate standard 70 mm perforated film, manufactured in accordance with Military Standard MS33525 (see Fig. 1). The sprocket has 56 teeth, a roll diameter of 3.3273 in. that will provide a mean diameter to the film (0.006 in. thick) of 3.3333 in.

In the design of this particular sprocket, the teeth are shaped to meet two different conditions: the film can leave the sprocket on a 1-in. dia roller, or the film can be pulled off in a tangential path. These two extremes call respectively for a tooth contour not exceeding an epicycloid or an involute using the roll diameter of the sprocket as a base circle, see Figs. 2*a* and 2*b*.

A tooth radius is chosen coincident with both curves at their base and lying just inside at the tip of the tooth. A radius of 0.280 in. with its center 0.010 in. below the roll diameter is found to be satisfactory. It is now necessary to compute data to design and make a templet that can be used on the Panto-Crush grinder. The templet is used for dressing a wheel to grind the space between the sprocket teeth as shown in Figs. 3 and 4. Note: the computations and process described here are not restricted to any specific kind of gear grinder. In fact, this information can also be used for any kind of a gear hobbing machine by simply making a hob to the same proportion and with the same profile as the grinding wheel shown in Fig. 7. It will be noticed that the tooth fillet radius *R* must originate below the roll diameter, and that the circumference of the wheel may be straight in section, i.e., parallel to the axis of the wheel.

One method of fabricating the sprocket is to make it from a 3-piece sandwich construction utilizing ground aluminum side plates of a diameter equal to the roll diameter. The center plate is made of steel equal in thickness to the sprocket tooth thickness, and hardened and ground from the crushed wheel, Fig. 4. The work piece is indexed 56 times to grind the gear tooth spaces; however this method is extremely time consuming for a production setup.

A better production method was devised as a result of a study conducted to determine if it is practical to grind the sprocket teeth on the Reishauer gear grinder. This method requires dressing the grinding wheel to fabricate a tooth form that would satisfy the conditions as stated above. The tooth

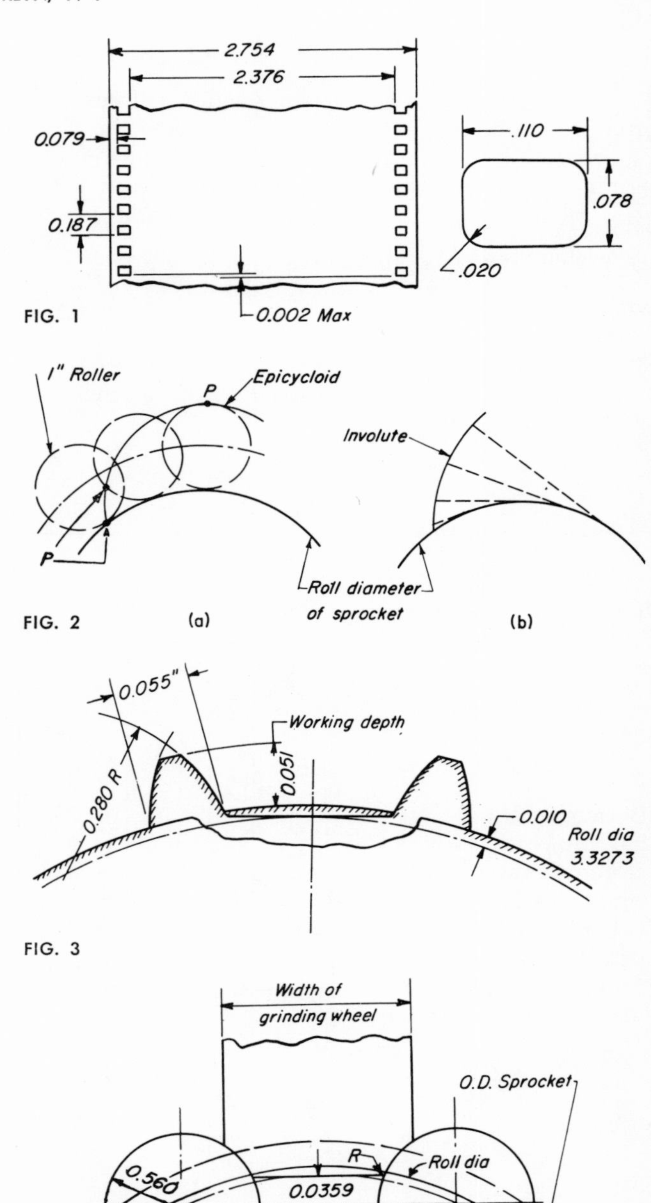

now has an involute form instead of only a radius.

The first step in this preferred production method is to establish the correct pressure angle for the involute tooth form. As indicated above, the tooth profile must be generated in such a manner that there will not be any interference with the epicycloid or involute take-off.

If the involute has a base circle below the roll

Design of Precision Sprockets . . II

diameter of the sprocket the tooth form will clear the perforations in the film while the film is being loaded or unloaded tangent to the roll diameter.

To help determine the correct pressure angle, it is necessary to establish how many degrees of rotation on the sprocket are needed to satisfy an epicycloid profile tooth. This information can be established as follows (See Fig. 5) and these computations:

R_o—Outside radii of sprocket teeth, 1.7146 in.

r —Mean radii of the film rolled on 1-in. dia roller, 0.503 in.

C —Center distance between the sprocket and roller with film in between, 2.1697 in.

R —Mean radii of film rolled on roll diameter of sprocket, 1.6667 in.

$$\cos\theta = \frac{r^2 + C^2 - R_o^2}{2\,r\,C}$$

$$\theta = \frac{0.503^2 + 2.1697^2 - 1.7146^2}{2 \times 0.503 \times 2.1697}$$

$$= 0.92567$$

$$\theta = 22.2302°$$

$$\cos\Psi = \frac{1.7146^2 + 2.1697^2 - 0.503^2}{2 \times 2.1697 \times 1.7146}$$

$$= 0.99382$$

$$\Psi = 6.37202°,\ \text{or}\ 0.1112\ \text{radians}$$

Because the roller r rolls on the radius R and does not slip, they both roll off an equal amount of their circumference. Therefore, their arcs AB and BD are equal. Employing the theorem—radius multiplied by the included angle expressed in radians equals the length of arc in the included angle; then because the two arcs are equal to each other, their equations are also equal to each other.

$$\frac{\theta\,\pi}{180} \times r = \frac{\delta\,\pi}{180} R$$

$$\therefore\ \delta = \frac{\theta\,\pi\,r}{180} \times \frac{180}{R\,\pi} = \frac{\theta\,r}{R}$$

$$\delta = \frac{22.2301 \times 0.503}{1.667}$$

$$= 6.709°,\ \text{or}\ 0.1171\ \text{radians}$$

$$\phi_E = \delta - \Psi = 0.0059\ \text{radians}$$

It is important to make certain that the pressure angle of the teeth at the outside diameter of the sprocket when generated is an involute greater than 14°47′, which would be the pressure angle of an involute tooth whose involute function is equal to ϕ_E. Because the pitch diameter of the gear in the following computations is just about equal to the outside diameter of the sprocket, a pressure angle of 15° 3½′ is selected because the base circle for this gear would then fall approximately 0.011 in. below the roll diameter. Furthermore, by providing a minimum radius of 0.004 in. on the wheel, the sprocket tooth will not be undercut.

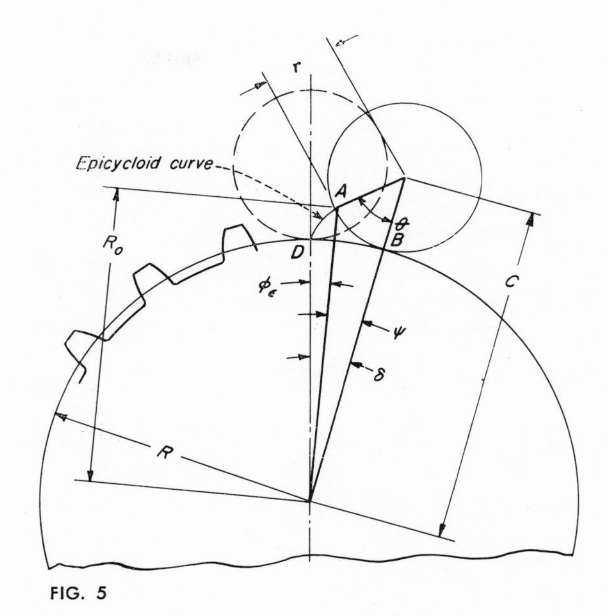

FIG. 5

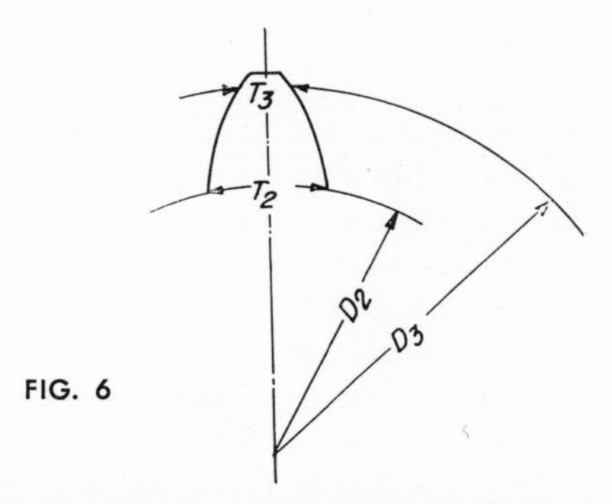

FIG. 6

This data is now converted into information similar to gear calculations in order to setup the Reishauer gear grinder, or any other gear-generating machine tool. From the Reishauer manual PZA 75 a gear train can be setup as follows:

$$\frac{12}{DP} = \frac{G_1}{G_2} \times \frac{G_3}{G_4}$$

From the selection of change gears a 16 1/3 DP is easily obtained.

$$\frac{12}{DP} = \frac{40}{70} \times \frac{54}{42}$$

$$\therefore\ DP = 16\ 1/3$$

Design of Precision Sprockets . . III

Computing the imaginary gear

No. of teeth	N	$= 56$
Diametral Pitch	P	$= 16\ 1/3$
Pressure angle	ϕ_2	$= 15°\ 3\tfrac{1}{2}'$
Pitch diameter	D_3	$= 3.4286$
Circular pitch	C	$= \frac{\pi}{P} = 0.1923$
Outside diameter	OD	$= \frac{N+2}{P} = 3.5510$
Whole depth	D	$= \frac{2.156}{P} = 0.132$
Root diameter	R D	$= 3.2870$

$$D_3 = \frac{N}{P} = 3.4286$$

Base dia. $= 0.9915 \times 3.3333 = 3.3048$

Referring to Fig. 3, the sprocket tooth shows a height of 0.051 in. and an undercut of 0.10 in. below roll diameter. Therefore, the wheel will penetrate 0.061 in. below the outside diameter of the sprocket. Also note that the tooth has a chordal thickness of 0.055 in. at the roll diameter. The arc tooth thickness is 0.055 in. at the point of contact with the mean thickness of the film. However, for the purpose of dimensioning the grinding wheel the arc tooth thickness must be determined at the pitch diameter of the imaginary gear.

T_2 = Arc tooth thickness of tooth at $D_2 = 0.055$
T_3 = Arc tooth thickness of tooth at D_3
ϕ_1 = Pressure angle at point where the mean diameter of the film makes contact with the tooth
D_2 = mean dia of film $= 3.3333$
D_3 = pitch dia $= 3.4286$

$$\text{Cos } \phi_2 = 15°\,3\tfrac{1}{2}' = 0.9639$$

$$\text{Cos } \phi_1 = \frac{D_3 \text{ Cos } \phi_2}{D_2} = \frac{3.4286 \times 0.9639}{3.3333} = 0.99145$$

$$\phi_1 = 7°30'$$

$$\text{Inv } \phi_1 = 0.00075 \quad \text{Inv } \phi_2 = 0.00622$$

$$T_3 = D_3 \left[\frac{T_2}{D_2} \times \text{Inv } \phi_1 - \text{Inv } \phi_2 \right]$$

$$T_3 = 3.4286 \left[\frac{0.055}{3.3333} + 0.00075 - 0.00622 \right] = 0.0343$$

The root diameter of the sprocket is equal to the roll diameter minus 0.020 in. as indicated in Fig. 3, or 3.3073 in. This figure is 0.1213 in. less than the pitch dia of the imaginary gear. Therefore, to determine the dimension for the width of the groove in the grinding wheel at the point of deepest penetration, Fig. 7, multiply 0.1213 in. by the tangent of ϕ_2 and add this value to T_3.
ie: $0.26904 \times 0.1213 + 0.0343 = 0.067$ in.

From this information the grinding wheel can now be dimensioned.

It should be noted that the dimensions given in Fig. 7 are normal to the tooth and not parallel to axis of wheel.

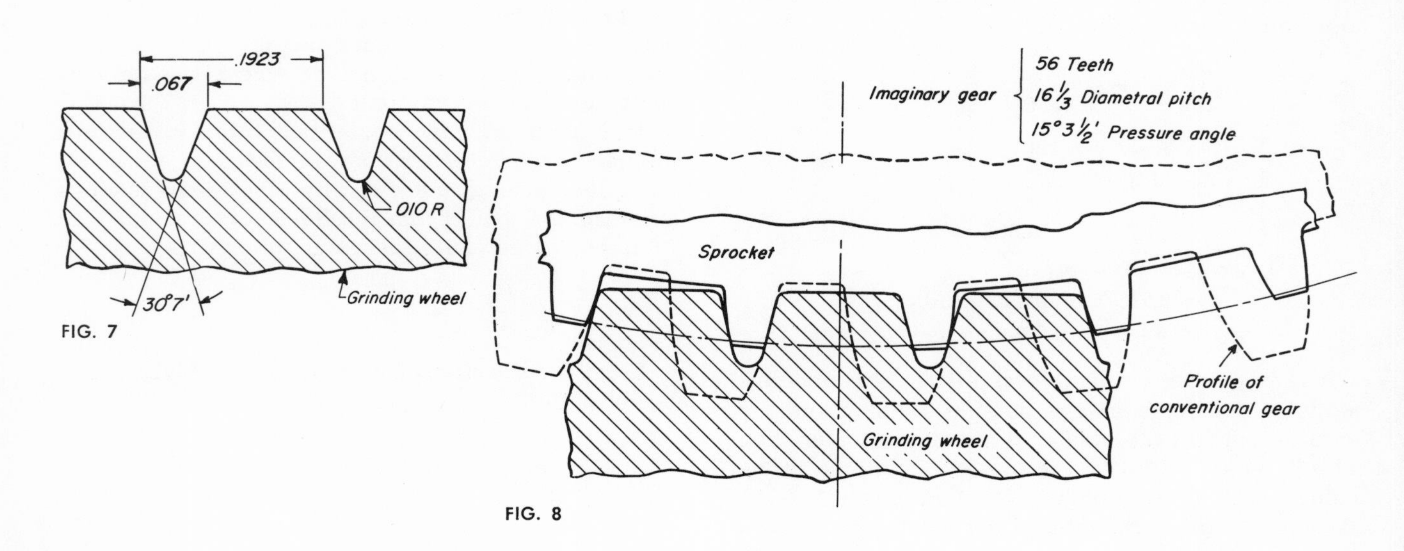

FIG. 7

FIG. 8

Chapter 4

Grinding

High-efficiency grinding

By increasing the cutting speed from 6000 to 12000 sfpm, grinding forces and surface roughness of ground work can be reduced while wheel life and work quality increase (Fig. 1). In this illustration, increase of 'life volume' refers to the width of grinding zone and the cubic inches removed at various speeds.

Grinding capacity (Fig. 2) is the amount of material removed per second by a 1 in. width of wheel. Note that in all of these tests a 60 grit aluminum oxide wheel is being used on AISI 1045 steel. Obviously, a significant increase in capacity is obtained by doubling the wheel speed and supplying oil under high pressure as a lubricant-coolant.

Air flow around the wheel can prevent coolant from penetrating to the grinding zone. In such cases, a sheet metal deflector can be applied. Grinding forces are smaller when using oil than emulsion.

The influence of wheel speeds up to 18,000 sfpm and the results of very high grinding capacity are illustrated by Fig. 3. Here the relations between cutting force, wheel speed and grinding capacity are shown. The resultant cutting force is expressed by means of its normal and tangential components.

Quality of the work in form and dimensional accuracy are affected by the amount of normal force applied. Power requirements for the drive units can be calculated from the tangential force. Grinding forces can be decreased without variation in grinding capacity simply by increasing the cutting speed.

Grinding forces are smaller when oil is used as a coolant instead of an emulsion (Fig. 4). Low grinding capacity and rough surface finish are the results when emulsion is used. High grinding capacity and very smooth surface finish are achieved with oil.

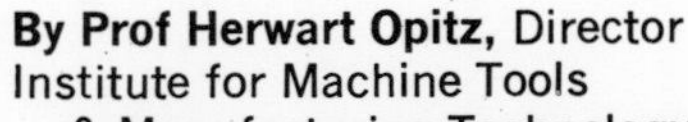

By Prof Herwart Opitz, Director
Institute for Machine Tools
& Manufacturing Technology
Technical University, Aachen, Germany

Source: 'Frontiers in Manufacturing Technology—Vol. III,' published by the Industrial Development Division, Institute of Science & Technology, University of Michigan, Ann Arbor, Mich. This volume is a compilation of papers devoted to important advances in manufacturing technology

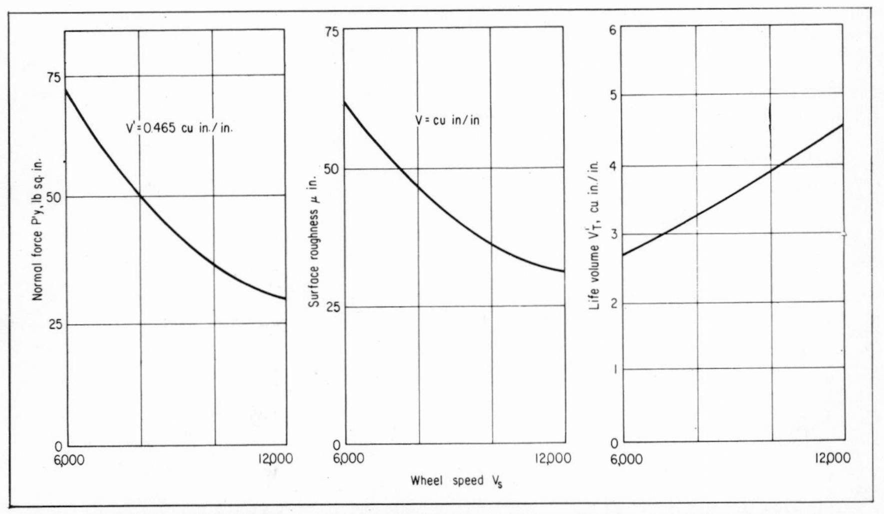

1. Grinding force and surface roughness decrease as wheel speed increases in grinding AISI 1045 with a 60-grit wheel. Stock removal: 0.465 cu in./in.

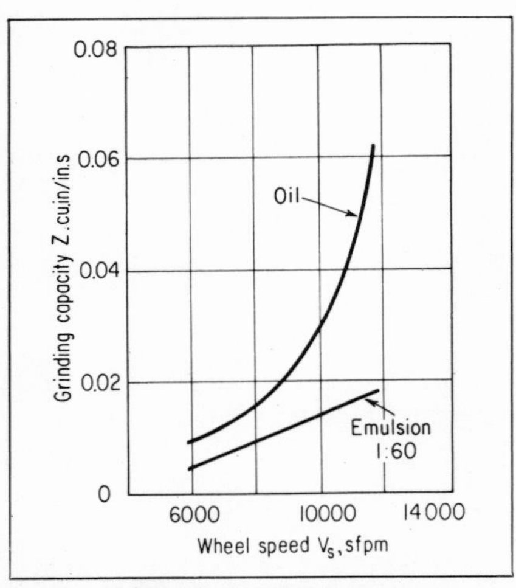

2. High-efficiency grinding should be done with oil for reasons of capacity, surface finish, and burning

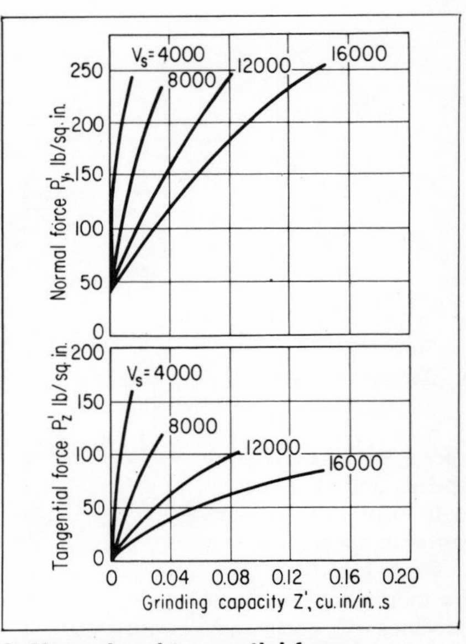

3. Normal and tangential forces vary with grinding capacity. Stock removal: 0.177 cu in./in.

The advantages of oil as a coolant are thus twofold: (1) grinding capacity is increased even with constant wheel speed, and (2) forces are decreased and a better surface finish is produced.

The influence of wheel speed on wheel life during the machining of various materials is shown in Fig. 5. According to the machining properties of the different materials, life volumes can be increased by maintaining grinding capacities. In the most unfavorable case (the 12-1-4 German highspeed steel), life volume can be increased tenfold. On the other hand, for the light machinable steels such as AISI 1045, a thousandfold increase is possible.

Highest capacities can be achieved with hard wheels, but a considerable temperature load can occur in the rim zone. At all capacities, temperatures increase with wheel speeds. The speed ratio (the ratio between wheel and work speeds) is important in high efficiency grinding. To avoid change in the microstructure of the surface zone of the workpiece, increase the work

High-efficiency grinding [Continued]

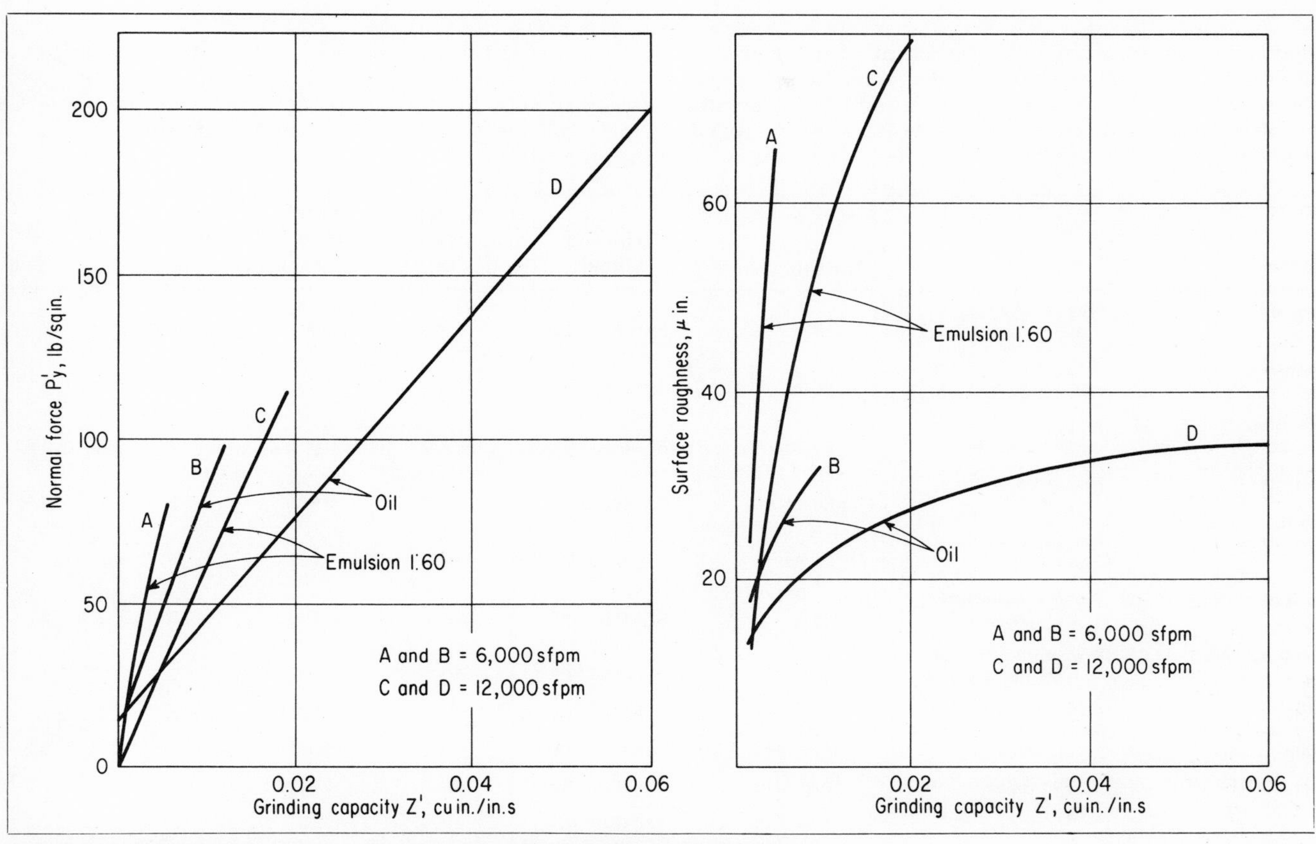

4. Coolant affects grinding forces and surface roughness. Speed ratio: q = 40

speed while increasing both the wheel speed and the capacity. Speed ratio will probably not exceed 60 under suitable speed and capacity conditions.

Real grinding time can be reduced by increasing the wheel speed so that a product of consistent quality can be produced at constant wear rates per workpiece. Fig. 6 shows a reduction in grinding time by 90 to 97%, depending on the type of wheel and work.

High-efficiency grinding makes it possible to solve grinding problems more quickly. Moreover, available methods are broadened. Certain machining operations preparatory to finish grinding can now be replaced by high efficiency grinding. For example, it is now possible to machine a part completely, even with a profile, in one operation from the rough stock to the finished form by plunge grinding. ■

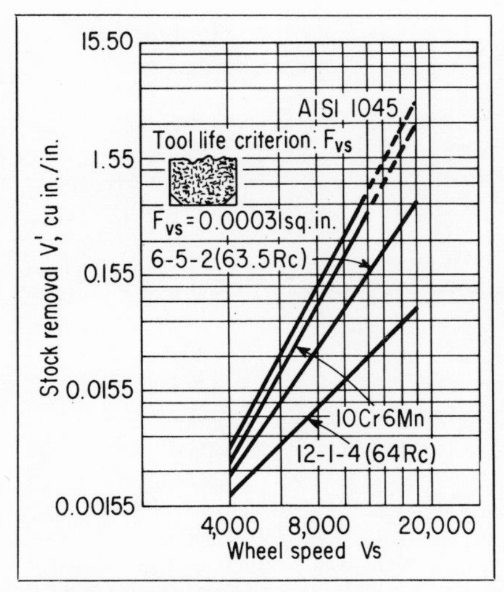

5. Tool life is affected by work material. Grinding capacity Z′ = 0.0155 cu in./in./sec

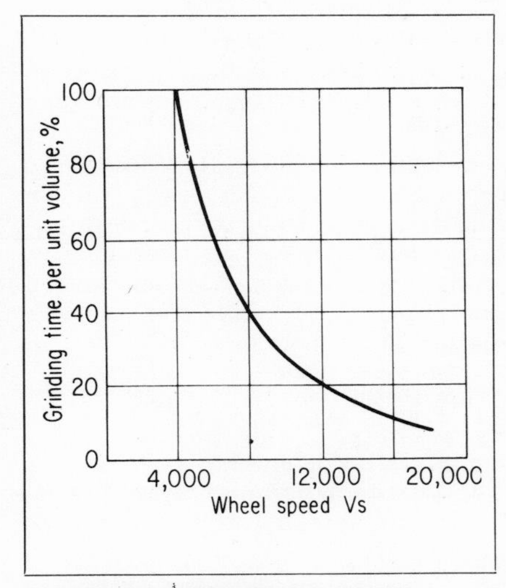

6. High wheel speeds can sharply reduce real grinding time without damage to microstructure

G-ratios in surface grinding

Source: 'Grinding Ratios for Aerospace Alloys;' Publication AFMDC-66-2, prepared under the auspices of the Manufacturing Technology Division, U S Air Force, Wright-Patterson Air Force Base, Ohio

Wheel wear is economically important in most grinding operations. Therefore, there has been much interest in a finite measure for it; namely, the grinding ratio G. This quantity is the volume of material removed divided by volume of wheel worn away.

Because of the importance of the refractory alloys—tungsten, molybdenum, columbium, tantalum and zirconium—in many aerospace applications, the G ratios for these materials are given for surface grinding operations.

Material	Condition	Hardness	Grinding wheel	Wheel speed ft./min.	Down feed in./pass	Grinding fluid	"G" ratio
Tungsten	Pressed & sintered 93% density	26 Rc	32A46N5VBE	2000	0.001	4	2.9
Tungsten	Pressed & sintered 93% density	26 Rc	32A46N5VBE	2000	0.001	1	2.8
Tungsten	Pressed & sintered 93% density	26 Rc	32A46N5VBE	2000	0.001	5	3.8
Tungsten	Pressed & sintered 93% density	26 Rc	32A46J8VBE	2000	0.001	5	2.5
Tungsten	Pressed & sintered 93% density	26 Rc	32A46K8VBE	2000	0.001	5	2.3
Tungsten	Pressed & sintered 93% density	26 Rc	32A46N5VBE	2000	0.0005	5	5.2
Tungsten	Pressed & sintered 93% density	26 Rc	32A46N5VBE	2000	0.002	5	3.6
Moly - .5 Ti	Hot rolled	220 BHN	32A46H8VBE	2000	0.001	2	1
Moly - .5 Ti	Hot rolled	220 BHN	32A46H8VBE	4000	0.001	2	3.2
Moly - .5 Ti	Hot rolled	220 BHN	32A46H8VBE	4000	0.0005	2	5.2
Moly - .5 Ti	Hot rolled	220 BHN	32A46J8VBE	4000	0.001	2	3.2
Moly - TZM	Hot rolled	248 BHN	32A46K8VBE	2000	0.001	5	13
Moly - TZM	Hot rolled	248 BHN	32A46K8VBE	4000	0.001	5	12
Moly - TZM	Hot rolled	248 BHN	32A46L8VBE	4000	0.001	5	13
Moly - TZM	Hot rolled	248 BHN	32A46N5VBE	2000	0.001	5	25
Moly - TZM	Hot rolled	248 BHN	32A46N5VBE	4000	0.001	5	12
Moly - TZM	Hot rolled	248 BHN	32A46N5VBE	4000	0.001	2	6
Moly - TZM	Hot rolled	248 BHN	32A46N5VBE	4000	0.001	3	5

Notes:

Table speed = 40 fpm and crossfeed = 0.050 in./pass in all cases.

Cutting fluid

1. Soluble oil (1:20)
2. Soluble oil (1:40)
3. Highly chlorinated oil
4. Highly sulfurized oil
5. Potassium nitrite KNO_2 (1:20)
6. Low viscosity sulfurized and chlorinated grinding oil

G-ratios in surface grinding (continued)

Material	Condition	Hardness BHN	Grinding wheel	Wheel speed ft./min.	Down feed in./pass	Grinding fluid	"G" ratio
Moly - TZM	Hot rolled	248	32A46N5VBE	4000	0.001	4	5
Columbium-unalloyed	Extruded	112	32A46J8VBE	4000	0.001	3	2.9
Columbium-unalloyed	Extruded	112	32A46K8VBE	4000	0.001	3	3.1
Columbium-unalloyed	Extruded	112	32A46K8VBE	4000	0.001	1	1.0
Columbium-unalloyed	Extruded	112	32A46K8VBE	4000	0.001	4	1.8
Columbium - D31	Extruded	217	GC46K6VP	2000	0.001	5	1.5
Columbium - D31	Extruded	217	32A46K8VBE	2000	0.001	5	5.0
Columbium - D31	Extruded	217	32A46K8VBE	4000	0.001	1	3
Columbium - D31	Extruded	217	32A46K8VBE	4000	0.001	4	2.5
Columbium - D31	Extruded	217	32A46K8VBE	4000	0.001	3	2.0
Columbium - D31	Extruded	217	32A46K8VBE	4000	0.001	5	4.8
Columbium - D31	Extruded	217	32A46J8VBE	4000	0.001	5	3.5
Columbium - D31	Extruded	217	32A46K8VBE	2000	0.0005	5	7.5
Tantalum - 10W	Forged	241	32A46J8VBE	2000	0.001	5	3.2
Tantalum - 10W	Forged	241	32A46J8VBE	4000	0.001	5	1.7
Tantalum - 10W	Forged	241	32A46J8VBE	4000	0.001	3	1.0
Tantalum - 10W	Forged	241	32A46G8VBE	2000	0.001	5	0.3
Tantalum - 10W	Forged	241	32A46K12VBEP	2000	0.001	5	2.2
Tantalum - 10W	Forged	241	6C60L6VP	2000	0.001	5	0.3
Zirconium	—	—	C46JV	3000	0.001	6	8.6
Zirconium	—	—	C46JV	6000	0.001	6	4.7
Zirconium	—	—	C46JV	3000	0.005	6	3.2
Zirconium	—	—	C46JV	6000	0.005	6	5.7

Notes:
Table speed = 40 fpm and crossfeed = 0.050 in./pass in all cases.

Cutting fluid
1. Soluble oil (1:20)
2. Soluble oil (1:40)
3. Highly chlorinated oil
4. Highly sulfurized oil
5. Potassium nitrite KNO_2 (1:20)
6. Low viscosity sulfurized and chlorinated grinding oil

Rules for surface grinding, 1

Four years of research will show the manufacturing engineer and foreman how to get the best results from use of segment wheels on vertical surface grinders

A constant bugaboo in all machining operations is to determine feeds and speeds that produce minimum total cost.

Fortunately, in vertical spindle surface grinding, changes in workpiece area, segment hardness, etc, have only a small effect on the minimum-cost, downfeed rate of the machine. Also, because the cost curve (Fig. 1) is generally flat in the area of minimum cost, small variations do not greatly effect the actual cost of the operation.

To use these curves, first decide what your labor cost is for a particular machine. Then, use your company's standard cost figures for machine-operating condition calculation. If you run your machine at downfeed rate indicated for your labor cost per hour, you will be operating very close to the point of minimum cost.

By Charles Pollock
Norton Co, Worcester, Mass

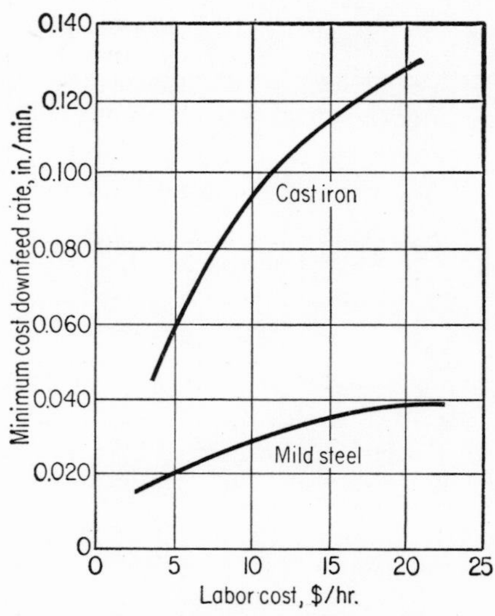

Fig. 1. Once you know the labor cost per hour, you can find the minimum-cost downfeed

Ourt tests were run on 1020 hot-rolled steel, because it is so commonly used in industry. The 100-hp vertical spindle surface grinder had a 54-in. dia work table and a 26-in. dia segment wheel, containing 12 segments. Wheel speed was 4900 fpm.

Fig. 2 shows a plot of metal removal rate against downfeed rate. Vertical spacing of the curves is roughly proportional to changes in workpiece area. Therefore, instead of downfeed rate we prefer to use volumetric feed rate, as in Fig. 3.

Loading the machine

The most significant result of our research at Norton Co is shown in Fig. 3. We found that the rate of stock removal on a surface grinder is dependent on the volumetric feed rate of the machine. Volumetric feed rate (the rate at which material is removed if the grinding wheel does not wear) is determined by multiplying the down-feed rate of the grinding machine times the area of the workpiece being ground.

Consequently, to increase output it is merely necessary to increase the volumetric feed rate. This can be done by either increasing the downfeed rate or by increasing the number of parts on the work table, or both together.

Which costs less—increasing the downfeed rate or increasing the number of pieces ground? Fig. 4 gives us the answer. Here we plot the wheel wear rate against volumetric feed rate. The result is a family of straight lines with different slopes. The smaller the workpiece area, ie, the fewer the number of parts on the work table, the steeper the line (see curve A) and, conversely, the larger the workpiece area the lower the slope (see curve C).

To answer the question of whether to increase downfeed rate or work area, let us look at a simple example. You will notice from curve B (Fig. 3) that an increase in volumetric feed rate from 15 cu in. per min to 30 cu in. per min gives an increase in output from 6 cu in. of material removed per minute to 8 cu in. of material removed per minute.

In Fig. 4 assume a volumetric feed rate of 15 with a 324 sq-in. workpiece area (see curve A). Then wheel wear rate will be approximately 2¼ cu in. per min. If we double the downfeed rate, so that volumetric feed rate is 30, the wheel wear rate increases to ap-

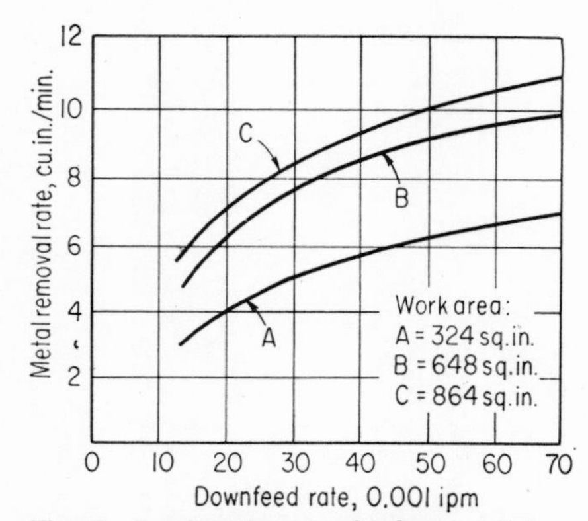

Fig. 2. In abrasive machining, metal removal is a function of downfeed rate or volumetric feed rate

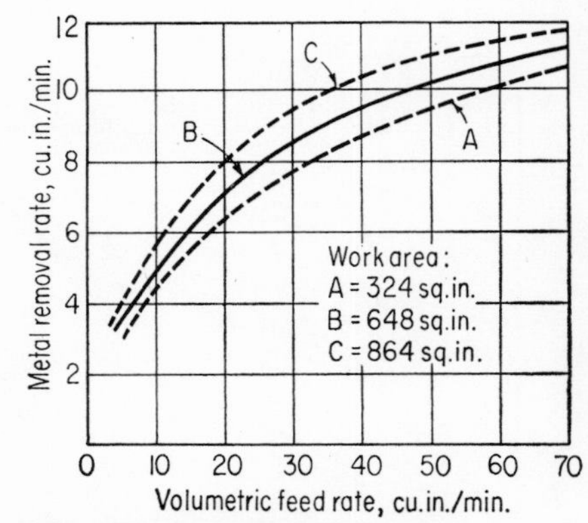

Fig. 3. To increase output, it is merely necessary to increase the down feed rate or parts on table

Rules for surface grinding [Continued]

proximately 5½ cu in. per min; a rather substantial increase. However, if we increase volumetric feed rate to 30 by doubling the number of pieces on the work table (648 sq in., or curve B) the wheel wear rate increases to only 2½ cu in. per min; a very small increase.

Wheel wear increases

Other measures of performance are of interest. Fig. 5 shows a graph of grinding ratio (the volume of metal removed per unit volume of wheel wear) versus the volumetric feed rate. In the previous example, the grinding ratio was 2.5 under the initial conditions (volumetric feed rate of 15, curve A.) When the feed rate is doubled, the value of G drops to 1.2, because wheel wear increases more rapidly than metal removal. If the work area is doubled, instead of feed rate, the value of G increases to 3.5.

Thus it is obvious that if we can obtain the same increase in output by doubling either the downfeed rate or the number of pieces on the work table, we will be better off doubling the work, because this results in only a slight increase in wheel wear rate.

Minimum cost

By increasing the work area we still operate at the lowest-cost downfeed rate. Hence by doing this we continue to operate at minimum cost and at the same time we have greatly increased output without significantly changing the rate of wheel wear. The net effect is to reduce costs from 5¢ per cu in. of material removed to 3 2/3¢. If we had obtained the same increase in productivity by doubling the downfeed rate, costs would have gone up to 8 2/3 cents.

Power goes up

We have not achieved something for nothing. Fig. 6 shows the price we pay is a sligthly increased power usage. From this illustration, we see that when grinding at a volumetric feed rate of 15 (curve A), we consume approximately 60 horsepower. By doubling the downfeed rate the power increases to 77 horsepower. However, if we double the work area, the machine will draw 83 hp at a volumetric feed rate of 30 cu in. per min.

The extra 6 horsepower is a low price for a 51% increase in productivity and a 60% decrease in wheel wear per part ground.

What does this mean to you? From this information we can derive two rules of thumb for operating vertical spindle grinders in an economical manner:

First, use the chart in Fig. 1 to select the downfeed rate that gives you minimum cost grinding.

Second, place as many parts on the work table as possible.

Keep one restriction in mind. Do not put parts in the center of the table, because they do not have an opportunity to get out from under the wheel and be cooled by grinding fluid. As a result, these parts are often overheated. They either burn or expand. If they expand, they will be undersize after cooling.

There is one limitation on the use of this information. The above concepts do not apply to grinding tool steel. Usually, the primary limitation on productivity is the high tendency of tool steels to burn and warp.

Results apply generally

The curves in Figs. 3, 4 and 6 were generated on one grinder, using one work material. Consequently, we would not expect that every grinding machine and every work material will give identical results. However, the shapes of the curves will be the same and the general relationships and rules of thumbs will apply for all types of vertical spindle surface grinders and for all mild steels and irons.

Curves for other jobs

In Fig. 4, the slope of the straight lines of wheel wear are also proportional to the ratio of the workpiece area divided by the grinding wheel area. Consequently, it is possible to generate a family of curves for two or three workpiece areas on one machine and then figure out where the lines would be for other workpiece sizes. ■

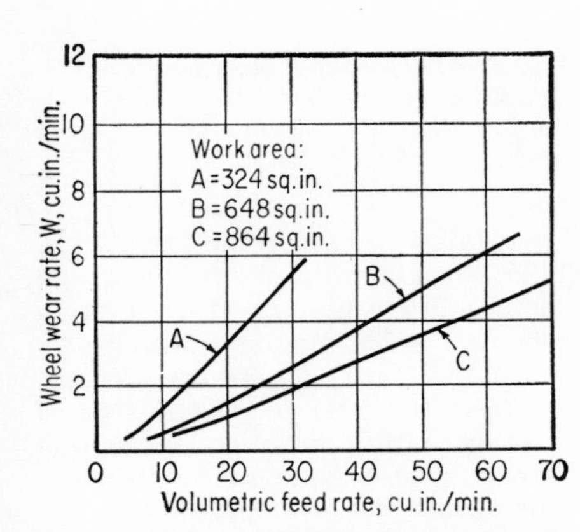

Fig. 4. An increase in work area is a more effective way of increasing output than increasing the downfeed

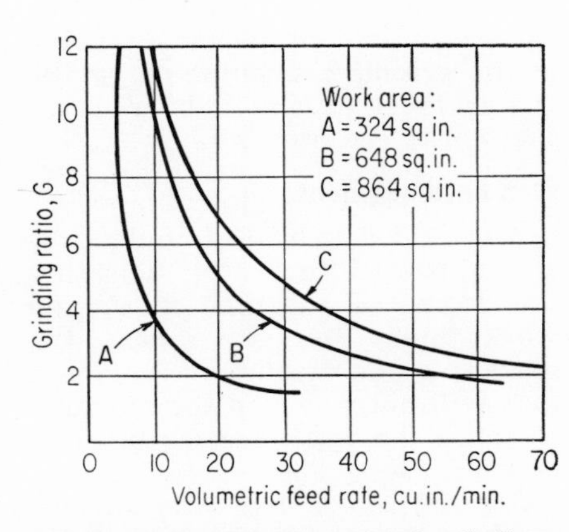

Fig. 5. If feed rate is doubled, grinding ratio drops, because wheel wear increases faster than metal removal

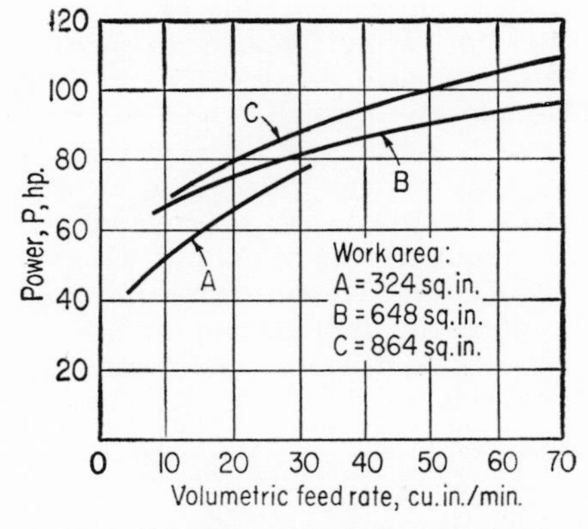

Fig. 6. Power consumption also favors doubling the work area instead of increasing the downfeed rate

Rules for surface grinding, 2

Four rules are: (1) select downfeed for minimum cost, (2) use largest effective table load, (3) apply 24- or 30 grit, and (4) use hardest feasible segments

This concluding installment will discuss selection of the proper segment-wheel grade for best grinding results, and also how to select the best abrasive grit size to assure that finish requirements are met.

In Fig. 7, metal removal rate is plotted against volumetric feed rate for four different grades of segments. (Each curve represents a curve such as the one developed in Fig. 3 of the preceding installment). This chart shows that for a given volumetric feed rate, output increases with the hardness of the segment. However, with each increase in hardness, the increase in metal removal rate is less.

Consequently, less benefit accrues from increased hardness when a hard segment grade is already being used than when hardness is increased from the softest grade.

Cost sets downfeed

In the preceding installment (**AM**—Jan 16 '67, p153) it was shown that the downfeed rate of a vertical spindle grinder, in order to cut operating costs to a minimum, depends primarily on the hourly time charge for the grinding machine and the material being ground.

In Fig. 8, for the same segment grades, we get a plot of straight lines for wheel wear rate against volumetric feed rate. These lines are closely spaced, showing that there is relatively little change in grinding wheel wear rate as the grade of the segment is changed. Consequently (Fig. 7 and 8) only the H grade segment might be used because the increase in metal removal rate (over an E grade segment) is relatively large, yet the decrease in wheel wear rate is relatively small. But as segment hardness increases the abrasive wears slower. Individual grains are held in place longer and consequently become somewhat duller before release from the matrix. Thus metal is removed faster by grinding with a hard wheel, but it is not as sharp as a soft wheel. More power is needed to push the abrasive through the metal.

By Charles Pollock
Norton Co, Worcester, Mass

Using the power

On the other hand, when grinding with a soft, sharp, E grade segment a large portion of the power is for metal removal.

Thus relatively large increases in horsepower must be expected when going from a G to an H grade segment, while achieving smaller increases in metal removal rate.

Fig. 9 confirms expectations. This chart shows that increasing segment hardness from E through H grades results in approximately equal increases in the horsepower necessary to remove material. These equal increases in power, coupled with the reduced increases in metal removal rate (shown in Fig. 7) confirm our thinking about the dull grains in hard wheels.

Thus, in selecting the best segment grade for a particular job, we are limited by power available at the grinding machine or by the power level at which burning of the workpiece occurs.

Pick hard segments

As a third rule of thumb, it is recommended that after selecting the downfeed rate and filling the work table, the user selects the hardest grade of abrasive segment that will not overload the machine or cause the workpiece to burn or warp.

Fig. 10 shows how the surface finish of a steel workpiece is closely related to abrasive grit size. This chart represents the average surface finish measured under two sets of grinding conditions: no spark-out, and 90-sec spark-out.

These finishes, representative of 'best' grinding practice, were obtained under specific conditions. The similarity of slopes of the two lines confirms the relative depend-

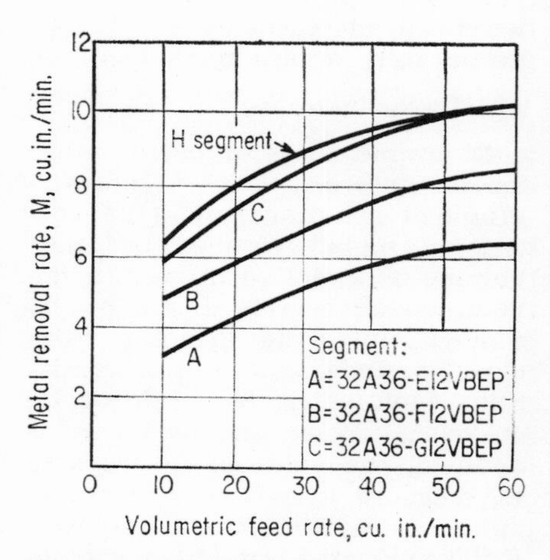

Fig. 7. Metal removal rate rises when one applies the harder segments to surface grinding

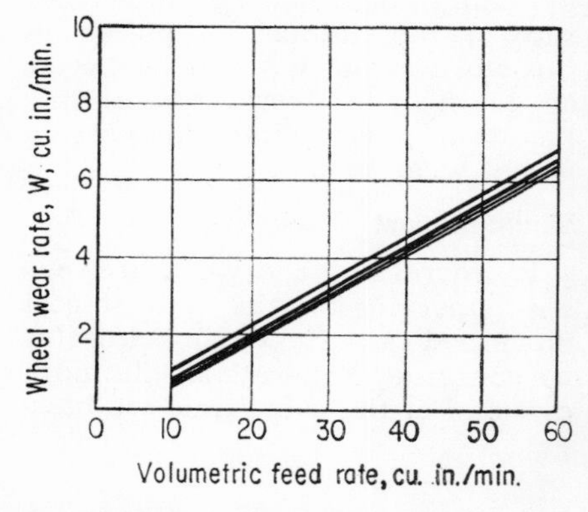

Fig. 8. Wheel wear rate is about the same for grades E to H, but harder segments will dull

Rules for surface grinding [Continued]

ence of finish upon grit size, providing the segment is not loaded or glazed.

This chart also indicates that proper spark-out period is more important in obtaining a good surface finish than selection of a fine grit abrasive. Unless surface finishes of less than 15 to 20 microinches AA are desired, a 24-grit segment, with proper spark-out, will generally do the job.

Because spark-out is more important in determining surface finish than grit size, a decision regarding the proper grit size should be based primarily on productivity of different abrasive sizes. Fig. 11 shows this relationship, and that either 24 or 30-grit size will give the best performance for general-purpose grinding.

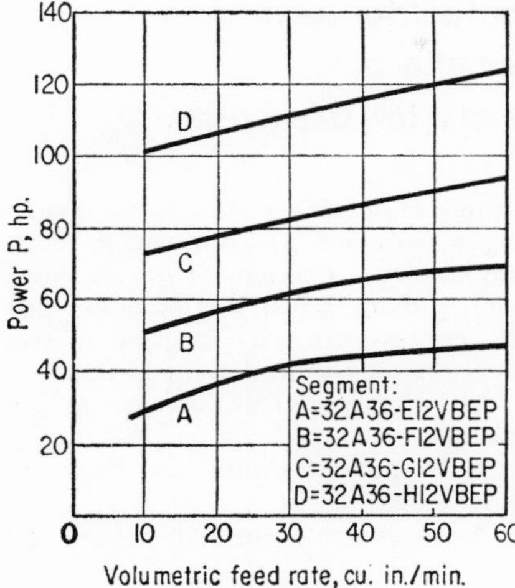

Fig. 9. Power consumption rises with segment hardness and may be beyond capacity of the grinder

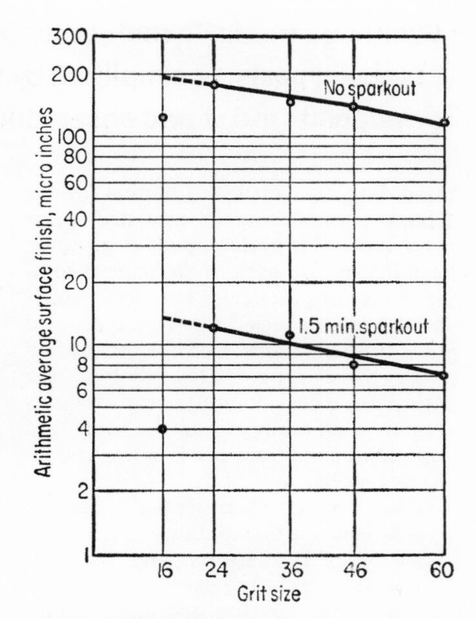

Fig. 10. Sparkout is always helpful in securing a better finish, even with a coarser grit

To boost output

Note also the loss of productivity between the peak of the curve (24 and 30 grit) and the low at 60-grit segments. For example, suppose that you are now using 80-grit segments and a 50 microinch finish is satisfactory. Output could be increased by as much as 75% by changing to either 24 or 30-grit size segments of the same grade.

The metal removal rate ratio shown in Fig. 11 is a means of comparing the performance of various grit sizes. It is the average ratio of the metal removal rate obtained for a particular grit size divided by the metal removal rate, at an equal power level, for 24-grit segments.

This ratio, then, says: To find the average metal removal rate, at a specific power level for an abrasive grit size, multiply the metal removal rate of a 24-grit-size segment by the metal removal rate ratio.

Example: Suppose that the test results show that at 100 hp and a feed rate of 0.030 ipm, a 32A24-H12VBEP segment has removed 10 cu in. of metal per minute. To find the metal removal rate of an 60-grit segment at 100 hp, and the same feed rate, multiply 10 cu in. per minute by the metal removal rate ratio from Fig. 11, for the 60-grit size. Thus, metal removal rate for 60 grit equals 10 x 0.76 = 7.6 cu in. per minute.

From the data presented in these two reference sheets we can develop four rules of thumb for obtaining the best possible results when abrasive machining steel or cast iron on vertical spindle surface grinders. These are:

Rule 1—Using Fig. 1, in the first reference sheet (**AM**—Jan 16 '67, p 153), select the downfeed rate that will give minimum cost. This downfeed rate is based on the hourly charge for the machine and the material being ground.

Rule 2—Place as many parts as possible on the grinding machine work table. Warning: Do not place parts over the center section of the table (approximately 12 to 18 in. in diameter).

Rule 3—Select either 24 or 30-grit segments, unless surface finish requirements are so restrictive that a finer grit segment is required. Remember that excellent surface finishes can be obtained by properly sparking out the workpiece.

Rule 4—Select the hardest possible grade of segment which will not overload the machine or cause the workpieces to burn or warp.

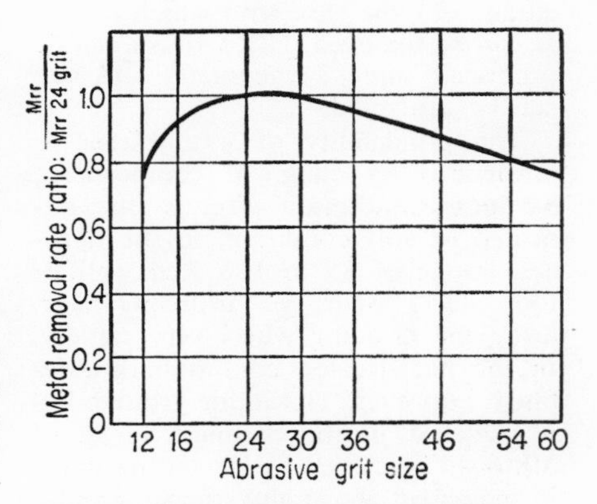

Fig. 11. A 24- or 30-grit segment will give the best metal-removal rate, exclusive of finish

Application of these rules will depend upon the size of the workpieces being ground. For instance, if you are grinding small rims on castings, because of the small total work area you will be able to use a very hard grade of segment. On the other hand, if you are grinding relatively large flat steel plates, where flatness is desired, you are likely to run into heat limitations and a soft segment grade will be required. ■

Grinding tool steels

Grinding of today's tool steels, whether for tools or precision parts, is simplified by use of a grindability index and observance of a few basic rules

Grinding is a critical function in the shaping and sizing of tool-steel parts and tools. To define specific grinding conditions for each trade-named tool-steel is impractical, but tool steels have been classified by AISI type, and the brand names can be identified with these types. This enables the grinding wheel manufacturer to recommend suitable grinding wheels and operating conditions for grinding specific tool steels.

The AISI classification can be worked into a Grindability Index (see table on p55) that compares the relative ease of grinding each steel and, in addition, defines the operating conditions that are most suitable for each classification.

Factors affecting grindability include: (1) the ease with which a metal can be removed, (2) a finish can be produced, and (3) geometry and size can be generated.

The grindability of a tool steel is influenced by chemical composition, hardness, and grain size. In the annealed or soft condition, in the hardness range of 85 to 100 Rockwell B, tool steels wear, or probably tear, away the grinding wheel very rapidly. In the heat-treated condition, grindability improves. Optimum grindability is achieved in the hardness range of 30 to 40 Rockwell C. As the hardness is increased above this range, grindability decreases.

Hard carbides, which are added to the steel to increase wear resistance, reduce its grindability. Finer grain size also makes a steel more difficult to grind.

Guiding principles given here are simply indicators of how to use grinding wheels and are not meant to be absolute specifications. They do, however, provide a fast way of preventing difficulties—not only in grinding tool steels but in all grinding operations.

- To remove stock, use coarser grit wheels.
- To produce finish, use finer grit wheels.
- To generate form, use finer grit wheels.
- To grind hard metals, use softer grade, and finer grit wheels.
- To grind soft metals, use harder grade, and coarser grit wheels.
- To grind large areas, use softer grade wheels.
- To grind small areas, use harder grade wheels.

Surface grinding

- To minimize heat and warpage, use fast table speeds, light downfeeds, and dress wheel before final size.
- To produce finish, dress the wheel to a fine finish.

Cylindrical grinding

- To remove stock, use a fast traverse speed and a slow work speed.
- To produce finish, use a slow traverse speed and a fast work speed.

Tool and cutter grinding

- To grind primary angles—small contact area, use a harder grade wheel of finer grit size.
- To grind secondary angles—large contact area, use a softer grade wheel of coarser grit size.
- The grinding wheel can be altered to act softer or harder by changing the width of the land on the wheel face. For harder action, widen the land; for softer action, narrow the land; if the wheel burns the work, narrow the land; and if the wheel wears away too rapidly, increase the land.

The wheel should be dressed frequently to prevent glazing and to provide an open surface. Bring the wheel to the work gradually, using less infeed.

Where complex forms must be ground and grinding fluids can be used, straight grinding oil will maintain form with fewer dressings than either dry grinding or grinding with a water-miscible fluid. On the other hand, some straight grinding oils may also produce compressive stress in the ground part.

Regardless of how smooth the tool surfaces are, the ground work surface will contain grinding scratches parallel to the direction of grinding. It has been proved that the direction of grinding marks has a lot to do with tool life, especially for such tools as deep-drawing dies and metals under high pressure. Tools with grinding marks parallel to the direction of movement of metal over the tools greatly outperform those with grinding marks at right angles to the direction of movement.

Economics

When grinding tool steels, the cost of the abrasive is insignificant as compared to the total cost. So several factors which would appear to increase abrasive consumption should be considered from the standpoint of cost. For example, frequent wheel dressings will insure freedom from burn, warp, and heat in the ground part. Softer grade wheels inhibit loading and glazing. The use of a friable abrasive grain increases productivity because it permits an inexperienced man to grind with a minimum likelihood of danger and makes it easier to generate geometry and size in hard-

By John A. Mueller, manager
Abrasives Systems Development
Carborundum Co., Niagara Falls, N.Y.

Source: "Tool Steel Trends," published by the Committee of Tool Steel Producers, American Iron & Steel Institute, New York

Grinding tool steels [Continued]

Grindability Index					
		Easy to Grind			Harder to Grind
		High Grindability	Medium Grindability	Low Grindability	Difficult to Grind
	Tool Steel Groups	**A** Cold Work — Air	**F** Carbon Tungsten	**M** High Speed	**A** 7
		H Hot Work **O** Cold Work — Oil	**D** Cold Work	**T** High Speed	**D** 7 **M** 4
		L Low Alloy **P** Mold	**H** 41	**D** 2	**M** 15
		S Shock Resisting	**H** 42		**T** 15
		W Water Hardening	**H** 43		
Surface Grinding	**Grinding Wheel Specification**				
	Grain Type	Semi Friable	Friable Aluminum Oxide	Friable Aluminum Oxide	Friable Aluminum Oxide
	Grit Size	46-60	46-60	46-60	80-100
	Grade	H-J	G-H	F-G	F
	Bond	Vitrified	Vitrified	Vitrified	Vitrified
	Speeds & Feeds				
	Wheel Speed	5500-6500 S.F.P.M.	5500-6500 S.F.P.M.	5500-6500 S.F.P.M.	5500-6500 S.F.P.M.
	Table Speed	50-75 F.P.M.	50-75 F.P.M.	Fast as possible	Fast as possible
	Crossfeed	1/32″-1/16″	1/32″-1/16″	1/32″-1/64″	Full face of wheel
	Infeed	.001″-.003″	.001″-.002″	.0005″-.001″	.0005″-.001″
Cylindrical Grinding	**Grinding Wheel Specification**				
	Grain Type	Semi-Friable Aluminum Oxide	Semi-Friable Aluminum Oxide	Friable Aluminum Oxide	Friable Aluminum Oxide or Friable Silicon Carbide
	Grit Size	60-80	60-80	60-80	80-100
	Grade	K-L	J-K	H-J	F-H
	Bond	Vitrified	Vitrified	Vitrified	Vitrified
	Speeds & Feeds				
	Wheel Speed	5500-6500 S.F.P.M.	5500-6500 S.F.P.M.	5500-6500 S.F.P.M.	3000 S.F.P.M.
	Work Speed				
	Roughing —	50-100 S.F.P.M.	50-100 S.F.P.M.	70-120 S.F.P.M.	70-120 S.F.P.M.
	Finishing —	100-150 S.F.P.M.	100-150 S.F.P.M.	120-150 S.F.P.M.	120-150 S.F.P.M.
	Infeed	.001″-.002″	.001″-.002″	.001″-.0015″	.0005″-.001″
	Table Speed				
	Roughing —	1/2-1/3 width of wheel per rev. of work	1/2-1/3 width of wheel per rev. of work	1/3-1/6 width of wheel per rev. of work	1/3-1/6 width of wheel per rev. of work
	Finishing —	1/3-1/6 width of wheel per rev. of work	1/3-1/6 width of wheel per rev. of work	1/6-1/12 width of wheel per rev. of work	1/6-1/12 width of wheel per rev. of work
Tool & Cutter Grinding	**Grinding Wheel Specification**				
	Grain Type	Semi-Friable Aluminum Oxide	Semi-Friable Aluminum Oxide	Friable Aluminum Oxide	Friable Aluminum Oxide
	Grit Size	46-60	46-60	46-60	60-80
	Grade	J-K	H-J	G-H	F-G
	Bond	Vitrified	Vitrified	Vitrified	Vitrified
	Speeds & Feeds				
	Wheel Speed	5500-6500 S.F.P.M.	5500-6500 S.F.P.M.	5500-6500 S.F.P.M.	5500-6500 S.F.P.M.
	Table Speed	6-10 F.P.M.	6-10 F.P.M.	6-10 F.P.M.	Fast as possible
	Infeed	.001″-.003″	.001″-.003″	.001″-.002″	.0005″-.001″
Grinding Fluid	**Straight Grinding**	Water Miscible Fluid	Water Miscible Fluid	Water Miscible Fluid	Water Miscible Fluid
	Complex Forms	Straight Oil	Straight Oil	Straight Oil	Straight Oil
	Note: Cutter grinding usually does not use grinding fluid.				
	Note: To produce better finish — use finer grit size wheels. To generate better form — use finer grit size wheels.				

to-grind metals or difficult shapes.

Diamond abrasive wheels

Diamond wheels can grind tool steels with acceptable rates of removal, but appear to be most effective in grinding the M and T types. Diamond-impregnated, epoxy-faced wheels are coming into use for hard materials. The use of these wheels, at present, appears to be in the surfacing of metals rather than in the grinding of forms and shapes.

Wheel specifications

Suggested starting recommendations both for grinding wheel specifications and speeds and feeds are shown in the table. Different steels in the same AISI group exhibit different grinding characteristics, so minor changes may have to be made. For example, D3 is harder to grind than D2. Types A7, D7, M4, M15, and T15 are in a class by themselves, because they are very resistant to abrasion and are difficult to grind.

Grinding wheels are tools of extreme versatility. A wide selection of wheels exists, from which optimum wheels may be chosen for specific operations. In addition, each grinding wheel itself can be manipulated by means of speeds, feeds, and dressing techniques to accommodate variations in the steel and in the grinding operation. The terms "hard acting" or "soft acting" should not be confused with the inherent hardness or softness of the wheel in respect to its grade. The measure of hardness or softness of the wheel in respect to grade is its ability to resist breakdown or shedding of grain.

Operating suggestions

Use grinding wheels that are best suited for the job, and maintain the machine in the best operating condition. Use the proper jigs and attachments to save both time in grinding and checking. Use correct clearance angles for each type of tool for each material. Maintain a bank of tools, ready to be placed on the job, and work out a schedule of tool replacement so that tools are resharpened at the proper intervals.■

Abrasive belt machining

Here are definite figures on stock removal rates for belt surface grinding, and comparative cost per cubic inch of metal removed by abrasive belt and bonded-wheel grinding

Abrasive belt machining can now attain stock removal rates of 10 lb/min and finishes to between 5 and 15 microinches, and hold tolerances of ±.001 in. for surface grinding and 0.00025 in. to 0.001 in. on centerless grinding. Grinding belts used for abrasive machining are resin-bond, coated with either aluminum oxide, silicon carbide or diamond.

Good abrasive maching operations consume only 4.5 to 7 hp per cu in. of low carbon steel removed per minute, versus 6 to 10 hp for comparable bonded-wheel operations.

Belts either operate at a lower, equal, or slightly higher cost than grinding wheels and other tools. They cost less to use on low tensile materials, are about equal on common steels, and slightly higher on harder materials. Abrasive cost is often a low percentage of overall cost in any operation—the main saving is in labor.

Speeds and feeds

Most abrasve machining is done between 2500 and 5500 fpm, with the average being 4500 fpm. Feeds vary from 1 in. per min for heavy slabbing cuts up to 240 ft per min for special precision centerless grinding applications.

Contact wheels

Most good abrasive belt machining applications use neoprene serrated contact wheels in the range of 90 to 95 durometer. Some polyurethane, cast iron, and steel contact wheels are used for special applications. Centerless grinding of small diameters is done with hard, plain-faced contact wheels. Hard contact wheels are used for heavy stock removal, high accuracy and best abrasive economics. Contact wheels should always be straddle-mounted for maximum rigidity.

Grinding aids

Abrasive belt machining is done wet, mostly with water-base fluids. Heavy-duty emulsions with high percentages of active sulfur, at a ratio of 8 to 1, are used most commonly in centerless grinding of steels. Surface grinding with belts is done more often with solution-type coolants so as to keep the work and carrier clean and visible. Some operations are done with straight grinding oils, requiring mist collection and automatic fire control systems. Filtration systems are essential for belt grinding and are often of the two-stage type. A drag conveyor is usually combined with a fine filter, since high stock removal and fine finishes are usually done on the same machine.

Centerless grinding

The first precision (±0.001 in.) centerless belt grinder was introduced in 1947 and featured 1.25 hp per in. of abrasive width.

In 1960, the world's largest centerless (belt) grinder was developed for grinding up to 9-in.-dia-by-30-ft steel bars. This machine had pushbutton setup of all functioning units, reverse work traverse, automatic loading and unloading, and a 75-hp motor to drive a 12-in. wide belt. This system, which weighed 75 tons, removed up to 0.032 in. on large bars to accuracies of 0.0005 in.

That same year, the first super-precision centerless belt grinder was shown to be capable, under

Stock removal rates K by belt surface grinding
(Hard serrated-contact wheels and 2 hp per in. of belt width)

Material	Grit 24	40	50	60	80	100	120
Aluminum	0.64	0.40	0.32	0.26	0.20	0.16	0.13
Magnesium	0.57	0.34	0.27	0.23	0.17	0.14	0.11
Bronze	0.44	0.27	0.21	0.18	0.13	0.10	0.08
Lead	0.35	0.21	0.17	0.14	0.10	0.08	0.06
Cast iron	0.24	0.16	0.14				
Titanium				0.006	0.005		
1020 steel		0.20	0.17	0.14	0.10	0.08	0.06
4140 steel		0.17	0.14	0.12	0.09	0.07	
Hardened steel 45 Rockwell C			0.14	0.12	0.09	0.06	

K = Constant (cu. in. per min. per inch abrasive) $\frac{K}{d} = f$

d = depth of cut in 0.001 in.

f = feed rate, ipm

Examples: 1020 CRS –0.010 in. stock removal. $\frac{K}{d} = f = \frac{0.20}{0.010} = 20$ ipm feed

If same job required 80-grit finish, add constants for 40 and 80 grit (0.20 + 0.10 = 0.30) and determine feed on a 2 head machine $\left(\frac{0.30}{0.010}\right. = 30$ ipm.

By James D Kreager
Sundstrand Machine Tool Division
Sundstrand Corp

Presented before American Society for Abrasive Methods in Chicago March 7, 1967

Abrasive belt machining [Continued]

specific conditions of operation, of grinding to 0.000050 in. This machine was developed and shown years before its time; most companies still work between 0.0005 and 0.001 in.

In 1963, the first precision multiple-head centerless bar grinder was installed. With it, one operator could do the work of three to four men by combining sizing and finishing in one fast pass. This machine featured a dialable unitized work rest common to all three heads, and a proprietary synchronizing regulating wheel drive. Today it is common to see two, three, four and five head machines throughout the basic metal industry. It is said that more rods and bars are ground by belt than by wheel in the steel industry.

Today's belt machines are sizing and finishing drawn or turned bars (1/16-9 in. OD) to close tolerances and superfinishes at rates equaling 5 to 8 bonded-wheel centerless grinders. Carbon steel, alloy steel, stainless steel, tool steel, and high speed steel are successfully ground.

Power input of 7.5 hp per in. width of abrasive is standard on large capacity machines. Stock-removal rates may be 0.25 cu in per in. of belt width on small size bars, and up to 1 cu. in. on large bars. By comparison, wide wheel grinders remove 0.6 to 0.75 cu in. per inch of wheel.

Abrasive belt cost per cu in of stock removal

Material	Grinding method Centerless $/cu in.	Surface $/cu in.
Steels		
1010	$0.023	$0.028
1045	0.026	0.032
4140 An.	0.030	0.035
9260 H.R.	0.031	NA
303 SS	0.039	0.048
304 SS-An.	0.060	NA
416 SS	0.028	0.037
M1	0.049	0.057
M3	0.057	0.063
Aluminum		
24 ST	0.008	0.010
6061	0.009	0.011
Magnesium	0.01	0.015
Titanium	0.210	0.290

Surface grinding

Abrasive belt machining of flat surface has long been done on precision platen grinders utilizing hydraulically controlled feed tables. Initially, this was done on nonferrous castings, but today almost all materials are processed by this method.

The precision platen grinder often works to 0.001 in. flatness and seldom produces work beyond 0.002 in. A good application should take only 10 to 30 sec., and gang grinding several parts within this time cycle is common. This procedure often replaces separate milling and grinding jobs.

Abrasive machining on continuous-feed, conveyorized surface grinders is not new, but heavy, rigid, high horsepower machines of this sort have only been available for about four years. Machines with multiple heads combine roughing, sizing and finishing in one pass on a continuous feed conveyorized production line. Applications range from printed circuits to chain-saw bars.

Hydraulic, reciprocating, magnetic-chuck surface grinders with multiple heads permit roughing and finishing in a single pass on parts such as shear blades and gear housings. Special versions of this type of machine are used for the grinding of tensile coupons in the aircraft industry.

Multiple-head, continuous-coil grinders are being used to grind both sides of rolled steel and aluminum, combining roughing and finishing in one fast pass. This type of machine has even been arranged to grind four sides of rectangular or square stock at one pass.

Cylindrical grinding

Abrasive-belt machining of rotors for motors is now common. Formerly, lathes and grinders were used to achieve stock removal and finish, but this is now done in one fast pass with improved accuracy. Often, several rotors are gang ground per load. Some companies grind as many as 1,000 small rotors per hour, and abrasive cost is considerably less than tool cost on this application.

Pushbutton automatic cycling machines range from 7.5 to 30 hp. Motor stators are also ground on a special chucking grinder to improve accuracy and increase production by gang grinding.

Chucking grinders have been used for years in the jet engine industry for sizing compressor stators and turbine rotors up to 50 in. in diameter. These are made of either stainless steel or titanium and the interrupted cut has proved difficult for bonded-wheel grinding. Special abrasive belt grinders remove 0.5 in. from the diameter in 8 minutes, versus 30 minutes for wheel grinders. Abrasive cost is $0.13 per part versus $0.22 for bonded wheel. Accuracy is identical and finish a bit better by belt. Widebelt grinders, which will plunge and oscillate grind, are steadily replacing narrow-bonded-wheel grinders, which must traverse across the face of a wide cylindrical part.

Special applications

An automatic cam grinder, using small contact wheels and a long belt in place of small diameter grinding wheels, has speeded the grinding of elliptically shaped cams while eliminating heat checking. This same type machine with wider belts is being used for grinding jet turbine blades to precision contours.

Outlook

Abrasive belt machining is growing faster than most other machining processes. While, it is no cure-all, it does offer substantial cost improvements in most fields of machining if properly applied. ■

Use of Diamond Grinding Wheels—I

By LEONARD I. SMITH, supervisor
Diamond Products Dept, Norton Co, Worcester, Mass

Grinding of tungsten carbide is one of the principal applications of diamond grinding wheels, and this use continues to expand each year. But because of the relatively high cost of diamond compared to conventional abrasives, it is important that diamond wheels be used in the most efficient manner.

What factors are to be considered in selection and use of diamond wheels to grind carbide efficiently, considering the wide variety of possible conditions?

These variables can all influence the cost of the diamond grinding operation to a greater extent than grinding with conventional abrasive wheels.

Let's consider the variables:

TYPE OF DIAMOND—Grain shape can range from a cube or sphere, which is classed as a strong shape, to the needle, sliver or plate shape, or a multi-crystal agglomerate as found in some manufactured diamond, all of which are classed as weak or friable.

The weaker or more friable shapes show a definite advantage in carbide grinding. Here, 75% of wheels are resinoid-bonded, 15% are vitrified bonded. Wheel life is 15 to 30% better than cubed-grain wheels. The more friable grain is more firmly anchored. Fracture at dulling renews sharpness. Major portions of the fractured grain remain in position in the wheel face to do more grinding. On the other hand, a strong-shape crystal has more tendency to pull out of the bond when it becomes dull.

The weaker-shape particles do not perform satisfactorily in metal-bonded wheels. The diamond is broken off at the surface of these wheels and no sharp protrusions are left to grind.

TYPE OF BOND—For "Fixed Feed" wet or dry grinding, the resinoid-bonded wheel is industry's choice as the fastest, free-cutting type of diamond wheel and it returns the best over-all grinding cost.

The vitrified-bonded diamond wheel is generally slower cutting but holds the diamond more effectively. It combines a reasonably good rate of cut with generally longer wheel life. It finds greatest use in offhand wet grinding or sharpening of carbide tools.

OFF-HAND TOOL SHARPENING

Metal-bonded diamond wheels are generally slower cutting than either resinoid or vitrified and find use principally in wet offhand tool sharpening where the rate of production is not important, and chip-breaker grinding.

GRIT SIZE—The selection of grit size is influenced by finish, formholding ability, and stock-removal rate.

As a general rule, the coarsest grit size commensurate with surface-finish requirements should be used.

Carbide tool-sharpening wheels range from 80 to 220 grit, depending on work area, feed, wheel-operating truth, number of die-out passes, etc. These grits, can be used to produce finishes from 5 to 20 micro-inches.

When certain radii or formholding requirements are to be met, the finer grit sizes such as 220, 320 or 400 are often used.

GRADE—This property is defined as the degree of bonding strength developed in manufacture of the wheel. Norton Co uses five grades —H, J, L, N, R, which are progressively "harder" (i.e. stronger). R grade is most prevalently used for carbide grinding. Softer grades find their principal use on some of the harder, more heat-sensitive carbides and in superfinishing operations. In a comparison of the effect of grade on grinding cost, grades J, N and R were used in 6-in. resinoid-bonded cup wheels, 120-grit, 100 concentration. At 0.001 in. downfeed per pass, the wheel costs per unit volume of carbide removed was J grade—$0.25; N grade—$0.21 and R grade—$0.17.

CONCENTRATION — Standard concentrations are 100, 75, 50, and 25, with 100 being the maximum.

Price should not be the criterion of selection. Grinding operations which involve a narrow or line contact (as obtained with straight wheels and most flaring cup and saucer wheels) are more economically performed with 100-concentration wheels in relation to the amount of carbide removed per carat of diamond wear.

BOND MODIFICATION—Various resinoid-bond modifications have been developed for special jobs:

1. Combinations of carbide and steel, as experienced in grinding lamination dies.
2. Severe pressures and intermittent ones.
3. Dry grinding as opposed to wet grinding.

In one test four bonds were used wet on a surface grinder at 0.050 in. crossfeed per table reversal, and 0.001 in. downfeed per crossfeed reversal. Details of the wheels were: 100 grit, 100 concentration, R grade. The G ratios (cubic inches of carbide removed per cubic inch of wheel wear) were found to be:

Bond	G-ratio
No. 1	52.8
No. 2	64.0
No. 3	94.6
No. 4	176.1

Use of Diamond Grinding Wheels—II

In dry grinding on a tool and cutter grinder at 0.002 in. infeed, B grade, 150 grit, 100 concentration, the ratios were:

Bond	G-ratio
No. 2	13.8
No. 4	19.6

Machine Condition — A grinding machine employing a diamond wheel should be kept in the best possible operating condition. This is particularly true in regard to spindle and flange truth.

On "Fixed-Feed" machines the feed-screw mechanism should be periodically checked to insure against excessive infeeds which can drastically affect diamond wheel wear.

Wherever possible the machine should be equipped for wet grinding. A flood coolant directed into the point of grinding contact can sharply reduce operating costs.

Some machines, particularly cutter grinders, are not equipped for wet grinding. On such machines, consideration should be given to spray coolant or use of a saturated wick of coolant held against the wheel face.

Wheel Size—There is a tendency, because of higher cost, to use smaller diamond wheels than conventional abrasive wheels on the same machines. This is satisfactory if the wheel speed is not permitted to become too low.

It should be pointed out, however, that larger diamond wheels generally operate at a lower carat loss per unit volume of carbide removed than a smaller wheel operated at the same conditions of speed and feed.

Wheel Speed — Resinoid-bonded diamond wheels operate most economically at speeds from 3000 to 5500 sfpm. If the wheel is operated at speeds either side of this range, the cost goes up quite sharply.

Rate of Production — Grinding costs are influenced by the Volumetric Feed Rate, V_f:

$V_f = C_f \times D_f \times$ traverse rate

Surface grinding tests were made at various downfeeds and at traverse rates of 300, 600 and 1200 ipm, the crossfeed being held constant at 0.050 in.

If the G ratio were the only criterion, we would recommend 0.0005 in. downfeed and 300 in./min. traverse rate (see chart). Here the G value of 212 is the highest of the three sets of data. The conditions also produced the lowest volumetric feed rate used in the tests:

$V_f = 0.0075$ cu. in./min.

To determine the most favorable operating procedure as it relates to wheel and labor costs within the limits established in the test conditions, cost curves, were plotted for the three tests. The lowest cost per cubic inch of carbide removed, namely $7.39, occurs at a volumetric feed rate of 0.030 cu. in./min., where the traverse rate is 300 in./min. and the downfeed 0.002 inch.

The same V_f (volumetric feed rate) of 0.030 cu. in./min. also occurs at 600/min. table speed. The total cost, however, is somewhat higher and increases at the higher traverse rates.

Several conclusions can be drawn from this study:

1. Carbide grinding costs can be changed significantly by relatively small increments of change in downfeed rates.
2. Wheel cost is less significant than labor cost when using a modern traverse rate (300 in./min.) in the generally used area of 0.0005 to 0.002 in. downfeed.
3. Wheel life alone does not reflect the true cost of using a diamond wheel.
4. It pays to study a grinding operation to find the most favorable grinding conditions.
5. There are many variables possible in making a diamond grinding wheel specification, and there are many variables possible in the method of using a diamond wheel.
6. The difference in grinding costs which can result from properf (or improper) selection and uses of a diamond wheel should influence the alert and cost-conscious user to seek the best diamond grinding wheel and the best method of using it. ●

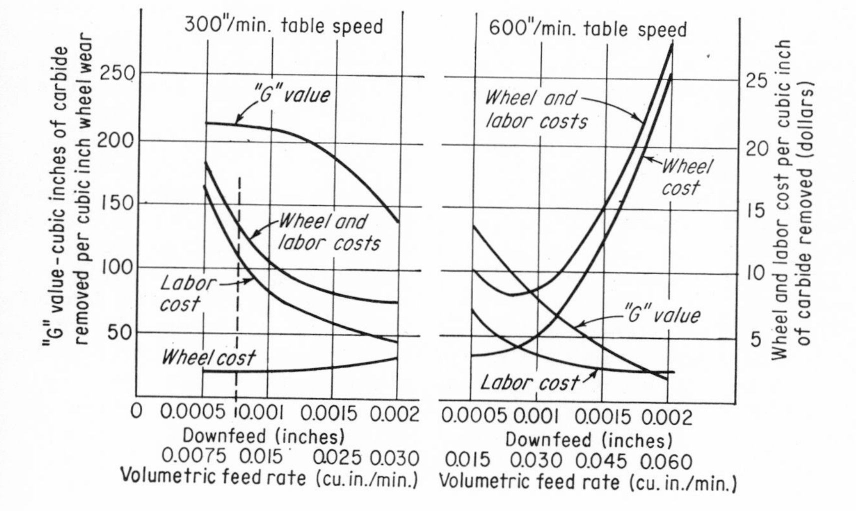

Grinding Concaved Rolls in a Lathe

By L KASPER

On an order to produce a pair of concaved rolls, hardened and ground, the dimensions given were the maximum and minimum diameters and the length of the chord Y, Fig. 1. Because the only critical dimensions were the diameter of the rolls and the diameter at the bottom of the concavity, and not having a suitable grinder available, we decided to complete the job on a lathe, selecting one with a taper attachment.

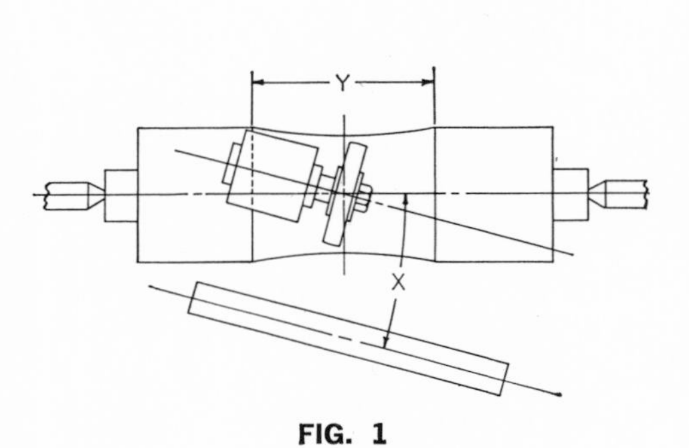

FIG. 1

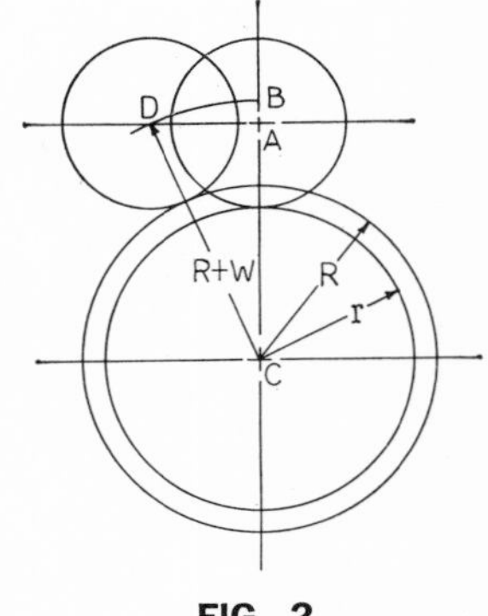

FIG. 2

Fig. 1 illustrates the setup, viewed from above, which satisfactorily completed the job. Before hardening, the concavities were rough finished to a sheet-metal templet. After hardening, the rolls were mounted on aligned centers of the lathe, and the maximum diameter was ground with a portable grinder mounted on the carriage. To grind the concaves, brackets were mounted on the carriage to locate the grinder above the roll. These brackets were provided with a slide, vertically adjustable, to carry the grinder.

The taper attachment was set with the guide bar at angle X with the axis of the roll. A formula for calculating X will be given later. The grinder is set with its axis parallel with the guide bar, and the carriage is positioned so that the center of the grinding wheel will be vertically over the axis of the roll at the midpoint of its length. As the lathe is operated, the carriage travels parallel with the line of centers, whereas the grinding wheel, which is mounted on the cross-slide, travels at an angle with the line of centers. This causes the cutting point of the grinding wheel to move toward and away from the axis of the roll rotation, so that the diameter of the cut is gradually increasing on both sides of the center position, until the cut runs out at the periphery of the roll.

The grinding wheel must be dressed to a radius which will permit point contact only at any point in its travel. The grinding wheel may be of any diameter but if the arc is to be of small radius, it will be necessary to use a thin wheel to prevent cutting by the side of the wheel.

In order to determine angle X, in Fig. 1, refer to Fig. 2. Two concentric circles are drawn, with radius R of the maximum diameter, and with radius r of the minimum diameter. Let W represent the radius of the wheel. From center C, locate point A using radius $r + W$. From A, with radius W, draw a circle tangent to r, to indicate the position of wheel at the bottom of the concave. From C, with radius $R + W$, locate point B, drawing an arc intersecting the upper horizontal centerline at D. From D with radius W, draw the tangent circle, indicating the position of the wheel as it emerges from the concave. The distance AD indicates the horizontal offset of the wheel relative to the center of the roll at this point. AD is determined by the equation:

$$AD = \sqrt{(CD)^2 - (CA)^2} = \sqrt{(R + W)^2 - (r + W)^2}$$

Then, $\tan X =$

$$\frac{\sqrt{(R + W)^2 - (r + W)^2}}{Y/2}$$

Simplifying: $\tan X =$

$$\frac{2\sqrt{(R + W)^2 - (r + W)^2}}{Y}$$

Because the grinding wheel rotates on an axis, which is at an angle with the roll axis, the grinding lines appear as short angular scratches. These lines would be noticeable on the material being rolled. To avoid this condition, the grinding wheel is allowed to "spark out" when the minimum diameter is within about 0.0005 in. from the required dimension. The grinder is then stopped and the spindle locked to prevent turning, and the carriage is fed back and forth, feeding the grinder into the roll a slight amount with each pass. This will set the grinding lines perpendicular to the roll axis, and they will appear lengthwise on the material being rolled. •

Chapter 5

Milling

Using miniature end mills

Efficient use requires skilled operators, proper equipment, attention to detail, and intelligent consideration of procedures

Metal cutting tool manufacturers have established industry wide physical standards for a relatively complete line of so-called 'miniature end mills.' (This title was assigned to all tools having 3/16-in. diameter straight shanks with cutting diameters 3/16 in. or less regardless of the number of flutes and end tooth configuration.)

Manufacturers of space hardware, electronic gear, instruments, and some appliance parts use small and very small diameter end mills to produce miniature components. Many shops have skilled hands available but are somewhat lacking in equipment that provides the necessary basic requirements to insure good small tool performance.

Spindles and tool holding devices must be accurate. Tools should operate as close to zero radial runout as possible. This is a must since 0.0003 to 0.0005 in. radial runout (not uncommon in many machine tools) can represent three to five times the chip load per tooth normally carried by miniature end mills. A tool that runs out in excess of its normal feed per tooth cannot operate efficiently. It will suffer premature failure. The importance of a balanced chip load per tooth on small end mills cannot be overemphasized.

Proper spindle speeds, feedrates

Tool room and production milling machines are available with spindle speeds in the range of 12,000 rpm. They have been designed for efficiently operating small diameter tools. To see why such high speeds are necessary, consider the following example:

An end mill ½ in. in diameter cutting mild steel at a surface speed of 90 sfpm requires a spindle speed of about 690 rpm. Speeds in this range are available on most milling machines. However, if a 1/32-in. diameter miniature end mill were to be used in the same steel at 90 sfpm, it would require a spindle speed of 11,000 rpm Most shops do not have spindles that can operate in this range.

If the 1/32-in. diameter miniature end mill is used in a machine with the more conventional top speed of approximately 3600 rpm, the cutting speed drops to about 30 sfpm. A drop in cutting efficiency and probably quick tool breakage are the penalties.

High rotational speeds tend to stabilize a small tool on its axis and minimize deflection. This stability is very important because of the physically frail nature of miniature end mills made of high speed steel and heat treated to the proper cutting hardness. Proper speeds are necessary if small end mills are to be operated successfully.

Feedrates for miniature end mills go hand in hand with spindle speeds. They are, perhaps, the most critical factor in avoiding tool breakage. Small diameter end mills are extremely limited as to the amount of feed per tooth they can handle without breaking. Feedrates of 0.000,01 to 0.0001 feet per tooth are common. Tools operating at lower than recommended rpm become extremely difficult to feed properly at these low feedrates.

A 3/32-in. slot is milled in a bronze aircraft pump part by a miniature (3/16-in. shank), two-flute, ball-end cutter

Smoothly operating power feed mechanisms should be used whenever possible. Miniature end mills are very sensitive to shock loading caused by uneven feedrates. High spindle speeds will minimize tool breakage on intricate parts that require hand feeding.

Operators must exercise good judgment when selecting feedrates. The machinability of the workpiece, the dimensions of the cut to be taken, and the diameter-to-length ratio of the tool should be considered.

Chip removal

Chips produced by low feedrates on small diameter miniature end mills are extremely fine and almost powdery in nature. It is important to good cutter life, however, that even these fine chips be removed from the cut.

One of the best methods for removing these chips is a mist coolant applicator which combines a small amount of liquid in a stream of air. The air jet can be directed to remove chips as they are cut. The liquid in the air stream will keep the tool lubricated and dissipate heat.

The direction in which the air stream strikes the tool can mean the difference between success and failure. This air can remove chips from the cut or blow chips into the cut. The latter condition will cause problems. Caution: Eye protection should be used by anyone operating a machine using air to remove chips.

A chart of suggested feed and speed ranges for miniature end mills follows. It is intended to serve as a guide to assist the planner or machine operator in coming reasonably close to a realistic point from which to begin operations. Because of the multitude of variable factors involved in every cutting application, feedrates are given in ranges rather than fixed figures. These feed ranges are for two-fluted tools; they should be doubled for four-fluted tools.■

By Thomas A Ribich, Director of Engineering
The Weldon Tool Co, Cleveland, Ohio

Feeds and speeds for miniature end mills

Cutting Diameter	Surface Feet Per Minute sfm	10	15	20	25	30	35	40	50	60	70	80	90	100
1/64	R P M	2445	3670	4890	6110	7335	8555	9780	12225	14670	17110	19560	22000	24445
	Table Feed Inches Per Minute	.002-1/64	.003-1/64	.005-1/32	.007-1/32	.010-1/32	1/64- 1/16	1/64- 1/16	1/32- 3/32	1/32- 7/64	3/64- 5/32	1/16-15/64	5/64- 5/16	7/64- 3/8
1/32	R P M	1225	1830	2445	3055	3665	4280	4890	6110	7335	8556	9780	11000	12225
	Inches Per Minute	.006-1/32	.010-3/64	1/64- 1/16	1/64- 5/64	1/32- 3/32	1/32- 1/8	3/64- 5/32	1/16-15/64	5/64- 5/16	7/64-13/32	5/32- 5/8	3/16- 3/4	1/4 1
3/64	R P M	815	1225	1630	2035	2445	2850	3260	4075	4890	5705	6520	7335	8150
	Inches Per Minute	1/64- 3/64	1/64- 5/64	1/32- 7/64	1/32- 9/64	3/64-11/64	3/64-15/64	5/64-17/64	3/32- 3/8	9/64- 1/2	3/16-45/64	1/4 -31/32	5/16 1-1/8	3/8 1-1/2
1/16	R P M	610	915	1225	1530	1835	2140	2445	3055	3665	4280	4890	5500	6110
	Inches Per Minute	1/64- 5/64	1/32- 7/64	3/64- 5/32	1/16- 7/32	5/64- 1/4	5/64-11/32	7/64-13/32	9/64- 1/2	7/32-13/16	9/32 1-1/16	3/8 1-1/4	1/2 1-7/8	5/8 2-1/2
5/64	R P M	490	735	980	1220	1465	1710	1955	2445	2935	3420	3910	4400	4885
	Inches Per Minute	1/64- 3/32	1/32- 9/64	3/64-13/64	5/64-17/64	3/32-11/32	7/64- 7/16	5/32- 9/16	15/64- 7/8	5/16 1-3/8	3/8 1-7/16	1/2 1-7/8	5/8 2-3/8	13/16 3-1/8
3/32	R P M	410	610	815	1020	1220	1425	1630	2040	2445	2850	3260	3665	4075
	Inches Per Minute	1/64- 7/64	3/64-11/64	1/16- 1/4	5/64-11/32	3/32- 7/16	1/8 - 9/16	5/32- 3/4	15/64-15/16	5/16 1-1/4	7/16 1-3/4	1/2 2-1/8	5/8 2-7/8	3/4 3-1/2
7/64	R P M	350	525	700	875	1050	1220	1395	1745	2095	2445	2795	3145	3490
	Inches Per Minute	1/32- 1/8	3/64- 7/32	5/64- 5/16	3/32-13/32	1/8- 1/2	5/32- 5/8	3/16- 3/4	1/4 1-1/8	11/32 1-3/8	7/16 1-7/8	9/16 2-9/16	3/4 3-1/8	7/8 3-7/8
1/8	R P M	305	460	610	765	915	1070	1220	1530	1835	2140	2445	2750	3055
	Inches Per Minute	1/32- 5/32	3/64- 1/4	5/64-11/32	7/64- 1/2	9/64-9/16	11/64- 3/4	7/32- 7/8	5/16 1-3/16	3/8 1-5/8	1/2 2-1/4	11/16 2-7/8	7/8 3-1/2	1 4-3/8
9/64	R P M	270	405	545	680	815	950	1085	1360	1630	1900	2175	2445	2715
	Inches Per Minute	3/64-11/64	1/16- 1/4	3/32- 3/8	1/8 - 1/2	5/32- 5/8	13/64-13/16	15/64 1	5/16 1-3/8	13/16 1-3/4	9/16 2-7/16	23/32 3-1/8	15/16 4	1-1/8 4-7/8
5/32	R P M	245	365	490	610	735	855	980	1220	1465	1710	1955	2200	2445
	Inches Per Minute	3/64- 3/16	5/64- 5/16	7/64- 7/16	9/64- 9/16	3/16- 3/4	1/4 - 7/8	1/4 1-1/16	3/8 1-7/16	1/2 1-7/8	5/8 2-1/2	7/8 3-3/8	1 4-1/4	1-3/8 5-1/8
11/64	R P M	220	335	445	555	665	780	890	1110	1335	1555	1780	2000	2220
	Inches Per Minute	1/16- 7/32	3/32-21/64	1/8 - 7/16	5/32-19/32	3/16- 3/4	1/4 - 7/8	1/4 1-1/8	3/8 1-1/2	1/2 2	5/8 2-9/16	7/8 3-5/8	1-1/16 4-1/2	1-7/16 6
3/16	R P M	205	305	405	510	610	715	815	1020	1220	1425	1630	1835	2035
	Inches Per Minute	1/16- 7/32	3/32-11/32	1/8 - 1/2	5/32- 5/8	7/32- 7/8	1/4 -1	9/32 1-3/16	3/8 1-5/8	9/16 2-1/4	3/4 2-7/8	7/8 3-3/4	1-1/16 5	1-3/8 6-1/2

Note: All feed ranges shown are for two-fluted tools. For four-fluted tools, double the ranges.

Application of Carbide End Mills

By Carl Neiderfringer, president, The Atrax Co, Newington, Conn

Rules for Use of Carbide End Mills

Materials

Solid-carbide end mills are recommended for milling all types of materials with the possible exception of soft steels.

Milling Set-Up

Climb milling with solid-carbide end mills is possible only with an extremely rigid set-up.

Eliminate Vibration

Hold end mills as close to cutting diameter as possible to prevent excessive overhang which can cause vibration, chatter and scored surfaces.

When to Sharpen

Carbide end mills should be sharpened as soon as appreciable wear lands are noticed.

How to Sharpen

A fine-grit, resinoid-bond diamond wheel is recommended. A true wheel spindle and rigid sharpening setup will determine the production that can be expected from the new cutting edge.

Relief When Sharpening

Too little relief causes the cutter to "drag" or "pick-up" on the heel of the land. Too much relief results in a weak tooth form.

Rake When Sharpening

When necessary to deepen flutes, the flute should be ground to conform to original shape for easy chip disposal. Incorrect flute forms cause clogging of chips, resulting in overheated workpiece and cutter, with eventual chipping and breakage of the tool.

Coolants

Most end-milling operations can be performed dry or with air blast to remove chips. A flow of coolant such as soluble oil or light oil is recommended in some cases, but it is necessary to use a heavy, continual flood to prevent intermittent heating and quenching which may result in hard spots in workpiece.

A heavy flow of coolant or strong air blast is recommended when milling aluminum to facilitate chip disposal, which is otherwise a problem.

A coolant high in cooling action but low in lubricating quality is recommended when milling stainless steels.

Recommended Applications of Solid-Carbide End Mills

Materials Group	Type of End Mill
Group 1:	
Brass and bronze Aluminum Zinc alloys Copper Hard rubber Fiber Plastics Magnesium	4-flute, right-hand spiral, square end
Group 2:	
Cast Iron	4-flute, straight-flutes, square end or 2-flute, straight-flutes, square end
Group 3: (also see Group 1):	
Aluminum Magnesium	2-flute, right-hand spiral
Group 4:	
Carbon steels Free-cutting steels Nickel Stainless steels Malleable iron Monel Metal Heat-treated steels Titanium	4-flute, slow spiral, right-hand spiral, square end

Speeds and Feeds for Solid-Carbide End Mills

		FEEDS, lpm		
MATERIAL	CUTTING SPEEDS Fpm	CUTTER DIA Up to 3/16	CUTTER DIA ¼ to ½	CUTTER DIA ½ and Over
Carbon steel	90-250	1-4	2-10	3-10
Free-cutting steels	100-300	1-4	3-15	4-15
Nickel	90-250	1-4	2-12	4-12
Nickel chrome	80-250	1-4	2-12	4-12
Stainless steels	70-250	1-4	2-10	4-12
Cast iron	60-200	1-5	3-15	4-15
Malleable iron	60-200	1-5	3-15	4-15
Brass and bronze	100-300	2-6	4-20	6-20
Aluminum and aluminum alloys	125-350	2-8	4-20	6-30
Zinc alloys	150-400	2-8	4-20	6-30
Copper	125-350	2-6	4-20	6-20
Hard rubber	150-500	3-12	4-30	6-40
Fiber	150-400	3-12	4-30	6-40
Plastics	200-600	3-15	6-40	10-50
Monel Metal	100-250	2-10	4-20	6-30

How to sharpen end mills with eccentric relief

If end mills are sharpened with eccentric peripheral relief, results are superior to the concave relief that is produced by a cup-wheel grind

By Alan Baker

Most end-mill manufacturers now grind eccentric peripheral relief on the primary cutting edges of their tools. Superior finish is obtained because the saw-tooth edge produced by a cup-wheel grind is not present, and the support behind the cutting edge is significantly greater.

There are, however, many end-mill user and even end-mill reconditioning vendors who still use the straight, cup-wheel method for sharpening end mills. The result is a weaker cutting edge and a relatively poor surface finish.

The grinding procedures required for eccentric peripheral relief are simple and should be used wherever end mills are sharpened. The helical flute of the mill, resting on the finger, is pulled past the dressed surface of the wheel, with the wheel angle such that the helical flute periphery is in contact with the wheel over a comparatively long area. The trailing edge of the peripheral primary contacts the wheel first, well below the centerline of the wheel. The leading edge of the peripheral primary contacts the wheel last, almost on the centerline of the wheel. The result is eccentric relief because the leading and trailing edges of the primary meet the wheel at different angles.

The grinding wheel used should be 6 or 7 inches in diameter, 60 or 80 grit, ¼ to ⅜ inches thick and of medium

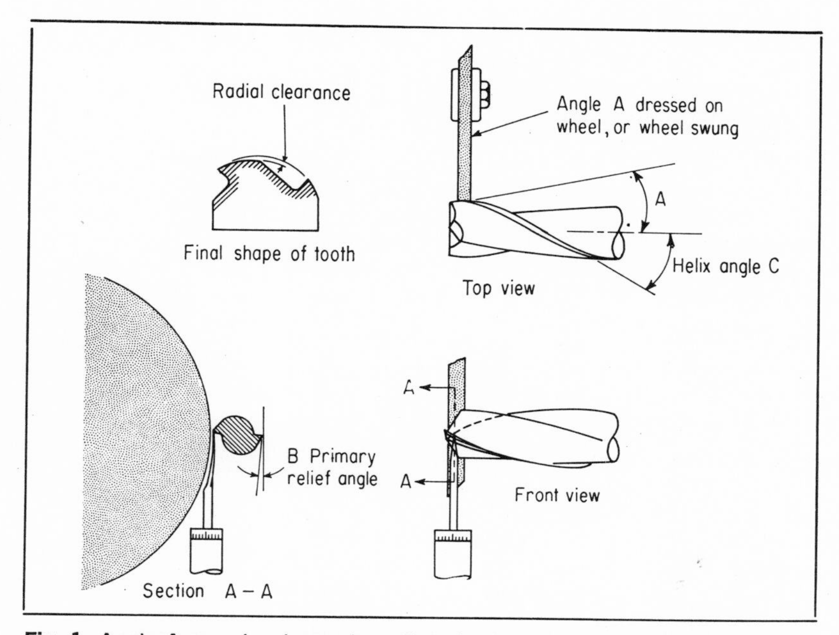

Fig. 1. Angle **A** may be dressed on the wheel or the wheelhead swung to **A**

Table 1. Set-up chart for eccentric relief

Tool dia	General-purpose 30° helix		Aluminum-cutting tool 37½° helix		Aluminum-cutting tool 45° helix	
	B	A	B	A	B	A
3/16-1/4	12°	7°00′	14°	10°50′	14°	14°
9/32-3/8	11°	6°24′	13°	9°39′	13°	13°
13/32-1/2	10°	5°48′	12°	9°16′	12°	12°
9/16-11/16	9½°	5°31′	11½°	8°52′	11½°	11½°
3/4-7/8	9°	5°13′	10½°	8° 6′	10½°	10½°
15/16-1-1/8	8°	4°38′	9½°	7°19′	9½°	9½°
1-1/4-2	7°	4° 3′	8°	6° 9′	8°	8°
2-1/4-2-1/2	6°	3°28′	7°	5°23′	7°	7°

hardness, perhaps letter I or J. The tooth-rest finger should be located almost in the center of the wheel width, slightly off-center toward the end-mill holding device. If available, an end-mill grinding fixture or other freely revolving, air-supported cylinder should be used to hold the end mill. The grinding wheel axis, end-mill axis and height of the rest finger should all be in line, as checked with an indicator. The finger should be contoured slightly to curve around the wheel periphery.

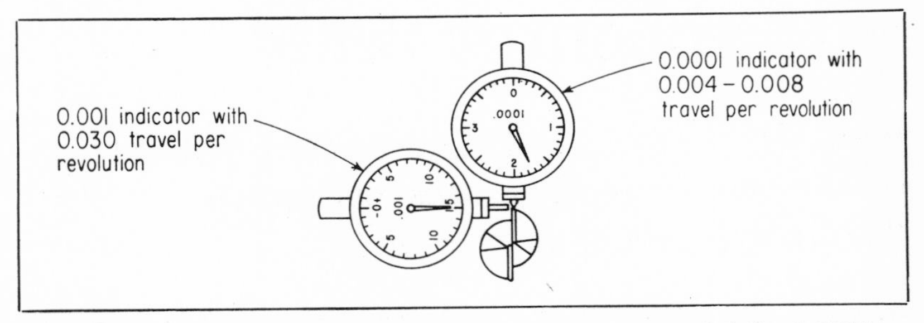

Fig. 2. In the two-indicator method of checking primary relief the 0.0001 indicator measures primary drop **X** and the 0.001 indicator measures travel **Y** as the end mill is revolved

To produce the eccentric relief desired, angle *A* must be dressed on the wheel, Fig. 1, or the wheelhead must be swung to angle *A*. Tangent of angle *A* equals the tangent of angle *B* (primary relief) times the tangent of angle C, (helix angle).

Objections overcome

When this method of grinding peripheral relief is first introduced to your personnel, there may be complaints about excessive burning of the cutting edge as the end mill is pulled against the smoothly dressed wheel. This objection is easily overcome by grinding the secondary relief first with a cup-wheel grind, to remove the old primary relief and leave a very small amount of stock to be removed during the eccentric primary grind which follows.

Table 1 gives suggested relief angles for various sizes of end mills. The table shows suggested primary relief angles for three standard end mills: general-purpose with a 30° helix, aluminum cutting with a 37½° helix and aluminum cutting with a 45° helix. If your end mills have some other helix angle, the tangent of angle *A* equals the tangent of the primary relief times the tangent of the helix angle. If *A* is increased the primary relief will be increased.

An inspection procedure for primary relief should accompany introduction of this eccentric grinding method in your shop. Table 2 shows an accurate, easy method for checking the amount of peripheral relief. Suggested relief angles and corresponding indicator drops are average for general-purpose and aluminum-cutting end mills on the market today. The end mill, held in a collet, is revolved in the horizontal plane over a surface plate causing the 0.001 indicator to travel the Y amount. The 0.0001 in. indicator should drop the approximate *X* amount. The 0.001 indicator button should be on the flute face at the periphery and as near the 0.0001 indicator button as possible without obstructing its travel. For normal end mill usage the tolerance on the required 0.0001 indicator drop, *X*, should be ±0.0003 in.

For difficult-to-machine materials or increased feed rates in non-ferrous materials, you may find it advantageous to change these relief angles and *X* dimensions to suit requirements. The changing of a relief angle is simple; just change *A*.

For information on grinding end teeth on end mills, refer to Special Report No. 518 (**AM**—Mar 19 '62, p121).

Table 2. Inspection chart for primary relief

Tool dia	General-purpose Angle B	General-purpose Drop X	Aluminum-cutting tools Angle B	Aluminum-cutting tools Drop X	Travel Y
3/16-1/4	12°	0.0021	14°	0.0024	0.010
9/32-3/8	11°	0.0019	13°	0.0023	0.010
13/32-1/2	10°	0.0026	12°	0.0031	0.015
9/16-11/16	9½°	0.0025	11½°	0.0030	0.015
3/4-7/8	9°	0.0031	10½°	0.0038	0.020
15/16-1-1/8	8°	0.0028	9½°	0.0033	0.020
1-1/4-2	7°	0.0037	7°	0.0042	0.030
2-1/4-2-1/2	6°	0.0031	8°	0.0037	0.030

HOW TO MILL A LARGER RADIUS THAN CUTTER'S

LAURENCE P HERRON

When a blueprint calls for milling a larger radius than possessed by any available cutter, the solution is to tilt the cutter. It's true that an ellipse, rather than a circular element, is generated. However, the error may not be important in many cases, and if it is the discrepancy can be corrected by working to a templet.

The tilt height of the cutter equals the cutter dia times the cutter dia divided by the dia of the required curvature.

Example: to approximate an arc of a 12-in. circle, using a 6-in. cutter.

Solution: tilt height = 6 x 6/12 = 3 in.

The accompanying table of ratios is based on the law of sines and will give the tilt angle when the ratio between cutter dia and curvature dia is known. In the above example, the ratio is 3/6 = 0.50. Tilt angle is then 30°, which checks with the example.

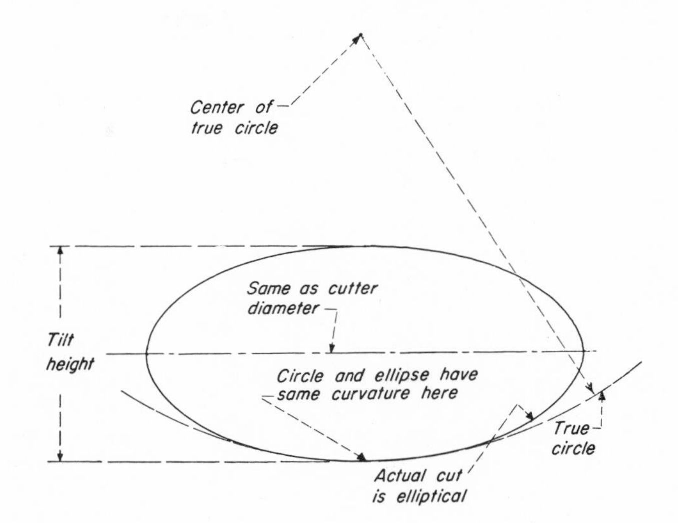

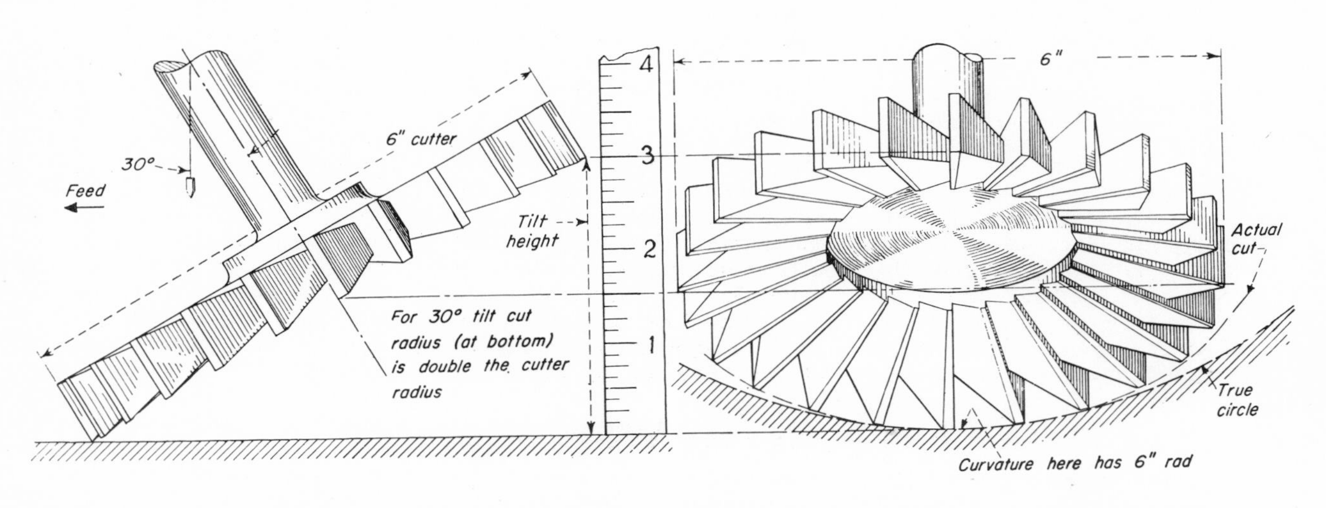

RATIO	ANGLE	RATIO	ANGLE	RATIO	ANGLE	RATIO	ANGLE	RATIO	ANGLE	RATIO	ANGLE
0.05	2.9	0.21	12.1	0.37	21.7	0.53	32.0	0.69	43.6	0.85	58.2
0.06	3.4	0.22	12.7	0.38	22.3	0.54	32.7	0.70	44.4	0.86	59.3
0.07	4.0	0.23	13.3	0.39	23.0	0.55	33.4	0.71	45.2	0.87	60.5
0.08	4.6	0.24	13.9	0.40	23.6	0.56	34.1	0.72	46.1	0.88	61.7
0.09	5.2	0.25	14.5	0.41	24.2	0.57	34.8	0.73	46.9	0.89	62.9
0.10	5.7	0.26	15.1	0.42	24.8	0.58	35.5	0.74	47.7	0.90	64.2
0.11	6.3	0.27	15.7	0.43	25.5	0.59	36.2	0.75	48.6	0.91	65.5
0.12	6.9	0.28	16.3	0.44	26.1	0.60	36.9	0.76	49.5	0.92	66.9
0.13	7.5	0.29	16.9	0.45	26.7	0.61	37.6	0.77	50.4	0.93	68.4
0.14	8.0	0.30	17.5	0.46	27.4	0.62	38.3	0.78	51.3	0.94	70.1
0.15	8 6	0.31	18.1	0.47	28.0	0.63	39.0	0.79	52.2	0.95	71.8
0.16	9.2	0.32	18.7	0.48	28.7	0.64	39.8	0.80	53.1	0.96	73.7
0.17	9.8	0.33	19.3	0.49	29.3	0.65	40.5	0.81	54.1	0.97	75.9
0.18	10.4	0.34	19.9	0.50	30.0	0.66	41.3	0.82	55.1	0.98	78.5
0.19	11.0	0.35	20.5	0.51	30.7	0.67	42.1	0.83	56.1	0.99	81.8
0.20	11.5	0.36	21.1	0.52	31.3	0.68	42.8	0.84	57.1	1.00	90.0

Milling Superalloys—I

By Carl T Olofson, principal metallurgist, and Francis W Boulger, chief, Metalworking Research Div, Battelle Memorial Institute, Columbus, Ohio

Milling Setup

The precautions described for turning (*AM/MM*—Jan 8, '62, p95) also apply in the more difficult operation of milling the superalloys. Climb milling is preferred in order to obtain a shorter tooth path in these difficult materials. It also promotes a thin chip as the cutter leaves the work.

Special designs of milling cutters may be needed, rather than those generally available in commercial stock sizes. Cutters should have helical teeth where possible to promote smooth cutting action. They should also have as many cutting edges or teeth as practical without sacrificing necessary chip space. This will partially offset the relatively low speeds at which cutters must operate in these materials.

Tool Materials

Both high-speed steel and insert-type carbide cutters can be used to mill superalloys. High-speed steel cutters are more reliable in terms of eliminating sudden failure. In addition, high-speed milling cutters give better short-time cutting speed but fall off in tool life faster than carbides do. For Rene 41, the crossover point is around 35 fpm and 25 min tool life. Hence, for general-purpose use and on small-lot production, high-speed steel cutters are preferred to carbide cutters. Types M-3 and T-15 high-speed steel have been used successfully for specific milling applications. An axial hole should be a part of the high-speed-steel cutter design to allow coolant injection at the site of cutting.

Carbide insert-type cutters can successfully mill superalloys when reasonable care is exercised. Carbide grades C-1, C-2, C-5a, and C-7a have been used for various rough-and-finish-machining operations.

Tool Geometry

An axial rake is usually recommended to reduce impact against the cutter and to provide a suitable shearing action. A large radial rake angle helps to reduce the cutting load, work hardening, and cutting temperature. The size of this angle will be limited by the strength of the tooth. Relief angles, although larger than standard, should be kept as low as possible—just enough to prevent the tool from dragging on the workpiece. When practical, a bevel angle should be used to protect the entering chamfer.

Conditions recommended for milling the superalloys are given in the table on page 69. Cutting speed is the most critical factor; excessive speed causes overheating of the cutting edges and subsequent rapid tool failure. When starting a new job, a cutting speed in the lower part of the recommended range should be used.

Lower feed values will minimize deflection in the workpiece, if this is a problem. It is best to maintain a uniform feed. Cutters should not idle in the cut, because the surface will work harden and cause rapid tool failure.

Whenever the cut or the machine tool permits, a climb cut (down milling) should be used. This should not be attempted, however, unless all lost motion is removed from the milling-machine-table feed mechanism.

Tool Angles for Milling Superalloys

Tool Materials	Geometry	Rake		Relief		Cutting Edge		Chamfer, degrees
		Axial	Radial	Face	Peripheral	Peripheral	Face	
HSS	FMA	0	10	5(c)	5(a)	45	5	0
"	FMA_1	15	10	2-4(c)	2-4(a)	0	1-3(a)	45
"	FMB	10-15	10-15	3-5(c)	3-5(a)	0	5-7	45 with 22½ bevel
"	FMB_1	+10	+10	5	5(a)	45	5	0
"	PMC	10-12(c)			6(a)			
"	PMD	10-12(c)	•		4-5(b)			
"	EME	35			12-15	45x0.030 in.		
"	EME_1	30	7		6			N.R. 0.060
"	SMF	10	10	10	10	45	5	N.R. 0.060
"	SMF_1	10	10	6	6			
"	G	+5	+5	6	6	15	8	
"	H	0	+5	6	6	15	8	
Carbide	FMI	+15	—10	5	5	0	5	0
"	FMJ	—5	—5	5	5	15	6	
"	FMK	0	0	7	7	15	15	N.R. 0.032R

(a) Add 5 deg. for secondary clearance values.

(b) Add 3 deg. for secondary clearance values.

(c) Where cutting edge is a helix (slab mill, end mill, etc.) axial rake=helix angle.

Note: The following cutter codes are used: FM=face mill; PM=peripheral mill; EM=end mill; and SM=side mill.

Milling Superalloys—II

Feeds and Speeds for Milling Superalloys

Alloy	Condition	Tool Material	Design (a)	Roughing Speed, fpm	Roughing Feed, ipt	Roughing Depth of Cut, in.	Finishing Speed, fpm	Finishing Feed, ipt	Finishing Depth of Cut, in.	Cutting Fluid
René 41	Aged	HSS (M-3)	FMA, FMB	—	—	—	15-20	.002-.006	.010-.015	Sulfur-base oils
			EME	—	—	—	15-20	.002-.004	.010-.015	
Hastelloy X	Sol. treated	HSS	—	15-25	.002-.006	.10-.20	—	—	—	Sulfur-base oil
Waspalloy M-252 J-1500 Udimet 500 René 41 (b)	Sol. treated	Carbide C-1	FMI	40-60	.006-.010	.125-.25	—	—	—	Sulfur-base or chemically active soluble oil
	Aged	Carbide C-2	FMI	—	—	—	60-100	.003-.006	.03-.125	
		HSS (T-15)	EME	30	.002	.25	—	—	—	
		HSS	FMH	—	—	—	35	.0025	—	
		HSS (T-15)	SMF	20	.010	.06	—	—	—	
		Carbide	FMJ	—	—	—	50	.002	—	
Inconel X and Nimonic 90	Sol. treated	HSS	FMA, FMB	—	—	—	25-35	.003-.006	—	Chemically active soluble oils, sulfur-base oils, or sulfur-base oils with 10-25% kerosene Sulfurized-chlorinated oils sometimes thinned with 10-25% kerosene
			EM	—	—	—	25-35	.001-.003	—	
			SM	—	—	—	25-35	.002-.004	—	
	Aged	HSS	FMA, FMB	—	—	—	15-30	.003-.006	—	
			EM	—	—	—	15-30	.001-.003	—	
			SM	—	—	—	15-30	.002-.004	—	
	Sol. treated	Carbide C-1	FMI	40-70	.006-.010	.125-.25	—	—	—	
	Aged	Carbide C-2	FMI	—	—	—	40-90	.003-.006	.03-.125	
Haynes Alloy No. 25	Sol. treated	HSS	—	15-17	.002-.006	.10-.20	—	—	—	Sulfochlorinated oil or chemically active soluble oil
		HSS	EME	30	.001	.25	—	—	—	
		HSS (T-15)	FMA	35	.005	.060	—	—	—	
J-1570 S-816 J-1650(e)	Aged	Carbide C-1	FMI	40-60	.006-.010	.125-.25	—	—	—	Sulful-base oil or chemically active soluble oil
		Carbide C-2	FMI	—	—	—	60-100(c)	.003-.006	.03-.125	
L-605	Annealed	Carbide C-1	FMK	50	.005-.01	.06-.08	—	—	—	Kerosene and sulfurized oil
		HSS (T-15)	EME_1	22	.0056	.060	—	—	—	
		HSS (M-2)	SMF_1	28	.0054	.060	—	—	—	
HS-21	Cast	Carbide C-2	—	—	—	—	100	.0011	—	(d)
J-1300 M-308	—	Carbide C-5a	FMI	50-75	.0060-.010	.125-.250	—	—	—	Waterbase soluble oil or sulfur-base oil
		Carbide C-7a	FMI	—	—	—	70-110	.003-.006	.03-.125	
J-155 S-590	Aged	HSS	—	15-25	.002-.006	.10-.20	—	—	—	Sulfur-chlorinated oil

(a) See table on page 68 for design data appropriate to the code shown.
(b) Aged only.
(c) Use 40-75 fpm for J-1650.
(d) Try the various cutting fluids listed in this table.

Simple Indexing...I

BY WM. H. SIEBRECHT 3rd, GENERAL PRECISION LABORATORY, INC.

The accompanying table avoids need for extended calculations to find the number of turns and crank travel required for any problem in simple, or plain, indexing. Further, instead of giving the angular distance between holes as divisions of a circle, the table is based on the distance between holes in the index plate in terms of degrees, minutes and seconds, as is common on engineering drawings. Therefore, the operator need not make a conversion. And a choice of index plates is frequently given in the tabulated figures, so that it is often unnecessary to change the plate already set up on the dividing head.

These tables are based on standard dividing heads with a 40:1 drive. With such heads, each turn of the crank moves the spindle through a 9° angle. Conversely, if spindle travel less than 9° is desired, the crank will have to move through less than one turn. How much it moves will be determined by the desired angle. All spindle angles less than 9° are expressed as crank travel in terms of index plate holes.

Values given in the tables are calculated from index plates with the following hole series: 15, 16, 17, 18, 19, 20, 21, 23, 24, 27, 28, 29, 30, 31, 33, 34, 37, 38, 39, 41, 42, 43, 46, 47, 49, 54, 58, 62 and 66.

The first column is the spindle travel in degrees, minutes and seconds. The second is the decimal equivalent. The last is the index plate and hole necessary to produce this angular displacement.

Example 1: Two holes 6.3000° apart are to be spotted on a circular piece of work. Referring to the table, this angle can be produced by using either the 14th hole in a plate with 20 holes or by the 21st hole in a plate with 30 holes.

Example 2: A plain cam is to be milled with a lobe that subtends an angle of 148° 40′ 00″. The number of turns of the crank is determined by dividing the angle by 9. The result is 16 turns plus 4° 40′ 00″. The 4° 40′ 00″ is read directly from the table as the 14th hole in a 27-hole plate or as the 28th hole in a 54-hole plate.

Angle	Decimal Equiv.	Hole and Plate
0° 8′ 11″	0.1364°	1/66
0° 8′ 43″	0.1452°	1/62
0° 9′ 19″	0.1552°	1/58
0° 10′ 00″	0.1666°	1/54
0° 11′ 1″	0.1837°	1/49
0° 11′ 29″	0.1915°	1/47
0° 11′ 44″	0.1957°	1/46
0° 12′ 33″	0.2093°	1/43
0° 12′ 51″	0.2143°	1/42
0° 13′ 10″	0.2195°	1/41
0° 13′ 51″	0.2308°	1/39
0° 14′ 13″	0.2368°	1/38
0° 14′ 36″	0.2432°	1/37
0° 15′ 53″	0.2647°	1/34
0° 16′ 22″	0.2727°	1/33
0° 16′ 22″	ditto	2/66
0° 17′ 25″	0.2903°	1/31
0° 17′ 25″	ditto	2/62
0° 18′ 00″	0.3000°	1/30
0° 18′ 37″	0.3103°	1/29
0° 18′ 37″	ditto	2/58
0° 19′ 17″	0.3214°	1/28
0° 20′ 00″	0.3333°	1/27
0° 20′ 00″	ditto	2/54
0° 22′ 2″	0.3673°	2/49
0° 22′ 30″	0.3750°	1/24
0° 22′ 59″	0.3830°	2/47
0° 23′ 29″	0.3913°	1/23

Angle	Decimal Equiv.	Hole and Plate
0° 23′ 29″	ditto	2/46
0° 24′ 33″	0.4091	3/66
0° 25′ 7″	0.4186°	2/43
0° 25′ 43″	0.4286	1/21
0° 25′ 43″	ditto	2/42
0° 26′ 8″	0.4355°	3/62
0° 26′ 20″	0.4390°	2/41
0° 27′ 00″	0.4500°	1/20
0° 27′ 42″	0.4615°	2/39
0° 27′ 56″	0.4655°	3/58
0° 28′ 25″	0.4737°	1/19
0° 28′ 25″	ditto	2/38
0° 29′ 11″	0.4865°	2/37
0° 30′ 00″	0.5000°	1/18
0° 30′ 00″	ditto	3/54
0° 31′ 46″	0.5294°	1/17
0° 31′ 46″	ditto	2/34
0° 32′ 44″	0.5454°	2/33
0° 32′ 44″	ditto	4/66
0° 33′ 4″	0.5510°	3/49
0° 33′ 45″	0.5625°	1/16
0° 34′ 28″	0.5745°	3/47
0° 34′ 50″	0.5806°	2/31
0° 34′ 50″	ditto	4/62
0° 35′ 13″	0.5870°	3/46
0° 36′ 00″	0.6000°	1/15
0° 36′ 00″	ditto	2/30
0° 37′ 14″	0.6207°	2/29

Angle	Decimal Equiv.	Hole and Plate
0° 37′ 14″	ditto	4/58
0° 37′ 40″	0.6279°	3/43
0° 38′ 34″	0.6429°	2/28
0° 38′ 34″	ditto	3/42
0° 39′ 31″	0.6585°	3/41
0° 40′ 00″	0.6666°	2/27
0° 40′ 00″	ditto	4/54
0° 40′ 54″	0.6818°	5/66
0° 41′ 32″	0.6923°	3/39
0° 42′ 38″	0.7105°	3/38
0° 43′ 33″	0.7258°	5/62
0° 43′ 47″	0.7297°	3/37
0° 44′ 5″	0.7347°	4/49
0° 45′ 00″	0.7500°	2/24
0° 45′ 57″	0.7660°	4/47
0° 46′ 33″	0.7759°	5/58
0° 46′ 57″	0.7826°	2/23
0° 46′ 57″	ditto	4/46
0° 47′ 39″	0.7941°	3/34
0° 49′ 5″	0.8182°	3/33
0° 49′ 5″	ditto	6/66
0° 50′ 00″	0.8333°	5/54
0° 50′ 14″	0.8372°	4/43
0° 51′ 26″	0.8571°	2/21
0° 51′ 26″	ditto	4/42
0° 52′ 15″	0.8710°	3/31
0° 52′ 15″	ditto	6/62
0° 52′ 41″	0.8780°	4/41

Angle	Decimal Equiv.	Hole and Plate
0° 54′ 00″	0.9000°	2/20
0° 54′ 00″	ditto	3/30
0° 55′ 6″	0.9184°	5/49
0° 55′ 23″	0.9231°	4/39
0° 55′ 52″	0.9310°	3/29
0° 55′ 52″	ditto	6/58
0° 56′ 51″	0.9474°	2/19
0° 56′ 51″	ditto	4/38
0° 57′ 16″	0.9545°	7/66
0° 57′ 27″	0.9574°	5/47
0° 57′ 51″	0.9643°	3/28
0° 58′ 23″	0.9730°	4/37
0° 58′ 42″	0.9783°	5/46
1° 00′ 00″	1.0000°	2/18
1° 00′ 00″	ditto	3/27
1° 00′ 00″	ditto	6/54
1° 00′ 58″	1.0161°	7/62
1° 2′ 47″	1.0465°	5/43
1° 3′ 32″	1.0588°	2/17
1° 3′ 32″	ditto	4/34
1° 4′ 17″	1.0714°	5/42
1° 5′ 10″	1.0862°	7/58
1° 5′ 27″	1.0909°	4/33
1° 5′ 27″	ditto	8/66
1° 5′ 51″	1.0976°	5/41
1° 6′ 7″	1.1020°	6/49
1° 7′ 30″	1.1250°	2/16
1° 7′ 30″	ditto	3/24

Simple Indexing...II

Angle	Decimal Equiv.	Hole and Plate
1° 8′ 56″	1.1489°	6/47
1° 9′ 14″	1.1538°	5/39
1° 9′41″	1.1613°	4/31
1° 9′ 41″	ditto	8/62
1° 10′ 00″	1.1666°	7/54
1° 10′ 26″	1.1739°	3/23
1° 10′ 26″	ditto	6/46
1° 11′ 3″	1.1842°	5/38
1° 12′ 00″	1.2000°	2/15
1° 12′ 00″	ditto	4/30
1° 12′ 58″	1.2162°	5/37
1° 13′ 38″	1.2272°	9/66
1° 14′ 29″	1.2414°	4/29
1° 14′ 29″	1.2414°	8/58
1° 15′ 21″	1.2558°	6/43
1° 17′ 9″	1.2857°	3/21
1° 17′ 9″	ditto	4/28
1° 17′ 9″	ditto	6/42
1° 17′ 9″	ditto	7/49
1° 18′ 23″	1.3065°	9/62
1° 19′ 1″	1.3171°	6/41
1° 19′ 25″	1.3235°	5/34
1° 20′ 00″	1.3333°	4/27
1° 20′ 00″	ditto	8/54
1° 20′ 26″	1.3404°	7/47
1° 21′ 00″	1.3500°	3/20
1° 21′ 49″	1.3636°	5/33
1° 21′ 49″	ditto	10/66
1° 22′ 10″	1.3696°	7/46
1° 23′ 5″	1.3846°	6/39
1° 23′ 48″	1.3966°	9/58
1° 25′ 16″	1.4211°	3/19
1° 25′ 16″	ditto	6/38
1° 27′ 6″	1.4516°	5/31
1° 27′ 6″	ditto	10/62
1° 27′ 34″	1.4595°	6/37
1° 27′ 54″	1.4651°	7/43
1° 28′ 10″	1.4694°	8/49
1° 30′ 00″	1.5000°	3/18
1° 30′ 00″	ditto	4/24
1° 30′ 00″	ditto	5/30
1° 30′ 00″	ditto	7/42
1° 30′ 00″	ditto	9/54
1° 30′ 00″	ditto	11/66
1° 31′ 55″	1.5319°	8/47
1° 32′ 12″	1.5366°	7/41
1° 33′ 6″	1.5517°	5/29
1° 33′ 6″	ditto	10/58
1° 33′ 55″	1.5652°	4/23
1° 33′ 55″	ditto	8/46
1° 35′ 18″	1.5882°	3/17
1° 35′ 18″	ditto	6/34
1° 35′ 48″	1.5968°	11/62
1° 36′ 26″	1.6071°	5/28
1° 36′ 55″	1.6154°	7/39
1° 38′ 11″	1.6364°	6/33
1° 38′ 11″	ditto	12/66
1° 39′ 11″	1.6531°	9/49
1° 39′ 28″	1.6579°	7/38
1° 40′ 00″	1.6666°	5/27
1° 40′ 00″	ditto	10/54
1° 40′ 28″	1.6744°	8/43
1° 41′ 15″	1.6857°	3/16
1° 42′ 10″	1.7027°	7/37
1° 42′ 25″	1.7069°	11/58
1° 42′ 51″	1.7143°	4/21
1° 42′ 51″	ditto	8/42
1° 43′ 24″	1.7234°	9/47
1° 44′ 31″	1.7419°	6/31
1° 44′ 31″	ditto	12/62
1° 45′ 22″	1.7561°	8/41
1° 45′ 39″	1.7609°	9/46
1° 46′ 22″	1.7727°	13/66
1° 48′ 00″	1.8000°	3/15
1° 48′ 00″	ditto	4/20
1° 48′ 00″	ditto	6/30
1° 50′ 00″	1.8333°	11/54
1° 50′ 12″	1.8367°	10/49
1° 50′ 46″	1.8462°	8/39
1° 51′ 11″	1.8529°	7/34
1° 51′ 43″	1.8621°	6/29
1° 51′ 43″	ditto	12/58
1° 52′ 30″	1.8750°	5/24
1° 53′ 1″	1.8837°	9/43
1° 53′ 14″	1.8871°	13/62
1° 53′ 41″	1.8947°	4/19
1° 53′ 41″	ditto	8/38
1° 54′ 33″	1.9091°	7/33
1° 54′ 33″	1.9091°	14/66
1° 54′ 54″	1.9149°	10/47
1° 55′ 43″	1.9286°	6/28
1° 55′ 43″	ditto	9/42
1° 56′ 45″	1.9459°	8/37
1° 57′ 23″	1.9565°	5/23
1° 57′ 23″	ditto	10/46
1° 58′ 32″	1.9756°	9/41
2° 00′ 00″	2.0000°	4/18
2° 00′ 00″	ditto	6/27
2° 00′ 00″	ditto	12/54
2° 1′ 2″	2.0172°	13/58
2° 1′ 13″	2.0204°	11/49
2° 1′ 56″	2.0323°	7/31
2° 1′ 56″	ditto	14/62
2° 2′ 44″	2.0454°	15/66
2° 4′ 37″	2.0769°	9/39
2° 5′ 35″	2.0930°	10/43
2° 6′ 00″	2.1000°	7/30
2° 6′ 23″	2.1064°	11/47
2° 7′ 4″	2.1176°	4/17
2° 7′ 4″	ditto	8/34
2° 7′ 54″	2.1316°	9/38
2° 8′ 34″	2.1429°	5/21
2° 8′ 34″	ditto	10/42
2° 9′ 8″	2.1522°	11/46
2° 10′ 00″	2.1666°	13/54
2° 10′ 21″	2.1724°	7/29
2° 10′ 21″	ditto	14/58
2° 10′ 39″	2.1774°	15/62
2° 10′ 54″	2.1818°	8/33
2° 10′ 54″	ditto	16/66
2° 11′ 21″	2.1892°	9/37
2° 11′ 42″	2.1951°	10/41
2° 12′ 15″	2.2041°	12/49
2° 15′ 00″	2.2500°	4/16
2° 15′ 00″	ditto	5/20
2° 15′ 00″	ditto	6/24
2° 15′ 00″	ditto	7/28
2° 17′ 52″	2.2979°	12/47
2° 18′ 8″	2.3023°	11/43
2° 18′ 28″	2.3077°	10/39
2° 19′ 5″	2.3182°	17/66
2° 19′ 21″	2.3226°	8/31
2° 19′ 21″	ditto	16/62
2° 19′ 39″	2.3276°	15/58
2° 20′ 00″	2.3333°	7/27
2° 20′ 00″	ditto	14/54
2° 20′ 52″	2.3478°	6/23
2° 20′ 52″	ditto	12/46
2° 21′ 26″	2.3571°	11/42
2° 22′ 6″	2.3684°	5/19
2° 22′ 6″	ditto	10/38
2° 22′ 56″	2.3824°	9/34
2° 23′ 16″	2.3878°	13/49
2° 24′ 00″	2.4000°	4/15
2° 24′ 00″	ditto	8/30
2° 24′ 53″	2.4146°	11/41
2° 25′ 57″	2.4324°	10/37
2° 27′ 16″	2.4545°	9/33
2° 27′ 16″	ditto	18/66
2° 28′ 4″	2.4677°	17/62
2° 28′ 58″	2.4828°	8/29
2° 28′ 58″	ditto	16/58
2° 29′ 22″	2.4894°	13/47
2° 30′ 00″	2.5000°	5/18
2° 30′ 00″	ditto	15/54
2° 30′ 42″	2.5116°	12/43
2° 32′ 18″	2.5385°	11/39
2° 32′ 36″	2.5435°	13/46
2° 34′ 17″	2.5714°	6/21
2° 34′ 17″	ditto	8/28
2° 34′ 17″	ditto	12/42
2° 34′ 17″	ditto	14/49
2° 35′ 27″	2.5909°	19/66
2° 36′ 19″	2.6053°	11/38
2° 36′ 46″	2.6129°	9/31
2° 36′ 46″	ditto	18/62
2° 37′ 30″	2.6250°	7/24
2° 38′ 3″	2.6341°	12/41
2° 38′ 17″	2.6379°	17/58
2° 38′ 49″	2.6471°	5/17
2° 38′ 49″	ditto	10/34
2° 40′ 00″	2.6666°	8/27
2° 40′ 00″	ditto	16/54
2° 40′ 32″	2.6757°	11/37
2° 40′ 51″	2.6809°	14/47
2° 42′ 00″	2.7000°	6/20
2° 42′ 00″	ditto	9/30
2° 43′ 15″	2.7209°	13/43
2° 43′ 38″	2.7272°	10/33
2° 43′ 38″	ditto	20/66
2° 44′ 21″	2.7391°	7/23
2° 44′ 21″	ditto	14/46
2° 45′ 18″	2.7551°	15/49
2° 45′ 29″	2.7581°	19/62
2° 46′ 9″	2.7692°	12/39
2° 47′ 9″	2.7857°	13/42
2° 47′ 35″	2.7931°	9/29
2° 47′ 35″	ditto	18/58
2° 48′ 45″	2.8125°	5/16
2° 50′ 00″	2.8333°	17/54
2° 50′ 32″	2.8421°	6/19
2° 50′ 32″	ditto	12/38
2° 51′ 13″	2.8537°	13/41
2° 51′ 49″	2.8636°	21/66
2° 52′ 20″	2.8723°	15/47
2° 53′ 34″	2.8929°	9/28
2° 54′ 12″	2.9032°	10/31
2° 54′ 12″	ditto	20/62
2° 54′ 42″	2.9118°	11/34
2° 55′ 8″	2.9189°	12/37
2° 55′ 49″	2.9302°	14/43
2° 56′ 5″	2.9348°	15/46
2° 56′ 20″	2.9388°	16/49
2° 56′ 54″	2.9483°	19/58
3° 00′ 00″	3.0000°	5/15
3° 00′ 00″	ditto	6/18
3° 00′ 00″	ditto	7/21
3° 00′ 00″	ditto	8/24
3° 00′ 00″	ditto	9/27
3° 00′ 00″	ditto	10/30
3° 00′ 00″	ditto	11/33
3° 00′ 00″	ditto	13/39
3° 00′ 00″	ditto	14/42
3° 00′ 00″	3.0000°	18/54
3° 00′ 00″	ditto	22/66
3° 2′ 54″	3.0484°	21/62

Simple Indexing...III

Angle	Decimal Equiv.	Hole and Plate
3° 3′ 50″	3.0638°	16/47
3° 4′ 23″	3.0732°	14/41
3° 4′ 44″	3.0789°	13/38
3° 6′ 12″	3.1034°	10/29
3° 6′ 12″	ditto	20/58
3° 7′ 21″	3.1224°	17/49
3° 7′ 50″	3.1304°	8/23
3° 7′ 50″	ditto	16/46
3° 8′ 11″	3.1364°	23/66
3° 8′ 22″	3.1395°	15/43
3° 9′ 00″	3.1500°	7/20
3° 9′ 44″	3.1622°	13/37
3° 10′ 00″	3.1666°	19/54
3° 10′ 35″	3.1765°	6/17
3° 10′ 35″	ditto	12/34
3° 11′ 37″	3.1935°	11/31
3° 11′ 37″	ditto	22/62
3° 12′ 51″	3.2143°	10/28
3° 12′ 51″	ditto	15/42
3° 13′ 51″	3.2308°	14/39
3° 15′ 19″	3.2553°	17/47
3° 15′ 31″	3.2586°	21/58
3° 16′ 22″	3.2727°	12/33
3° 16′ 22″	ditto	24/66
3° 17′ 34″	3.2927°	15/41
3° 18′ 00″	3.3000°	11/30
3° 18′ 22″	3.3061°	18/49
3° 18′ 57″	3.3158°	7/19
3° 18′ 57″	ditto	14/38
3° 19′ 34″	3.3261°	17/46
3° 20′ 00″	3.3333°	10/27
3° 20′ 00″	ditto	20/54
3° 20′ 19″	3.3387°	23/62
3° 20′ 56″	3.3488°	16/43
3° 22′ 30″	3.3750°	6/16
3° 22′ 30″	ditto	9/24
3° 24′ 19″	3.4054°	14/37
3° 24′ 33″	3.4091°	25/66
3° 24′ 50″	3.4138°	11/29
3° 24′ 50″	ditto	22/58
3° 25′ 43″	3.4286°	8/21
3° 25′ 43″	ditto	16/42
3° 26′ 28″	3.4412°	13/34
3° 26′ 49″	3.4468°	18/47
3° 27′ 42″	3.4615°	15/39
3° 29′ 2″	3.4839°	12/31
3° 29′ 2″	ditto	24/62
3° 29′ 23″	3.4898°	19/49
3° 30′ 00″	3.5000°	7/18
3° 30′ 00″	ditto	21/54
3° 30′ 44″	3.5122°	16/41
3° 31′ 18″	3.5217°	9/23
3° 31′ 18″	ditto	18/46
3° 32′ 9″	3.5357°	11/28

Angle	Decimal Equiv.	Hole and Plate
3° 32′ 44″	3.5454°	13/33
3° 32′ 44″	ditto	26/66
3° 33′ 9″	3.5526°	15/38
3° 33′ 29″	3.5581°	17/43
3° 34′ 8″	3.5690°	23/58
3° 36′ 00″	3.6000°	6/15
3° 36′ 00″	ditto	8/20
3° 36′ 00″	ditto	12/30
3° 37′ 45″	3.6290°	25/62
3° 38′ 18″	3.6383°	19/47
3° 38′ 34″	3.6429°	17/42
3° 38′ 55″	3.6486°	15/37
3° 40′ 00″	3.6666°	11/27
3° 40′ 00″	ditto	22/54
3° 40′ 24″	3.6735°	20/49
3° 40′ 54″	3.6818°	27/66
3° 41′ 32″	3.6923°	16/39
3° 42′ 21″	3.7059°	7/17
3° 42′ 21″	3.7059°	14/34
3° 43′ 3″	3.7174°	19/46
3° 43′ 27″	3.7241°	12/29
3° 43′ 27″	ditto	24/58
3° 43′ 54″	3.7317°	17/41
3° 45′ 00″	3.7500°	10/24
3° 46′ 3″	3.7674	18/43
3° 46′ 27″	3.7742°	13/31
3° 46′ 27″	ditto	26/62
3° 47′ 22″	3.7895°	8/19
3° 47′ 22″	ditto	16/38
3° 49′ 5″	3.8182°	14/33
3° 49′ 5″	ditto	28/66
3° 49′ 47″	3.8298°	20/47
3° 50′ 00″	3.8333°	23/54
3° 51′ 26″	3.8571°	9/21
3° 51′ 26″	ditto	12/28
3° 51′ 26″	ditto	18/42
3° 51′ 26″	ditto	21/49
3° 52′ 46″	3.8793°	25/58
3° 53′ 31″	3.8919°	16/37
3° 54′ 00″	3.9000°	13/30
3° 54′ 47″	3.9130°	10/23
3° 54′ 47″	ditto	20/46
3° 55′ 10″	3.9194°	27/62
3° 55′ 23″	3.9231°	17/39
3° 56′ 15″	3.9375°	7/16
3° 57′ 4″	3.9512°	18/41
3° 57′ 16″	3.9545°	29/66
3° 58′ 14″	3.9706°	15/34
3° 58′ 36″	3.9767°	19/43
4° 00′ 00″	4.0000°	8/18
4° 00′ 00″	ditto	12/27
4° 00′ 00″	ditto	24/54
4° 1′ 17″	4.0213°	21/47
4° 1′ 35″	4.0263°	17/38

Angle	Decimal Equiv.	Hole and Plate
4° 2′ 4″	4.0345°	13/29
4° 2′ 4″	ditto	26/58
4° 2′ 27″	4.0408°	22/49
4° 3′ 00″	4.0500°	9/20
4° 3′ 52″	4.0645°	14/31
4° 3′ 52″	ditto	28/62
4° 4′ 17″	4.0714°	19/42
4° 5′ 27″	4.0909°	15/33
4° 5′ 27″	ditto	30/66
4° 6′ 31″	4.1087°	21/46
4° 7′ 30″	4.1250°	11/24
4° 8′ 6″	4.1351°	17/37
4° 9′ 14″	4.1538°	18/39
4° 10′ 00″	4.1666°	25/54
4° 10′ 15″	4.1707°	19/41
4° 10′ 43″	4.1786°	13/28
4° 11′ 10″	4.1860°	20/43
4° 11′ 23″	4.1897°	27/58
4° 12′ 00″	4.2000°	7/15
4° 12′ 00″	ditto	14/30
4° 12′ 35″	4.2097°	29/62
4° 12′ 46″	4.2128°	22/47
4° 13′ 28″	4.2245°	23/49
4° 13′ 38″	4.2272°	31/66
4° 14′ 7″	4.2353°	8/17
4° 14′ 7″	ditto	16/34
4° 15′ 47″	4.2632°	9/19
4° 15′ 47″	ditto	18/38
4° 17′ 9″	4.2857°	10/21
4° 17′ 9″	ditto	20/42
4° 18′ 16″	4.3043°	11/23
4° 18′ 16″	ditto	22/46
4° 20′ 00″	4.3333°	13/27
4° 20′ 00″	ditto	26/54
4° 20′ 41″	4.3448°	14/29
4° 20′ 41″	ditto	28/58
4° 21′ 17″	4.3548°	15/31
4° 21′ 17″	ditto	30/62
4° 21′ 49″	4.3636°	16/33
4° 21′ 49″	4.3636°	32/66
4° 22′ 42″	4.3784°	18/37
4° 23′ 5″	4.3846°	19/39
4° 23′ 25″	4.3902°	20/41
4° 23′ 43″	4.3953°	21/43
4° 24′ 15″	4.4043°	23/47
4° 24′ 29″	4.4082°	24/49
4° 30′ 00″	4.5000°	8/16
4° 30′ 00″	ditto	9/18
4° 30′ 00″	ditto	10/20
4° 30′ 00″	ditto	12/24
4° 30′ 00″	ditto	14/28
4° 30′ 00″	ditto	15/30
4° 30′ 00″	ditto	17/34
4° 30′ 00″	ditto	19/38

Angle	Decimal Equiv.	Hole and Plate
4° 30′ 00″	ditto	21/42
4° 30′ 00″	ditto	23/46
4° 30′ 00″	ditto	27/54
4° 30′ 00″	ditto	29/58
4° 30′ 00″	ditto	31/62
4° 30′ 00″	ditto	33/66
4° 35′ 31″	4.5918°	25/49
4° 35′ 45″	4.5957°	24/47
4° 36′ 17″	4.6047°	22/43
4° 36′ 35″	4.6098°	21/41
4° 36′ 55″	4.6154°	20/39
4° 37′ 18″	4.6216°	19/37
4° 38′ 11″	4.6364°	17/33
4° 38′ 11″	ditto	34/66
4° 38′ 43″	4.6452°	16/31
4° 38′ 43″	ditto	32/62
4° 39′ 19″	4.6552°	15/29
4° 39′ 19″	ditto	30/58
4° 40′ 00″	4.6666°	14/27
4° 40′ 00″	ditto	28/54
4° 41′ 44″	4.6957°	12/23
4° 41′ 44″	ditto	24/46
4° 42′ 51″	4.7143°	11/21
4° 42′ 51″	ditto	22/42
4° 44′ 13″	4.7368°	10/19
4° 44′ 13″	ditto	20/38
4° 45′53″	4.7647°	9/17
4° 45′ 53″	ditto	18/34
4° 46′ 22″	2.7727°	35/66
4° 46′ 32″	4.7755°	26/49
4° 47′ 14″	4.7872°	25/47
4° 47′ 25″	4.7903°	33/62
4° 48′ 00″	4.8000°	8/15
4° 48′ 00″	ditto	16/30
4° 48′ 37″	4.8103°	31/58
4° 48′ 50″	4.8140°	23/43
4° 49′ 17″	4.8214°	15/28
4° 49′ 45″	4.8293°	22/41
4° 50′ 00″	4.8333°	29/54
4° 50′ 46″	4.8462°	21/39
4° 51′ 54″	4.8649°	20/37
4° 52′ 30″	4.8750°	13/24
4° 53′ 29″	4.8913°	25/46
4° 54′ 33°	4.9091°	18/33
4/ 54′ 33″	ditto	36/66
4° 55′ 43″	4.9286°	23/42
4° 56′ 8″	4.9355°	17/31
4° 56′ 8″	ditto	34/62
4° 57′ 00″	4.9500°	11/20
4° 57′ 33″	4.9592°	27/49
4° 57′ 56″	4.9655°	16/29
4° 57′ 56″	ditto	32/58
4° 58′ 25″	4.9737°	21/38
4° 58′ 43″	4.9787°	26/47

Simple Indexing...IV

BY WM. H. SIEBRECHT 3rd, GENERAL PRECISION LABORATORY, INC.

Angle	Decimal Equiv.	Hole and Plate	Angle	Decimal Equiv.	Hole and Plate	Angle	Decimal Equiv.	Hole and Plate	Angle	Decimal Equiv.	Hole and Plate
5° 00′ 00″	5.0000°	10/18	5° 28′ 42″	ditto	28/46	6° 00′ 00″	ditto	12/18	6° 25′ 43″	ditto	30/42
5° 00′ 00″	ditto	15/27	5° 29′ 16″	5.4878°	25/41	6° 00′ 00″	ditto	14/21	6° 25′ 43″	ditto	35/49
5° 00′ 00″	ditto	30/54	5° 30′ 00″	5.5000°	11/18	6° 00′ 00″	ditto	16/24	6° 27′ 23″	6.4565°	33/46
5° 1′ 24″	5.0233°	24/43	5° 30′ 00″	ditto	33/54	6° 00′ 00″	ditto	18/27	6° 27′ 42″	6.4615°	28/39
5° 1′ 46″	5.0294°	19/34	5° 30′ 37″	5.5102°	30/49	6° 00′ 00″	ditto	20/30	6° 29′ 18″	6.4884°	31/43
5° 2′ 44″	5.0454°	37/66	5° 30′ 58″	5.5161°	19/31	6° 00′ 00″	ditto	22/33	6° 30′ 00″	6.5000°	13/18
5° 2′ 56″	5.0488°	23/41	5° 30′ 58″	ditto	38/62	6° 00′ 00″	ditto	26/39	6° 30′ 00″	ditto	39/54
5° 3′ 45″	5.0625°	9/16	5° 32′ 18″	5.5385°	24/39	6° 00′ 00″	ditto	28/42	6° 30′ 38″	6.5106°	34/47
5° 4′ 37″	5.0769°	22/39	5° 33′ 11″	5.5532°	29/47	6° 00′ 00″	ditto	36/54	6° 31′ 2″	6.5172°	21/29
5° 4′ 50″	5.0806°	35/62	5° 33′ 32″	5.5588°	21/34	6° 00′ 00″	ditto	44/66	6° 31′ 2″	ditto	42/58
5° 5′ 13″	5.0870°	13/23	5° 34′ 17″	5.5714°	13/21	6° 3′ 6″	6.0517°	39/58	6° 31′ 56″	6.5323°	45/62
5° 5′ 13″	ditto	26/46	5° 34′ 17″	ditto	26/42	6° 3′ 40″	6.0612°	33/49	6° 32′ 44″	6.5454°	24/33
5° 6′ 00″	5.1000°	17/30	5° 35′ 10″	5.5862°	18/29	6° 3′ 55″	6.0652°	31/46	6° 32′ 44″	ditto	48/66
5° 6′ 29″	5.1081°	21/37	5° 35′ 10″	ditto	36/58	6° 4′ 11″	6.0698°	29/43	6° 34′ 3″	6.5676°	27/37
5° 7′ 14″	5.1207°	33/58	5° 35′ 27″	5.5909°	41/66	6° 4′ 52″	6.0811°	25/37	6° 35′ 7″	6.5854°	30/41
5° 8′ 34″	5.1429°	12/21	5° 35′ 41″	5.5946°	23/37	6° 5′ 18″	6.0882°	23/34	6° 36′ 00″	6.6000°	11/15
5° 8′ 34″	ditto	16/28	5° 37′ 30″	5.6250°	10/16	6° 5′ 48″	6.0968°	21/31	6° 36′ 00″	ditto	22/30
5° 8′ 34″	ditto	24/42	5° 37′ 30″	ditto	15/24	6° 5′ 48″	ditto	42/62	6° 36′ 44″	6.6122°	36/49
5° 8′ 34″	ditto	28/49	5° 39′ 4″	5.6512°	27/43	6° 6′ 26″	6.1071°	19/28	6° 37′ 4″	6.6176°	25/34
5° 10′ 00″	5.1666°	31/54	5° 39′ 41″	5.6613°	39/62	6° 7′ 40″	6.1277°	32/47	6° 37′ 54″	6.6316°	14/19
5° 10′ 13″	5.1702°	27/47	5° 40′ 00″	5.6666°	17/27	6° 8′ 11″	6.1364°	45/66	6° 37′ 54″	ditto	28/38
5° 10′ 54″	5.1818°	19/33	5° 40′ 00″	ditto	34/54	6° 8′ 47″	6.1463°	28/41	6° 38′ 34″	6.6429°	31/42
5° 10′ 54″	ditto	38/66	5° 40′ 26″	5.6739°	29/46	6° 9′ 28″	6.1579°	13/19	6° 39′ 8″	6.6522°	17/23
5° 12′ 38″	5.2105°	11/19	5° 41′ 3″	5.6842°	12/19	6° 9′ 28″	6.1579°	26/38	6° 39′ 8″	ditto	34/46
5° 12′ 38″	ditto	22/38	5° 41′ 3″	ditto	24/38	6° 10′ 00″	6.1666°	37/54	6° 40′ 00″	6.6666°	20/27
5° 13′ 33″	5.2258°	18/31	5° 41′ 38″	5.6939°	31/49	6° 11′ 15″	6.1875°	11/16	6° 40′ 00″	ditto	40/54
5° 13′ 33″	ditto	36/62	5° 42′ 00″	5.7000°	19/30	6° 12′ 25″	6.2069°	20/29	6° 40′ 21″	6.6724°	43/58
5° 13′ 57″	5.2326°	25/43	5° 42′ 26″	5.7073°	26/41	6° 12′ 25″	ditto	40/58	6° 40′ 39″	6.6774°	23/31
5° 15′ 00″	5.2500°	14/24	5° 43′ 38″	5.7272°	21/33	6° 12′ 51″	6.2143°	29/42	6° 40′ 39″	ditto	46/62
5° 16′ 6″	5.2683°	24/41	5° 43′ 38″	ditto	42/66	6° 13′ 51″	6.2308°	27/39	6° 40′ 54″	6.6818°	49/66
5° 16′ 33″	5.2759°	17/29	5° 44′ 29″	5.7414°	37/58	6° 14′ 31″	6.2419°	43/62	6° 41′ 32″	6.6923°	29/39
5° 16′ 33″	5.2759°	34/58	5° 44′ 41″	5.7447°	30/47	6° 14′ 42″	6.2449°	34/49	6° 41′ 52″	6.6977°	32/43
5° 16′ 57″	5.2826°	27/46	5° 46′ 9″	5.7692°	25/39	6° 15′ 39″	6.2609°	16/23	6° 42′ 8″	6.7021°	35/47
5° 17′ 39″	5.2941°	10/17	5° 47′ 9″	5.7857°	18/28	6° 15′ 39″	ditto	32/46	6° 45′ 00″	6.7500°	12/16
5° 17′ 39″	ditto	20/34	5° 47′ 9″	ditto	27/42	6° 16′ 22″	6.2727°	23/33	6° 45′ 00″	ditto	15/20
5° 18′ 28″	5.3077°	23/39	5° 48′ 23″	5.8065°	20/31	6° 16′ 22″	ditto	46/66	6° 45′ 00″	ditto	18/24
5° 19′ 5″	5.3182°	39/66	5° 48′ 23″	ditto	40/62	6° 16′ 45″	6.2791°	30/43	6° 45′ 00″	ditto	21/28
5° 19′ 36″	5.3265°	29/49	5° 49′ 25″	5.8235°	11/17	6° 18′ 00″	6.3000°	14/20	6° 47′ 45″	6.7959°	37/49
5° 20′ 00″	5.3333°	16/27	5° 49′ 25″	ditto	22/34	6° 18′ 00″	ditto	21/30	6° 48′ 18″	6.8049°	31/41
5° 20′ 00″	ditto	32/54	5° 50′ 00″	5.8333°	35/54	6° 19′ 9″	6.3191°	33/47	6° 48′ 39″	6.8108°	28/37
5° 21′ 5″	5.3514°	22/37	5° 50′ 16″	5.8378°	24/37	6° 19′ 28″	6.3243°	26/37	6° 49′ 5″	6.8182°	25/33
5° 21′ 26″	5.3571°	25/42	5° 51′ 00″	5.8500°	13/20	6° 20′ 00″	6.3333°	19/27	6° 49′ 5″	ditto	50/66
5° 21′ 42″	5.3617°	28/47	5° 51′ 38″	5.8605°	28/43	6° 20′ 00″	ditto	38/54	6° 49′ 21″	6.8226°	47/62
5° 22′ 15″	5.3710°	37/62	5° 51′ 49″	5.8636°	43/66	6° 21′ 11″	6.3529°	12/17	6° 49′ 39″	6.8276°	22/29
5° 24′ 00″	5.4000°	9/15	5° 52′ 10″	5.8696°	15/23	6° 21′ 11″	ditto	24/34	6° 49′ 39″	6.8276°	44/58
5° 24′ 00″	ditto	12/20	5° 52′ 10″	ditto	30/46	6° 21′ 43″	6.3621°	41/58	6° 50′ 00″	6.8333°	41/54
5° 24′ 00″	ditto	18/30	5° 52′ 39″	5.8776°	32/49	6° 21′ 57″	6.3659°	29/41	6° 50′ 52″	6.8478°	35/46
5° 25′ 52″	5.4310°	35/58	5° 53′ 48″	5.8966°	19/29	6° 22′ 30″	6.3750°	17/24	6° 51′ 26″	6.8571°	16/21
5° 26′ 31″	5.4419°	26/43	5° 53′ 48″	ditto	38/58	6° 23′ 14″	6.3871°	22/31	6° 51′ 26″	ditto	32/42
5° 26′ 51″	5.4474°	23/38	5° 55′ 16″	5.9211°	25/38	6° 23′ 14″	ditto	44/62	6° 52′ 6″	6.8684°	29/38
5° 27′ 16″	5.4545°	20/33	5° 55′ 37″	5.9268°	27/41	6° 23′ 41″	6.3947°	27/38	6° 52′ 56″	6.8824°	13/17
5° 27′ 16″	ditto	40/66	5° 56′ 10″	5.9362°	31/47	6° 24′ 33″	6.4091°	47/66	6° 52′ 56″	ditto	26/34
5° 27′ 51″	5.4643°	17/28	5° 57′ 6″	5.9516°	41/62	6° 25′ 43″	6.4286°	15/21	6° 53′ 37″	6.8936°	36/47
5° 28′ 42″	5.4783°	14/23	6° 00′ 00″	6.0000°	10/15	6° 25′ 43″	ditto	20/28	6° 54′ 00″	6.9000°	23/30

Simple Indexing...V

Angle	Decimal Equiv.	Hole and Plate
6° 54′ 25″	6.9070°	33/43
6° 55′ 23″	6.9231°	30/39
6° 57′ 16″	6.9545°	51/66
6° 58′ 4″	6.9677°	24/31
6° 58′ 4″	ditto	48/62
6° 58′ 47″	6.9796°	38/49
6° 58′ 58″	6.9828°	45/58
7° 00′ 00″	7.0000°	14/18
7° 00′ 00″	ditto	21/27
7° 00′ 00″	ditto	42/54
7° 1′ 28″	7.0244°	32/41
7° 2′ 36″	7.0435°	18/23
7° 2′ 36″	ditto	36/46
7° 3′ 15″	7.0541°	29/37
7° 4′ 17″	7.0714°	22/28
7° 4′ 17″	7.0714°	33/42
7° 5′ 6″	7.0851°	37/47
7° 5′ 27″	7.0909°	26/33
7° 5′ 27″	ditto	52/66
7° 6′ 19″	7.1053°	15/19
7° 6′ 19″	ditto	30/38
7° 6′ 46″	7.1129°	49/62
7° 6′ 59″	7.1163°	34/43
7° 7′ 30″	7.1250°	19/24
7° 8′ 17″	7.1379°	23/29
7° 8′ 17″	ditto	46/58
7° 8′ 49″	7.1471°	27/34
7° 9′ 14″	7.1538°	31/39
7° 9′ 48″	7.1633°	39/49
7° 10′ 00″	7.1666°	43/54
7° 12′ 00″	7.2000°	12/15
7° 12′ 00″	ditto	16/20
7° 12′ 00″	ditto	24/30
7° 13′ 38″	7.2272°	53/66
7° 14′ 21″	7.2391°	37/46
7° 14′ 28″	7.2439°	33/41
7° 15′ 29″	7.2581°	25/31
7° 15′ 29″	ditto	50/62
7° 16′ 36″	7.2766°	38/47
7° 17′ 9″	7.2857°	17/21
7° 17′ 9″	7.2857°	34/42
7° 17′ 35″	7.2931°	47/58
7° 17′ 50″	7.2973°	30/37
7° 18′ 45″	7.3125°	13/16
7° 19′ 32″	7.3256°	35/43
7° 20′ 00″	7.3333°	22/27
7° 20′ 00″	ditto	44/54
7° 20′ 32″	7.3421°	31/38
7° 20′ 49″	7.3469°	40/49
7° 21′ 49″	7.3636°	27/33
7° 21′ 49″	ditto	54/66
7° 23′ 5″	7.3846°	32/39
7° 23′ 34″	7.3929°	23/28
7° 24′ 12″	7.4032°	51/62
7° 24′ 42″	7.4118°	14/17
7° 24′ 42″	ditto	28/34
7° 26′ 5″	7.4348°	19/23
7° 26′ 5″	ditto	38/46
7° 26′ 54″	7.4483°	24/29
7° 26′ 54″	ditto	48/58
7° 27′ 48″	7.4634°	34/41
7° 28′ 5″	7.4681°	39/47
7° 30′ 00″	7.5000°	15/18
7° 30′ 00″	ditto	20/24
7° 30′ 00″	ditto	25/30
7° 30′ 00″	7.5000°	35/42
7° 30′ 00″	ditto	45/54
7° 30′ 00″	ditto	55/66
7° 31′ 50″	7.5306°	41/49
7° 32′ 6″	7.5349°	36/43
7° 32′ 26″	7.5405°	31/37
7° 32′ 54″	7.5484°	26/31
7° 32′ 54″	ditto	52/62
7° 34′ 44″	7.5789°	16/19
7° 34′ 44″	ditto	32/38
7° 36′ 12″	7.6034°	49/58
7° 36′ 55″	7.6154°	33/39
7° 37′ 50″	7.6304°	39/46
7° 38′ 11″	7.6364°	28/33
7° 38′ 11″	ditto	56/66
7° 39′ 00″	7.6500°	17/20
7° 39′ 34″	7.6596°	40/47
7° 40′ 00″	7.6666°	23/27
7° 40′ 00″	ditto	46/54
7° 40′ 35″	7.6765°	29/34
7° 40′ 59″	7.6829°	35/41
7° 41′ 37″	7.6935°	53/62
7° 42′ 51″	7.7143°	18/21
7° 42′ 51″	ditto	24/28
7° 42′ 51″	ditto	36/42
7° 42′ 51″	7.7143°	42/49
7° 44′ 39″	7.7442°	37/43
7° 45′ 31″	7.7586°	25/29
7° 45′ 31″	ditto	50/58
7° 46′ 22″	7.7727°	57/66
7° 47′ 2″	7.7838°	32/37
7° 48′ 00″	7.8000°	13/15
7° 48′ 00″	ditto	26/30
7° 48′ 57″	7.8158°	33/38
7° 49′ 34″	7.8261°	20/23
7° 49′ 34″	ditto	40/46
7° 50′ 00″	7.8333°	47/54
7° 50′ 19″	7.8387°	27/31
7° 50′ 19″	ditto	54/62
7° 50′ 46″	7.8462°	34/39
7° 51′ 4″	7.8511°	41/47
7° 52′ 30″	7.8750°	14/16
7° 52′ 30″	ditto	21/24
7° 53′ 53″	7.8980°	43/49
7° 54′ 9″	7.9024°	36/41
7° 54′ 33″	7.9091°	29/33
7° 54′ 33″	ditto	58/66
7° 54′ 50″	7.9138°	51/58
7° 55′ 43″	7.9286°	37/42
7° 56′ 28″	7.9412°	15/17
7° 56′ 28″	7.9412°	30/34
7° 57′ 13″	7.9535°	38/43
7° 59′ 2″	7.9839°	55/62
8° 00′ 00″	8.0000°	16/18
8° 00′ 00″	ditto	24/27
8° 00′ 00″	ditto	48/54
8° 1′ 18″	8.0217°	41/46
8° 1′ 37″	8.0270°	33/37
8° 2′ 9″	8.0357°	25/28
8° 2′ 33″	8.0426°	42/47
8° 2′ 44″	8.0454°	59/66
8° 3′ 9″	8.0526°	17/19
8° 3′ 9″	ditto	34/38
8° 4′ 8″	8.0690°	26/29
8° 4′ 8″	ditto	52/58
8° 4′ 37″	8.0769°	35/39
8° 4′ 54″	8.0816°	44/49
8° 6′ 00″	8.1000°	18/20
8° 6′ 00″	ditto	27/30
8° 7′ 19″	8.1220°	37/41
8° 7′ 45″	8.1290°	28/31
8° 7′ 45″	ditto	56/62
8° 8′ 34″	8.1429°	19/21
8° 8′ 34″	ditto	38/42
8° 9′ 46″	8.1628°	39/43
8° 10′ 00″	8.1666°	49/54
8° 10′ 54″	8.1818°	30/33
8° 10′ 54″	ditto	60/66
8° 12′ 21″	8.2059°	31/34
8° 13′ 3″	8.2174°	21/23
8° 13′ 3″	ditto	42/46
8° 13′ 27″	8.2241°	53/58
8° 14′ 3″	8.2340°	43/47
8° 15′ 00″	8.2500°	22/24
8° 15′ 55″	8.2653°	45/49
8° 16′ 13″	8.2703°	34/37
8° 16′ 27″	8.2742°	57/62
8° 17′ 22″	8.2895°	35/38
8° 18′ 28″	8.3077°	36/39
8° 19′ 5″	8.3182°	61/66
8° 20′ 00″	8.3333°	25/27
8° 20′ 00″	ditto	50/54
8° 20′ 29″	8.3415°	38/41
8° 21′ 26″	8.3571°	26/28
8° 21′ 26″	ditto	39/42
8° 22′ 20″	8.3721°	40/43
8° 22′ 46″	8.3793°	27/29
8° 22′ 46″	ditto	54/58
8° 24′ 00″	8.4000°	14/15
8° 24′ 00″	ditto	28/30
8° 24′ 47″	8.4130°	43/46
8° 25′ 10″	8.4194°	29/31
8° 25′ 10″	ditto	58/62
8° 25′ 32″	8.4255°	44/47
8° 26′ 15″	8.4375°	15/16
8° 26′ 56″	8.4490°	46/49
8° 27′ 16″	8.4545°	31/33
8° 27′ 16″	ditto	62/66
8° 28′ 14″	8.4706°	16/17
8° 28′ 14″	ditto	32/34
8° 30′ 00″	8.5000°	17/18
8° 30′ 00″	ditto	51/54
8° 30′ 49″	8.5135°	35/37
8° 31′ 35″	8.5263°	18/19
8° 31′ 35″	ditto	36/38
8° 32′ 4″	8.5345°	55/58
8° 32′ 18″	8.5385°	37/39
8° 33′ 00″	8.5500°	19/20
8° 33′ 40″	8.5610°	39/41
8° 33′ 52″	8.5645°	59/62
8° 34′ 17″	8.5714°	20/21
8° 34′ 17″	ditto	40/42
8° 34′ 53″	8.5814°	41/43
8° 35′ 27″	8.5909°	60/66
8° 36′ 31″	8.6087°	22/23
8° 36′ 31″	8.6087°	44/46
8° 37′ 1″	8.6170°	45/47
8° 37′ 30″	8.6250°	23/24
8° 37′ 58″	8.6327°	47/49
8° 40′ 00″	8.6666°	26/27
8° 40′ 00″	ditto	52/54
8° 40′ 43″	8.6786°	27/28
8° 41′ 23″	8.6897°	28/29
8° 41′ 23°	ditto	56/58
8° 42′ 00″	8.7000°	29/30
8° 42′ 35″	8.7097°	30/31
8° 42′ 35″	ditto	60/62
8° 43′ 38″	8.7272°	32/33
8° 43′ 38″	ditto	64/66
8° 44′ 7″	8.7353°	33/34
8° 45′ 24″	8.7568°	36/37
8° 45′ 47″	8.7632°	37/38
8° 46′ 9″	8.7692°	38/39
8° 46′ 50″	8.7805°	40/41
8° 47′ 9″	8.7857°	41/42
8° 47′ 27″	8.7907°	42/43
8° 48′ 16″	8.8043°	45/46
8° 48′ 31″	8.8085°	46/47
8° 48′ 59″	8.8163°	48/49
8° 50′ 00″	8.8333°	53/54
8° 50′ 41″	8.8448°	57/58
8° 51′ 17″	8.8548°	61/62
8° 51′ 49″	8.8636°	65/66

Chapter 6

Reaming

How to ream stainless steels

Stainless steels can be reamed successfully if the shop adheres to the narrow range of conditions that produce optimum results and basic precautions are heeded

Rigidity of setup, sharp tools and positive feed (to prevent riding without cutting) are basic requirements when reaming the stainless steels.

A problem in machine reaming can occur if the axis of the spindle is not in exact alignment with the axis of the reamer. Causes for misalignment may originate from wear in the ways, wear or dirt in the sleeve or tool clamp, or poor leveling of the machine. These shortcomings are indicated when a rigidly mounted reamer produces poor finishes and oversized or eccentric holes, especially noticeable at the start of the reaming operation.

To ream precisely and to minimize bell-mouth holes, guide bushings are recommended. Where reamers must guide themselves into the hole, a parallel floating holder to compensate for any parallel or angular misalignment is suggested. Smooth, straight holes are required for reaming to high-dimen-

Table 1. Feeds, speeds and coolants for five types of stainless

Alloy Group	Alloy Number	Alloy condition*	Tool material Grade	Tool material Design†	Cutting Speed** fpm	Cutting fluid	Feed for reamer diameter shown, ipr — Nominal reamer diameter, in. ⅛	¼	½	1	1½	2
					Stainless steel alloys							
	Straight-chromium grades											
B4	405	Ann (150-200 Bhn)	T1, M1	A	40	I, IIa	0.004	0.007	0.010	0.016	0.020	0.025
			C-2	C	140	I, IIa	0.004	0.007	0.010	0.016	0.020	0.025
B5	410	Ann (160-220 Bhn)	T1, M1	A	45	I, IIa	0.003	0.005	0.010	0.016	0.020	0.025
			C-2	C	140	I, IIa	0.003	0.005	0.010	0.016	0.020	0.025
F5	410	HT (300-350 Bhn)	T5, M2	D	20	IIa, IIIa, IV	0.002	0.005	0.010	0.016	0.020	0.025
			C-2	D	80	IIa, IIIa, IV	0.002	0.005	0.010	0.016	0.020	0.025
D5	440B	Ann (215-260 Bhn)	T5, M2	A	25	I, IIa	0.002	0.005	0.010	0.016	0.020	0.025
			C-2	C	100	I, IIa	0.002	0.005	0.010	0.016	0.020	0.025
G5	440B	HT (375-440 Bhn)	T5, M2	D	15	IIa, IIIa, IV	0.001	0.003	0.007	0.010	0.012	0.015
			C-2	D	60	IIa, IIIa, IV	0.001	0.003	0.007	0.010	0.012	0.015
	Precipitation-hardenable grades											
D6	17-7 PH	Ann (160-180 Bhn)	T5, M2	C	25	I, IIa	0.003	0.005	0.010	0.016	0.020	0.025
			C-2	C	100	I, IIa	0.003	0.005	0.010	0.016	0.020	0.025
F6	17-7 PH	HT (380-440 Bhn)	T15	D	15	IIa, IIIa, IV	0.002	0.003	0.004	0.007	0.010	0.012
			C-2	D	60	IIa, IIIa, IV	0.002	0.003	0.004	0.007	0.010	0.012
	Chromium-nickel grades											
D7	347	Ann (160-220 Bhn)	T1, M1	A	30	I, IIa	0.002	0.005	0.010	0.016	0.020	0.025
			C-2	C	120	I, IIa	0.003	0.005	0.010	0.016	0.020	0.025
					Austenitic stainless steel superalloys							
	Nonheat-treatable grades											
D8	19-9DL	ST (180-220 Bhn)	T15	C	15	IIa, IIIa, IV	0.002	0.003	0.005	0.008	0.010	0.010
			C-2	C	65	IIa, IIIa, IV	0.002	0.003	0.005	0.008	0.010	0.015
E8	Timken 16-25-6	ST		C	20	IIa, IIIa, IV	0.002	0.003	0.005	0.008	0.010	0.010
				C	70	IIa, IIIa, IV	0.002	0.003	0.005	0.008	0.010	0.010
	Age-hardenable grades											
E9	A-286	STA (320 Bhn)		C	20	IIa, IIIa, IV	0.002	0.003	0.005	0.008	0.010	0.010
				D	70		0.002	0.003	0.005	0.008	0.010	0.010

*Ann = annealed; HT = heat treated; ST = solution treated; STA = solution treated and aged.
†See Table 3 for tool angles.
**For quality surface finishes on stainless steels, speeds of 20 to 40 fpm are recommended for high-speed steel reamers. If hole size is more important, speeds between 40 and 120 fpm are sometimes used.

Source: 'Machining and Grinding of Ultra High-Strength Steels and Stainless Steel Alloys'—NASA Handbook SP-5084

How to ream stainless steels [Continued]

sional accuracy and surface finish.

Cutting speeds

Recommended speeds, Table 1, vary among different materials and will depend on the need for dimensional tolerances or smooth finishes, or perhaps both. Where tolerance of the hole is the main factor, higher speeds can be used. Where smooth finishes are required, lower speeds are recommended. High-speed steel reamers operate in these high-strength materials within a range of 15 to 55 fpm depending on the alloy and heat-treated condition. Carbide reamers operate in the higher speed range of 25 to 160 fpm.

Feeds for reamers

Feeds are based on the type of material, depth of cut, finish required, and design of reamer being used. Small feeds are usually required to produce acceptable holes, and may range between 0.002 and 0.020 ipr depending on the reamer diameter and the alloy. Too low a feed will result in chatter, glazing, and excessive wear.

Depth of cut

Continuous cutting action is necessary in reaming; otherwise, burnishing, or work hardening of the work surface may result. Reamer life is also shortened when insufficient stock is removed. If the drill size is too close to the reamer size or if it drills oversize, the reamer will not have enough bite. Thus, the reamer may wedge rather than cut, resulting in excessive wear or even breakage. As the diameter of the hole increases, the amount allowed for reaming also should be increased.

Types of reamers

Fluted reamers are available as standard items. Spiral-fluted reamers can be identified as those usually possessing right-hand cuts, along with positive axial (helix) and radial rake angles. Both types of reamers afford a smoother cutting action, better chip disposal, and are less likely to dig in or chatter than straight-fluted reamers, although the straight-fluted styles are usually preferred when extreme accuracy is required. Some people recommend a left-hand spiral reamer for a right-hand cut. Although right-hand spiraling of the flutes helps the tool to cut more freely in this situation, it also tends to make the reamer feed into the work too rapidly.

The use of end-cutting, straight-fluted right-hand-cut reamers is recommended for maraging steels except when reaming a tapered hole, or when reaming under conditions where the tool tends to jump ahead as a result of uneven feed. Under these conditions reamers with the hand of spiral opposite to the hand of cut, such as left-hand spiral, right-hand cut are recommended. Tapered-shank reamers are preferred since they give less runout than the straight-shank reamers do.

Reamers for the stainless steels should be ordered with narrow land widths to reduce rubbing, thus minimizing work hardening of the work material. Reamers with margins about 0.010 in. wide produce acceptable holes. Scoring is a problem with wider margins, and excessive chatter is likely to occur when margins are less than 0.005 in.

All reamers should be as short as possible, not only for maximum rigidity, but also to lessen the possibility of reaming bell-mouthed or tapered holes. In addition, the chamfer must be concentric with all flutes of the reamer.

Tool materials

AISI M2 and M10 high-speed steel reamers can be used on stainless steel alloys. Sometimes a surface treatment will improve the resistance of reamers to wear and abrasion.

Carbide-tipped reamers also can be used and operate at somewhat higher speeds and have much better tool life. Type C-50 and C-2 carbide tools are recommended for hot-worked or annealed maraging steel and Type C-2 or Type C-70 are recommended for reaming maraged material.

Reamers should be ground in fixtures or on tool and cutter grinders. The quality of finish produced by reaming depends to a great extent on the proper width of lands and on the finish of the cutting edges. A polished finish on a reamer will produce a better finish on the part. Coarse grinding marks on the reamer will transfer their patterns to the finished hole. Furthermore, stoning the edges of reamers after grinding materially lengthens the life of the tool.

Reamers should be handled and stored carefully. This usually means storage in individual racks or partitioned boxes. If the reamer is dropped on hard surfaces or hit by other tools, unprotected cutting edges may be nicked. When not in use, reamers should be protected with a coating of oil to prevent rusting. A small spot on the cutting edge will start a pit. A deep nick can spoil a reamer. ■

Table 2. Cutting fluids for stainless steels

Symbol	Cutting fluid
I	Water-base coolant, soluble-oil type, or chemical type
IIa	Sulfurized oil
IIIa	Chlorinated oil
IV	Sulfochlorinated oil

Table 3. Tool angles found best for reaming stainless

Reamer information	Stainless steels	Maraging steels Diameter of hole, in.			pH Stainless steels 180/250 bhn
		Up to ½	½ to 1	1½ to 2	
	High-speed steel reamers				Carbide
Tool angles, deg.					
Helix angle	7	—	—		5 to 8
Radial rake	3 to 5	0 to 5	0 to 5	—	7 to 10
Relief	6 to 8	10 to 15	8 to 10	0 to 5	—
Primary chamfer	30 to 35	45	45	6 to 8	—
Secondary chamfer (lead)	—	—	—	45	2
Chamfer relief	—	12	8 to 12	—	—
Margin data				7 to 9	
Margin width, in.	—	0.004 to 0.010	0.010 to 0.015	0.015 to 0.020	0.005 to 0.010

Application of Carbide Reamers

By Carl Neiderfringer, president, The Atrax Co, Newington, Conn

The following recommendations apply to use of solid-carbide reamers:

For Longer Tool Life

1. Elimination of chatter will protect cutting edges, insure better hole finish.
2. To eliminate chatter, reduce rpm.
3. If reduced rpm does not eliminate chatter, increase feed as rpm is reduced.
4. Carbide bushings are recommended. In cases where bushings are necessary they should fit reamer closely (approx. 0.0002) with just enough clearance to avoid freezing.
5. Use power feeds on all materials up to Rockwell C-55.
6. Use hand feed on materials harder than Rockwell C-55.
7. Be sure workpiece is firmly held.
8. Be sure spindle is in excellent condition.
9. Never stop reamer while engaged in cut.
10. Handle, store, or ship reamer with care.

Stock Removal

Stock left in hole before reaming must be sufficient to engage reamer properly. Too little stock will cause excessive reamer wear, too much stock can cause a chip problem resulting in reamer freezing or breaking.

Hole Finish

Solid-carbide reamers, when used properly, produce superior hole finishes, especially in tough materials such as hardened steels up to Rockwell C-60. Excellent finishes can also be produced in aluminum, brass, bronze, stainless, etc.

Hole Accuracy

Solid-carbide reamers, like other solid-carbide tools, because of their hardness, hold precise tolerances many times longer than comparable high-speed-steel tools in any material.

Maintenance

When resharpening, use 320-grit or finer diamond wheel of resinoid-bond type.

Reamer Rake Angles

Most reamers are supplied with a radial to one degree positive rake. Unless materials to be reamed require special rake angles, standard reamers with radial rake will work well.

Steel sometimes requires a shear cutting action. This is obtained by using a radial-rake reamer having right-hand spiral flutes.

Left-hand spiral reamers have a negative shear cutting action at the lead or cutting chamfer.

Reamer Back Taper

The back taper of a reamer is generally 0.0002 to 0.0003 per in. on chucking reamers and 0.0008 to 0.0010 on rose-type reamers, which have no other relief on the diameter.

Odd-Fluted Reamer

Odd-fluted reamers offer advantages where extremely fine finishes and close-tolerance, round holes must be produced. Special construction eliminates opposing flutes and incorporates unequal spacing of flutes.

Spiral Flutes

Right-hand spiral flutes produce better finishes in certain materials, because of shearing action on lead end. In some materials, however, where a negative shear is desired, a left-hand spiral will obtain better results.

Taper Holes

Again, a rigid setup and good spindle are necessary. A minimum amount of stock, sufficient to clean up, should be left in the hole. If a relatively large amount of stock must be removed, use a straight-fluted reamer as a rougher, followed by a spiral-fluted finisher.

For close tolerance taper holes, semi-finish with standard spiral taper and finish with high-spiral reamer.

Cutting Speeds

MATERIAL	SFPM
Steel (All Types)	
Rockwell C 60 or harder	8-20
Rockwell C 50 to 60	15-30
Rockwell C 40 to 50	20-40
Rockwell C 30 to 40	35-65
Under Rockwell C 30	60-90
Cast iron and malleable iron	50-85
Non-Ferrous	
Aluminum	
Brass	
Bronze	
Copper	90-175
Fiber	
Plastic	
Hard Rubber, etc.	

FEEDS	IPR
Steels	
Rockwell C 50 or harder	0.002-0.004
Rockwell C 30 to 50	0.004-0.008
Cast iron, malleable iron	0.005-0.010
Non-ferrous materials	0.005-0.012

Coolants Required

MATERIAL	LUBRICANTS
Steel harder than Rockwell C 50	Light oil
Steels softer than Rockwell C 50	Light oil for good finishes or Soluble oil and water
Cast iron and malleable iron	Soluble oil and water
Non-ferrous materials	Soluble oil and water

NOTES:

Use a heavy flow of lubricant for best reamer performance, especially on tough materials.

Kerosene gives superior finishes on many non-ferrous metals—aluminum, bronze, etc. Exercise caution with flammable lubricants.

Stock Removal

REAMER DIA	STOCK REMOVAL
Up to 1/16 Incl.	0.003 to 0.005
Over 1/16 to 1/8 "	0.004 to 0.008
" 1/8 to 1/4 "	0.006 to 0.012
" 1/4 to 3/8 "	0.008 to 0.014
" 3/8 to 1/2 "	0.010 to 0.015
" 1/2 to 3/4 "	0.012 to 0.018

Chapter 7

Threads and Threading

Thread series designations, 1

When a shop drawing or specification gives only the designation of a thread, here is how to find the thread series wanted, and also the applicable ASA standard

Designation	Thread Series	References: American Standards	References: Handbook H28 (1957), Section No.
ACME—C	Acme threads, centralizing	B1.5	XII.
ACME—G	Acme threads, general purpose (See also "STUB ACME")	B1.5	XII.
AMO	American Standard microscope objective threads	B1.11	XIV
ANPT	Aeronautical National Form taper pipe threads[d]	- - - -	- - - -
F—PTF	Dryseal fine taper pipe thread series	B2.2	Appendix 7.
M	Metric standard threads, (all except "S" threads)	- - - -	Appendix 14.
N BUTT	National Buttress threads	B1.9	XVI
8N	American National 8-thread series[c]	B1.1	Appendix 1.
12N	American National 12-thread series[c]	B1.1	Appendix 1.
16N	American National 16-thread series[c]	B1.1	Appendix 1.
NC	American National coarse thread series[c]	B1.1	Appendix 1.
NEF	American National extra-fine thread series[c]	B1.1	Appendix 1.
NF	American National fine thread series[c]	B1.1	Appendix 1.
NGO	National gas outlet threads[b]	B57.1	IX.
NGS	National gas straight threads	B57.1	IX.
NGT	National Gas taper threads (See also "SGT")	B57.1	IX.
NH	American National hose coupling and firehose coupling threads[g]	B26, B33.1	X.
NPSC	American Standard straight pipe threads in pipe couplings	B2.1	VII.
NPSF	Dryseal American Standard fuel internal straight pipe threads	B2.2	VIII.
NPSH	American Standard straight pipe threads for loose-fitting mechanical joints for hose couplings	B2.1, B33.1	VII, X.
NPSI	Dryseal American Standard intermediate internal straight pipe threads	B2.2	VIII.
NPSL	American Standard straight pipe threads for loose-fitting mechanical joints with locknuts	B2.1	VII.
NPSM	American Standard straight pipe threads for free-fitting mechanical joints for fixtures	B2.1	VII.
NPT	American Standard taper pipe threads for general use	B2.1	VII.
NPTF	Dryseal American Standard taper pipe threads	B2.2	VIII.
NPTR	American Standard taper pipe threads for railing joints	B2.1	VII.
NR	American National thread with a 0.108p to 0.144p controlled root radius[e]	- - - -	- - - -
NS	American National threads of special diameters, pitches, and length of engagements[c]	B1.1	Appendix 2.
PTF—SAE, SHORT	Dryseal SAE short taper pipe threads	B2.2	VIII.
PTF—SPL, SHORT	Dryseal special short taper pipe threads	B2.2	Appendix 7.
PTF—SPL, EXTRA SHORT	Dryseal special extra short taper pipe threads (See also "SPL—PTF")	B2.2	Appendix 7.
- - - - - -	Surveying instrument mounting threads	- - - -	XVII.

Thread series designations, 2

Designation	Thread Series	References	
		American Standards	Handbook H28 (1957), Section No.
S	Standard coarse metric threads to 5mm incl.	- - - -	Appendix 14.
SGT	Special gas taper threads	B57.1	IX.
SPL—PTF	Dryseal special taper pipe threads	B2.2	Appendix 7.
STUB ACME	Stub Acme threads	B1.8	XIII.
UN	Unified constant-pitch thread-series	B1.1	III.
UNC	Unified coarse thread series	B1.1	III.
UNEF	Unified extra-fine thread series	B1.1	III.
UNF	Unified fine thread series	B1.1	III.
UNJ	Unified constant-pitch thread series with a 0.150 11p to 0.180 42p controlled root radius[f]		(Proposed)
UNJC	Unified coarse thread series with a 0.150 11p to 0.180 42p controlled root radius[f]		(Proposed)
UNJEF	Unified extra-fine thread series with a 0.150 11p to 0.180 42p controlled root radius[f]		(Proposed)
UNJF	Unified fine thread series with a 0.150 11p to 0.180 42p controlled root radius[f]		(Proposed)
UNR	Unified constant-pitch thread series with a 0.108p to 0.144p controlled root radius		(Proposed)
UNRC	Unified coarse thread series with a 0.108p to 0.144p controlled root radius		(Proposed)
UNRF	Unified fine thread series with a 0.108p to 0.144p controlled root radius		(Proposed)
UNREF	Unified extra-fine thread series with a 0.108p to 0.144p controlled root radius		(Proposed)
UNM	Unified miniature thread series	B1.10	V.
UNS	Unified threads of special diameters, pitches, or lengths of engagement	B1.1	IV.

[a]Methods of designating multiple threads are shown in ASA B1.5, Acme Screw Threads, and Part III, Section XII of Handbook H28 (1957).

[b]All threads, except NGO, are right hand, unless otherwise designated. For NGO threads, designations "RH" or "LH" are required.

[c]Superseded by the Unified thread series.

[d]As published in Military Specification MIL-P-7105.

[e]As published in Military Specification MIL-B-7838.

[f]As published in Military Specification MIL-S-8879.

[g]Under consideration in Sectional Committee B2 is the adoption of "NFH" for firehose coupling threads and "GH" for garden hose threads

*Handbook H-28 was compiled by the Federal Screw Thread Commission,, at the direction of the National Bureau of Standards, and is available from the U.S. Government Printing Office, Washington.

Common Thread-Fit Problems

This helpful guide to the specific causes of thread misfits, their symptoms and means of detection was compiled jointly by the three SPS Screw Thread Metrology Laboratories at Jenkintown, Cleveland, and Santa Ana, Calif, where problems on screw thread fits are solved for industry

By John B Twining, chief metrologist, Screw Thread Metrology Laboratory, Standard Pressed Steel Co, Jenkintown, Pa

Possible Cause	Detection	Accuracy Direct Reading to:
	Bolts Loosen in Tapped Hole	
(a) External thread PD is too low	(a), (c) Measure pitch dia on 3-roll gage	0.0001
(b) Internal hole is oversize	(a), (c) Check major dia with micrometer or supermike	0.0001
(c) Tapered threads	(b), (c) Check tapped hole with thread plug gage or internal thread comparator	0.0001
(d) Incorrect preload	(b) Check minor dia with plain plug or adj. internal micrometer	Plus functional gage fit
	(d) Check proper installation	
	Binding in a Tapped Hole	
(a) PD or major diameter oversize	(a), (c), (d) Measure PD on 3-roll gage set up (PD & functional)	0.0001
(b) Excessive lead error	(a), (d) Check major dia with micrometer or supermike	0.0001
(c) Improper thread form	(b) Check lead on lead tester	0.00001
(d) Tapered thread	(c), (d) Thread form and lead by optical comparator	To 100X mag.
	(c) Use form and radii charts	To 100X mag.
	Thread Will Not Start	
(a) Distorted lead thread	(a) Check on optical comparator	To 100X mag.
(b) Nicked lead thread	(a) Check with ring thread—also visual	Functional check
(c) Mating parts not same class or size of thread	(b) Visual check and use of ring thread	Functional check
	(b) Use of microscope	40X
	(c) Use of magnetic plating thickness gage, 3-roll gage or internal thread comparator	0.0001
	Loose or Sloppy Fit	
(a) Undersize PD	(a), (d) Use PD & functional 3-roll gage	0.0001
(b) Mating parts not same class or size of thread	(b), (d) Internal thread comparator	0.0001
	(b) Ring thread and thread plug gages	Functional check
(c) PD of tapped hole is oversize	(c) Thread plugs—internal thread comparators	Functional check
(d) Tapered thread	(c) Three-wire check with mold of tapped hole	0.0001
	Low Torque When Used With Locknut	
(a) Locking feature of nut won't meet torque requirements (on standard studs)	(a) Use of torque wrench and standard stud	inch oz.
	(a) Use of thread plug gage	Functional check
(b) PD of bolt is undersize	(b) Use of PD and functional 3-roll gage	0.0001
	(b) Three-wire check	0.00001
	Excessive Torque When Used With Locknut	
(a) Locking feature of nut won't meet torque requirements (on standard studs)	(a) Use of torque wrench and standard stud	inch oz.
	(a) Use of thread plug gage	Functional check
(b) PD of bolt is oversize	(b) Use of PD and functional 3-roll gage	0.0001
(c) Lack of lubricant on locknut	(b) Three-wire check	0.00001
	(c) Visual inspection	Visual
	Galling Threads	
(a) Burrs	(a) Microscope check	40X
	(a) Use of ring thread	Functional check
(b) Excessive lead errors	(b) Use of lead tester	0.00001
	(b) Optical comparator	100X
(c) Materials prone to galling—not properly plated	(c), (e) Chemical test (etching)	(To magnify laps)
	(c) Magnetic plating thickness gage	0.0001
(d) Poor surface finish on threads	(d) Use of surface roughness checker	1 micro inch
(e) Lapped threads	(d), (e) Use of microscope	40X

Threading Speeds for Various Materials

COURTESY LANDIS MACHINE COMPANY

NATIONAL COARSE THREADS

	MATERIALS											
	TOOL STEEL (CARBON)		VANADIUM STEEL STAINLESS STEEL FORGED STEEL OPEN HEARTH MONEL METAL		MACHINE STEEL DRAWN STEEL TOBIN BRONZE BESSEMER SCREW STOCK		MALLEABLE IRON CAST IRON		CAST BRASS BAR BRASS PHOS. BRONZE COPPER		FIBER BAKELITE ALUMINUM LEAD	
DIA.	RPM	SURFACE SPEED	RPM	SURFACE SPEED	RPM	SURFACE SPEED	RPM	SURFACE SPEED	RPM	SURFACE SPEED	RPM	SURFACE SPEED
1/4	306	20	535	35	764	50	917	60	1070	70	1222	80
5/16	245		428		611		733		856		978	
3/8	183	18	306	30	458	45	560	55	662	65	764	75
7/16	157		262		393		481		568		656	
1/2	137		191	25	306	40	382	50	459	60	535	70
9/16	102	15	170		272		340		407		475	
5/8	92		153		245		306		367		428	
3/4	76		102	20	178	35	229	45	279	55	330	65
7/8	65		87		153		196		241		285	
1	46	12	76		134		172		210		258	
1 1/8	41		68		119		153		187		221	
1 1/4	37		55	18	92	30	123	40	153	50	183	60
1 1/2	31		46		76		102		127		153	
1 3/4	22	10	39		66		87		109		131	
2	19		29	15	48	25	67	35	86	45	105	55
2 1/4	17		25		42		59		76		94	
2 1/2	15		23		38		53		69		84	
2 3/4	11	8	21		35		49		62		77	
3	10		13	10	26	20	38	30	51	40	64	50

NATIONAL FINE SERIES

DIA.	RPM	SURFACE SPEED	RPM	SURFACE SPEED	RPM	SURFACE SPEED	RPM	SURFACE SPEED	RPM	SURFACE SPEED	RPM	SURFACE SPEED
1/4	382	25	611	40	764	50	917	60	1222	80	1222	80
5/16	306		489		611		733		978		978	
3/8	255		357	35	458	45	611		815		815	
7/16	175	20	306		393		524		656	75	699	
1/2	153		268		344		459		573		611	
9/16	136		238		306		373	55	475	70	543	
5/8	122		214		276		337		428		489	
3/4	92	18	153	30	203	40	254	50	357		381	75
7/8	79		131		153	35	219		285	65	329	
1	69		115		134		191		258		287	
1 1/8	61		85	25	119		153	45	204	60	255	
1 1/4	46	15	76		107		137		183		214	70
1 1/2	38		64		89		115		153		178	
1 3/4	33		44	20	66	30	87	40	120	55	153	
2	29		38		57		76		105		124	65
2 1/4	20	12	25	15	42	25	59	35	85	50	111	
2 1/2	18		23		38		54		76		99	
2 3/4	17		21		35		49		69		83	60
3	15		19		32		45		64		76	

CUTTING SPEEDS LISTED ARE FOR ANNEALED STOCK ONLY. WHEN HEAT TREATED MATERIAL IS THREADED, DECREASE CUTTING SPEED 20 TO 30 %.

Wire Measurement of Screws

By Robert T Parsons, president, The Van Keuren Co, Watertown, Mass

The following form, which is worked out with figures for a 5-40 NC Go thread gage, allows the simple calculation for three-wire measurements of threads where the effect of lead angle is considered. Only tables of Axial Angle Screws containing Best Size Wire and Wire Constants are needed. Complete thread tables for all thread sizes are contained in the Van Keuren Handbook No. 36, pp 207-211.

The inaccuracies which arise from neglecting the effect of lead angle are becoming more and more critical with the narrower tolerances and increased accuracies which technological advance is bringing.

DATA

1. Number of Starts $N_s = 1$
2. Pitch Diameter $D = .1088''$
3. No. Threads per in. $n = 40$
4. Axial pitch (1/n) $p_x = .025''$ (Also $= L/N_s$)
5. Diametral pitch P or $P_n = .$
6. Lead $L = .025''$
7. Axial Pressure angle $\phi_x = 30°$ (½ Included angle)
8. Normal Pressure angle $\phi_n = .°$
9. Find the ratio of lead to pitch dia. $\dfrac{L}{D} = \dfrac{0.025}{0.1088} = 0.2298$

DATA FROM TABLES

10. Refer to the table for the specified number of starts for Axial Angle Screws or Normal Pressure Angle Screws. Enter in the following spaces the next *lower* and *higher* values of ratio (L/D) and corresponding 1″ pitch values for (d_{pin}) and (C).

	L/D	Diff	1″ AXIAL PITCH VALUES d_{pin}	Diff	C	Diff
		(c)	(d)	(e)	(f)	(g)
Next lower value (a)	0.2252	0.0055	0.57567	0.0008	0.86309	0.00014
Next higher value (b)	0.2307		0.57559		0.86295	

11. Using the values in step (10), interpolate to get the Best Size Wire (d_{pin}) and Wire Constant (C), for a 1″ Axial pitch screw having the same L/D ratio as the given screw.

L/D value (9) $= 0.2298$
L/D value (a) $= 0.2252$
Diff (h) $= 0.0046$

$$d_{pin} = \underset{(d)}{0.57567} - \frac{\overset{(h)}{0.0046}}{\underset{(c)}{0.0055}} \times \underset{(e)}{0.00008}$$

$$= 0.57560''$$

$$C = \underset{(f)}{0.86309} - \frac{\overset{(h)}{0.0046}}{\underset{(c)}{0.0055}} \times \underset{(g)}{0.00014}$$

$$= 0.86297''$$

12. Find the Best Wire Size for the specified screw.

$$d_{pin} = \frac{d_{pin} \text{ for } 1'' \text{ pitch}}{n}$$

$$= \frac{0.57560}{40} = 0.01439''$$

13. Find wire constant for specified screw

$$C = \frac{C \text{ for } 1'' \text{ pitch}}{n}$$

$$= \frac{0.86297}{40} = 0.02157''$$

Note: For metric or for diametral pitch screws multiply the 1″ pitch values by the linear axial pitch.

14. Find Measurement over wires. (Zero backlash)

$M = D + C = 0.1088'' + 0.02157''$
$M = 0.13037''$

If the actual measurement over the wires is under the computed value, the pitch diameter of the screw is undersize by approximately the same amount.

BACKLASH SCREWS

Worm threads, which are to be cut with thin threads, to provide a certain amount of backlash (Δt_x) may be considered as being undersize on the pitch diameter by an amount (ΔD).

15. Amount to make the pitch dia. undersize for (Δt_x) backlash.

$\Delta D = \Delta t_x \cot \phi_x = . - - - \times . - - -$
$= . - - - -$

16. Undersize pitch dia. for . - - -″ backlash

$D_1 = D - \Delta D$

This value should be used in items (2) and (9) for calculating (d_{pin}) and (C)

USING SUBSTITUTE WIRES

If Exact Best Size Wires cannot be obtained, slightly larger or smaller wires may be used with approximate results.

17. Best Size Wire from (12) $d_{pin} = 0.01439''$
18. Available size $\dfrac{0.57735}{n}$ $G = 0.01443''$
19. Wire difference $\Delta G = +0.00004''$
20. Wire Factor* $(1 + \operatorname{cosec} \phi_n)$ $2H = 3.00000$ See page 234 Cat. No. 36

Conversion Formulas

$$\tan \lambda = \frac{L}{3.141593\, D}$$

From trig tables, $\lambda = . - - - °$
$\cos \lambda = 0. - - -$
$\tan \phi_n = \tan \phi_x \cos \lambda$
$= . - - - \times . - - -$
$\tan \phi_n = . - - -$
From tables, $\phi_n = . - - - °$

21. Increase or reduction with substitute wires

$\Delta M = 2H \times \Delta G = 3.0 \times .00004''$
$= 0.00012''$

22. Measurement over substitute wires

$M_1 = M \pm \Delta M$
$= 0.13037'' + 0.00012''$
$= 0.13049''$

RESULT OF MEASUREMENT

23. Actual measurement with . - - - ″ wires $M_A = . - - - ''$
24. Computed measurement with . - - - ″ wires M or $M_1 = . - - - - ''$
25. Amount screw is over or under theoretical size $\Delta M = M_A - M$ or M_1
26. Amount teeth are thick or thin $\Delta t_x = \Delta M \tan \phi_x$
27. Constant for .01443″ diameter wires:

$C = M_1 - D$
$= 0.13049'' - 0.1088''$
$= 0.02169''$

* $(1 + \operatorname{cosec} \phi_x)$ may be used for 60° threads when lead angle is less than 5°, thread is single start and ΔG is .0020″ or less.

Simplified Measurement Over Wires for National Acme Threads

By D E SWEET, Tool Engineer, UNITED SHOE MACHINERY CORP

To save a great deal of labor in calculating M, the measurement over wires, for National Acme threads, I have devised the simplified formula shown in the table and constants 1 and 2 to be used therewith. For each size of thread, three wire diameters are listed. Thus, it is likely that a shop will have at least one appropriate wire size on hand.

The complete formula for measurement of National Acme threads is:
$M = E - \cot a \times p/2 + G\,(1 + \operatorname{cosec} a')$
where M = measurement over wires

E = pitch dia
$a = 14\frac{1}{2}°$
p = pitch
G = wire dia
a' = ½ normal thread angle

And to use this formula one must first solve ½ the normal thread angle.

These involved calculations can be avoided by the use of the simplified formula.

Example: Assume a 2-in—4 NA single thread—Class 2G fit. From tables of thread parts, we find that the pitch dia is 1.8637 in. From the accompanying table, we will select a wire size of 0.12831 in. Appropriate constants are: No. 1—0.1574319, and No. 2—0.0015206.

Then $M = 1.8637 + 0.1574319 + 0.0015206 \div (1.8637)^2$
$= 2.0215696$

If we were to calculate the value of M by the exact formula, the value of $M = 2.0215742$. The difference is of no importance in normal manufacturing and inspection of a screw thread. Hence, these constants save time and labor.

WIRE MEASUREMENT CONSTANTS FOR 29° NATIONAL ACME THREADS

Wire measurement M = Pitch dia E + constant 1 + constant 2 ÷ E^2

NOM DIA & THDS PER IN.	WIRE DIA IN.	CON-STANT 1	CONSTANT 2		NOM DIA & THDS PER IN.	WIRE DIA IN.	CON-STANT 1	CONSTANT 2		NOM DIA & THDS PER IN.	WIRE DIA IN.	CON-STANT 1	CONSTANT 2	
			SINGLE THD	DOUBLE THD				SINGLE THD	DOUBLE THD				SINGLE THD	DOUBLE THD
	.03206	.0392706	.0000237	.0000948		.10498	.1375913	.0007955	.0031820		.17000	.2035158	.0035794	.0143176
1/4-16	.03228	.0503693	.0000238	.0000952	1 - 5	.10329	.1291515	.0007827	.0031308	2 1/4 - 3	.17215	.2152528	.0036247	.0144988
	.03608	.0593372	.0000266	.0001064		.11547	.1899777	.0008750	.0035000		.19245	.3166295	.0040521	.0162084
	.03602	.0417846	.0000348	.0001392		.10498	.1375913	.0007956	.0031824		.17000	.2035158	.0035803	.0143212
5/16-14	.03689	.0461293	.0000356	.0001424	1 1/8 - 5	.10329	.1291515	.0007828	.0031312	2 1/2 - 3	.17215	.2152528	.0036255	.0145020
	.04123	.0678030	.0000398	.0001592		.11547	.1899777	.0008751	.0035004		.19245	.3166295	.0040531	.0162124
3/8-12	.04441	.0606674	.0000584	.0002336		.10498	.1375913	.0007957	.0031828		.17000	.2035158	.0035812	.0143248
	.04304	.0538258	.0000566	.0002264	1 1/4 - 5	.10329	.1291515	.0007829	.0031316	2 3/4 - 3	.17215	.2152528	.0036265	.0145060
7/16-12	.04811	.0791450	.0000632	.0002528		.11547	.1899777	.0008753	.0035012		.19245	.3166295	.0040541	.0162164
	.05250	.0688456	.0000994	.0003976		.12831	.1574319	.0015194	.0060786		.25600	.3117676	.0121278	.0485112
1/2-10	.05164	.0645509	.0000978	.0003912	1 3/8 - 4	.12911	.1614271	.0015289	.0061156	3 - 2	.25822	.3228541	.0122330	.0489320
	.05773	.0949639	.0001093	.0004372		.14423	.2374347	.0017091	.0068364		.27500	.4066523	.0130279	.0521116
	.06418	.0788408	.0001899	.0007596		.12831	.1574319	.0015196	.0060784		.25600	.3117676	.0121359	.0485436
5/8- 8	.06456	.0809386	.0001911	.0007644	1 1/2 - 4	.12911	.1614271	.0015291	.0061164	4 - 2	.25822	.3228541	.0122412	.0489648
	.07216	.1186924	.0002136	.0008544		.14423	.2374347	.0017093	.0068372		.27500	.4066523	.0130367	.0521468
	.08250	.0897732	.0004342	.0017368		.12831	.1574319	.0015201	.0060804		.25600	.3117676	.0124571	.0498284
3/4- 6	.08608	.1076515	.0004531	.0018124	1 3/4 - 4	.12911	.1614271	.0015295	.0061180	5 - 2	.25822	.3228541	.0125651	.0502604
	.09622	.1582899	.0005065	.0020260		.14423	.2374347	.0017098	.0068382		.27500	.4066523	.0133816	.0535264
	.08250	.0897732	.0004343	.0017372		.12831	.1574319	.0015206	.0060824					
7/8- 6	.08608	.1076515	.0004532	.0018128	2 - 4	.12911	.1614271	.0015300	.0061200					
	.09622	.1582899	.0005065	.0020260		.14423	.2374347	.0017104	.0068416					

Normal Thread Angles and Bottom Flats (Acme) . . . I

DONALD SWEET

Normal thread angle is the angle on the cutter blade that produces the sloped side of the thread. As the cutter works, it is aligned at the helix angle, twisted from the plane perpendicular to the thread center-line, so the normal thread angle is even less than the nominal 29° of the Acme thread.

The following tables give the normal thread angle to the nearest minute for helix angles from 0° to about 22°. In addition, the width of cutter at the tip, where the thread bottom is cut, is given for Acme threads from 4 to 16 tpi, including helix angles from 0° to about 24°.

ANGLES

Consider a 1-in. Acme thread, 0.2-in. pitch, 0.8-in. lead, 3G fit: Pitch dia would be (E), $\frac{0.8940}{0.8849}$ in., and minor dia would be $\frac{0.7800}{0.7664}$ in.

Tan ½ normal thread angle
= Tan 14°30′ (half 29°) x cos helix angle
= 0.2586176 x 0.95905
= 0.24801

½ normal thd angle
= 13°56′15″

From the table, this angle would be read 13°56′, within a minute of the exact figure.

ANGLES . . . A

Helix Angle From Deg	Min	Helix Angle to Deg	Min	Normal Thread Angle, A Deg	Min	Helix Angle From Deg	Min	Helix Angle to Deg	Min	Normal Thread Angle, A Deg	Min
0	00	1	25	14	30	15	16	15	32	14	00
1	25	2	24	14	29	15	32	15	47	13	59
2	24	4	24	14	28	15	47	16	02	13	58
4	24	5	14	14	27	16	02	16	17	13	57
5	14	5	58	14	26	16	17	16	31	13	56
5	58	6	35	14	25	16	31	16	46	13	55
6	35	7	10	14	24	16	46	17	00	13	54
7	10	7	42	14	23	17	00	17	14	13	53
7	42	8	12	14	22	17	14	17	28	13	52
8	12	8	40	14	21	17	28	17	41	13	51
8	40	9	06	14	20	17	41	17	54	13	50
9	06	9	32	14	19	17	54	18	08	13	49
9	32	9	56	14	18	18	08	18	22	13	48
9	56	10	20	14	17	18	22	18	34	13	47
10	20	10	42	14	16	18	34	18	47	13	46
10	42	11	04	14	15	18	47	19	00	13	45
11	04	11	24	14	14	19	00	19	12	13	44
11	24	11	45	14	13	19	12	19	24	13	43
11	45	12	05	14	12	19	24	19	37	13	42
12	05	12	25	14	11	19	37	19	49	13	41
12	25	12	44	14	10	19	49	20	01	13	40
12	44	13	02	14	09	20	01	20	13	13	39
13	02	13	20	14	08	20	13	20	24	13	38
13	20	13	38	14	07	20	24	20	36	13	37
13	38	13	55	14	06	20	36	20	47	13	36
13	55	14	12	14	05	20	47	21	00	13	35
14	12	14	28	14	04	21	00	21	11	13	34
14	28	14	44	14	03	21	11	21	22	13	33
14	44	15	01	14	02	21	22	21	33	13	32
15	01	15	16	14	01	21	33	21	44	13	31

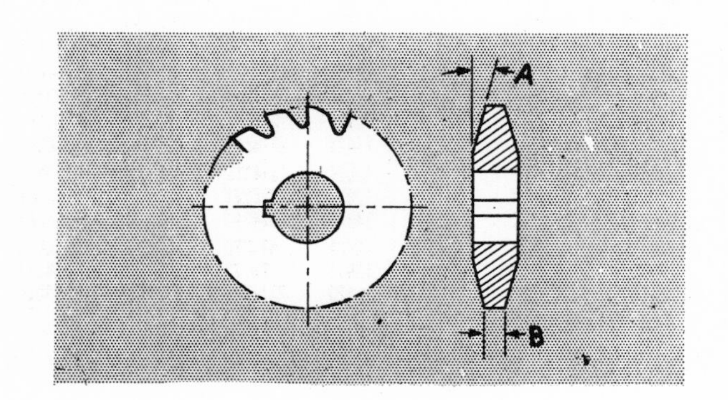

Normal Thread Angles and Bottom Flats (Acme) . . . II

FLATS

For the same 1-in. thread, the flat would be:

$X = 0.100 - 2(0.2586176 \times 0.057)$
$= 0.0705$ in.

From the table, the flat would read 0.0660 in., which is less than maximum normal flat to allow for cutter wear. Flat dimensions in the table go in increments of 0.0005 in., but can in no cases fall outside the limit of the root dia.

FLATS . . . B

Helix Angle From Deg	Helix Angle From Min	Helix Angle to Deg	Helix Angle to Min	Normal Bottom Flat, (in.) B	Tpi	Helix Angle From Deg	Helix Angle From Min	Helix Angle to Deg	Helix Angle to Min	Normal Bottom Flat, (in.) B	Tpi
0	0	9	45	0.0205	16	12	01	14	15	0.0550	6
9	45	16	20	0.0200	16	14	15	16	11	0.0545	6
16	20	20	46	0.0195	16	16	11	17	55	0.0540	6
20	46	24	43	0.0190	16	17	55	19	30	0.0535	6
0	0	8	28	0.0240	14	19	30	20	58	0.0530	6
8	28	14	29	0.0235	14	20	58	22	21	0.0525	6
14	29	18	29	0.0230	14	0	0	3	27	0.0690	5
18	29	21	57	0.0225	14	3	27	8	20	0.0685	5
21	57	23	51	0.0220	14	8	20	10	54	0.0680	5
0	0	5	23	0.0283	12	10	54	12	54	0.0675	5
5	23	13	02	0.0278	12	12	54	14	39	0.0670	5
13	02	17	00	0.0273	12	14	39	16	12	0.0665	5
17	00	20	12	0.0268	12	16	12	17	38	0.0660	5
20	12	22	58	0.0263	12	17	38	18	58	0.0655	5
0	0	5	23	0.0320	10	18	58	20	12	0.0650	5
5	23	12	15	0.0315	10	20	12	21	22	0.0645	5
12	15	15	59	0.0310	10	0	0	3	04	0.0875	4
15	59	18	59	0.0305	10	3	04	7	24	0.0870	4
18	59	21	35	0.0300	10	7	24	9	38	0.0865	4
21	35	23	54	0.0295	10	9	38	11	26	0.0860	4
0	0	4	28	0.0410	8	11	26	13	00	0.0855	4
4	28	10	49	0.0405	8	13	00	14	23	0.0850	4
10	49	14	07	0.0400	8	14	23	15	39	0.0845	4
14	07	17	45	0.0495	8	15	39	16	49	0.0840	4
17	45	19	02	0.0390	8	16	49	17	55	0.0835	4
19	02	21	05	0.0385	8	17	55	18	57	0.0830	4
21	05	22	57	0.0380	8	18	57	19	56	0.0825	4
0	0	3	49	0.0565	6	19	56	20	52	0.0820	4
3	49	9	13	0.0560	6	20	52	21	46	0.0815	4
9	13	12	01	0.0555	6	21	46	22	38	0.0810	4

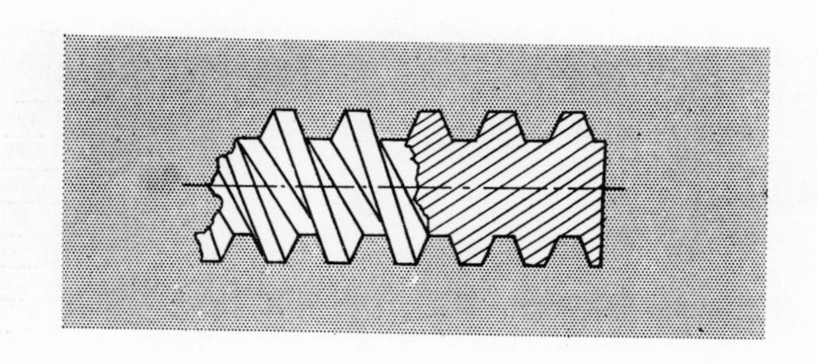

END PLAY OF ACME SCREW THREADS . . . I

D SWEET, tool engineer

All general-purpose Acme threads are designed to allow a certain lateral movement when assembled. This end play is related to the class of fit.

Maximum end play is usually more important than minimum end play. To find the maximum end play, subtract the minimum pitch diameter of the screw from the maximum diameter of the nut, and multiply the result by tan 14° 30′. See diagram.

The nomograms that will appear in an early number make it possible to determine the correct fit to afford the desired end play. It is necessary only to lay a straightedge from the screw size to the number of threads per inch and read the maximum end play on the center scale.

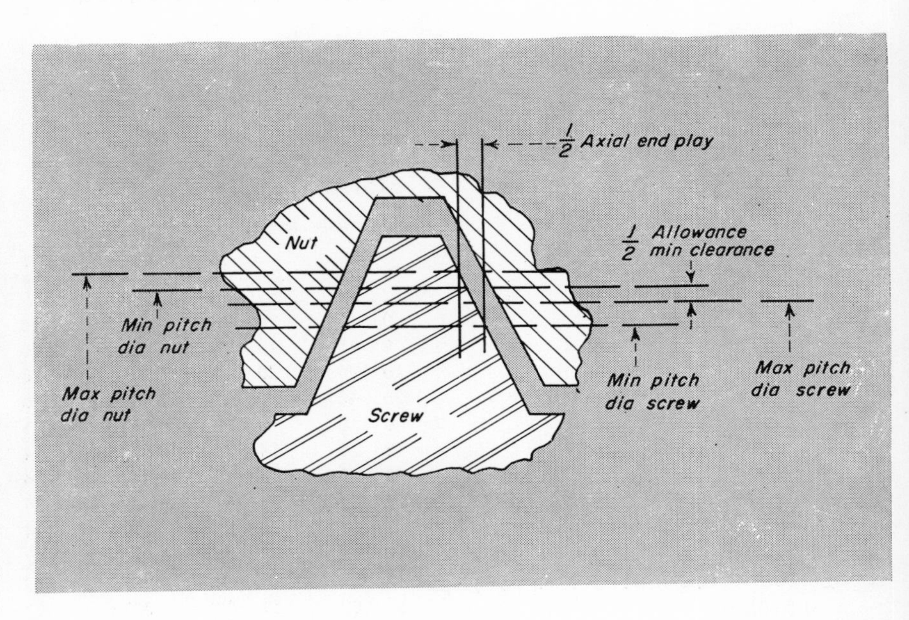

EXAMPLE:

An old-style ⅝-8 Acme screw was designed for an end play of 0.005 in. max, when assembled. What class of fit should the designer specify for a ⅝-8 Acme thread (ASA B1.5—1952) to obtain the same end play?

Max end play = Nut PD max − Screw *PD* min times tan 14° 30′

= 0.5697 − 0.5506

= 0.00494 in.

By application of a straight edge to the nomogram (future issue) for a Class 3G fit, it will be seen that the end play scale is cut at 0.0049 in.

MINIMUM END PLAY

One chart, reproduced herewith, can be used to find the minimum end play for all classes of fit. The value of the minimum end play depends only on the nominal size or diameter of the screw.

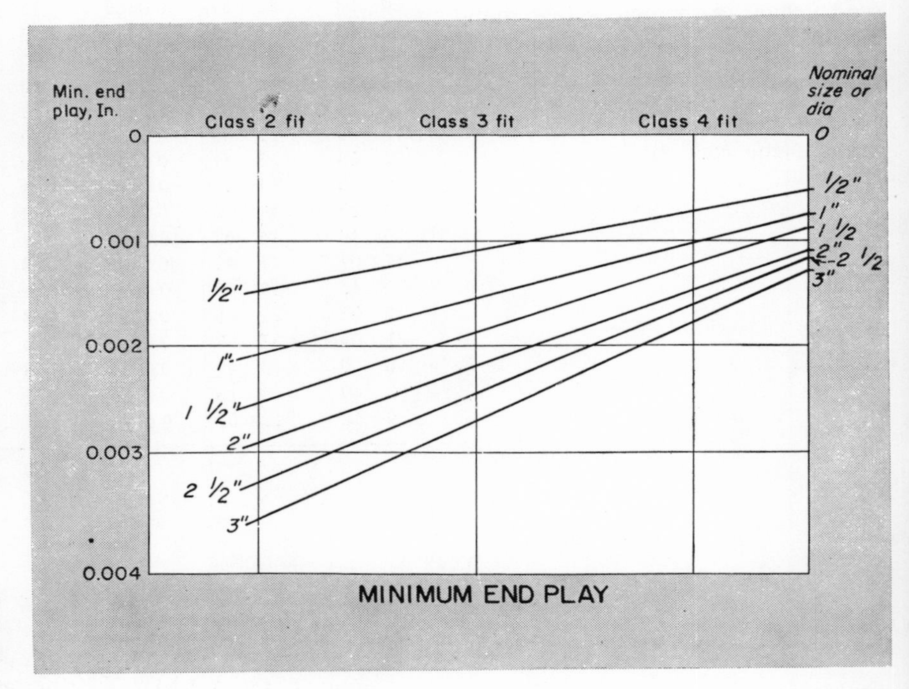

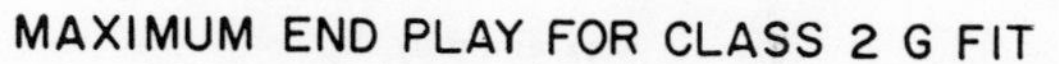

END PLAY OF ACME SCREW THREADS . . . II

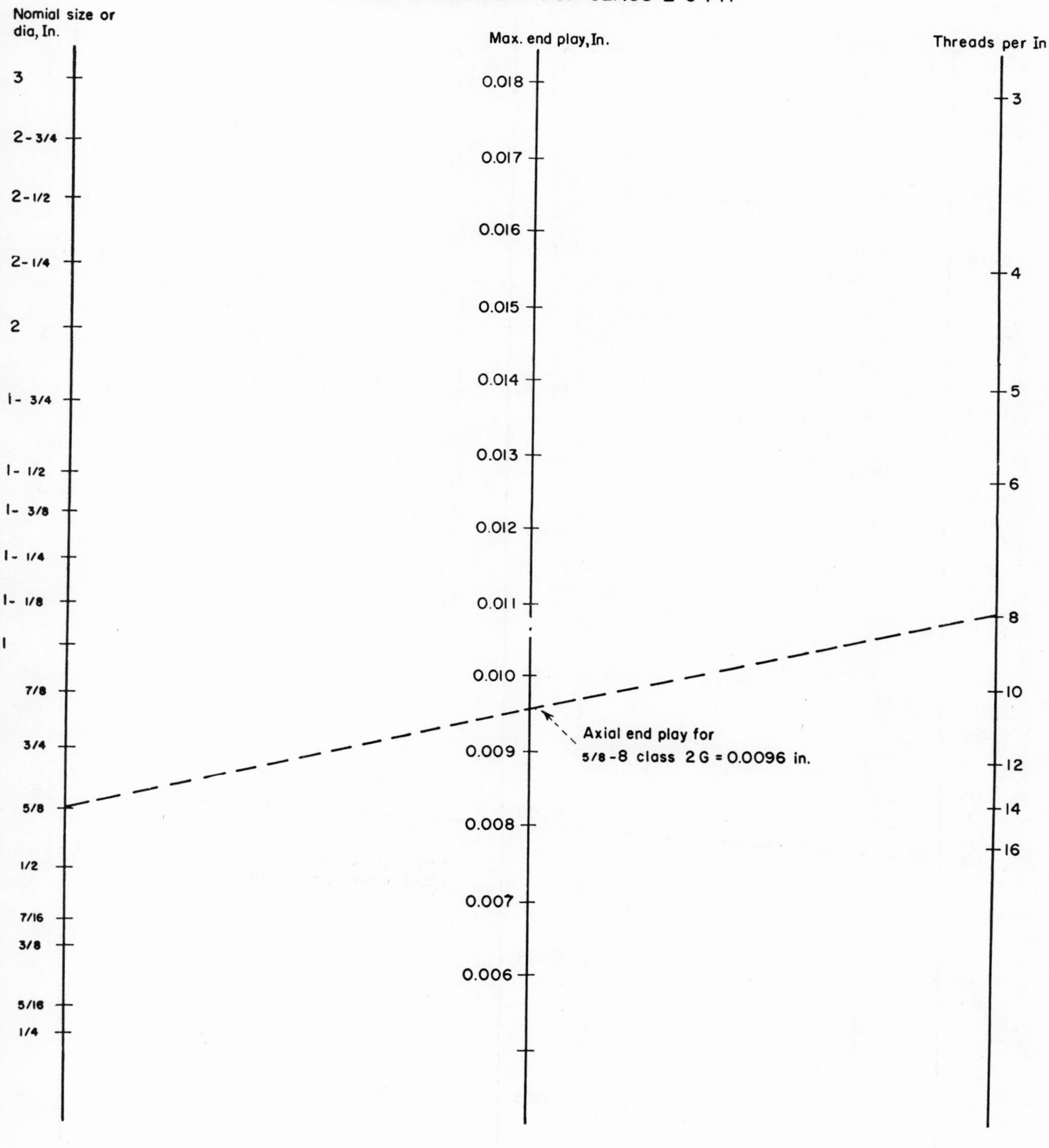

END PLAY OF ACME SCREW THREADS . . . III

MAXIMUM END PLAY FOR CLASS 3G FIT

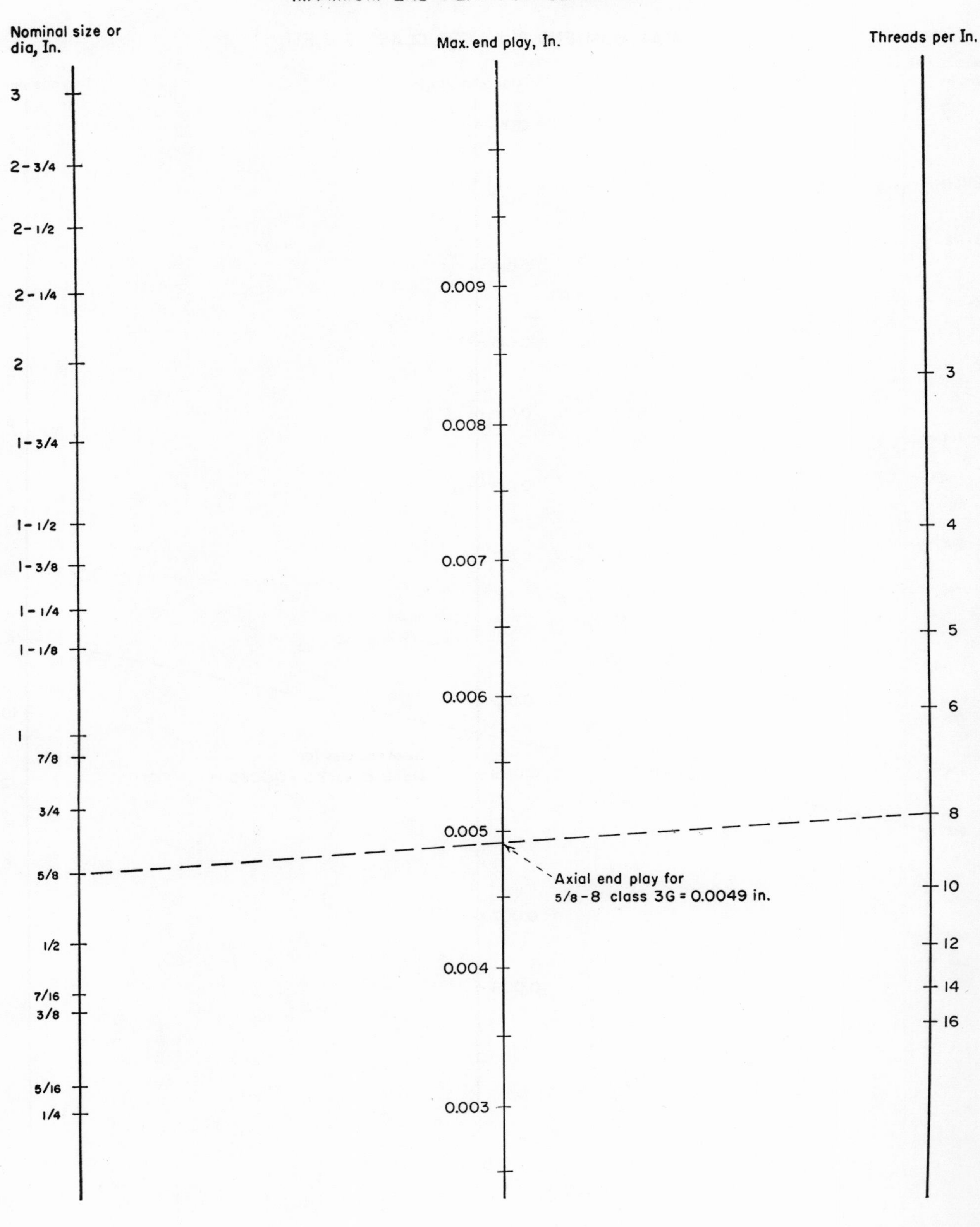

END PLAY OF ACME SCREW THREADS . . . IV

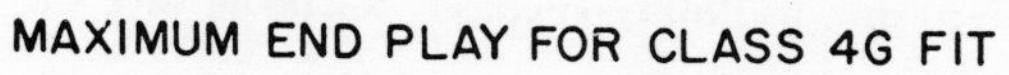

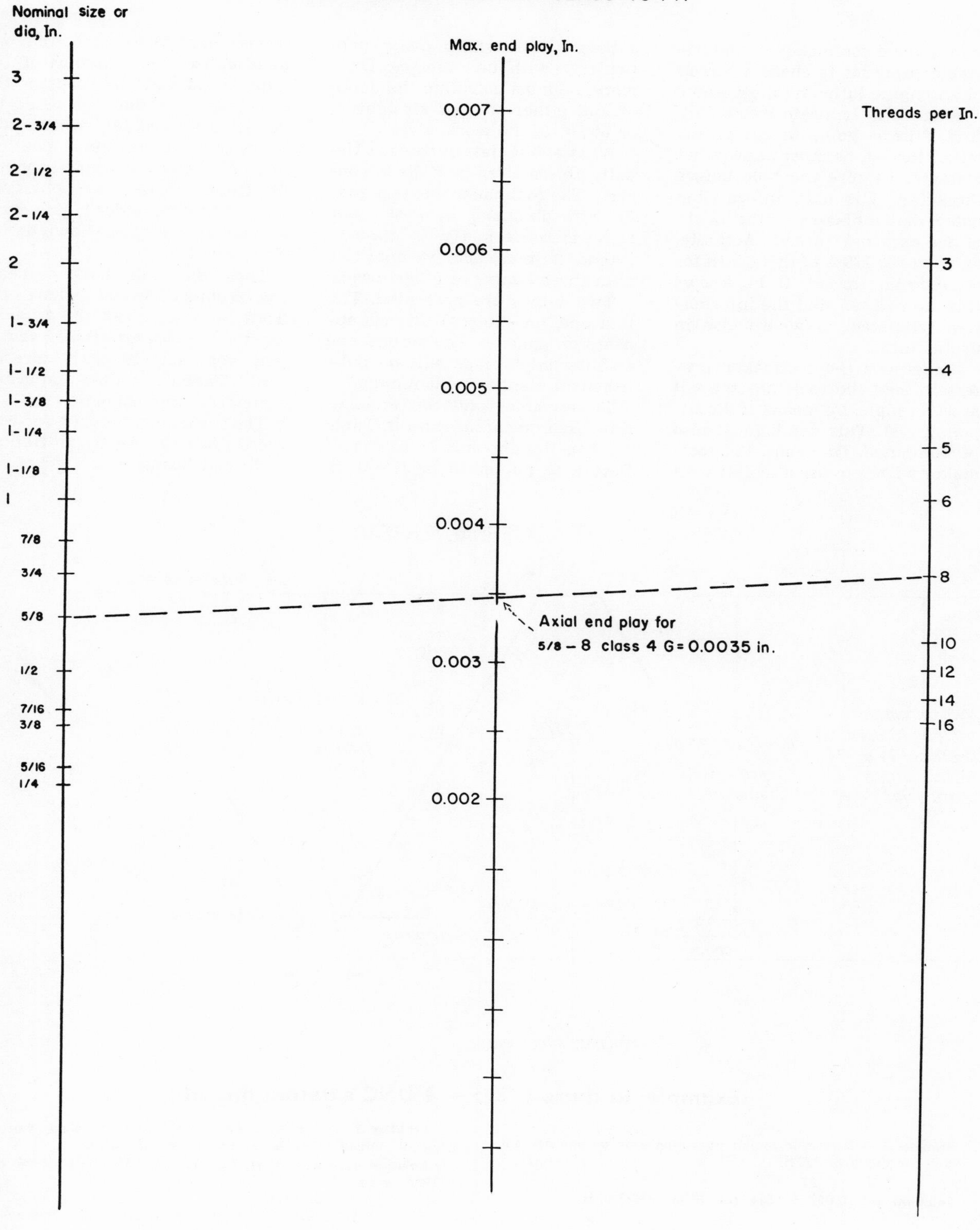

Single-Point Thread Chasing—I

By Warren G Ogden, Jr, manufacturing feasibility engineer, Raytheon Mfg Co, Lowell, Mass

Whenever a toolmaker or model-maker prepares to chase a thread in an engine lathe, he must know how much to truncate the toolbit. And if he is going to cut an internal thread, he must know what diameter to bore the hole before threading. The only information listed in handbooks is the depth of an external thread. Actually, he wants 83 1/3% of this depth for an internal thread. If he knows this, he will not find the information tabulated anywhere else in useful form.

Sometimes the toolmaker prefers to feed the tool into the cut at a 29° angle, by means of a compound rest. This depth must also be computed. Or again, the toolmaker wishes to use a toolbit with a theoretically sharp point, particularly with fine pitches. Once more, he must calculate the depth of cut, either to feed straight in or at 29° to the work.

What about class of thread? Usually this matter is of little concern. The toolmaker wants a good fit, with no shake, in most cases. If he is concerned with class of thread, he must achieve final size with thread gages or other means.

Two tables are presented. The first one, on external threads appears on page 93. The second one on internal threads will be published in a subsequent number.

These tables and the sketches are based on thread data in Table 111.1 in Handbook H 28 (1957) — Part 1, as published by the U. S. Government. The sketches aid one to visualize the meanings of the table headings. Of course the tables and sketches are based on design-size profilé, where the screw and nut have one identical pitch diameter, or metal-to-metal fit. Under these conditions, you have "Ogden's value" (for want of a better name) as shown on the sketches.

These diameters have no bearing on class of thread and are only 0.005 in. away from the true diameter for coarser pitches and as you approach 80-pitch threads, the "Ogden's value" approximates the true diameter.

The tables can be used for both the Unified and American National thread forms.

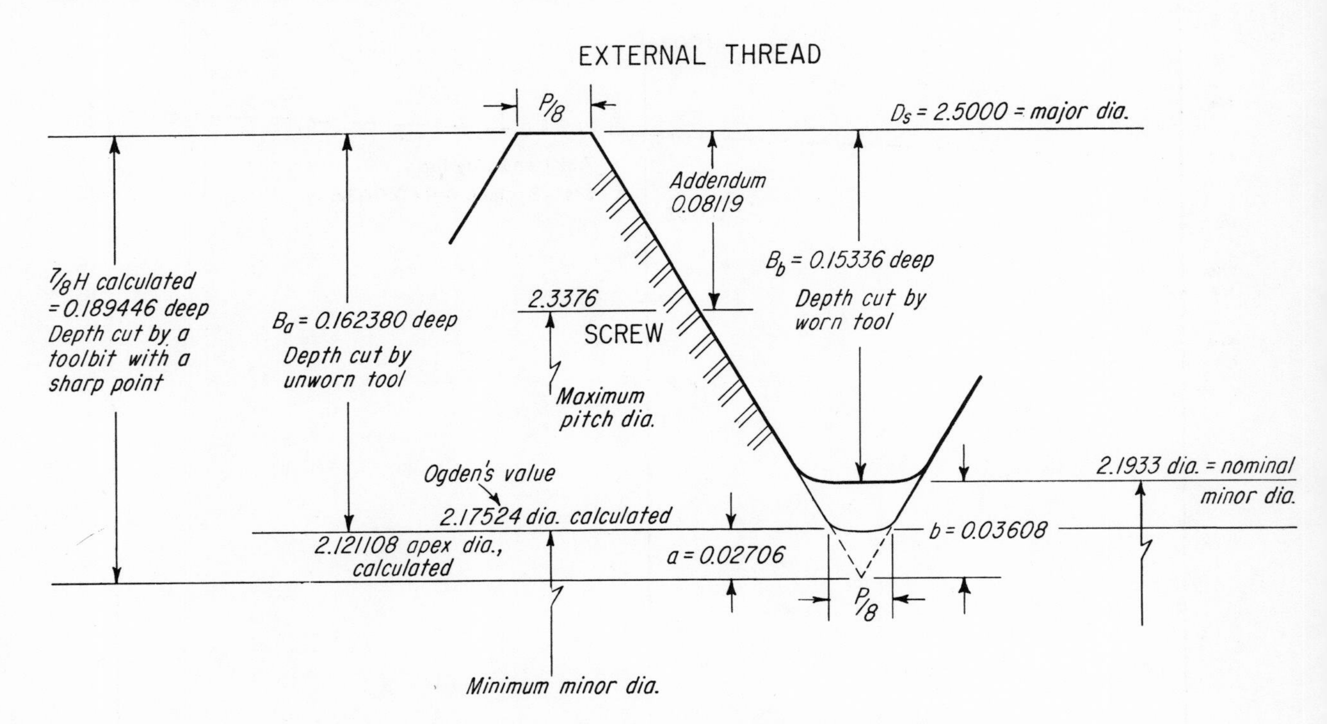

Example: to chase a 2½ — 4 UNC external thread

Problem 1 — Determine toolbit truncation a on an unworn tool, and b for depth of $\frac{17}{24}$ H

Solution: $a = 0.02706$ (See col. 3 in table) $= \frac{H}{8}$

$b = 0.03608 = \frac{H}{6}$, where $H =$ height of sharp V thread $= 0.86603 \div n = 0.21650$ for $n = 4$

Problem 2 — Given maximum major dia of screw, calculate the 90° infeed of the toolbit for the variants above.

Solution $B_a = h_b = \frac{3}{4}\ H$, Fig 111.1, col 13 of Handbook H 28 (1957) — part 1.

$= 0.162380$

$B_b = h_s = \frac{17}{24}\ H$, Fig 111.1, col 12 of Handbook.

$= 0.15336$, or the minimum depth or maximum material condition.

Single-Point Thread Chasing—II

SINGLE POINT THREAD CHASING IN A TOOLROOM LATHE EXTERNAL THREADS ONLY UNIFIED THREAD FORM				IF TOOLBIT IS NOT TRUNCATED Theoretical sharp point	
THREADS PER INCH n	INFEED at 90° to the MINIMUM MINOR DIAM.	INFEED at 29° to the MINIMUM MINOR DIAM.	DEPTH of TOOLBIT TRUNCATION	INFEED at 90° to apex of fundamental triangle	INFEED at 29° to apex of fundamental triangle
	inch	inch	inch	inch	inch
80	0.008119	0.00928	0.00135	0.00948	0.010825
72	.009021	.01031	.00150	.01053	.012028
64	.010149	.01161	.00169	.01184	.013532
56	.011599	.01326	.00193	.01354	.015465
48	.013532	.01547	.00226	.01578	.018042
44	.014762	.01688	.00246	.01722	.019682
40	.016238	.01857	.00271	.01894	.021651
36	.018042	.02063	.00301	.02105	.024056
32	.020297	.02320	.00338	.02368	.027063
28	.023197	.02651	.00387	.02706	.030929
27	.024056	.02751	.00401	.02807	.032075
24	.027063	.03094	.00451	.03157	.036084
20	.032476	.03714	.00541	.03789	.043301
18	.036084	.04125	.00601	.04210	.048113
16	.040595	.04641	.00677	.04736	.054127
14	.046394	.05304	.00773	.05413	.061859
13	.049963	.05712	.00833	.05829	.066617
12	.054127	.06189	.00902	.06315	.072169
11½	.056480	.06458	.00941	.06590	.075307
11	.059047	.06752	.00984	.06889	.078730
10	.064952	.07426	.01083	.07577	.086603
9	.072169	.08252	.01203	.08420	.096225
8	.081190	.09283	.01353	.09472	.108253
7	.092788	.10609	.01546	.10826	.123718
6	.108253	.12377	.01804	.12630	.144338
5	.129904	.14852	.02165	.15156	.173205
4½	.144338	.16503	.02406	.16839	.192450
4	.162380	.18566	.02706	.18945	.216506

Single-Point Thread Chasing—III

Warren G Ogden, Jr, manufacturing feasibility engineer, Raytheon Mfg Co, Lowell, Mass

As stated in the previous installment (AM/MM—May 1 '61, p97), a toolmaker must know how much to truncate a toolbit when he prepares to chase a thread. In the case of an internal thread, he must also know the diameter of the hole to be bored before threading.

The sketch below is based on thread data in Table 111.1 of Handbook H 28 (1957)—Part I, as published by the U S Government. This sketch and the table on page 95 are based on the design-size profile, when the screw and nut have one identical pitch diameter, or metal-to-metal fit. The "Ogden's value" is the calculated root diameter established by an unworn tool.

The table on page 95 gives seven columns of data. These enable the toolmaker to find size to bore the hole, the depth of toolbit truncation for an unworn tool, the infeed at 90° and at 29°. The table also gives the infeed at 90° and 29° when the toolbit is not truncated.

Example: to chase a 2½—4 UNC internal thread
To find:

a—truncation for unworn tool
b—truncation for worn tool
D_a—infeed, at 90° for unworn tool
D_b—infeed, at 90° for worn tool
Bore size of nut

Solution:

$a = 0.00902$, col 5 of table. Truncation a also equals $H/8 - H/12 = H/24$, where $H = 0.216506$ for $n = 4$. Also $H = 0.866025p$, where $p = 1/n$.

$b = H/8 = 0.02706$

$D_a = 0.15336$, col 3 of table $= 17/24\ H$

$D_b = ⅝H = 0,13532$

Bore size of nut = minimum major dia minus twice the height of design-size thread = 2.500 — $2D_b$ = 2.2294. Or bore size = 2.500 — 0.27063 (Col 2 of table) = 2.2294.

Infeed at 90° for a sharp V thread is 0.16238, Col 7 in table, or ¾ H.

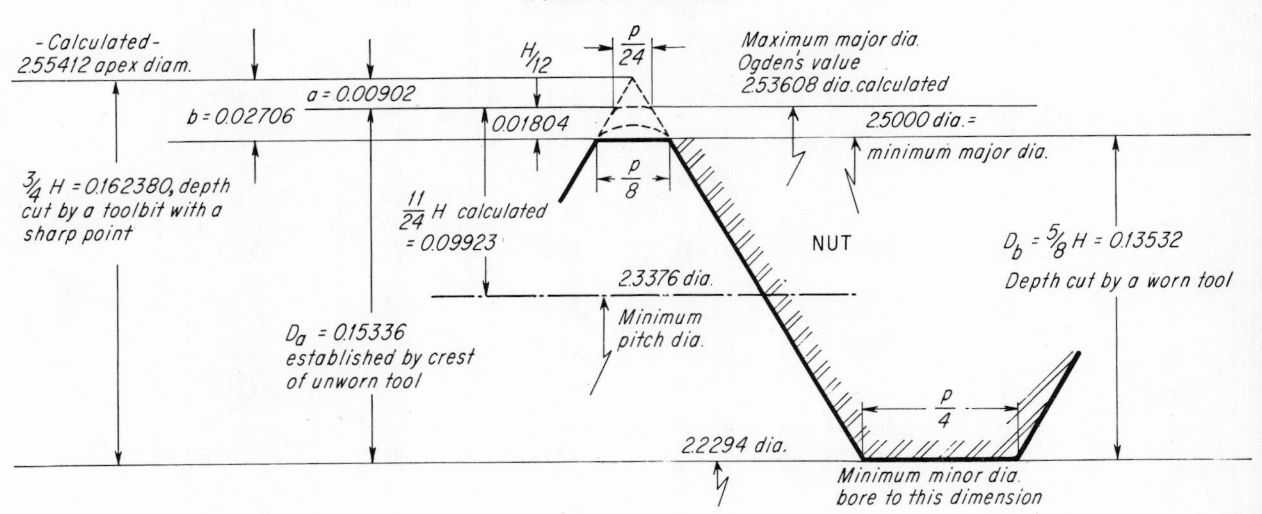

Single-Point Thread Chasing—IV

THREADS PER INCH n	SINGLE POINT THREAD CHASING IN A TOOLROOM LATHE INTERNAL THREADS ONLY UNIFIED THREAD FORM				IF TOOLBIT IS NOT TRUNCATED Theoretical sharp point	
	CONSTANT to determine bore diam. before threading	INFEED at 90° to the MAXIMUM MAJOR DIAM.	INFEED at 29° to the MAXIMUM MAJOR DIAM.	DEPTH of TOOLBIT TRUNCATION	INFEED at 90° to apex of fundamental triangle	INFEED at 29° to apex of fundamental triangle
	inch	inch	inch	inch	inch	inch
80	0.01353	0.00767	0.00877	0.00045	0.008119	0.00928
72	.01504	.00852	.00974	.00050	.009021	.01031
64	.01691	.00958	.01095	.00056	.010149	.01161
56	.01933	.01095	.01252	.00064	.011599	.01326
48	.02255	.01278	.01461	.00075	.013532	.01547
44	.02460	.01394	.01594	.00082	.014762	.01688
40	.02706	.01534	.01754	.00090	.016238	.01857
36	.03007	.01704	.01948	.00100	.018042	.02063
32	.03383	.01917	.02192	.00113	.020297	.02320
28	.03866	.02191	.02505	.00129	.023197	.02651
27	.04009	.02272	.02598	.00133	.024056	.02751
24	.04511	.02556	.02922	.00150	.027063	.03094
20	.05413	.03067	.03507	.00180	.032476	.03714
18	.06014	.03408	.03897	.00200	.036084	.04125
16	.06766	.03834	.04384	.00226	.040595	.04641
14	.07732	.04382	.05010	.00258	.046394	.05304
13	.08327	.04719	.05395	.00278	.049963	.05712
12	.09021	.05112	.05845	.00301	.054127	.06189
11½	.09413	.05334	.06099	.00314	.056480	.06458
11	.09841	.05577	.06376	.00328	.059047	.06752
10	.10825	.06134	.07013	.00361	.064952	.07426
9	.12028	.06816	.07793	.00401	.072169	.08252
8	.13532	.07668	.08767	.00451	.081190	.09283
7	.15465	.08763	.10019	.00515	.092788	.10609
6	.18042	.10224	.11690	.00601	.108253	.12377
5	.21651	.12269	.14028	.00722	.129904	.14852
4½	.24056	.13632	.15586	.00802	.144338	.16503
4	.27063	.15336	.17534	.00902	.162380	.18566

Subtract constant from major diameter of external thread.
If crest of tool is rounded it should be tangent to p/24 flat

Change gears for metric threads

By John H Wilson, president, Wilson Mfg Co, Inc., Wichita Falls, Texas

The accompanying table gives the change gears required to cut metric threads on a lathe built to cut U S standard threads. With these gears, practically any metric thread can be cut. In some of the less common sizes, it is necessary to change gears *A* or *D* (see sketch and table). One note of caution: the leadscrew should not be disengaged after starting the cut. Either the lathe must be reversed, or one must use the leadscrew reversing attachment if the lathe is so equipped.

Gears *A* run at same speed as lathe spindle.

Where the notation $A = D$ appears, it means that any convenient size can be used as long as *A* and *D* have the same number of teeth.

On the table where gears are listed $B = C$ or $C = B$, a single idler gear of any convenient size can be used between the 50 and 127 tooth gears.

If for any reason the 120 to 127 gears cannot be used, 68 to 72 tooth gears (or any multiple of 17 to 18) can be used instead for a close approximation of correct pitch. The error with these gears is less than 0.0005 in. in each inch of length of thread.

If for any reason the 50 to 127 tooth gears cannot be used, 37 to 94 tooth gears may be substituted in their place. With these gears the error will be less than 0.0002 in. in each inch length of thread. Threads made with these approximate gears are ordinarily sufficient for most ordinary purposes, but with the gears listed in the table the pitch of the threads produced will be as accurate as the leadscrew on the lathe.

The figures are correct for all lathes that use a one-to-one ratio from spindle to feed box. If the ratio is not one-to-one the change gears *A-B* must be changed or *C-D*, whichever is convenient.

The feed box setting can be changed in place of gears *A-B* or *C-D* if the ratio is not one-to-one. For example, if one was cutting a 1 m/m pitch, the table indicates the feed box is set for 10 threads. Now if the lathe has a one-to-two reduction from the spindle to the feed box, one would set the feed box for 5 threads.

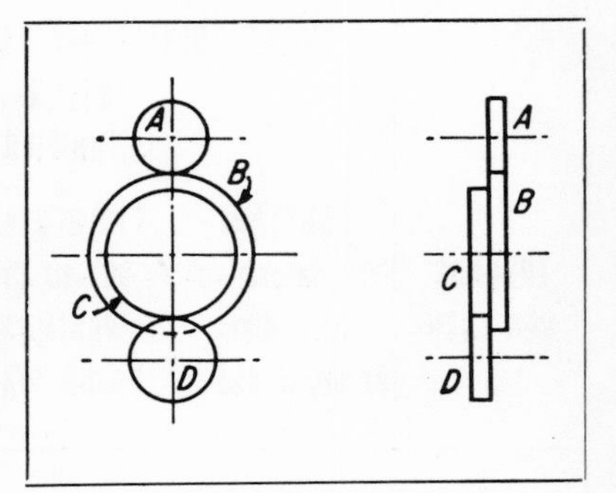

Pitch Rqd in mm	Set Gearbox to Cut USS Thrd	Use These Change Gears (See sketch above)			
		A	B	C	D
0.25	80	50	127	120	60
0.3	80	A=D	127	120	D=A
0.35	80	70	127	120	60
0.4	72	60	127	120	50
0.45	72	54	127	120	40
0.5	20	50	B=C	C=B	127
0.6	40	A=D	127	120	D=A
0.7	40	70	127	120	60
0.75	40	75	127	120	60
0.8	40	60	127	120	45
0.9	40	75	127	120	50
1.0	10	50	B=C	C=B	127
1.25	8	50	B=C	C=B	127
1.5	16	A=D	127	120	D=A
1.75	16	70	127	120	60
2.0	5	50	B=C	C=B	127
2.5	4	50	B=C	C=B	127
3.0	8	A=D	127	120	D=A
3.5	8	70	127	120	60
4.0	6	A=D	127	120	D=A
4.5	6	45	127	120	40
5.0	2	50	B=C	C=B	127
5.5	2	50	50	55	127
6.0	4	A=D	127	120	D=A
6.5	4	65	127	120	60
7.0	4	70	127	120	60
8.0	3	A=D	127	120	D=A
9.0	3	45	127	120	40
10.0	2	50	40	80	127
12.0	2	A=D	127	120	D=A

Chapter 8

Taps and Tapping

Breaking torque of taps

Using breaking torques of 4-flute taps as 'standard', it is possible to establish release settings for tap drivers, using 2- and 3-flute taps up to ½-in. thread

Manufacturing people often ask how much torque a tap will withstand before breaking. Most often this information is needed to determine safe settings for releasing tap drivers, so that they will slip before tap breakage occurs. Here are some guides for estimating tap-breaking torque in the common tap size range up to ½-in.

Relative breakage torques of various tap designs
(4-flute = 100%)

Tap type	Tap size #10-32	¼-28	5/16-24	3/8-24	½-20
3-flute chip driver	—	—	—	135	100
2-flute chip driver	130	130	135	—	—
4-flute, %	100	100	100	100	100
4-flute, in-lb	90	200	340	625	1500
2-flute	90	95	100	—	—
3-flute, 30° helix	—	—	70	95	100
2-flute, 30° helix	85	95	—	—	—
3-flute	85	80	65	75	65
3-flute fast helix	70	80	70	70	75

Calculation of breakage torque:
Example: 3-flute ¼-28 tap; rating 80% of 4-flute tap.
Then torque = 80% of 200 in.-lb. (for 4-flute tap.)
= 160 in.-lb.

Factors affecting breakage

Because taps are made of hardened tool steel they are relatively brittle compared with mechanical parts. High hardness does increase the basic strength level, but also tends to reduce impact strength. Sharp corners and other irregularities may become the starting point for cracks which can lead to breakage.

Specifically, the factors affecting tap breaking strength are:

Hardness—which increases static strength but reduces impact strength.

Thread height—coarse-pitch threads are deeper, may reduce strength, and require higher tapping torque.

Thread roots—sharp thread roots increase the tendency for breakage, especially under impact loading conditions.

Grinding damage—burned and checked surfaces usually reduce the required breaking torque.

Surface treatments — hard-case surface treatments improve wear resistance and tap life but increase brittleness, and this may cause breakage under poor operating conditions.

Misalignment—this condition, or excessive tap runout, applies a bending load to the tap, increasing the possibility of breakage.

Drive smoothness — non-uniform rotation of the tapping spindle causes impact loading and resultant easy breakage.

Under good operating conditions, most tap breakage stems from overloading of the tap in torsion, as caused by excessive dulling, by chip packing in the flutes or hole bottom, by an excessive percentage of thread height, or by attempting to tap difficult materials.

Difficult and unusual tapping operations often require special tap designs and operating procedures.

Breaking-torque test

To determine the average breaking torque for taps, National-Winter Research Laboratory, mounted taps in the normal manner on a leadscrew tapping machine. The workpiece was drilled with blind holes of the proper tap drill size and then mounted on a torque dynamometer. The output of the dynamometer was fed to a recording device.

The tapping machine was operated in a normal manner, except that the tap was allowed to bottom in the blind hole. This caused a large increase in torque which broke the tap. A very low speed was used to minimize severe impact affects. The nature of the breakage was a good approximation of normal tap breakage.

The size range checked was representative of more than 90% of all usage. Actual test sizes were #10-32 (0.190 dia.), ¼-28, 5/16-24, ⅜-24, and ½-20, all in the fine-pitch series. A few coarse-pitch taps were also checked to determine the effect of deep threads.

The basic design selected for comparison was the 4-flute, since this is standard in all sizes. Additional types checked were 2- and 3-flute taps, (where available), 2- and 3-flute chip driver taps (where available), 2- and 3-flute 30° helical flute taps (where available), and 3-flute fast helix (52°) taps. All of the taps were made of high speed steel, and had normal hardness.

Torque strength of taps

Tap design does have a marked influence on tap torsional strength,

Source: 'Metal Cuttings' published by National Twist Drill & Tool Co, Rochester, Mich.

Breaking torque of taps [Continued]

as illustrated in the table. Here the breaking torque of the standard 4-flute tap set is 100%. The chip driver type is stronger than the four-flute design by about 30-35% in all sizes except ½-in. The ½-in. tap has reduced shank diameter, so breakage occurrred in the shank for all except the weakest designs. The three-flute types (except chip driver) were generally the weakest in all sizes. In some sizes there was a nearly two-to-one spread between the weakest and the strongest designs.

Operating safety factors

If the breaking torque of a tap is divided by the tapping torque to be expected with a dull tap, the result is the safety factor. For the four-flute design, the safety factor is seen to range between about 5 for the No. 10-32 size to about 15 for the ½-20 size.

Effect of thread pitch

Tapping torques for coarse-pitch threads are appreciably higher than those for fine-pitch threads, and range between 60% and 100% for the size range considered here. Moreover, taps for coarse-series threads have breaking torques 10-20% lower than those for fine-pitch threads. When this is combined with the effect of the higher tapping torques, the net safety factor for coarse-series threads is only about 40% to 50% of that for fine-series threads according to the test results.

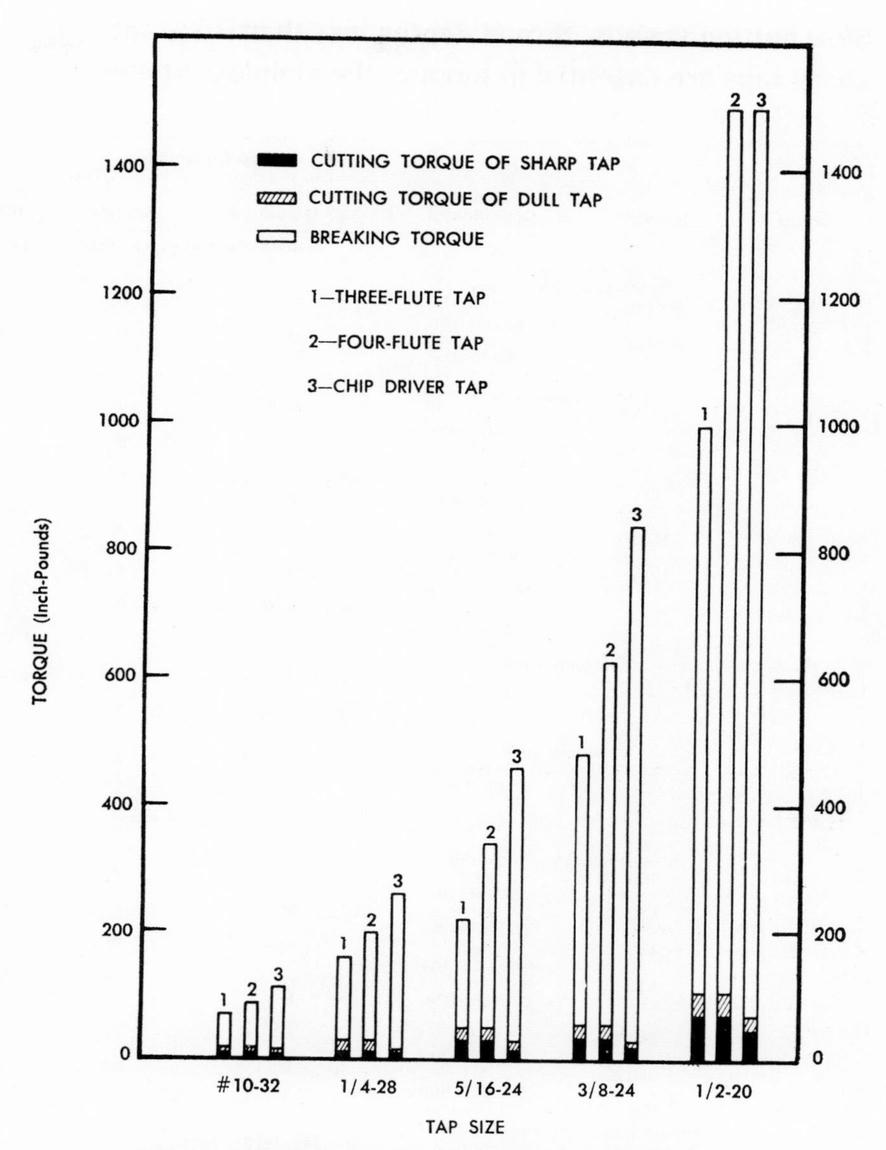

Breaking torques are many times the torques required when taps are sharp. Tests based on cutting steel of 200 Brinell

Effect of surface treatments

Hard-case surface treatments often appreciably extend tap life. Unfortunately, however, such treatments are also quite brittle, and this increases susceptibility to breakage by impact or overloading.

To check this effect, several of the test taps were given a simple 'nitride' surface treatment and were then broken in the same test setup used in the earlier tests. In every case, the breaking torque was severely reduced to between ¼ and ½ that of untreated taps.

Summary

Most commonly used taps have an adequate safety factor when used in ordinary engineering materials, and this safety factor increases with increasing tap size. Taps for coarse-pitch threads have less reserve than those for fine-pitch threads. Thus, small diameter coarse-pitch threads are more difficult to tap, an observation which is borne out in practice. Particular care is required in the use of hard-case surface treatments in marginal operations. ■

Tapping stainless Steels

Slow cutting speeds, thread depths less than 75% and sharp taps are essential in tapping the stainless steels

Alloy Group	Alloy Number	Alloy Condition[a]	Tap Designation: HSS Grade	Tap Designation: Tap Style[b]	Percent thread	Cutting speed, fpm	Cutting Fluid[c]
Nonstainless alloy steels							
	Cr-Mo Low-alloy steels						
A1	A-4130	Ann 150-230 Bhn	M2	A	70	30	IIa, IIIa, IIIb, IV
	A-4130	HT 315-370 Bhn	M2	B	60	15	IIa, IIIa, IIIb, IV
	Cr-Ni-Mo low-alloy steels						
C1	A-4340	Ann 150-230 Bhn	M2	A	70	30	IIa, IIIa, IIIb, IV
	A-4340	HT 315-370 Bhn	M2	B	60	15	IIa, IIIa, IIIb, Iv
	A-4340	HT 510-560 Bhn	M10 (cyanided) M2	B	60	5	IIIc
D2	H-11	Ann 160-220 Bhn	M2	A	70	20	IIa, IIIa, IIb, IV
	H-11	HT 340-370 Bhn	M2	B	60	15	IIa, IIIa, IIIb, IV
	H-11	HT 515-560 Bhn	M10 (cyanided) M2	B	60	5	IIIc
	Maraging steels						
D3	18 Ni Co Mo	HR	M10 (nitrided)	A	70-75	15	
		HR-Ann Mar.	M10 (nitrided)	B, C, D	50-55	7	
	Cr-Ni-Mo steel						
E1	D6a	HT 52 Rc	M10 (nitrided)	C	65	5	IIIc
Stainless steel alloys							
	Straight-chromium grades						
B4	405	Ann 150-200 Bhn	M2	A	< 75	10-25	IIb, IIIa, IIIb, IV
C4	442	Ann 150-200 Bhn	M2	A	< 75	10-25	IIb, IIIa, IIIb, IV
B5	410	Ann 160-220 Bhn	M2	A	< 75	10-25	IIb, IIIa, IIIb, IV
	410	HT 300-350 Bhn	M10 (nitrided) M2	B2	60-70	15-20	IIb, IIIa, IIIb, IIIc, IV
D5	440 B	Ann 215-260 Bhn	M2	A	< 75	20	IIb, IIIa, IIIb, IV
	440 B	HT 300-350 Bhn	M2	B2	< 75	15	IIb, IIIa, IIIb, IV
	Precipitation-hardenable grades						
D6	17-7 PH	Ann 160-180 Bhn	M2, M10	A		20	IIa, IIIa, IIIb, IV
	17-7 PH	HT 380-440 Bhn	M2, M10	B2	75	5-10	IIa, IIIa, IIIb, IV
	Chromium-nickel grades						
D7	347	Ann 160-220 Bhn	M2	A	65	10-25	IIb, IIIa, IIIb, IV
Austenitic stainless steels superalloys							
E9	A286	STA 320 Bhn	M10	A	75	30	IIIb

[a] Ann = annealed; HT = heat treated; HR = hot rolled; Mar. = maraged; STA = solution treated and aged.

[b] A = spiral pointed or spiral fluted
B = taper
C = plug
D = bottoming.

[c] Cutting fluids for stainless steels

I	Water-base coolant, soluble-oil type or chemical type
IIa	Sulfurized oil
IIb	Sulfurized oil plus 10 to 25% kerosene
IIIa	Chlorinated oil
IIIb	Highly chlorinated oil
IIIc	Highly chlorinated oil plus inhibited trichlorethane
IV	Sulfochlorinated oil

Tapping problems...causes and cures—I

COMPILED BY BAY STATE TAP & DIE, DIVISION OF THE CLEVELAND TWIST DRILL CO

You just can't fool around with taps the way you can with some other rotary cutting tools. That is to say there is considerably less margin for error in such variables as speed, lubrication, rigidity of setup, and tool geometry.

Thus the time to prevent tapping problems is *before* the machining starts. For example, when the engineering department designs a product that requires internal threading, tapping problems can be minimized by selecting (if possible) a material that is relatively easy to tap, by avoiding unreasonable lengths of engagement and high percentages of thread, by allowing adequate clearance at the bottom of blind holes, and by establishing realistic pitch-diameter tolerances.

For any given thread specification tapping difficulties may be further minimized by following a rule so basic that it almost goes without saying: Pick the right tap, use it in the right machine with a good lubricant, at reasonable speed. Important, too (except in completely automatic cycles) is the skill of the operator.

Tapping difficulties fall into seven basic categories: (1) tap breakage and chipping, (2) oversize threads, (3) undersize threads, (4) unsatisfactory tap life, (5) rough or torn threads, (6) wavy threads and (7) loading. One or more of these conditions can cause difficulties on a job.

TAP BREAKAGE AND CHIPPING

This is probably the most common difficulty encountered in tapping. The major causes are:

- **Hitting the bottom of the hole,** or hitting chips packed in the bottom of the hole. The drilled hole must be deep enough to permit the tap to produce a full thread to the desired depth—in addition to the chamfer length on the tap. Reasonable additional clearance should be allowed beyond this depth to make sure that forward rotation can be stopped and the tap reversed before the tap strikes the bottom of the hole (or the chips packed in the bottom).
- **Undersize hole**—Here a higher-than-specified percentage of thread is produced, causing unnecessary strain on the tap and a greater volume of chips. To prevent this, consult a tap drill chart which shows optional-size drills, and select a larger tap drill. The drilled hole size should be measured, because some materials tend to "close in" after the drill is withdrawn.
- **Incorrect tap style**. Different tap styles produce differences in chip action and variances in tap strength. Most tap manufacturers' catalogs show the application for which each tap is designed.

Select the tap design with relation to the type of material, hole size, the depth of the tapping. Various types of taps are available for special purposes. These tools differ in number of flutes, flute contours, and other features.

- **Chip packing in flutes.** This is responsible for many broken taps, and not only in the tapping of blind holes. One plant faced with this difficulty found that a spiral-pointed tap, which drives the chips ahead of the cutting ac-

Tapping a $7/8$-14 thread in a motor bracket requires optimum cutting action and positive chip removal

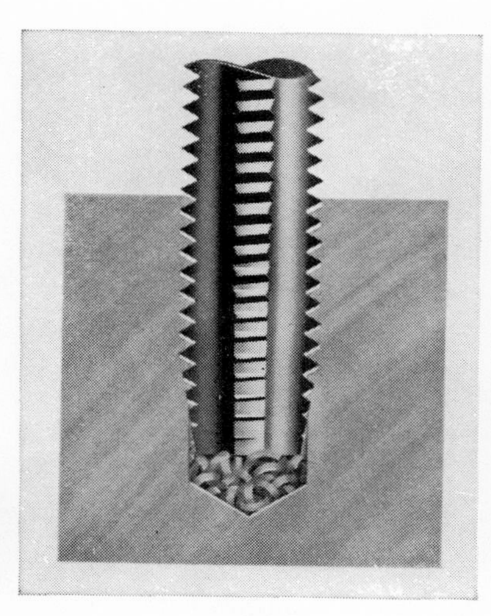

Chip packing may be overcome by a shorter thread depth or a lower percentage of thread

If a tap hits chips packed in blind hole, excessive torque is developed and the tap will break

Tapping problems...causes and cures—II

tion, is a better choice than other types.

• **Improper chamfer.** To offset this, be careful in your choice of the taper, plug or bottoming chamfer (as furnished by tap manufacturers), making sure that they are the right ones for your specific threading requirements.

When a dull tap is reground, the original chamfer angle should be duplicated, and make sure that all sections are ground uniformly. Only in unusual situations should the manufacturer's chamfer be altered, and then only by highly experienced grinding personnel.

• **Misalignment.** If this occurs between the tap and the work, uneven thrust and chip load can result in chipped teeth and broken taps.

• **Insufficient lubrication.** Because of the comparatively slow speeds at which taps operate, and because of the heavy chip load per tooth in the chamfer, lubricity is the basic requirement of any cutting fluid used. The cutting fluid must be capable of preventing loading, must aid in chip removal, and must insure cool operation.

• **Improper speed.** In other cutting tools, speed and feed can be changed independently. Not so with taps, because the feed per revolution is governed by the lead. For example, a single-lead tap having 20 tpi must have a feed of 1/20 (0.050) inch.

These are the factors to be considered in selecting tapping speed: composition and hardness of the material, efficiency of the cutting fluid, condition of the equipment, position of tapping operations (horizontal or vertical), type of hole (through, blind or bottoming), depth of hole, and percentage of thread. Tap manufacturers' recommended speeds are thus only starting points.

Taps with long chamfers may be run at higher speeds than taps with shorter chamfers. This is because the long chamfer reduces the chip load per tooth.

For the same reason, coarse-thread taps should be run somewhat slower than fine-thread taps of like diameter. The greater the percentage of thread, the heavier the chip load per tooth—resulting in need for a slower tapping speed. Fewer flutes mean increased chip load per tooth.

Another rule: speed must be decreased as the length of the thread to be tapped is increased —because frictional heat is generated by the increased chip congestion.

• **Reversing the tap.** Some materials tap undersize because of "flow back" immediately after the tapping operation. The material tends to seize and break the tap when the direction of operation is reversed. This same problem occurs in tapping any stringy material — because the heel of each tooth, when reversing, must cut through the chip that was still attached when the following tooth stopped cutting.

In some instances, grinding a hook or cutting face in the heel side of the land helps in reversing the tap.

Tap Chamfers

Taper—8 to 10 threads

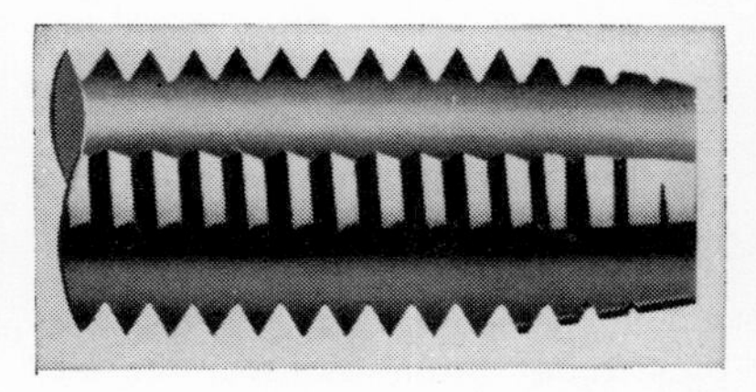

Plug—3 to 5 threads

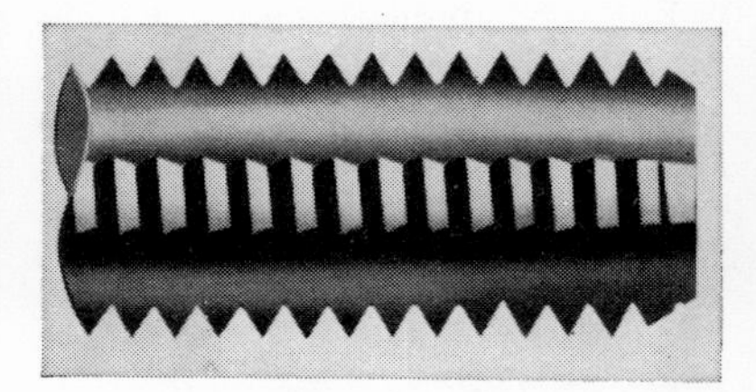

Bottoming—1½ to 2 threads

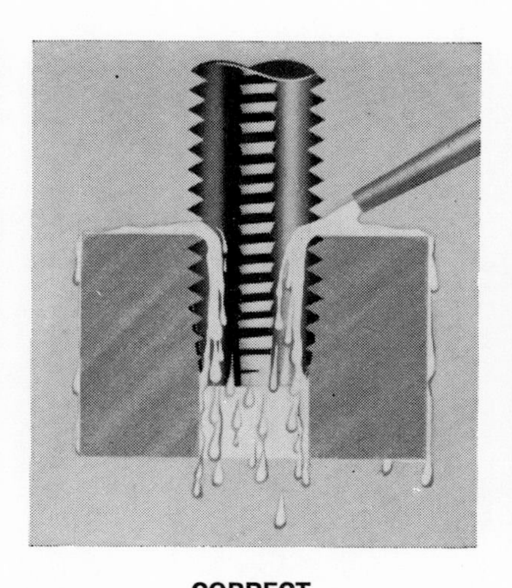

CORRECT
Coolant should be directed to flow, in copious amounts, so that it will reach the cutting faces of the tap.

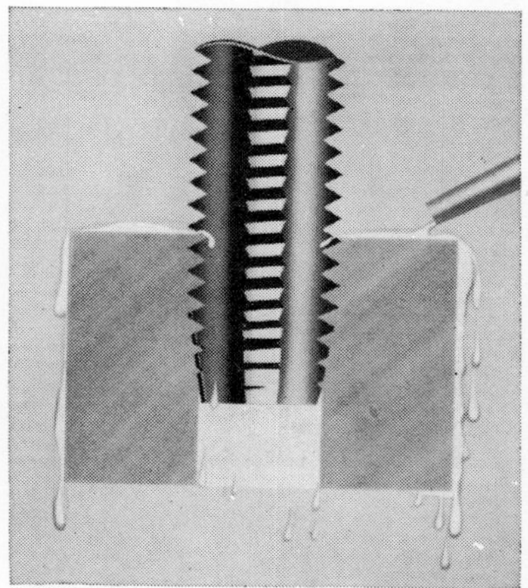

INCORRECT
Proper flow (left) provides cooling and lubrication and, in most cases, aids in chip removal

Tapping problems . . . causes and cures—III

By Bay State Tap and Die, Division of Cleveland Twist Drill Co

Parts I and II (AM/MM—Nov 25 '63, p 101) began discussion of the seven most common problems encountered in the use of taps, and what to do about them. These problems, and their cures are concluded here

OVERSIZE THREADS

Because oversize threads result in a loose fit, the fastener lacks the necessary holding power and, consequently the desired tightness between mating parts is lacking. Here are some contributing causes and how to correct them:

- **Misalignment** occurs, of course, when the tap and part are not lined up properly. The correction here is obvious: check and align the spindle, fixture, and work.
- **Tap too large**—frequently a high-limit tap is chosen when a lower limit one should be used for the desired type of thread.
- **Dull taps.** Taps that are worn or dull not only tend to produce oversize threads but are likely to chip, break and produce rough or poor threads. When the taps are sharpened, they must *not* be ground unevenly, otherwise they may still cut oversize.
- **Worn or loose spindles**—these or worn holders in the equipment can produce oversize threads. Replacement or repair of the spindles and/or the holders should correct this problem.
- **Chip packing in flutes** can produce an oversize thread in some instances. Choose a different style of tap, one that can prevent this condition.
- **Excessive feed pressure** whether manually or mechanically controlled, may cause shaving of the thread flanks and result in oversize threads. This applies both to the tapping of the hole and subsequent withdrawal of the tap.
- **Too much radial relief on tap**—If the tap is too free cutting it tends to cut an out-of-round or oversize thread. Here, a tap with less—or no—radial relief gives results that are much better.

UNDERSIZE THREADS

There are five major reasons why a tap cuts undersize threads:

- **Wrong size tap.** The tendency here is to select a lower-limit tap instead of a higher-limit one, especially in the tapping of materials that tend to contract.
- **Excessive speed.** In some materials, excessive speed generates excessive heat, causing the workpiece to expand during tapping. After the tap is removed the part contracts as it cools, resulting in an undersize thread.
- **Burr on the thread.** Threads are often thought to be undersize because the go gage will not enter. Actually, a burr is formed on the crest of the internal thread, preventing the go gage from moving. A tap with a higher limit can eliminate this.
- **Wrong allowance for plating.**

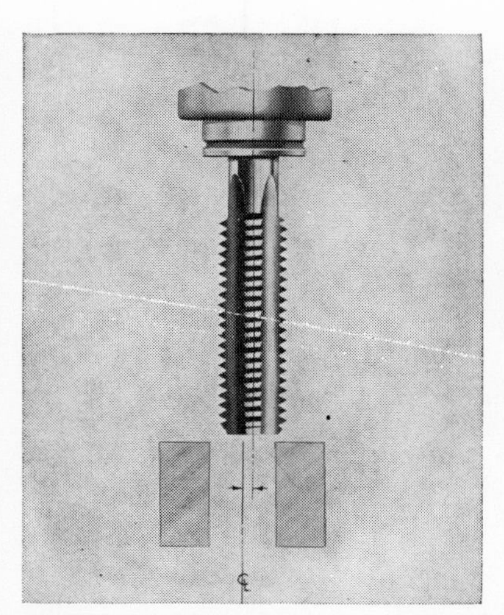

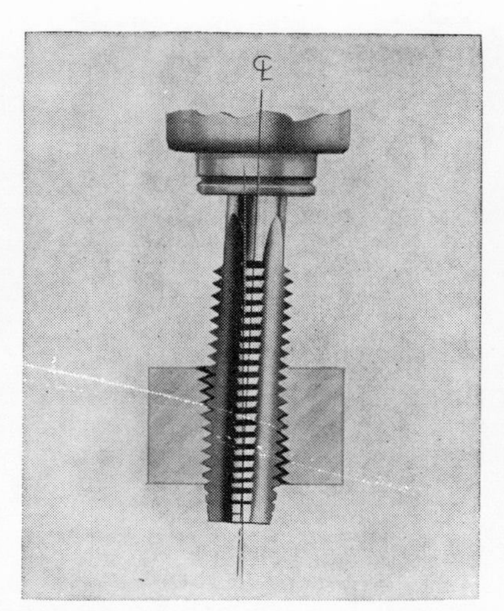

MISALIGNMENT between tap and workpiece may break the tool, because of uneven thrust and chip load, or produce oversize and bellmouth threads

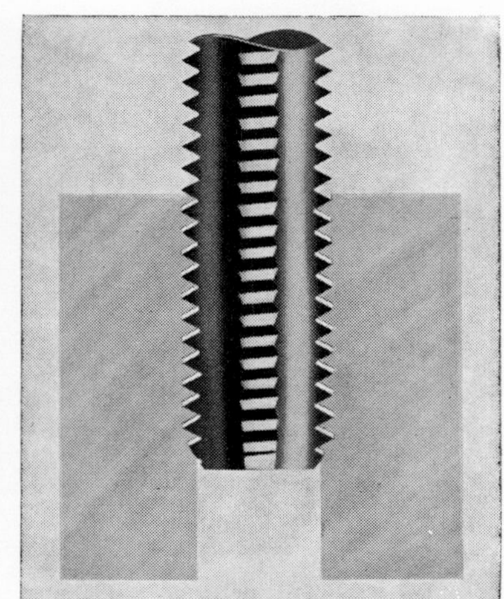

A TAP CUTS oversize when excessive pressure is applied to the spindle

Tapping problems . . . causes and cures—IV

Failure to make the right allowance for plating or heat treatment of parts after tapping can cause a high percentage of rejection. A tap with a higher limit prevents this.

• **Thin wall material.** Use a tap that cuts freely when tapping a thin-wall material. Usually, an oversize tap, or a tap with more hook or shear, helps. Lubrication should be adequate.

If a thin-wall part is held in a clamping device, the clamp should grip the part so that the material is not "squeezed" unduly.

UNSATISFACTORY TAP LIFE

Here are six reasons why tap life is unsatisfactory:

• **Improper tap style.** This is the first consideration when tap life is too short. Try switching to another type or different surface treatment.

• **Hard material.** Nitriding or flash chrome plating can extend tap life, and taps with longer chamfers are also helpful in machining hard material. As little thread engagement as is permissible is desirable, and often absolutely necessary. Reduced hook or rake angles improve tap life and thread finish in hard materials.

• **Abrasive material** often substantially reduces the wear life of taps. Here, use of nitrided or chrome-plated taps improves performance. Recommended, too, is reaming of cored holes to eliminate the draft angle and the abrasive surface.

• **Work-hardened hole.** This condition is largely dependent upon the work-hardening property of the material, and can be minimized by proper drilling techniques, reaming, or lower rake angles.

• **Insufficient lubrication.** A more suitable lubricant or more copious amounts may be the solution here.

• **Excessive speeds** can cause excessive heat, loss of tool life.

ROUGH OR TORN THREADS

These are caused by an improper cutting face on the tap, or insufficient hook. In some cases the hole is too small to be properly prepared. Dull taps do not cut smoothly and freely resulting in chip congestion and also production of poor threads.

WAVY THREADS

These can be caused by loose spindles, incorrect chamfer, incorrect thread relief or an out-of-round hole. Corrective measures: align spindle or fixture and work, correct chamfer grind (usually this means reducing the relief), and correct the out-of-round hole condition.

LOADING

Material loading or buildup on the chamfer tooth, or between the flanks of the thread, is usually caused by friction-pressure between the tap and the material being threaded. Contributing factors which tend to increase this friction-pressure are: incorrect tap geometry, insufficient lubrication, excessive speed, and improper hole preparation. Investigate these conditions in a thorough, precise manner to isolate the problem. •

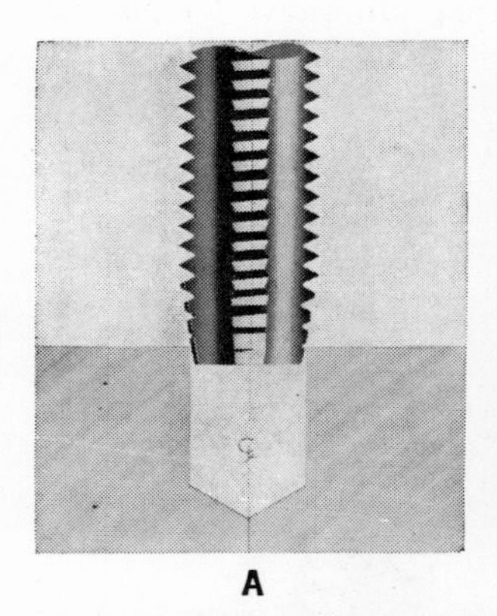

A

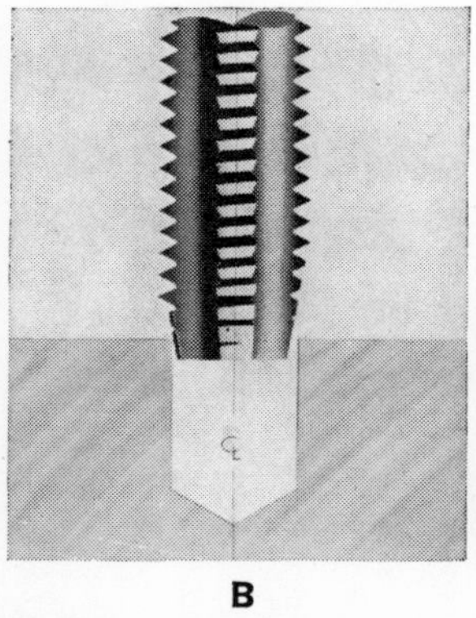

B

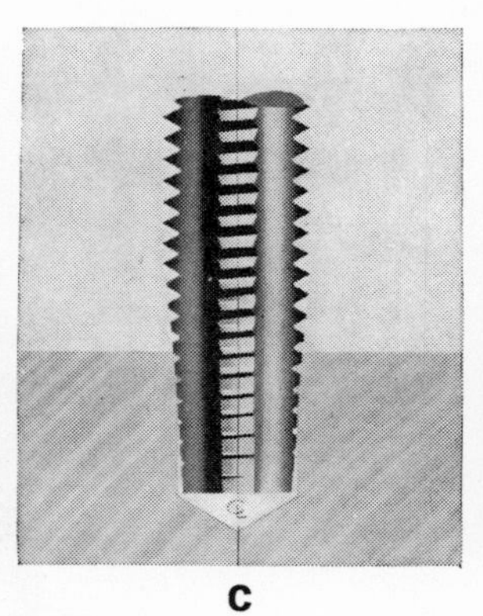

C

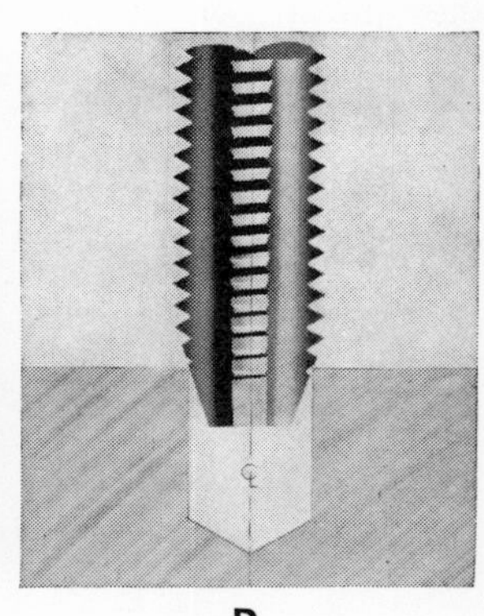

D

HOW CHAMFER AFFECTS TAPPED-HOLE QUALITY— (A) Correct chamfer cuts quality threads and provides long tool life, (B) Uneven chamfer produces rough, oversize threads and increases tap breakage, (C) When the chamfer is too long, the tap will hit bottom before a full thread is formed, and (D) Chamfer ground with a point too small and incorrect angle reduces number of effective teeth, increases chip load

Chapter 9

Turning and Boring

Carbide-tool breakage

Actual breakage, not wear or cratering, is one of the most perplexing causes of carbide-tool failure. This can occur shortly after the tool is put into service, or much later

Common causes of breakage shortly after tool starts to cut

Tool too small for job

Cutting forces cause breakage, bending or vibration in the tool shank. Use heavier shank.

Overhang is too great

Cutting forces cause breakage, bending or vibration in the tool shank. The tool overhang should not exceed the height dimension of the tool shank.

Vibration in the set up

Even though tool is rigidly clamped and overhang is kept to a minimum, vibration may occur. Check slides and spindle bearings for looseness. If the workpiece is hollow, fill it with a vibration-dampening material such as lead for small pieces or a material such as rubber for large pieces.

Carbide is too hard

If tool design changes, such as lead angle, negative rake land, or honing do not help, change to a tougher grade of carbide.

Carbide damaged by grinding

Damage can be caused by rough grinding the carbide tool at the cutting edge. A suggested grinding setup for preservation of grinding wheels and preventing damage to carbide is:

(a) Grind shank to 10°-12° clearance with aluminum-oxide wheel. Do not touch the carbide.

(b) Rough grind the carbide to 8°-10° clearance with a silicon-carbide wheel or coarse-grit diamond wheel. Do not touch the cutting edge with this wheel.

(c) Finish grind carbide to 6°-7° clearance with a medium or fine-grit diamond wheel. A 1/16- to ⅛-in. land is sufficient unless a piece is chipped out of the carbide.

Cutting-edge design is too weak

Sometimes the cutting edge is too weak, especially on interrupted cuts. Here are suggestions:

(a) Try a 45° side cutting edge angle on the tool.

(b) Grind a negative land 3° to 15° on the cutting edge.

(c) Hone cutting edge.

If none of the above steps prevent breakage, the feed should be reduced or a softer, tougher grade of carbide should be used.

Cutting edge too dull

As a result of excessive honing, the cutting edge may be dull. If light honing does not prevent breakage, then a negative land should be ground on the top of the cutting edge. A 3° to 15° negative land will not increase cutting pressures as much as excessive honing.

Edge breaking or spalling of chipbreaker

Such conditions are usually caused by short, broken chips resulting in severe hammering against the chipbreaker shoulder.

The chipbreaker is too deep or the shoulder of the chipbreaker is too steep or too close to the cutting edge. Decrease depth of chipbreaker and use 45° shoulder instead of the usual 90° shoulder. General chipbreaker recommendations are 0.015 in. deep by 5 to 8 times as wide as the feed per revolution.

Build up on cutting edge

Build up results from slow speeds and severe localized cutting pressures. This will cause chipping. Increase speed until build up disappears.

By V G Yost, chief service engineer
Tungsten Division
Firth Sterling, Inc, McKeesport, Pa

Carbide-tool breakage [continued]

Stopping the machine

On older machines or large workpieces, it is not enough to disengage the feed before stopping the machine. After stopping the feed, the tool should be permitted to clean up the cut or it should be retracted from contact with the workpiece.

Common causes of breakage after extended cutting

Worn carbide tool

Excessively worn carbide tools cause heavy buildup of cutting pressures. When these pressures exceed the strength of the carbide, the tip will break. This is particularly noticeable on the harder grades. A wear land of more than 0.030 in. should not be permitted to develop on a tool tipped with a general-purpose grade. Tools tipped with a medium-duty grade should be changed before the wear land exceeds 0.020 in.

Use of chipped tool

A tool with a chipped tip will develop heavy localized pressures at the chipped area, resulting in a crack across the carbide. The tip, if chipped, will eventually break unless it is reground before the crack develops.

Improper application of coolant

Interrupted quenching of the cutting edge may cause a crack which will frequently result in a broken tip after continued use.

Heavy-duty dry cutting with a brazed tool

Breakage can occur in a well-supported tool on a new machine. Brazed tool shanks are usually made from carbon steel and the heat and pressure from some jobs can cause deformation or mushrooming of the shank under the front of the tip. The tip will not deform but will break because it has lost its support.

To prevent breakage, change to an alloy steel shank or to the use of a water coolant or spray mist applied to the under side of the shank.

Heavy-duty cutting with copious coolant

A wide chip passing over the top of the tool will prevent the coolant from getting to the cutting edge and cause the cutting edge to become very hot.

When the chip breaks, the coolant is permitted to splash on the tool and quenches the carbide tip. This occurs many times and results in heat checks along the cutting edge. Usually the heat checks are so deep that regrinding will not remove them. Eventually these heat checks become so deep that a piece will break out of the tip. This may happen after the initial grind or after several regrinds.

Coolant should be applied to the underside of the shank to draw the heat away from the tip; it must not be permitted to impinge on the cutting tip. If this does not prevent heat checks and breakage, a softer, tougher grade should be used.

Throwaway inserts under heavy cuts

The problem is similar to the preceding one. With throwaway inserts, a transverse fracture occurs after the insert has been used on all corners of one side and is turned over. If the conditions are extremely severe, the insert will break even before it is turned over. Inspection of the insert will show numerous heat checks.

Breakage of the nose or corner

Sometimes the nose or corner will break when the tip is subjected to a steady non-interrupted cut. When cutting steel and some alloy irons, all grades will crater. Some will crater more than others. The heaviest crater is at the nose of the tool. The crater will start back of the cutting edge and will gradually widen. If the tool is not changed or reground soon enough, the edge of the creater will intersect the cutting edge and a piece of the tip or insert will break off, because of the careless handling. ■

Carbide tooling for alloy castings

Shops that machine a growing volume of high-alloy castings for the process industries are turning to modern machine tools and carbides to reduce their costs

Details of successful single-point carbide tools for machining alloy castings containing a minimum of 8% nickel and/or chromium are:

Side rake—4-8°.
Back rake—4-8°.
Side relief—4-7°.
End relief—7-10°.
End cutting angle—8-15°.
Side cutting angle—10-15°.
Nose radius—1/16-1/8 in.

Because of the increased speed and depth of cut possible with carbide tooling, horsepower requirements are greater than for conventional tooling. Toolholders must be extremely rigid to prevent chattering or rubbing of the tool on the metal surface. Such action, particularly on the austenitic alloys, forms irregular work-hardened patches leading to greatly increased tool breakage.

Chips of many high alloy grades are characteristically tough and stringy. Therefore, chip curlers and breakers are recommended whenever the size of the tool permits. They are especially important on the austenitic grades.

Interrupted cuts should be avoided whenever possible, so as to prevent work-hardening the alloy at the point where the tool leaves or enters.

To provide sufficient cooling and lubrication, the work must be flooded at all times. This is particularly true when a shallow cut is being made, since most of the heat generated in the tool will be localized in a very small volume of material at the tip. In general, a water-soluble coolant is best.

In selecting the best combination of speed and feed, speed should be reduced first and then feed.

Because of the greater initial expense and higher cost of retipping, drilling with carbide tooling is generally not economically justified except at fairly high production rates. To prevent breakage, holes below about 0.090 in. should be avoided whenever design considerations permit. If a hole below this diameter must be made with a carbide drill, use manual feed so that force can be adjusted by 'feel.'

Wherever possible, use plenty of coolant to flood the workpiece and the drill. An included angle of 135-140° with a clearance point angle of about 8-10° is generally best.

Milling and reaming can also be performed successfully. With milling, one can start at about 80 sfm.

For reaming, speed is approximately the same as for drilling, and the depth of cut is generally from 0.002 to 0.004 inch per reamer tooth.■

By W F Helber
Reprinted from '8 Plus' published by Alloy Casting Institute, New York

Table 1. Machining of cast corrosion-resistant high alloys

ACI type	Rough turning Speed sfm	Rough turning Feed ipr	Finish turning Speed sfm	Finish turning Feed ipr	Drilling Speed sfm	Drilling Feed ipr
CA-15*	150-300	.060-.125	200-300	.010-.030	50-80	a
CA-40*	150-250	.060-.090	200-300	.010-.030	40-60	a
CB-30	150-200	.060-.090	200-250	.010-.030	40-60	a
CB-7Cu	150-250	.060-.125	200-250	.010-.030	50-80	a
CC-50	150-200	.060-.090	200-250	.010-.030	50-80	a
CD-4MCu	150-250	.060-.125	200-300	.010-.020	50-80	a
CE-30	150-200	.060-.090	200-250	.005-.020	50-80	a
CF-3, CF-8	150-350	.060-.125	200-350	.010-.030	50-80	a
CF-20	150-300	.060-.125	200-300	.010-.030	50-80	a
CF-3M, CF-8M	150-300	.060-.090	200-300	.005-.020	50-80	a
CF-8C	150-300	.060-.090	200-300	.005-.020	30-60	a
CF-16F	150-300	.060-.090	200-300	.005-.020	30-60	a
CG-8M	100-250	.060-.090	200-250	.005-.020	40-60	a
CH-20	100-250	.060-.090	200-250	.005-.020	40-60	a
CK-20	100-250	.030-.060	150-200	.005-.020	40-60	a
CN-7M	100-200	.030-.060	150-250	.005-.020	30-60	a

Table 2. Machining of cast heat-resistant high alloys

ACI type	Rough turning Speed sfm	Rough turning Feed ipr	Finish turning Speed sfm	Finish turning Feed ipr	Drilling Speed sfm	Drilling Feed ipr
HA	150-300	.060-.125	200-300	.010-.030	50-100	a
HC	100-200	.030-.060	150-250	.005-.010	40-60	a
HD	100-200	.030-.060	150-250	.005-.010	40-60	a
HE	100-150	.020-.050	150-200	.005-.010	30-60	a
HF	100-200	.020-.050	150-200	.005-.010	30-60	a
HH	100-150	.020-.050	100-200	.005-.010	30-40	a
HI	100-150	.020-.050	100-150	.005-.010	20-40	a
HK*	100-150	.020-.090	150-250	.005-.015	25-50	a
HL*	90-150	.020-.060	150-200	.005-.015	25-60	a
HN*	90-150	.015-.060	150-200	.005-.015	30-60	a
HT*	100-150	.020-.090	150-250	.005-.015	30-60	a
HU*	100-150	.020-.060	150-200	.005-.015	30-60	a
HW*	80-100	.030-.090	100-180	.005-.015	25-30	a
HX*	80-100	.030-.090	100-180	.005-.015	30-60	a

[a] Drilling feeds

Drill dia., in.	Feed, ipr	Drill dia., in.	Feed, ipr
Under ⅛	.002—.004	½—1	.007—.015
⅛—¼	.004—.008	over 1	.020—.030
¼—½	.004—.008		

* Use chip curler

Turning with ceramic tools

Source: 'Machining Data Handbook' published by Metcut Research Associates, Inc., Cincinnati, Ohio

Material	Hardness BHN	Condition	Depth of cut in.	Speed fpm	Feed ipr
Free machining plain carbon steels Resulphurized steels 1108 1119 1144 1109 1120 1145 B1111 1126 1146 B1112 1132 1151 B1113 1137 1211 1113 1138 1212 1115 1139 1213 1117 1140 1213 + Te 1118 1141 Leaded steels 10L18 12L13 10L20 12L14	100 to 150	Hot rolled or as cast, normalized, annealed or cold drawn	Under 0.062	1000 to 2200	0.004 to 0.010
			Over 0.062	600 to 1600	0.005 to 0.020
	150 to 250	Hot rolled or as cast, normalized, annealed or cold drawn	Under 0.062	1000 to 2000	0.004 to 0.010
			Over 0.062	500 to 1500	0.005 to 0.020
	275 to 325	Quenched and tempered	Under 0.062	600 to 1600	0.004 to 0.010
			Over 0.062	400 to 1100	0.005 to 0.020
	325 to 375	Quenched and tempered	Under 0.062	300 to 800	0.004 to 0.010
			Over 0.062	250 to 650	0.005 to 0.015
	375 to 425	Quenched and tempered	Under 0.062	250 to 700	0.004 to 0.010
			Over 0.062	200 to 600	0.005 to 0.015
Plain carbon steels 1006 1023 1040 1064 1008 1024 1041 1065 1009 1025 1042 1066 1010 1026 1043 1070 1012 1027 1045 1074 1015 1030 1046 1078 1016 1033 1049 1080 1017 1034 1050 1084 1018 1035 1052 1085 1019 1036 1055 1086 1020 1037 1060 1090 1021 1038 1062 1095 1022 1039	85 to 175	Hot rolled or as cast, normalized, annealed or cold drawn	Under 0.062	650 to 1600	0.004 to 0.010
			Over 0.062	450 to 1300	0.005 to 0.020
	175 to 275	Hot rolled or as cast, normalized, annealed or cold drawn	Under 0.062	450 to 1300	0.004 to 0.010
			Over 0.062	350 to 900	0.005 to 0.020
	275 to 325	Quenched and tempered	Under 0.062	400 to 1200	0.004 to 0.010
			Over 0.062	300 to 900	0.005 to 0.020
	325 to 375	Quenched and tempered	Under 0.062	300 to 800	0.004 to 0.010
			Over 0.062	250 to 650	0.005 to 0.015
	375 to 425	Quenched and tempered	Under 0.062	250 to 700	0.004 to 0.010
			Over 0.062	200 to 600	0.005 to 0.015

Applicable ceramic tools are:
CCT 707—The Carborundum company
O-30—General Electric Company, Metallurgical Products Department
VR-97—VR/Wesson Company

Turning Superalloys—I

By C T Olofson, principal metallurgist, and Francis W Boulger, chief,
Metalworking Research Div, Battelle Memorial Institute, Columbus, Ohio

Superalloys are a group of complex alloys for use in the service-temperature range of 1100-1800 F. Chromium, nickel and iron are the common elements in these materials, but cobalt, molybdenum, tungsten, titanium and aluminum may be present. Superalloys are considerably more difficult to machine than ordinary constructional materials. As compared to B1112 steel at 100, these alloys have machinability ratings of 6-11.

The low machinability ratings of the superalloys generally result from the tendency of these materials to weld to the cutting edge of the tool as a built-up edge, and from the high work-hardening rates of the alloys. The built-up edge periodically spalls, pulling out parts of the cutting edge with it. The work-hardened chip produced during machining is highly abrasive to the tool face, while the remaining work-hardened machined surface may notch the tool at the depth-of-cut line during the next cut.

Cemented carbide and high-speed steel of maximum red hardness are usually employed for cutting tools. They generally have zero to positive rakes, and just enough clearance to prevent rubbing.

Superalloys can be machined in the solution-treated, partially aged, or fully aged conditions. The choice depends on the strength of the alloy, the surface finish desired, and the operation.

Solution-treated metal generally is gummy and shows a greater tendency to weld to the tool as a built-up edge. It also work hardens rapidly. Because high strain hardening and a built-up edge formation can cause poor surfcae finish, the solution-treated condition is often rated as having a low order of machinability. Nevertheless, it is sometimes preferred for drilling, tapping, threading, and rough-machining.

Aged alloys, although stronger than solution-treated alloys, will often machine easier because of lower work-hardening capabilities and higher yield strength-tensile strength ratios. Finer grain size and the presence of microscopic and submicroscopic precipitated metallic compounds are also thought to contribute to better machinability. The higher strength, lower strain-hardening coefficients, and lower ductilities of aged alloys will also contribute to better surface finishes.

A copious quantity of cutting fluid, properly directed, is an effective means of minimizing welding tendencies of these alloys.

Superalloys, in addition to their welding tendencies, are particularly vulnerable to work hardening. During the machining process, plastic deformation work hardens both the machined surface and the chip, the latter becoming quite abrasive to the tool. The amount of work hardening imparted to the work and chip is related to the cutting temperatures developed.

Cutting temperatures can be minimized by using proper tools, machining procedures, and cutting fluids. Sharp cutters with positive rake angles, adequate relief angles, and good surfaces keep cutting and frictional forces low. Slower speeds and lighter feeds than those used for stainless steel are also desirable. Furthermore, the depth of cut should be sufficient to prevent burnishing or glazing. Second cuts also should be avoided to prevent work hardening the previously machined surface.

Special mention should be made about the rubbing of tools against the workpiece. This action also contributes to excessive cutting temperatures and consequent work hardening.

Tool Angles for Turning Superalloys

Tool Material	Geometry	Tool Angles, deg.						Nose Radius, in.	Chip Breaker
		Rake		Relief		Cutting Edge			
		Back	Side	Side	End	Side	End		
Cobalt HSS	A	8-12	6-8	5-8	7-10	10-15	10-15	—	—
"	B	7-10	5-8	5-8	7-10	10-15	10-15	—	—
"	C	5-8	5-8	5-8	7-9	10-15	8-15	—	Yes
"	D	7-10	5-8	7-10	7-10	10-15	8-15	—	—
"	E	0	15	4-5	4-5	0	5	1/32	—
"	F	0-8	8-15	4-5	4-5	15-25	10-20	1/32	Yes
"	G	5-10	5-10	5-8	7-10	10-15	10-15	—	Yes
"	H	7-10	5-10	5-8	7-10	10-15	10-15	—	—
"	I	4-6	4-6	4-6	4-6	—	—	—	—
"	J	4-6	15-20	4-8	3-5	8-12	8-12	—	—
Castalloy	F	0-8	8-15	4-5	4-5	15-25	10-20	1/32	—
"	K	5-10	5-8	5-8	7-10	10-15	8-15	—	—
Carbide, C-1 to C-8	L	0-3	6	6	6	0	6	0.02	—
"	M	0-3	6	7	7	45	—	—	—
"	N	0-8	8-15	4-5	4-5	15-25	10-20	<1/32	Yes
"	O	5-8	8-10	5-7	5-7	15-30	—	1/32	—
"	P	15	10-15	6	6	—	—	—	—
"	Q	−6	9	6	6	30	9	—	—
"	R	+2	6	6	6	8-12	8-12	1/32	—
"	S	8-12	4-7	3-6	3-6	6-10	4-8	1/32	—
"	T	0	0	7	7	15	15	1/32	—
"	U	+5	NA	7	7	45	45	1/32	—
"	V	0	7-10	7-10	7-10	45	5	1/32	—

Turning Superalloys—II

Feeds and Speeds for Turning Superalloys

Alloy		Tool		Roughing			Finishing			
Designation	Condition	Material (a)	Design (b)	Speed, fpm	Feed. ipr	Depth of Cut, in.	Speed, fpm	Feed, ipr	Depth of Cut, in.	Cutting Fluid
Rene 41	Aged	M-3, HSS	?	—	—	—	20	.002-.006	.01-.05	Chemically active soluble oils
		Carbide C-2	L, M, N, O	35-60	.01-.015	.05-.125	—	—	—	
		Carbide C-3	U, V	—	—	—	60-120	.004-.010	.015-.15	
Hastelloy X	Sol. treated	HSS	C	18-21	.007-.011	.025-.045	18-21	.003-.008	.008-.011	Sulfur-base oil
		Carbide	O	68-88	.008-.016	.05-.10	68-88	.006-.009	.011-.015	
Waspaloy	Aged	HSS	F	10-20	.015	.09	—	—	—	Chemically active soluble oils, sulfur-base oils, or sulfur-base oils w/10-25% kerosene
		CA	F	40	.015-.025	To .25	—	—	—	
		Carbide	N	40-75	.008-.018	.045-.18	75-125	.005-.010	.015-.045	
	Sol. treated	HSS	F	25-35	.015	.09	—	—	—	
Inconel X	Aged	Carbide	N	35-50	.010-.015	.015-.125	60-90	.005-.010	.015-.045	
Nimonic 90 Inconel 700(c) Udimet 500(d) M-252(e) J-1500	Sol. treated	HSS	F	20-35	.015	.093 max	—	—	—	Heavy sulfur-base or sulfochlorinated oils
		CA	F	40	.015-.025	.25 max	—	—	—	
		Carbide	M, N, O	30-45	.008-.013	under .093	—	—	—	
	Aged	HSS	E, F	10-20	.01-.015	.093 max	—	—	—	
		Carbide	N	60	.008-.018	.187 max	60-90	.005-.010	.015-.045	
		Carbide	N	40-75	.010-.015	.05-.125	30-60	.003-.008	.03	
		Carbide	M, N, O	70-80	.010-.015	.015-.045	75-125	.005-.010	.015-.045	
Haynes Alloy No. 25	Sol. treated	HSS	E	25	.009	.062	—	—	—	Soluble oil and lubricant
		Carbide	N or O	30-45	.010-.015	.060-.10	30-45	.003-.005	.008-.010	
		Carbide C-6	N	100	.009	.10	—	—	—	
J-1570	Aged	Carbide	M, N, O	40-75 30-40	.010-.015 .008-.013	.063-.125	55-100	.004-.009	.031	Soluble oils or sulfur-base oils
S-816(f) HS-31 Alloy X-40	Aged	HSS	C	10-15	.008-.009	.25	—	—	—	Sulfur or sulfochlorinated oil
		Carbide	M, N, O	25-35	.009-.012	.25	—	—	—	
		Carbide	L	—	—	—	30-60	.005-.008	.015-.05	
		Carbide	N	45-65	.010-.015	.05-.125	60-80	.005-.008	.015-.05	
HS-21	Cast	Carbide	M, N, O	10-40	.009-.012	.25	65	.005-.008	.005	General-purpose cutting oil and kerosene
L-605	Annealed	Carbide	T	150	.008	.06	—	—	—	
J-1650	Sol. treated	Carbide	N	30-50	.015-.020	09-.25	—	—	—	Sulfur-base oils
	Aged	Carbide	N	40-75	.010-.015	.05-.125	75-100	.005-.01	.015-.05	Soluble oils or sulfur-base oils
J-1300		Carbide C-5a	M, N	60-80	.015-.020	.09-.25	—	—	—	
		Carbide C-7a	M, N	80-100	.010-.015	.05-.125	100-125	.005-.01	.015-.045	
N-155 S-590	Aged	HSS	C	18-20	.007-.011	.025-.045	18-20	.003-.008	.008-.011	Sulfochlorinated oils
		CA	K	20-40	.007-.011	.025-.045	20-40	.003-.008	.008-.011	
		Carbide	O	50-72	.008-.013	.050-.10	68-90	.006-.009	.011-.015	

(a) HSS=high-speed steel; CA=cast alloy.
(b) See table on page 110 for design data appropriate to the code shown.
(c) Machinable in all conditions.
(d) Machines best in solution-treated condition.
(e) Machines best in partially aged condition.
(f) Machines best in the aged condition.

Speeds for multiple cuts

When a multiple-turning attachment is fitted to a turret lathe, the most efficient compromise for turning speed can be found by use of the tables

When a multiple turning head is used on a turret lathe, the selected spindle speed is usually a compromise between the recommended speed for the cutter nearest to the workpiece centerline, and the speed for the cutter at the greater distance from the centerline.

Sometimes the speed is based on either the smallest workpiece diameter to be turned or the largest. This speed selection procedure occurs because the operator does not bother to compute a likely starting speed, based on the variable workpiece diameters. The end result is usually anything but a proper cutting speed to suit the machining situation at hand.

To aid the operator in selecting proper starting spindle speeds for multiple-diameter workpiece turning, and to counteract the problem mentioned above, precalculated spindle speeds, in chart and/or table form, can be useful. Tables 1, 2, 3 and 4 are presented for these purposes:

Table No. 1 gives recommended starting spindle speeds for rough turning of cast iron with multiple diameters; Table No. 2 is set up for the finish starting speeds.

Table No. 3 gives a constant for commonly used metals, relating their respective turning speeds, percentage-wise, to the values in Tables 1 and 2. Table 4 provides the actual cutting speed in surface feet per minute for the metals found in Table 3.

When using the tables to select starting speeds for turning a given workpiece, the operator enters the horizontal and vertical diameter values that correspond to those of the workpiece to be turned. Then he reads within the table body the recommended starting spindle speed directly in rpm at the intersection of the two values. The starting speed is then set in accordance with the nearest available spindle speed on the machine.

For materials other than soft

Table 1. Spindle speeds for rough turning cast iron
(Carbide tooling at 420 sfpm)

First Diameter	Second diameter														
0	1	2	3	4	5	6	7	8	9	10	11	12	13	14	15
1	1604	1069	802	642	535	458	401	357	321	292	267	247	229	214	201
2	1069	802	642	535	458	401	357	321	292	267	247	229	214	201	189
3	802	642	535	458	401	357	321	292	267	247	229	214	201	189	178
4	642	535	458	401	357	321	292	267	247	229	214	201	189	178	169
5	535	458	401	357	321	292	267	247	229	214	201	189	178	169	160
6	458	401	357	321	292	267	247	229	214	201	189	178	169	160	153
7	401	357	321	292	267	247	229	214	201	189	178	169	160	153	146
8	357	321	292	267	247	229	214	201	189	178	169	160	153	146	140
9	321	292	267	247	229	214	201	189	178	169	160	153	146	140	134
10	292	267	247	229	214	201	189	178	169	160	153	146	140	134	128
11	267	247	229	214	201	189	178	169	160	153	146	140	134	128	123
12	247	229	214	201	189	178	169	160	153	146	140	134	128	123	119
13	229	214	201	189	178	169	160	153	146	140	134	128	123	119	115
14	214	201	189	178	169	160	153	146	140	134	128	123	119	115	111
15	201	189	178	169	160	153	146	140	134	128	123	119	115	111	107

Table 2. Spindle speeds for finish turning cast iron
(Carbide tooling at 490 sfpm)

First Diameter	Second diameter														
0	1	2	3	4	5	6	7	8	9	10	11	12	13	14	15
1	1875	1248	936	749	624	535	468	416	374	340	312	288	268	250	234
2	1248	936	749	624	535	468	416	374	340	312	288	268	250	234	220
3	936	749	624	535	468	416	374	340	312	288	268	250	234	220	208
4	749	624	535	468	416	374	340	312	288	268	250	234	220	208	197
5	624	535	468	416	374	340	312	288	268	250	234	220	208	197	187
6	535	468	416	374	340	312	288	268	250	234	220	208	197	187	178
7	468	416	374	340	312	288	268	250	234	220	208	197	187	178	170
8	416	374	340	312	288	268	250	234	220	208	197	187	178	170	163
9	374	340	312	288	268	250	234	220	208	197	187	178	170	163	156
10	340	312	288	268	250	234	220	208	197	187	178	170	163	156	150
11	312	288	268	250	234	220	208	197	187	178	170	163	156	150	144
12	288	268	250	234	220	208	197	187	178	170	163	156	150	144	139
13	268	250	234	220	208	197	187	178	170	163	156	150	144	139	134
14	250	234	220	208	197	187	178	170	163	156	150	144	139	134	129
15	234	220	208	197	187	178	170	163	156	150	144	139	134	129	125

Frank L Rush
Columbus, Ohio

Speeds for multiple cuts [continued]

cast iron, the starting speed is determined by use of the percentage constants found in Table 3.

Example: Consider the case of turning a soft cast-iron workpiece with multiple diameters of 9-in. and 11-in., respectively. By entering the proper diameter values within the table, we find a recommended rough turning starting speed of 160 rpm (Table 1), and a finish turning starting speed of 187 rpm (Table 2). These values, or the nearest found on the machine, are then set according to the operation to be performed.

Production, quality and earnings stem from the amount of work passing the cutter. Hence it is important to come close to the ideal speed at first setting. The accompanying tables aid in determining these ideal values, by giving appropriate starting speeds for multiple-diameter workpieces, when using commercially available multiple turning heads on the turret lathe. Tables of the nature shown can benefit the user by reducing downtime for calculating starting speeds at the machine, and hence, contributing to better overall production efficiency.

Table 3. Spindle constants for carbide tooling

Material	Turning speed constants Rough	Turning speed constants Finish
C-1010	100%	100%
C-1020	100	100
C-1040	76	86
C-10045	76	86
C-1050	76	86
C-1070	52	47
C-1090	52	47
B-1112	117	114
C-1118	117	114
C-1137	100	100
C-1141	100	100
E-3310	58	65
A-4140	76	86
A-4150	76	86
E-4160	58	65
A-4340	76	86
A-4615	76	86
E-52100	52	47
Cast Steel	58	65
Cast Iron, Hard	76	86
Malleable Iron	76	86
Brass (Naval 73)	100	100
Brass (Free Machining)	Max.	Max.
Bronze (Phos. 64)	117	114
Bronze (Tobin)	100	100
Bronze (Alum. 68)	76	86
Stainless #416	100	100
Stainless #302	76	86
Stainless #304	76	86
Stainless #440C	58	65
Monel	52	47
Aluminum	Max.	Max.

Table 4. Machinability of materials (Carbide Tooling)

Material	Speed sfpm Rough	Speed sfpm Finish
C-1019	420	490
C-1020	420	490
C-1040	320	420
C-1045	320	420
C-1050	320	420
C-1070	220	230
C-1090	220	230
B-1112	490	560
C-1118	490	560
C-1137	420	490
C-1141	420	490
E-3310	245	320
A-4140	320	420
A-4150	320	420
E-4160	245	320
A-4340	320	420
A-4615	320	420
E-52100	220	230
Cast Steel	245	320
Cast Iron, Hard	320	420
Malleable Iron	320	420
Brass (Naval 73)	420	490
Brass (Free Machining)	Max.	Max.
Bronze (Phos. 64)	490	560
Bronze (Tobin)	420	490
Bronze (Alum. 68)	320	420
Stainless #416	420	490
Stainless #302	320	420
Stainless #304	320	420
Stainless #440C	245	320
Monel	220	230
Aluminum	Max.	Max.

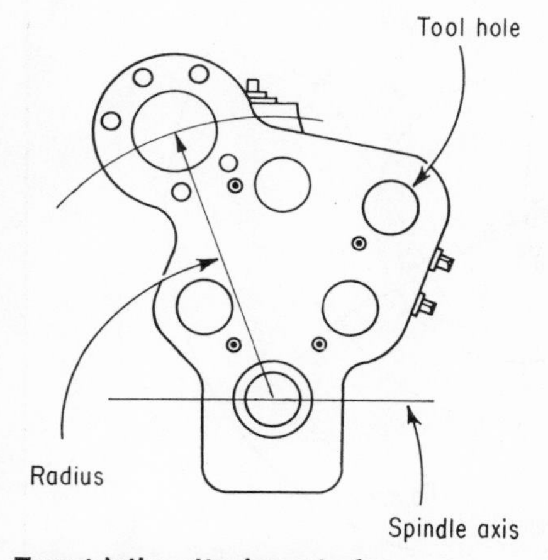

Turret lathe attachments for multiple turning can have tool holes with radiuses that differ widely from the spindle axis

Ed note: The spindle speed is definitely a compromise. In every case, the cutting speed of the larger diameter will be faster than recommended, and the cutting speed of the smaller diameter will be slower than recommended.

To take an extreme case, the table for rough turning shows that the spindle speed for turning a 1-in. diameter and a 15-in. diameter should be 201 rpm. At this speed, the 15-in. diameter will have a surface speed of 790, about twice as much as the recommended 420 rpm shown in the table.

At the same time, the 1-in. diameter will have a surface speed of 53 sfpm, which is about one-eighth of 420. Thus, we would expect the tooi on the 15-in. diameter to have a high wear rate, and in certain materials the tool on the 1-in. diameter might have difficulty in cutting at such a slow speed.

Nevertheless, most combinations of diameters experienced in actual practice will not be this extreme, and the closer any pair of diameters is to the same size, the less variance there will be between the actual cutting speeds and the recommended cutting speeds. ■

Facing tools for chuckers

When the tool feeds along an arc for facing cuts, special attention must be paid to the effect of cross-arm rotation on operating rake and relief angles

If you don't know what's going to happen to the rake and relief angles when the cutting tool is feeding along an arc, you may get poor tool performance, and you may even have smashups if the relief angle becomes negative.

One way to avoid the smashups, at least, is to locate the tip of the tool on an arc passing through the center of the workpiece. But this method causes large changes in operating rake and relief angles during long facing cuts, and these changes may force the use of undesirably small lip angles which compromise tool performance and prevent the use of disposable toolbits.

There is, however, a better method for cuts that do not go all the way to the center of the work. With the method described here, relatively long facing cuts can be made without introducing large changes in operating rake and relief angles.

This method can be used to advantage for short facing cuts because it reduces the number of tool blocks required for parts of various diameters and enables one facing tool block to handle various diameters without reshimming of tool holders for height or resetting of toolholder back-up screws for projection height.

Nomenclature:

R1 is radius of max diameter cut.
R2 is radius of min diameter cut.
R3 is distance from center of work spindle to axis of swinging cross-arm.
R4 (to be determined) is the distance from the axis of swinging cross-arm to tool tip.
A is axis of work spindle.
B is axis of swinging cross-arm.
AB is distance from axis of work spindle to axis of swinging cross-arm and is equal to R3.
Δ (to be determined) is maximum change in operating rake and relief angles during the course of the cut.

Layout method

1. Lay out points A and B.
2. Draw circles and arcs, as indicated, for R1, R2, R1-R2, and R3.
3. Locate point C at intersection of arcs R3 and R1-R2, and project line AC to point D, the position of the tool tip at the largest diameter cut.
4. Similarly, locate point E and project line EA to point F, the position of the tool tip at the smallest diameter cut.
5. About point B, draw arc DF, the path of the tool tip as it moves through the cut. This also checks steps 3 and 4. For counterclockwise spindle rotation, the outline of the toolbit may be drawn in at this time. For inverted tools (clockwise spindle rotation), refer to step 8.
6. Observe that the operating rake and relief angles are equal at D and F. To determine the maximum change (Δ) in these angles, proceed as follows:
7. Construct line BG perpendicular to and bisecting line AE and intersecting arc AE at point H.
8. Construct line HJ perpendicular to AB and intersecting arc AE at H and arc DF at K. The outline of the toolbit at K may be drawn at this time. Angle AKH is Δ, the maximum change in operating rake and relief angles.

Note that, if the tool is inverted for

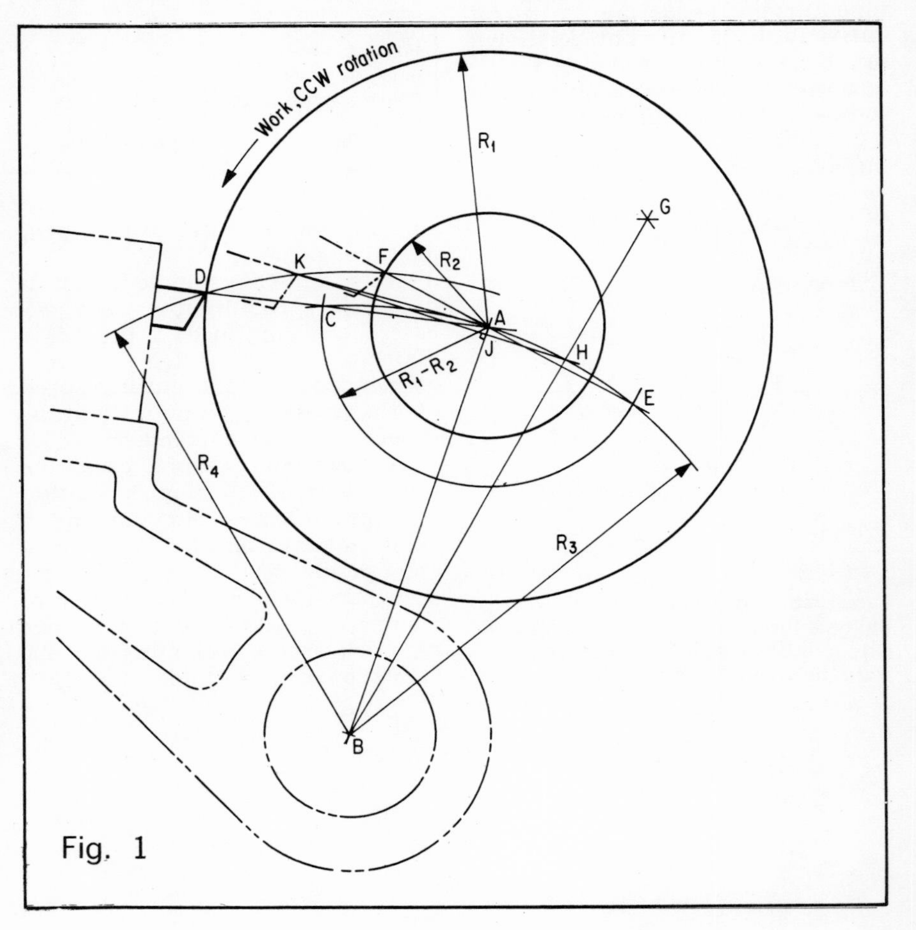

Fig. 1

By Warren A. Phinney
Senior industrial engineer
Teledyne Wisconsin Motor
Milwaukee

Facing tools for chuckers [Continued]

clockwise spindle rotation, operating relief at K will be less than at D and F. For this condition, the basic tool alignment may be drawn in at point K rather than at point F and the change in rake and relief angles checked at points D and F rather than at point K. This will avoid introducing, during the cut, operating relief angles less than the angles provided on the toolbit or toolholder.

Mathematical method

Tool tip location:

$$(1)\quad R4 = \sqrt{AB^2 + R1 \times R2}$$

Change in rake and relief angles for tool tip feeding along R4:

$$(2)\quad \Delta = \sin^{-1}\frac{AB}{R4} - \cos^{-1}\frac{R1 + R2}{2R4}$$

Change in rake and relief angles for tool tip feeding along R3:

$$(3)\quad \Delta = \cos^{-1}\frac{R2}{2AB} - \cos^{-1}\frac{R1}{2AB}$$

Note that, for clockwise spindle rotation, a component of tool motion is opposed by work rotation. Conversely, for counterclockwise spindle rotation, a component of tool motion is in the direction of work rotation, so climb-cutting conditions may exist. Make sure that the feed works are suitable for these conditions.

Examples of use

This method was developed because of the need to face a 10-in.-OD by 5½-in.-ID flywheel rim on a chucking lathe with a rotating cross-arm axis located 14¾ in. from the work-spindle centerline. Allowing ⅛ in. for tool approach and overtravel:

$R1 = 5\frac{1}{8}$
$R2 = 2\frac{5}{8}$
$R3 = AB = 14.75$

$$(1)\quad R4 = \sqrt{AB^2 + R1 \times R2} = 15.199$$

$$(2)\quad \Delta = \sin^{-1}\frac{AB}{R4} - \cos^{-1}\frac{R1 + R2}{2R4}$$
$$= \sin^{-1}\frac{14.750}{15.199} - \cos^{-1}\frac{5.125 + 2.625}{2 \times 15.199}$$
$$= \sin^{-1} 0.9704 - \cos^{-1} 0.2449$$
$$= 76°01' - 75°14'$$
$$= 0°47'$$

Had the tool tip been located conventionally on an arc passing through the center of the work spindle, the change in operating rake and relief angles would have been:

$$(3)\quad \Delta = \cos^{-1}\frac{R2}{2AB} - \cos^{-1}\frac{R1}{2AB}$$
$$= \cos^{-1}\frac{2.625}{2 \times 14.750} - \cos^{-1}\frac{5.125}{2 \times 14.750}$$
$$= \cos^{-1} 0.08898 - \cos^{-1} 0.17372$$
$$= 84°54' - 80°0'$$
$$= 4°54'$$

In round numbers, the change in operating rake and relief angles was reduced from 5 deg to 1 deg.

To clarify the change in operating rake and relief angles for alternate (not successive) tool tip locations, Fig. 2 shows the tool tip stationary and the arc of work-center travel moving relative to the toolbit instead of, as in Fig. 1, the toolbit moving relative to the work center. In addition, the ratio of the work radii relative to AB is increased, perhaps beyond practical limits.

The change in operating rake and relief angles is equal to the angle originating at the tool tip and subtending the arc A to A2. It can also be seen that, if the tool tip is positioned as shown at D2 and D3, the change in operating rake and relief angles increases as compared with that for position D.

To minimize the change in operating rake and relief angles occurring during the course of the cut and to equalize operating rake and relief angles at the beginning and end of the cut, it is necessary to calculate side BD, equal to R4, of triangle ABD by Equation (1). Then the change in operating rake and relief can be found with Equation (2).

To calculate the change in operating rake and relief angles for a tool tip located on an arc passing through the center of work, refer to Fig. 3, where A to N is the circular arc of work-center travel relative to the tool tip. The change in rake and relief angles is found with Equation (3). ■

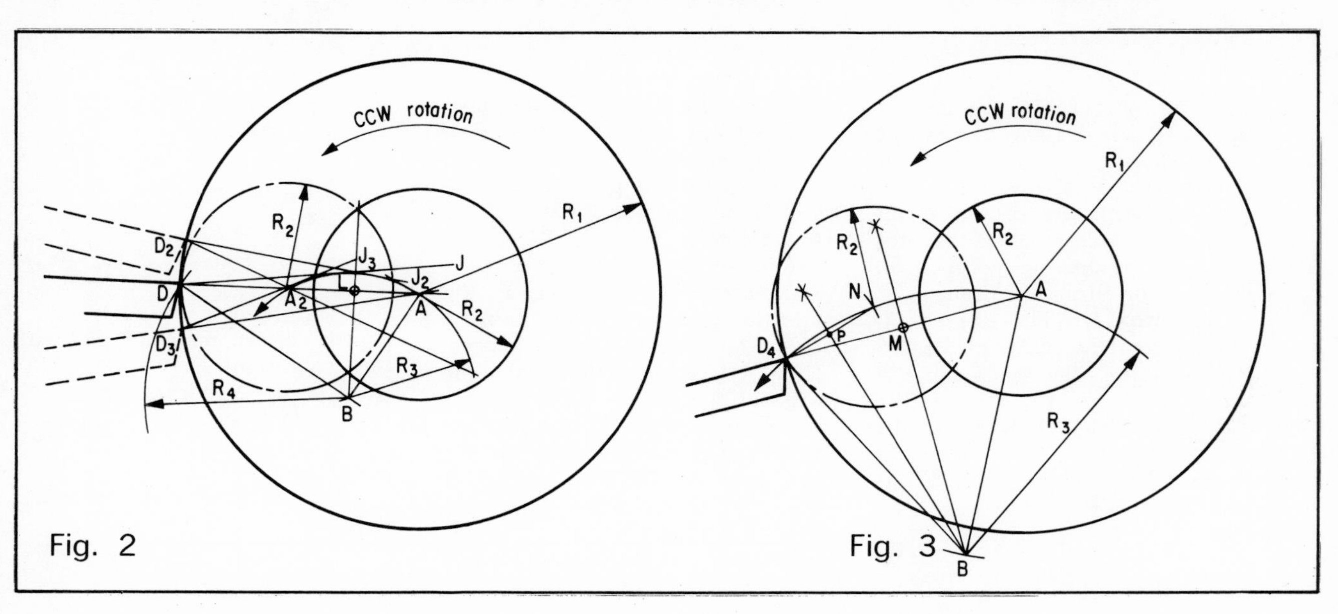

Fig. 2

Fig. 3

Shaving-Tool Corrections

I ZORICH, COLLEGE PARK, SOUTH AUSTRALIA

NOMENCLATURE:

c = step on work
d_1 = smallest work dia shaved
d = any work dia for which tool step is to be corrected
D = dia of supporting roller
R = Distance from work center to roller center
p = tool travel past center
X = depth of tool to be corrected
y = length of cutting edge to form profile
$Z = c - \Delta Z$
ΔZ = vertical tool movement of tool edge, when floating toolholder passes center by distance p
α = top rake angle
β = clearance angle
ψ = angle between tool top rake and a line drawn from intersection of top rake with work diameter to center of work

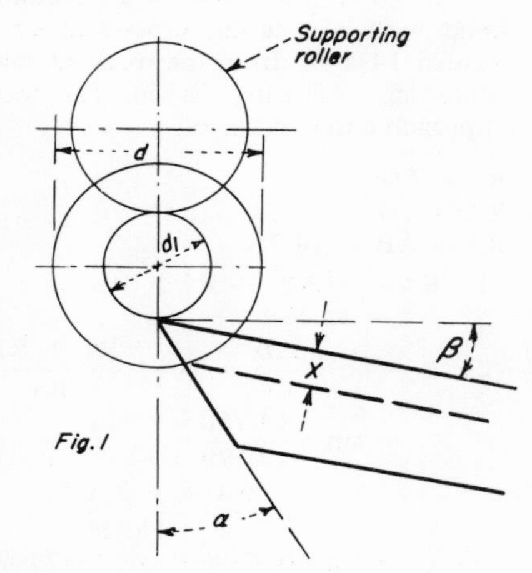

Fig. 1

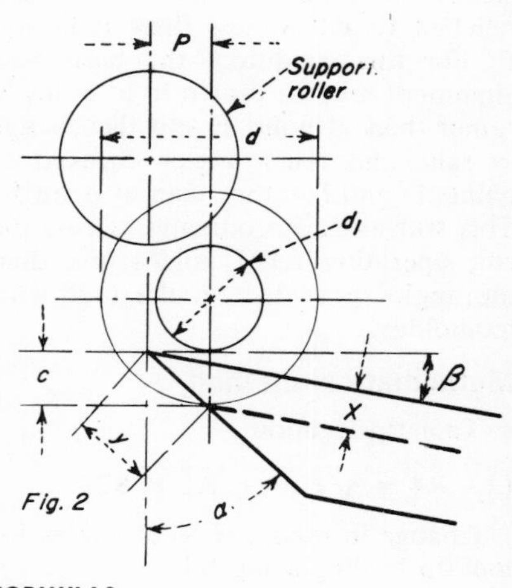

Fig. 2

FORMULAS

For Fig 2

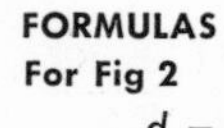

$$c = \frac{d - d_1}{2}$$

$$p = c \times \tan \alpha$$

$$y = \frac{c}{\cos \alpha}$$

$$X = y \times \sin 90° - (\alpha + \beta)$$

$$= y \times \cos (\alpha + \beta)$$

$$= c \times \frac{\cos (\alpha + \beta)}{\cos \alpha} \qquad (3)$$

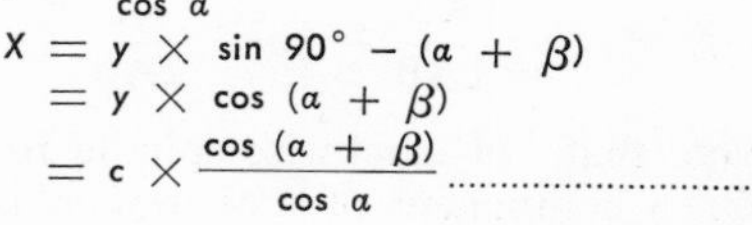

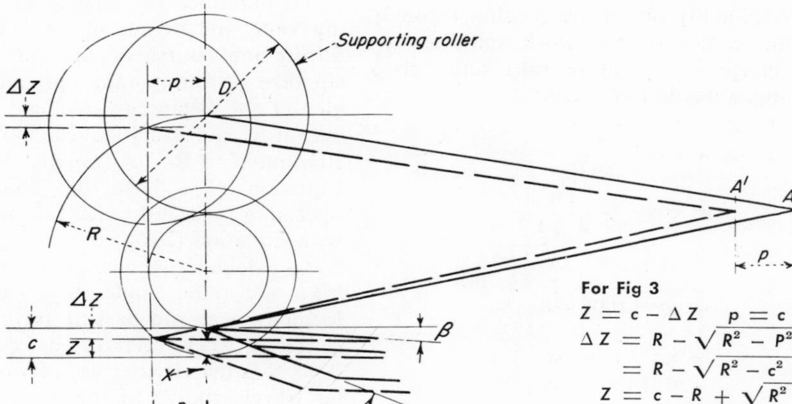

Fig. 3

For Fig 3

$$Z = c - \Delta Z \qquad p = c \times \tan \alpha$$

$$\Delta Z = R - \sqrt{R^2 - p^2}$$

$$= R - \sqrt{R^2 - c^2 (\tan \alpha)^2}$$

$$Z = c - R + \sqrt{R^2 - c^2 (\tan \alpha)^2}$$

$$X = \frac{Z \times \cos (\alpha + \beta)}{\cos \alpha}$$

$$= \left[c - R + \sqrt{R^2 - c^2 (\tan \alpha)^2}\right] \times \frac{\cos (\alpha + \beta)}{\cos \alpha} \qquad (4)$$

Shaving operations are finding increased use on automatics and turret lathes. The edge of the shaving tool is always on a vertical line with the center of the supporting roller. Here are the formulas for shaving-tool tool corrections for three conditions:

CONDITION 1—The edge of the shaving tool and center of the supporting roller just reach the center of the work, Fig 1.

Formulas for tool correction are:

$$X = \frac{d \sin (\alpha - \psi) \cos (\alpha + \beta)}{2 \sin \alpha} \qquad (1)$$

$$\sin \psi = \frac{d_1}{d} \sin \alpha \qquad (2)$$

CONDITION 2—Edge of the shaving tool and centerline of the supporting roller pass the center of the work but the tool step cutting the other diameter just reaches the center of the work, Fig 2. The toolholder is nonfloating.

CONDITION 3—The setup in Fig 3 is the same as in Fig 2, except that the toolholder is floating. Here A is the position when the edge of the shaving tool reaches work center and A' is the position when the center is passed and the tool step reaches the center at the other diameter.

EXAMPLE (Fig 3):

What is the dimension of step X when $c = 0.25$, $R = 2$, $\alpha = 15°$ and $\beta = 5°$?

$$X = \left[0.25 - 2 + \sqrt{4 - 0.0625 (\tan 20°)^2}\right] \times \frac{\cos 20°}{\cos 15°}$$

$$= 0.224$$

Under certain material and cutting conditions:

$$(\tan \alpha)^2 = \text{a constant } K_1$$

$$\frac{\cos (\alpha + \beta)}{\cos \alpha} = \text{constant } K_2$$

Then:

$$X = \left[c - R + \sqrt{R^2 - c^2 \times K_1}\right] \times K_2$$

Surface treatments for cutting tools

Operating details are given for five different methods that are used by the automotive industry and stamping plants to improve the life of various cutting tools

Fine-finish grinding

Under the microscope, a tool sharpened with a 46- to 60-grit wheel has a comparatively rough surface. Slight honing to remove the burr does not create a straight keen edge. But finishing with a 320-grit, shellac- or resinoid-bonded wheel produces a fine finish and sharp, straight edge, if 0.0005 to 0.001 in. of material is removed. The cost of fine sharpening is more than offset by reduction in number of sharpenings required.

Nitriding, single-treatment

Immersion of the tools (fine-finish ground for best results) in a liquid salt bath increases the hardness of hardened HSS tools, provides marked improvement in performance on all cuts, except heavy intermittent cutting of tough, hard metal. Nitriding temperature should not exceed the tempering temperature previously used. Tools should be air cooled after nitriding to avoid cracks.

Maintain the bath (see composition at right) in molten condition at about 1050 F for 12 hr before use. Normal operating temperature is 1025–1050 F. Time in the bath depends on size and type of tool:

¼–⅜-in. tap	12–15 min.
¼–⅜-in. reamer	15–18 min.
¼–⅜-in. thick cutter	15–20 min.
¼–⅜-in drill	20-25 min.
½-in. drill and over	30–40 min.

The work should be clean and dry. Nitride after each grind, because the case is generally less than 0.001 in. deep. Tools that might crack or warp should be preheated to 500-600 F before nitriding.

Nitriding, double-treatment

Forming punches and dies subject to heavy wear but not impact can be improved by using M-2 HSS steel (5% molybdenum, 4 chromium, 2 vanadium and 6 tungsten). Treat in salt as above, but hold in pot for 6 hr. Then place in gas nitriding furnace, purge with ammonia, nitride at 975 F for 24 hr with ammonia dissociation less than 60%. Tool surface may be discolored dull gray, but do not remove, because this surface is most resistant to pickup, scoring, seizing and similar troubles.

Hard chromium plating

Taps, reamers, broaches, cutters, drills, files and dies will give longer service if hard chromium-plated. Fine finish grinding is important prior to plating.

Before plating, clean the work thoroughly to remove oil or grease. Then place the work in the plating bath and allow to come to the bath temperature of 135-140 F. Apply reverse current to the work (work connected as anode) for a time sufficient to produce a suitable etch. On sharp-edged tools for cutting hard metals, the plating thickness should range from a flash up to not over 0.002 in. If necessary to relieve hydrogen embrittlement, heat-treat the tool for ½–¾ hr at 300–400 F.

Batch composition is 40–45 oz CrO_3 and 0.55–0.65 oz H_2SO_4 per U.S. gal. Plating current may vary from 100 to 1000 amp per sq ft.

Steam oxidizing

To produce a thin, black iron-oxide (Fe_3O_4) coating on properly hardened and tempered and ground HSS tools, they are placed in a special furnace containing a circulating fan and fitted with steam inlet and exhaust connections. The coating holds lubricant, resists chip pickup and wear. Preheat tools in an air furnace to 650-750 F to avoid formation of rust. Purge furnace with steam at 2 to 5 in. water pressure for ½ hr. Then raise temperature to or below the tempering temperature for tool. A 2-hr treatment at 1050 F will give about 0.0001 in. coating, a 5-hr treatment a 0.0002-in. coating. If work shows a red or velvet-brown surface, the furnace was not properly purged or the steam contained air.

Nitriding salt for tools

Sodium cyanide	30% min.
Sodium and/or potassium carbonate	25% max
Other active ingredients	4% max
Moisture	2% max
Sulfur	0.5% max
Impurities	1% max
Potassium chloride	Balance
Melting temperature	950 F max

Carbide Boring Bits—I

These data represent application information developed for production use of solid carbide boring bits

By Carl Neiderfringer,
president, The Atrax Co,
Newington, Conn

Speeds and Feeds for Ferrous Materials

Stainless Steels (Average Work)

MATERIAL TO BE CUT	1/16 TO 1/8 CUT .010—.030 FEED		1/64 TO 1/16 CUT .007 TO .020 FEED		.007 TO 1/64 CUT .004 TO .012 FEED	
AISI TYPE	SPEED FPM	CARBIDE GRADE CLASS	SPEED FPM	CARBIDE GRADE CLASS	SPEED FPM	CARBIDE GRADE CLASS
302	140-240	C-7	150-275	C-7	150-300	C-3
303	175-325	C-6	200-375	C-7	200-400	C-8
304	140-240	C-7	150-275	C-7	150-300	C-3
309	140-240	C-7	150-275	C-7	150-300	C-3
316	140-240	C-7	150-275	C-7	100-300	C-3
321	140-240	C-7	150-275	C-7	150-300	C-3
347	140-240	C-7	150-275	C-7	150-300	C-3
403	175-300	C-6	200-350	C-7	200-400	C-8
410	175-300	C-6	200-350	C-7	200-400	C-8
416	175-300	C-6	200-350	C-7	200-400	C-8
420	125-200	C-7	150-225	C-7	150-250	C-3
430	175-300	C-6	200-350	C-7	200-400	C-8
431	140-250	C-7	150-300	C-7	150-350	C-3
440 A, B, C	125-200	C-7	150-225	C-7	150-350	C-3
440 F	125-175	C-7	150-250	C-7	150-350	C-3
446	140-250	C-7	150-350	C-7	150-350	C-3

Steels (Average Work)

MATERIAL TO BE CUT		5/64 TO 5/32 CUT .010—.020 FEED		1/64 TO 1/16 CUT .008—.012 FEED		.005 TO 1/64 CUT .002—.008 FEED	
	S.A.E. No.	SPEED FPM	CARBIDE GRADE CLASS	SPEED FPM	CARBIDE GRADE CLASS	SPEED FPM	CARBIDE GRADE CLASS
CARBON	1010-1025	275-475	C-7	375-675	C-7	500-1500	C-8
	1030-1095	200-400	C-7	300-500	C-7	400-1000	C-8
FREE CUTTING	1112-1120	275-475	C-7	375-625	C-7	500-1500	C-8
	X1314-X1340	250-450	C-7	350-550	C-7	400-1200	C-8
MN.	T1330-T1350	150-350	C-7	250-450	C-7	300-800	C-8
NICKEL	2015-2320	250-450	C-7	350-550	C-7	400-1200	C-8
	2330-2515	200-400	C-7	300-500	C-7	400-1000	C-8
NICKEL	3115-3140	200-400	C-7	300-500	C-7	400-1000	C-8
CHROME	3145-3450	150-350	C-7	250-450	C-7	300-800	C-8
MO.	4130-4820	150-350	C-7	250-450	C-7	300-800	C-8
CR.	5115-52100	150-350	C-7	250-450	C-7	300-800	C-8
CR. V.	6115-6195	150-350	C-7	250-450	C-7	300-800	C-8
CAST STEEL		200-400	C-7	300-500	C-7	300-800	C-8

Irons (Average Work)

MATERIAL TO BE CUT		5/64 TO 5/32 CUT .010—.020 FEED		1/64 TO 1/16 CUT .008—.012 FEED		.005 TO 1/64 CUT .002—.008 FEED	
		SPEED FPM	CARBIDE GRADE CLASS	SPEED FPM	CARBIDE GRADE CLASS	SPEED FPM	CARBIDE GRADE CLASS
CAST IRON	HARD NO ALLOY	225	C-2	250	C-2	275	C-4
	MEDIUM NO ALLOY	250	C-2	275	C-2	300	C-4
	SOFT NO ALLOY	275	C-2	300	C-2	325	C-4
	HARD ALLOY	200	C-2	225	C-2	250	C-4
	MEDIUM ALLOY	225	C-2	250	C-2	275	C-4
	SOFT ALLOY	250	C-2	275	C-2	300	C-4
	CHILLED ROLLS	20	C-1	20	C-2	25	C-2
SEMI STEEL	UP TO 25% SEMI STEEL	250	C-2	275	C-2	300	C-4
	OVER 25% SEMI STEEL	200	C-2	225	C-2	250	C-4
MAL IRON	HARD	200	C-2	225	C-2	250	C-2
	MEDIUM	225	C-2	250	C-2	275	C-2
	SOFT	250	C-2	275	C-2	300	C-2

Carbide Boring Bits—II

Speeds and Feeds for Non-Ferrous and Non-Metallic Materials

MATERIAL TO BE CUT		5/64 TO 5/32 CUT .010—.020 FEED		1/64 TO 1/16 CUT .008—.012 FEED		.005 TO 1/64 CUT .002—.008 FEED	
		SPEED FPM	CARBIDE GRADE CLASS	SPEED FPM	CARBIDE GRADE CLASS	SPEED FPM	CARBIDE GRADE CLASS
BRASS & BRONZE	HARD	150-350	C-2	200-400	C-2	300-600	C-2
	SOFT	200-400	C-2	300-500	C-2	350-1000	C-2
ALUMINUM	CASTING	250-600	C-2	300-1000	C-2	400-1500	C-2
	BAR STOCK	250-600	C-2	300-1000	C-2	400-1500	C-2
ZINC ALLOYS		250-500	C-2	300-600	C-2	400-1000	C-2
RUBBER	HARD	300-500	C-2	350-600	C-2	400-800	C-3
	SOFT	400-800	C-2	500-1000	C-2	600-1200	C-3
COPPER		200-500	C-2	300-600	C-2	400-1000	C-2
FIBRE		300-500	C-2	300-800	C-2	350-1000	C-2
PLASTICS		300-800	C-2	400-1000	C-2	500-1500	C-2
MONEL METAL		200-275	C-6	200-300	C-6	225-325	C-6

Selection of Chart of Relief and Rake Angles

The angles given are intended to apply to jobs where the tool shank is set perpendicular to the work with the cutting edge on the centerline.

Material	Side Rake (Depth of cut less than ¼)	Back Rake	Front Relief	Side Relief
SOFT STEEL UNDER 200 BRINELL	8	0	7	7
MEDIUM SOFT STEEL 200-275 BRINELL	8	0	7	7
MEDIUM STEEL 276-350 BRINELL	8	0	7	7
MEDIUM HARD STEEL 351-425 BRINELL	6	0	7	7
HARD STEEL OVER 425 BRINELL	4	0	5	5
CAST IRON	8	0	7	7
BRASS & BRONZE	8	0	7	7
ALUMINUM	8	0	7	7
DIE CASTINGS	8	0	7	7
HARD RUBBER	8	10	7	7
SOFT RUBBER	15	15	10	10
COPPER	15	0	7	7
FIBER	10	0	7	7
PLASTICS	10	0	7	7

"Blending Angles" for Precision Boring Tools

F Caplan, Kaiser Engineers, Oakland, Calif

In precision boring, clearance angles vary with bore diameter, work material, and the height of the tool above center. When a toolbit with side angles L is ground with bore-clearance-angle C, the side-clearance angle $F = \tan^{-1}$ ($\cos L$ times $\tan C$). This normal relief or blending angle is the angle to which the tool is tilted during lapping. Reference: New American Machinist's Handbook, p 12-61.

Example: What is the side-clearance angle F if the tool has side angles $L = 10°$, and the bore clearance angle $C = 18°$.

Solution: Align $C = 18°$ with $L = 10°$ and read $F = 17° 45'$.

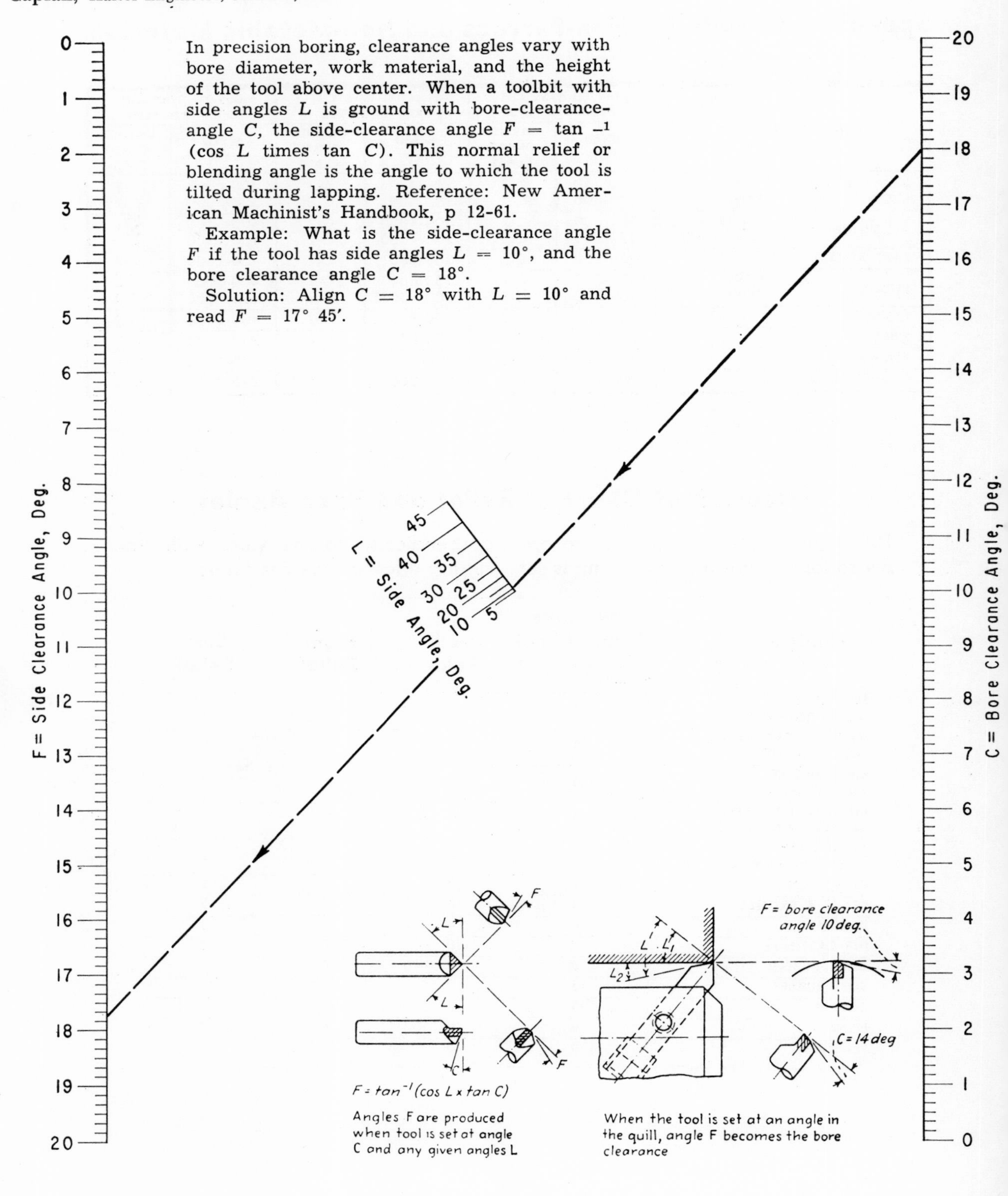

Chapter 10

Screw-machine Work

Design of
Screw-machine products, 1

If designers of screw machine products will follow practical 'standards' for certain features, they won't affect function but will save plenty

There are several aspects of drafting practice that need special attention when screw-machine products are involved. These details are often neglected, or the parts designer does not understand what is practical. From long association with such problems, we recommend the following "rules."

1. Sharp corners and corner breaks

External corners. Never specify "Sharp Corners." These do not exist. Corners must be dimensioned. (See sketches in Fig. 1.) A drawing note that is practical, economical, and permits easy machining at minimum cost is: "break corners 0.010 in. x 45°." If sharper corners are needed, specify them; for example, "0.002-in. R max." This is as close to a sharp corner as can be obtained. Remember that to achieve this optimum sharpness, considerable cost is incurred.

By S A Cappon, Manager
Production Methods Div,
National Acme Co, and
Herbert A Eichstaedt
Technical co-ordinator
National Screw Machine Products
Association, Cleveland

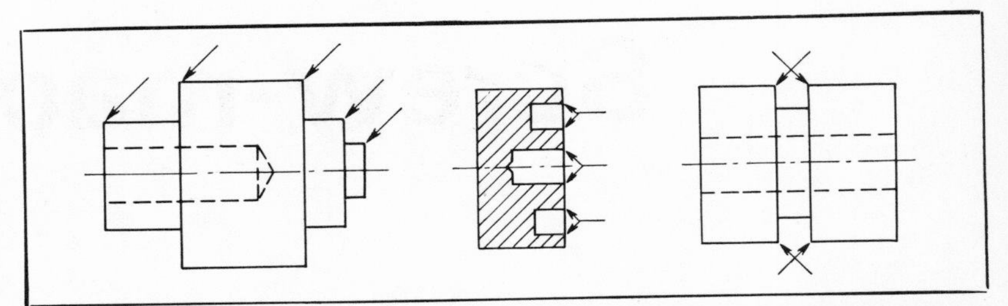

Fig. 1. Corners should be dimensioned as liberally as possible. A practical drawing note is: "break corners 0.010 in. by 45°"

Breaking a corner as suggested ".010 in. x 45°" is sometimes expressed as "max. R allowed:" This is costlier than an actual chamfer, because it is necessary to blend the radius with a diameter or a face.

The sharp-corner condition in Fig. 2 A must also be dimensioned. Although an external feature, it is equivalent to an internal corner. A practical dimension is: "0.010 in. x 45°", or possibly "max. 0.010R".

Use an undercut if possible. Thus, a theoretically sharp corner is achieved. This feature is necessary when the smaller diameter must be ground and

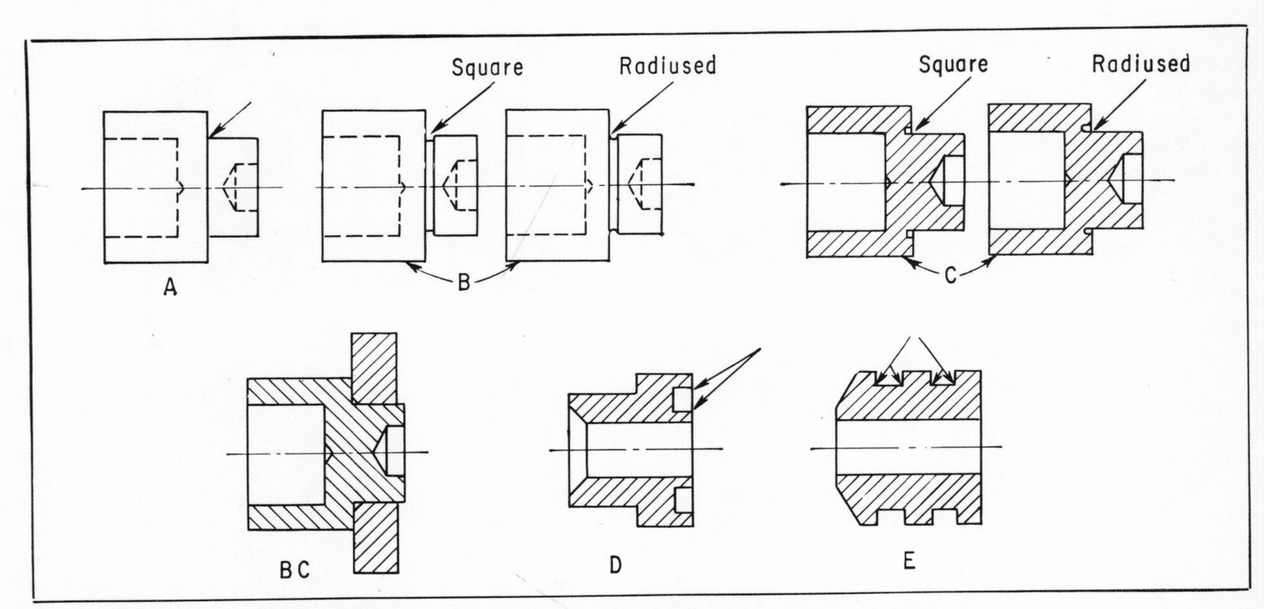

Fig. 2. Undercuts are often given a radius to avoid crack development in heat treatment

Design of
Screw-machine products, 2

grinding relief is required, or a mating part must fit flush against the shoulder face. The relief shown in Fig. 2 B is easy to machine, because it can be established by a cut-off blade or flat tool bit from the cross-slide, which is easier than Fig. 2 C. Dimensioning of such an undercut for a minimum cost is: "0.062 in. wide x 0.015 in. deep."

A square-corner undercut, Fig. 2 B or Fig. 2 C, is sometimes changed to a radiused undercut, particularly if there is concern about crack development after the product undergoes heat treatment.

The condition in Fig. 2 D has only one solution. To achieve minimum cost, the dimensions for such a sharp corner should not be less than "0.008 in. x 45°," or 0.008 in. R. Anything smaller results in excessive tool sharpening and downtime.

In the cases shown in Figs. 2 B and 2 C, if there is an assembly problem with a mating part that has to be flush with the shoulder face, an internal chamfer on the mating part will overcome some of the problems.

Internal corners. Here again there's no such thing as a sharp corner. The most economical dimension for the sketches that are shown in Fig. 3 is not less than 0.010 in. x 45°.

The condition shown in Fig. 3 B can be resolved from an economical standpoint with a dimension not less than 0.010 in. x drill-point angle (standard 118° or special angle).

Further variations of this condition are shown in Fig. 3 C and 3 D. For an economical dimensioning 0.010 in. x 45° is recommended. To establish Fig. C 3, a recessing tool with the above corner break can easily be ground. For Fig. 3 D trepanning blades ground in the same manner will achieve these results (0.010 in. x 45°).

If a sharper corner is required for Fig. 3 C, a blade-type tool will be needed. Both the tool and a reamer will require frequent regrinding, but a 0.003 in. R can be obtained if necessary. Similarly, for Fig. 3 D blade-type tools and end-cutting counterbore-reamers can be used.

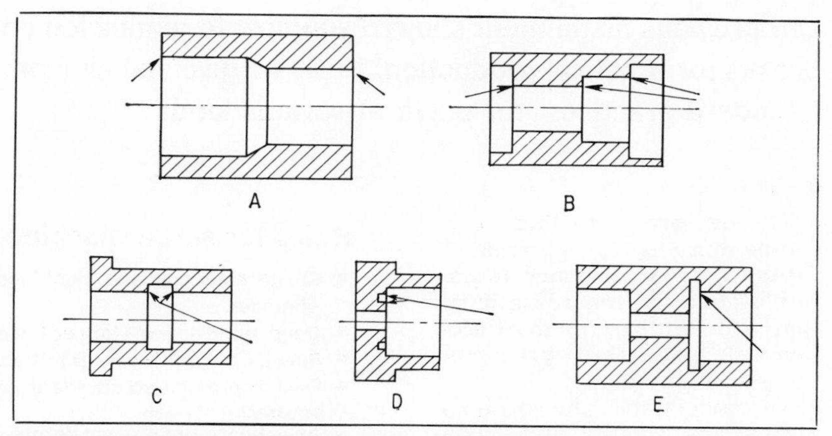

Fig. 3. Internal corners should be dimensioned "0.010 in. by 45°." In examples C, D and E, a side wall angle in the recesses of ½ to 2° is helpful for tool clearance. The same applies to Fig. 4 B and C

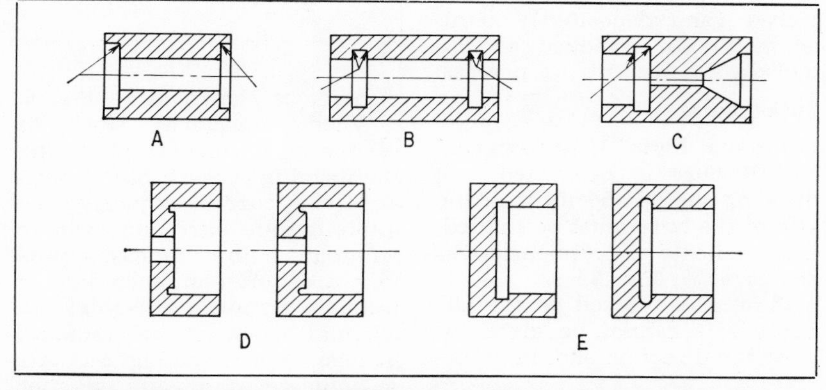

Fig. 4. Corners may be dimensioned as an 0.010 in. fillet or radius, or 0.010 in. x 45°. Radius under cuts, examples D and E, are preferable to those with a square corner

In the cases shown in Fig. 3 C and 3 D, a ½° to 2° angle is recommended for the sides of the recesses. This provides a natural tool relief at withdrawal.

If corner sharpness requirements are 0.005 in. or less in Fig. 4 A, use one of the alternatives, Fig. 4 D or Fig. 4 E. The dimensioning of such undercuts for practical and economical reasons should be "0.062 in. wide x 0.015 in. deep."

The radius-type undercut is more desirable and depends on what other operations are required on the part. However, the drawing should specify 0.062 in. wide x 0.015 in. deep and allow optional selection of a square corner or a radius relief.

The conditions shown so far should not be considered as "undercuts." They are merely "reliefs" to satisfy "sharp corner" situations. Although like undercuts in nature, they are not true undercuts.

Design of

Screw-machine products, 3

Our previous installment showed you how to dimension corner breaks for practical production. Now, we give you similar 'standard practices' for burrs of various kinds

2. Burrs

Through proper tooling, burrs can be minimized or eliminated. Certainly it is necessary to apply "standards" regarding burrs that will result in the most economical method of producing the screw product.

To understand the problem better, we should first know something about the kinds of burrs produced, and their nature. We can consider a burr, except the external cut-off type, to be nothing more than displaced (and deposited) metal, or metal not removed by the tool during the cutting process.

Rules for screw-machine part design

- Cut threads can't be machined closer than 2 threads from a shoulder.
- Deep blind holes (tapped) require space for chips.
- Specify cut off burrs. Don't say "Small cut off burr permitted."
- Design product so standard bars, shapers, and tolerances can be used.
- Investigate pre-shaped forms.
- Use standard threads and holes. Avoid special tooling expenses.
- Extra close tolerances are expensive. Diameters are held closer than lengths.
- Specify materials with maximum machinability characteristics.
- Specify corner breaks and radii.

Cut-off burrs

External burrs. If an external cut-off burr is permitted, the drawing should specify it. If the size of the burr must be limited, then indicate the dimension as shown at A, Fig. 5.

If burr is allowed on the OD but a size cannot be given, at least the direction should be indicated, as shown at B, Fig. 5.

When hexagonal stock is used, a burr is generated during the forming process, but removal of this burr is expensive. To minimize the burr and even to eliminate it in some cases, provide a chamfer on each side of the part section that retains the original stock shape (square, hexagon, etc.). See Fig. 5 C. Also, indicate specifically whether these burrs are allowed or not.

Generally speaking, to avoid generating external burrs it is necessary to provide chamfers or radiuses whenever you have intersecting surfaces.

Internal burrs. Such defects are generated when cutting off into a hole. As shown in A, Fig. 6, a burr will always be generated at the point shown. This defect is detrimental if the part must slide over a mating part. Also, gaging will be difficult.

If permitted from a functional standpoint, allow a generous chamfer as at B, Fig. 6. This feature can be established through a recessing operation before the part is cut off. In such cases a burr will be generated but it cannot be detrimental to assembly or gaging. And if this is the case, specify on the drawing that a burr is allowed.

Another application of the above idea occurs when the part must be cut off into a tapped

By S A Cappon, Manager
Production Methods Div,
National Acme Co, and
Herbert A Eichstaedt
Technical Co-ordinator
National Screw Machine Products
Association, Cleveland

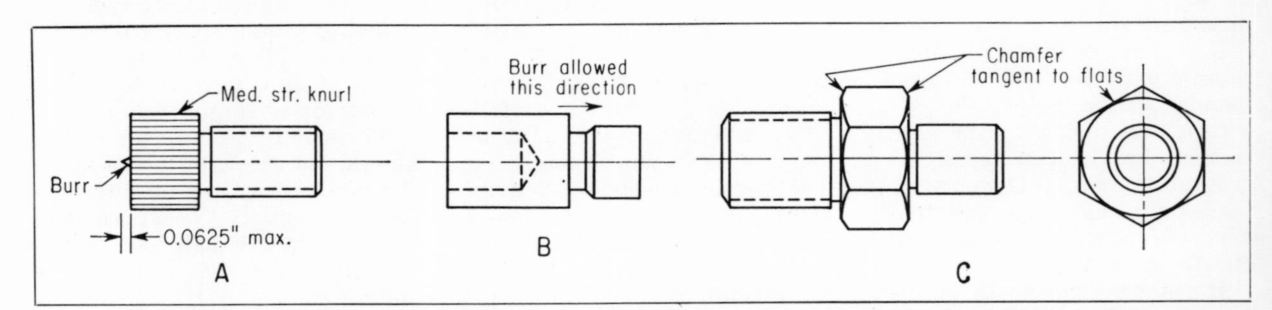

Fig. 5. External burrs should be dimensioned, as at A, controlled in direction as at B, or removed by chamfering

Design of
Screw-machine products, 4

hole. In addition to a chamfer for relocation of an acceptable burr, it is also important to allow for an ample recess for chip accumulation.

Intersecting surfaces 1, as at Fig. 6 C, will generate burrs. If such burrs are permitted, define them accurately on the drawing. Also clarify the permitted direction of the burr. Such instructions facilitate the method of production. Or, if permissible, avoid square shoulders on points 1.

Secondary operations like cross drilling, milling, tapping, etc. produce heavy burrs. Fig. 7 A shows a typical case.

If a slot is milled, Fig. 7 B, and a burr is permitted on the OD, that fact should be specified. Then the shop will produce the part with either a cutter rotating towards the outside, or will try to have a milling cutter travel from the center of the part to the outside (if a radius is not allowed on the bottom of the slot).

If a burr is not allowed on the OD or only a very small one, and none on the inside, offer an optional V-groove at the bottom of the slot, Fig. 7 C. This groove permits accumulation of a burr, while maintaining a clean OD, avoiding another operation.

In fact, there are some special cases. For example:

1. Avoid performing any operation on a threaded area. In designing a part, this can sometimes be avoided. See Fig. 7 D.

2. In broaching a blind hole, heavy burrs are generated. Provide enough depth for the hole prior to broaching. Otherwise, production of the part will involve excessive tool cost. ■

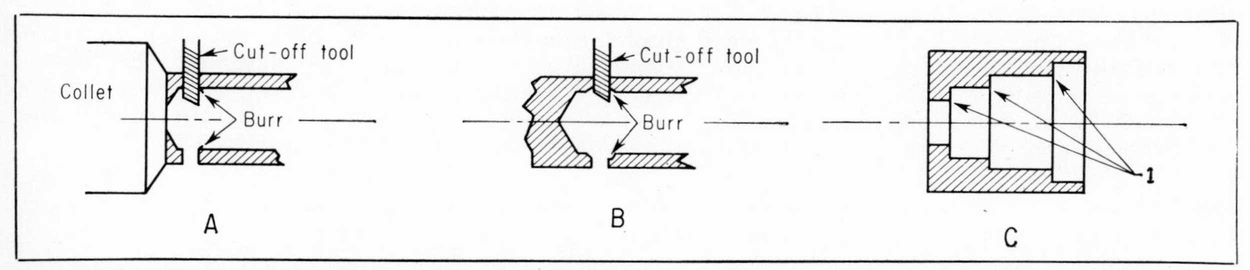

Fig. 6. Internal burrs can be minimized by forethought to such matters as tool shape or providing recesses or radii

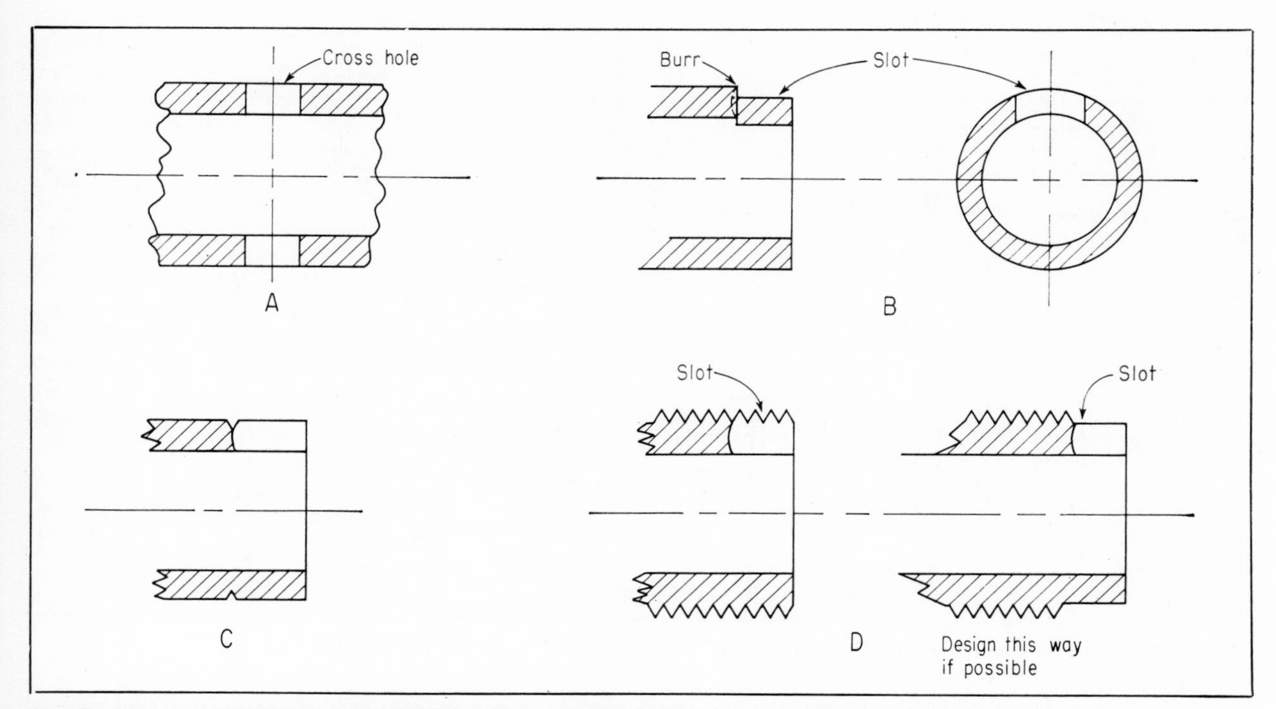

Fig. 7. Slots and crossholes can produce heavy unwanted burrs. A groove as at C or a turned down section as at D will avoid burrs that interfere with assembly

Design of

Screw-machine products, 5

The previous installment showed you how to handle burrs for various parts. Here we give you practical information on another area important to cost reduction

4. Dimensioning and finish

In addition to the standards established by large companies and associations, there are a few points that need emphasis:

(a) In dimensioning a screw product be guided by the probable gaging problems. As a rule, if the part can be easily gaged, it can also be easily manufactured.

(b) Use decimal dimensions. Standardize the block reference for tolerances as follows:

Two decimals	±0.010 in.
Three decimals	±0.005 in.
Angle dim.	±1°
Finish spec.	125 mu-in. RMS

Unless otherwise specified

(c) 'Unless otherwise specified'. Call out closer tolerances only when absolutely necessary. The most economical way is to complete the part in one operation. In this case, normal inside and outside operations can qualify the part to ±0.002 in. on diameters, but not recesses or total length. Any closer limits may require second operations, and this could make the cost of the product excessive.

Surface finish should be realistically related to function. If the finish specified is 125 microinches RMS, the screw-products manufacturer must tool up for 63 microinches RMS (or one-half the spec). Also think in terms of gaging the part or checking the finish. Sketch A in Fig. 8 shows an impossible condition, both internally and externally.

By S A Cappon, manager
Production Methods Div
National Acme Co
and **Herbert A Eichstaedt**
technical co-ordinator
National Screw Machine Products Association, Cleveland

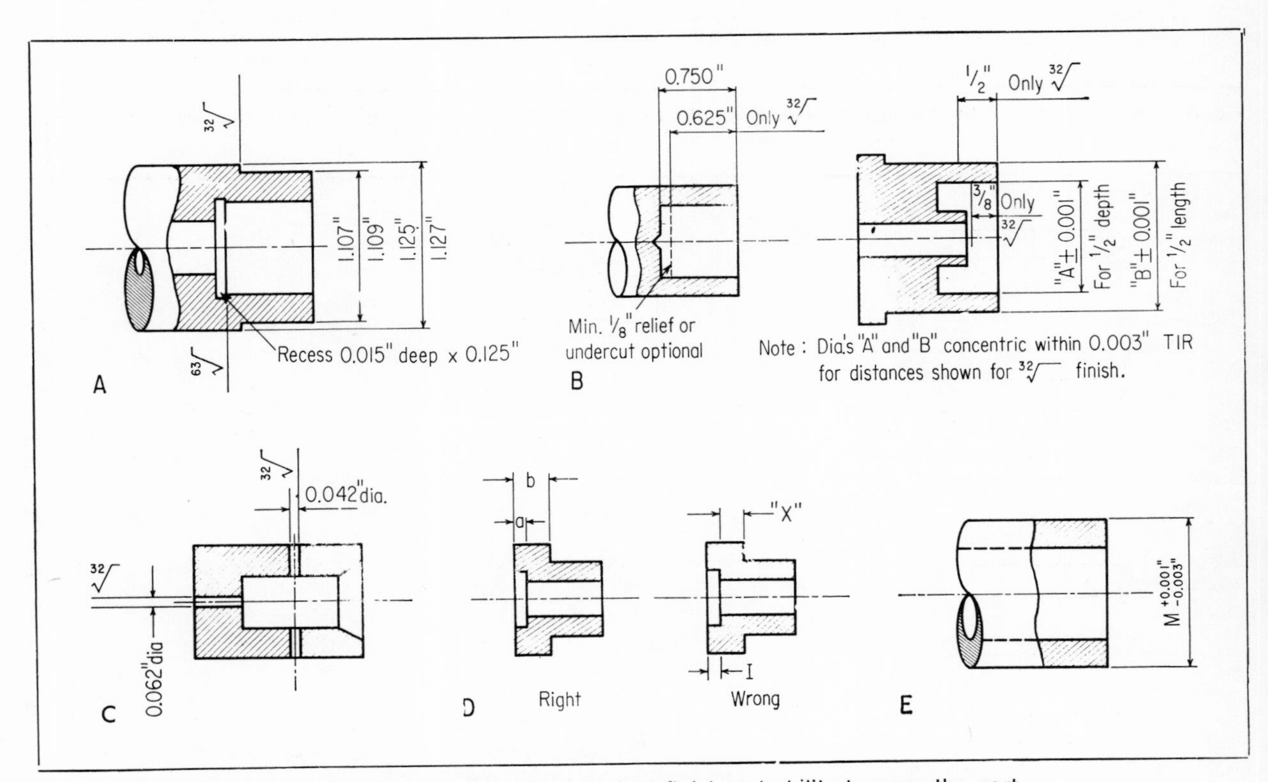

Fig. 8. Dimensioning is important to production of desired finish and ability to gage the part

Design of

Screw-machine products, 6

If a hole is blind, the specified finish cannot go all the way to the bottom. The first of the sketches at B, Fig. 8, shows a suitable specification. When special finish is required for a specific length, that length should be dimensioned. The same considerations apply to special close tolerances on diameter or concentricity.

Another situation difficult to measure and to produce is shown at C, Fig. 8. By sectioning a part one can measure the finish, but the result is not conclusive, nor is it accurate, for all parts in a batch or run.

(d) 'Special situations'. These should be avoided. In most cases, dimensioning can be done so that internal features are not tied together. Sketches at D, Fig. 8 show the right and wrong ways. By proper balancing of limits for dimensions *a* and *b*, one can determine what is required for dimension *X*.

Sometimes close limits must be held in one direction but not in the other (See Sketch E, Fig. 8). On the plus side only 0.001 in. from the nominal can be used, but on the minus side one can take advantage of 0.003 in. This is important, for it more or less indicates the functional aspect of dimension M.

If the part is to be plated or heat treated, or both, then the dimensions that are specified must be identified. The best method of dimensioning is to show dimensions "before" and "after."

In the case of a tapered part (Fig. 9), one of the circled dimensions must be omitted. For ease in gaging, it is preferable to leave out angle V. Otherwise, be guided by functional aspects.

(e) Concentricity between various diameters (internal and external) can be expected within 0.004 in. TIR, except when it relates to stock diameter.

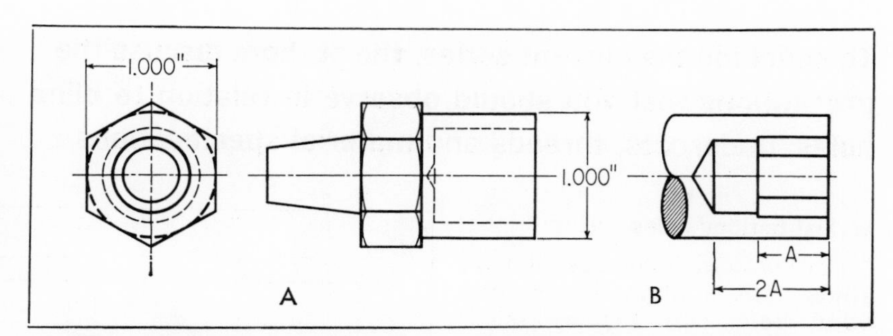

Fig. 10. The 1.000-in. dimension may not clean up to a cylinder. (left). Always allow sufficient chip clearance when broaching a hole

Closer limits can be obtained if necessary.

The degree of concentricity, perpendicularity, and parallelism required of the part should be amply described as a note rather than by symbols.

(f) Miscellaneous. When a part is made from hex or any multilateral stock, beware of an outside diameter that must be equal to the distance between opposite flats (See Sketch A, Fig. 10). Because of stock tolerances between flats, this diameter may not come out as a continuous circle at all points.

A broached hexagonal hole is established by first drilling a hole of a specific depth. In the case of a blind hole, allow enough extra depth for chips or burrs (Sketch B, Fig. 10). As a rule of thumb, allow additional depth equal to the broached depth. The diameter of the drilled hole should leave only minimum stock.

Do not specify "no concavity or convexity allowed" on a drawing. Give a specific figure: "Concavity (or convexity) to be within 0.003 in.

Tool clearance

On internal or external surfaces, where a straight shoulder or face must be established, tool clearance is important, particularly at withdrawal. In such cases the external face or the

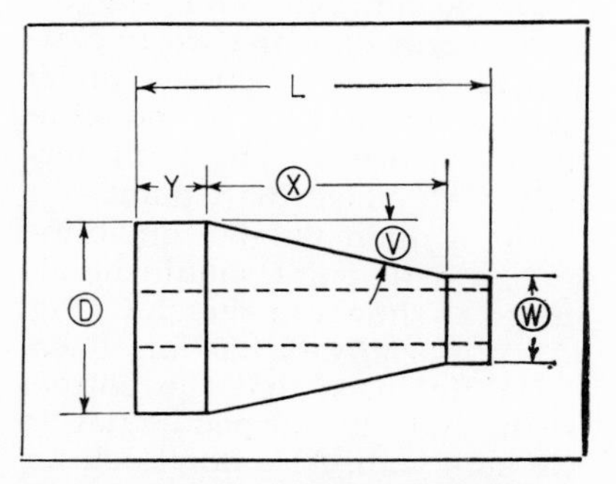

Fig. 9. Taper is customarily written as the difference between diameters divided by the distance between them, or 'total taper.' Taper can also be written as the angle measured from the centerline

sides of a recess should be dimensioned so that a ½ to 2° clearance angle is provided.

This will achieve two things. First, the tool will have a natural clearance, so that at the end of its travel and at the moment of withdrawal it is completely relieved from the work. And toolmarks are avoided.

When long operations, like facing, cutoff, or deep-hole drilling, are required, indicate by notes whether steps are permitted. If steps will not interfere with the function of the part, mention the dimensions of the steps in the notes or put them on the print. A step of 0.005 in. is sometimes beneficial.

Design of

Screw-machine products, 7

To conclude the current series, the authors discuss the precautions that you should observe in relation to blind holes, undercuts, threads and material specifications

5. Flat-bottom holes

Flat-bottom holes are of two kinds: (1) a plain flat-bottom blind hole, and (2) a 2-step hole with a squared corner on the bottom of larger hole, as shown at A and B, Fig. 11.

The conditions and suggestions regarding the corner itself have been discussed under "sharp corners." If possible, however, use an angle, preferably a standard drill-point.

For a plain flat-bottom blind hole, a drill point should be allowed as shown in sketch C, Fig. 11, preferably 0.125 in. deep.

Do not insist that the bottom of the hole be perpendicular to the hole wall. Also don't ask for excessively fine surface finish.

Limit tapping depth and surface-finish requirements to a set distance from the bottom of the hole. A recess solves this problem. If the hole is tapped the recess should be equivalent to two threads. See Sketch D, Fig. 11. But a recess is not the "final answer." Another solution is shown in Sketch E, Fig. 11, but here the distance to the bottom of the hole should be 2½ threads at least.

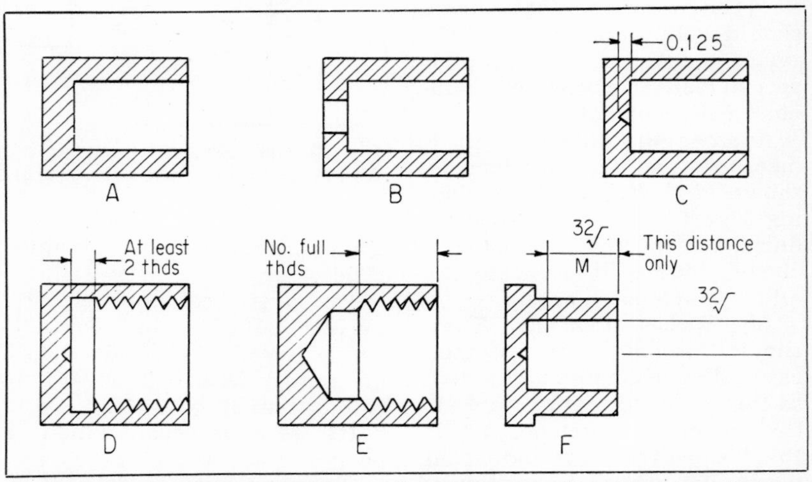

Fig. 11. Blind holes should allow a drill point, preferably at least 0.125 in. If the hole must be tapped a recess two threads wide minimizes important burrs. If a drilled hole is tapped, E, show the number of usable threads

By S A Cappon, Manager
Production Methods Div,
National Acme Co, and
Herbert A Eichstaedt
Technical Co-ordinator
National Screw Machine Products
Association, Cleveland

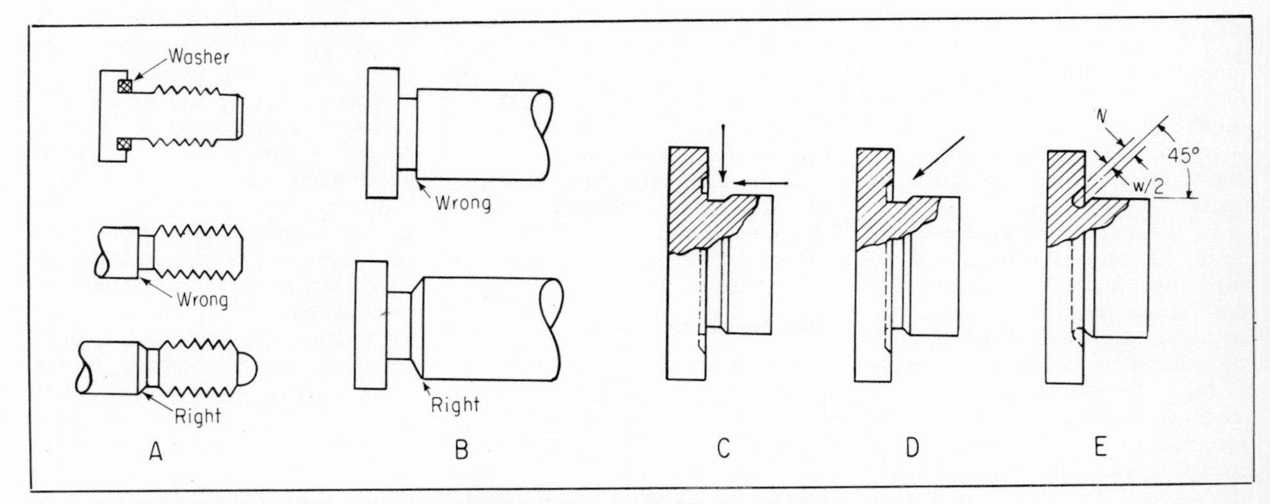

Fig. 12. Undercuts are used for many purposes. Use a chamfer as shown at B. Angular undercuts D and E should be avoided if possible

Design of
Screw-machine products, 8

If a special surface finish is required, see "Dimensioning." Sketch F shows depth M for the required surface finish, starting at least ⅛ in. from the bottom. If this much relief interferes with the function of the part, allow at least 0.032 in.

6. Undercuts

Undercuts are used for a large variety of purposes:

(a) Thread relief (internal or external)

(b) As a substitute for a sharp corner (internal or external)

(c) Broach relief

(d) Grinding relief

(e) Assembly relief

(f) Burnishing relief

Also, some undercuts are used in special applications. Examples: (1) an undercut for a packing or flue washer, Sketch A, Fig. 12, or (2) an undercut required in bearing races for shield grooves.

It is difficult to standardize dimensions for undercuts, except in the case of a thread (internal or external). Here the width of the undercut equals 2 to 2½ threads, and depth equals the depth of the thread at least. Whenever possible, specify a chamfered undercut rather than a straight one. (See B, Fig. 12).

An angular undercut, internal or external, is difficult to produce and should be avoided whenever possible. Usually when the need for such an undercut exists, a relief is required in both directions.

A typical undercut in both directions is shown at C, Fig. 12. This undercut can be established with two separate tools.

Sketch D, however, shows an angular undercut that must be machined with a special attachment, and sometimes obtained as a second operation. If a second operation is used, a burr will develop. Thus additional cost is incurred.

If an angular undercut is necessary due to some functional requirement of the part, then apply it as shown in Sketch E. Give the 45° angle, and specify width W and the maximum or minimum depth required.

If an undercut is primarily a manufacturing aid, or relief, the only one that must have a specific dimension is thread relief (external or internal). All others can vary in size, type and direction. Therefore they can be left entirely to the discretion of the screw-products manufacturer by an option on the drawing.

Any limitations as to size, kind and direction should be stated on the drawing. Also, if a relief or undercut is permitted, it should not be less than 0.032 in. wide by 0.005 in. deep.

7. Materials

There are several aspects of materials that must be specified on the print:

(a) Analysis. Obviously the analysis of the product material is closely related to its function or the methods of manufacture. You should note whether the material should be of welding quality, or cold forming quality if it is to be crimped or heat-treated if certain strength requirements are mandatory.

(b) Specify stock size. For example, if you specify a diameter equal to stock size, you let the supplier know that you will accept standard stock surface imperfection.

(c) Imperfections. If on sulphurized steels you require the part to be free of surface seams, pipes or impurities, then so specify. In this case larger stock will be required to allow removal of the layer of stock containing these blemishes.

(d) End use. In the case of stainless steel parts, indicate on print the end use of part. Thus the supplier is advised that proper annealing may be required. In case of aluminum it is important to specify the end use of the part. This will indicate what analysis and temper should be supplied.

(e) Stock shape. If the geometry of the part will allow use of tubing, specify if tubing is objectionable.

(f) Material treatment. Be specific in the case of carbon or alloy steels to state whether use of HR material is objectionable.

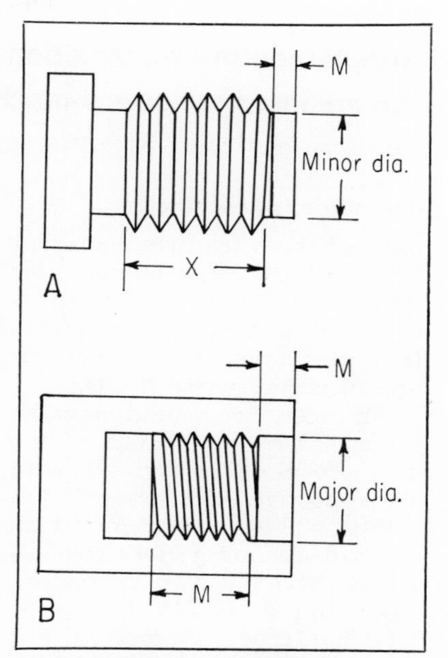

Fig. 13. Always dimension thread length X. The pilot or clearance M may be marred by the threading tool and is less than the minor diameter or the major diameter

8. Threads

The usable length of thread should be dimensioned in all cases, unless the end of the thread runs into an undercut.

Sometimes a pilot is provided ahead of the thread (internally or externally). In this case, make sure that either the pilot diameter M is less than the root diameter of the thread or specify on print that area M can be marred by the threading tool. See sketches A and B, Fig. 13.

Design of

Screw-machine products, 9

Supplementary information has been developed on dimensioning of screw-machine parts.

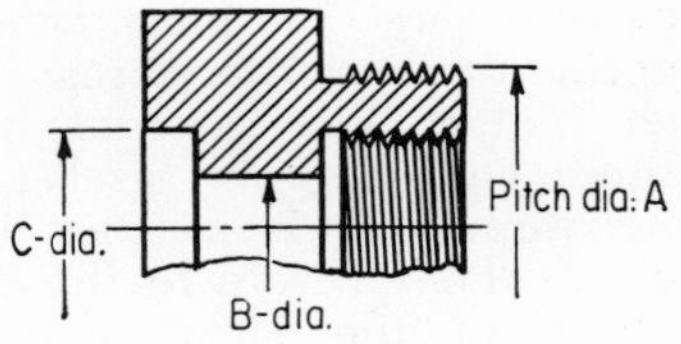

Concentricity-Eccentricity

Both of these features represent relationships between circular surfaces of a part: A concentricity notation designates surfaces that must have a common center within certain close limits.

An eccentricity notation is the distance between centers of one or more circular surfaces with its respective tolerance.

Fig. 14 shows a part for which we will assume a close concentricity between two diameters, say within 0.004 TIR. Let us relate counterbore diameter *C* and pitch diameter *A* in this manner. For good reasons, this part will be produced on an automatic with the threaded end out. A second operation will be required to machine the counterbore diameter C, using diameter B as a pilot. Obviously the concentricity specifications will be met easily in this case. In fact, any screw machine should maintain a 0.004 in. TIR between circular surfaces on parts with a 1:1 ratio of diameter and length. And special tooling will hold a closer concentricity limit.

Often a concentricity requirement does not extend for the entire length of a surface. This situation is seldom noted on the part-drawing as it should be. A suitable method is shown in Fig. 15. Possibly a note should be added to identify that X is being related to the concentricity between diameters A and B.

A drawing properly dimensioned with respect to such features provides an understanding of the functional characteristics of the part. In turn, functional aspects govern the manufacturing method and tooling.

Zero concentricity can be obtained between diameters A and B, Fig. 16. For this job it is possible to use a counterbalanced floating reamer holder with an overhanging turning tool and holder to cut diameter B.

Proper dimensioning of surfaces perpendicular to the part

By S A Cappon, Manager
Production Methods Div,
National Acme Co, Cleveland

Fig. 14

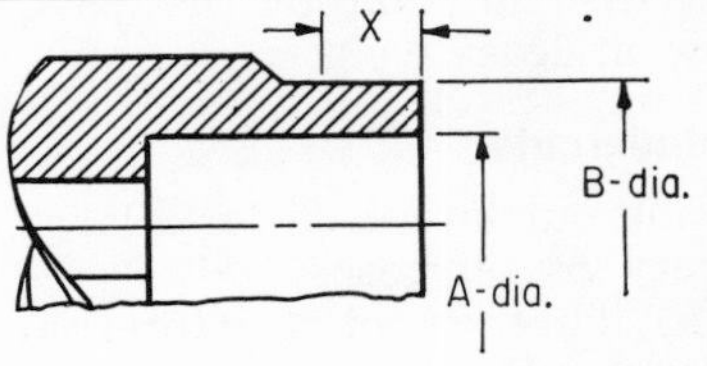

Fig. 15

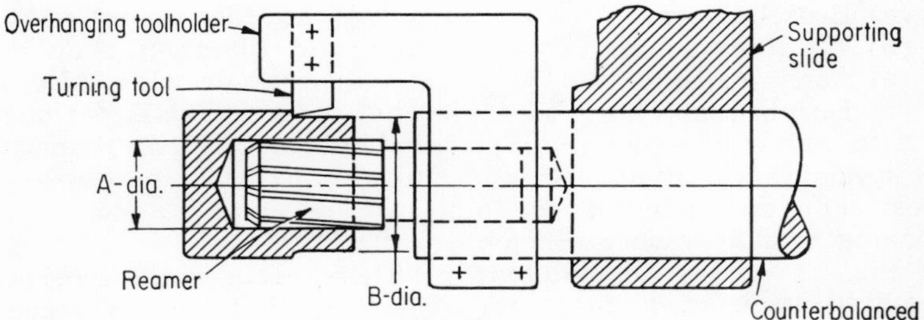

Fig. 16

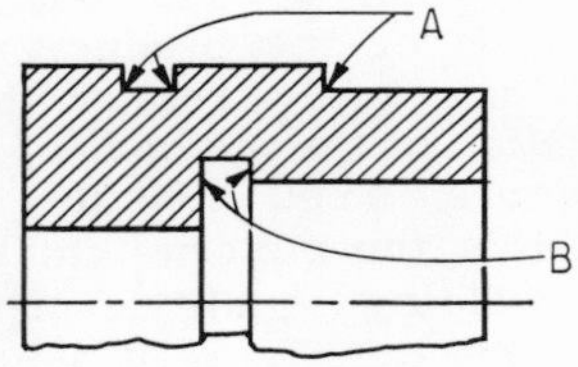

Fig. 17

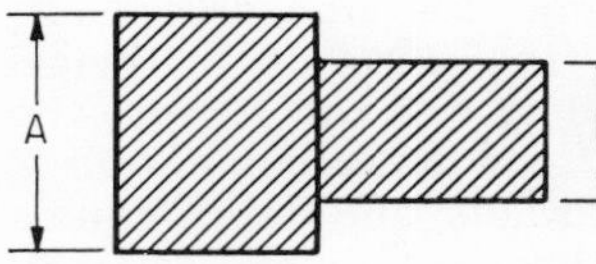

Fig. 18

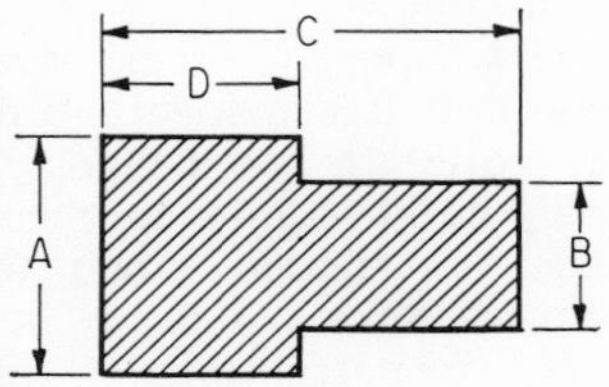

Fig. 19

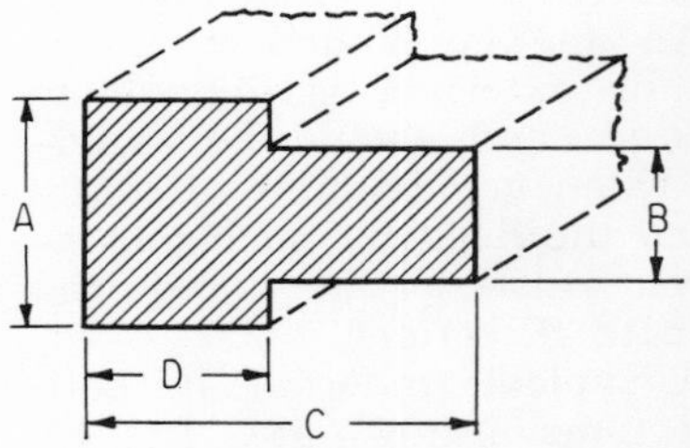

Fig. 20

Design of

Screw-machine products, 10

axis represents a chance for real economies. If perpendicularity is not functional, as at A and B, Fig. 17, consider allowing a slight angle, say ½° at least. Thus, a natural tool clearance is provided and surfaces A and B are established with minimum tool wear and are entirely unmarred. If, however, a specific perpendicularity is needed, then it should be so specified, even if a second operation appears essential.

Centerlines

It is often difficult to determine whether a dimension is a diameter or not. For example, in Fig. 18, if the note "dia" does not appear alongside of dimension A, it will be difficult to prove that A and B are truly diameters. A drawing should not leave anything to interpretation. Proper dimensioning avoids ambiguities and misunderstandings.

Perhaps the customary interpretation of Fig. 19 might be that A and B are diameters while C and D are lengths. However, a second possible interpretation is that the sketch represents a cross section of an extruded shape, Fig. 20. By adding a centerline, Fig. 21, we definitely and correctly identify A and B as diameters and C and D as lengths.

However, there is still a possibility for misunderstanding. Only by showing centerlines properly can we clearly define the part. The use of two centerlines, Fig. 22, now shows us that A, B and D are diameters and C is a length dimension.

Now suppose we consider Fig. 23 without centerlines. We can interpret the dimensions as:

(a) The part is an extruded shape as above.

(b) Diameters A and C are eccentric to each other.

(c) A is a diameter; B and D are lengths, and C is a truncated diameter.

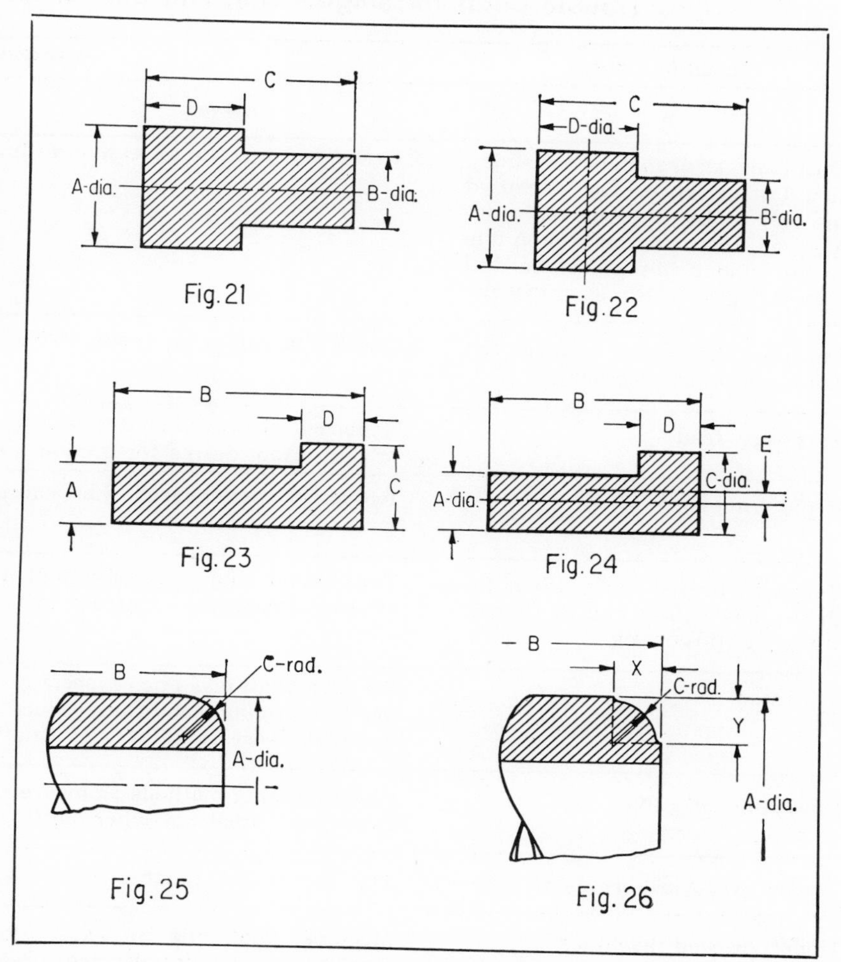

With centerlines properly shown the sketch does not allow for misinterpretation.

In Fig 24 the note "dia" appears where a diameter is the case and also dimension E is introduced to show the amount of eccentricity.

A symbol, ⌀, can be used in place of "dia" to indicate diameter, but diameters must be properly identified along with clearly defined centerlines.

Radii

Consideration should be given to radii. For instance in Fig. 25, if one wants a perfect blend, the radius C must be tangent to A and B. But if this blend is not necessary or functional, it should be replaced with a generous chamfer, to reduce manufacturing cost.

Good appearance is the only valid reason for a radiused corner. Only occasionally does a radius ease assembly or avoid damage to mating surfaces; for example, an outer bearing race. The race is normally ground on the OD and face after heat treatment, and grinding stock is left on these surfaces. Fig. 26 shows a method of dimensioning such surfaces, with the expectation that radus C will blend with the ground surfaces.

Screw Machine Work on Aluminum—I

By M Bryan Baker. Kaiser Aluminum & Chemical Corp, Oakland, Calif.

Source: Revised edition of "Machining Kaiser Aluminum With Automatic Screw Machines"

Trouble Chart for Single-Point, Flat and Circular Form Tools

Possible Causes	Corrections
Breakage	
Dull tools. Deflection of toolslides, turret or workpiece can be caused by dull tools; if other tools are engaged, they may be broken due to the sudden impact as the dull tool engages or leaves the work	Resharpen prior to excessive dulling
Improper sharpening methods can cause checking or cracking	Check sharpening methods, avoid overheating, quick cooling
Excessive feed	Reduce feed rate to enable tool to stand up under the work pressure, or revise tool design for greater strength
Rapid traverse	Reduce traverse speed that causes tools to "hog-in" or enter the metal too quickly
Shims in brazed tools, when too thick, may deflect end of carbide tip by mushrooming	Try thinner shim; excessive tool deflection may necessitate reduction of overhang and/or increase in tool support rigidity
Vibration	Be sure that machine, especially bearings, toolholders and fixtures are in first-class condition; reduce overhang and relief angles to minimum; increase rigidity (see "Chatter").
Clamping difficulty	Check holding methods to insure they do not induce stresses that can cause tool breakage when cutting
Wrong grade of carbide	Try less brittle, tougher grade. (Consult manufacturer)
Underpowered machines	Machines that will not cause slowing or hesitation of the tool or workpiece will minimize tool breakage
Intermittent coolant application	Check for worn pump, improper operation; decrease pressure so hot tool will not be subjected to sudden cooling
Inclusions, hard spots	Consult manufacturer, if excessive or consistent
Careless handling	Improper handling can cause nicks, start tool failure; use care in handling, storage; coat tools with grease when not in use
Chipping of Cutting Edge	
Excessive relief; tool jabbing	Too much relief weakens cutting edge; note recommendations
Clearances, chipbreakers improperly ground	Use clearances specified or less; see if chip breakers are smooth, carefully ground; avoid hasty additions of curlers, chip breakers
Careless handling	Check above under "Breakage"
Excessive vibration	Condition, rigidity of machine, holders and fixtures should be checked; determine if relief angles are excessive

Screw Machine Work on Aluminum—II

Trouble Chart for Single-Point, Flat, and Circular Form Tools—*Continued*

Cause	Remedy
Clogging of chips can cause tools to jam against the workpiece or impair cutting action	Altering lead angle will change nature of chips—increased angle causes thinner chip, less angle a thicker one, increase coolant flow; try liquid with more coolant properties. (Refer to "Chip clogging").
Chatter	
Worn, loose spindles, bearings; lack of rigidity in setup; speed too slow; depth of cut excessive; tool angles too great	Check setup components, machine units for play; increase (occasionally decrease) speed; reduce depth of cut; decrease tool angles to those recommended, or less
Dimensional Runout	
Excessive feeds; varying depths of cut may result from bent, burred, dirty collet, sleeve, or toolholder; built-up edge on tool	Speed or feed may have to be reduced. Check for damaged or dirty collets, holders; shank should be flat, properly seated; hold-down screws should be flat-bottomed, square with tool axis; keep tools sharp, honed
Excessive Pressure	
Chip-breaker too narrow; dull tools	Wide, shallower chip-breaker indicated; check clearances, increase likely; sharpen tools before dullness becomes noticeable
Rubbing	
Relief or clearance insufficient; tool held crooked	Increase clearance or relief until tool does not rub or drag; check for proper mounting, holding
Built-up Edge on Tool	
Rough tool finish; improper or insufficient coolant	Tool surface should be honed after grinding to remove all grinding marks; if coolant or lubricant is insufficient or of wrong type, heating and consequent metal build-up will result
Burning of Cutting Edge	
Cutting fluid inadequate; excessive speed; overheating during sharpening	Direct coolant flow to cutting edge; increase flow; change to fluid having more cooling properties as well as lubrication; reduce surface speed to where tool edge is not burned; tool material of greater red-hardness may be necessary
Rough Finish	
Improper tool angles; dull tool; improper cutting fluid or insufficient flow; too high feed rates; too low surface speed; build-up on tool cutting edge	Reduce tool angles; do not allow tool to become excessively dull; try cutting fluid with more lubricating quality and/or increase surface speed; hone cutting edge
Cratering	
Rake angle insufficient; feed and/or speed too high	Higher rake angles will reduce friction of chip against tool; reduce speed and/or feed; use tool material that is more resistant to abrasion
Chip Clogging	
Cutting fluid flow insufficient; chips not broken up sufficiently	Direct flow of cutting fluid to wash away chips; increase flow and pressure; grind chip breaker or curler into tool

Screw Machine Work on Aluminum—III

By M Bryan Baker, Kaiser Aluminum & Chemical Corp, Oakland, Calif

Trouble Chart for Drilling

Possible Causes	Corrections
Breakage	
Dull drill; lack of lip clearance; drill diving into work; forcing; excessive feed	Resharpen to proper angles; increase clearance behind cutting edges up to 17° at periphery with increase toward center until point angle of 135° or more is formed. "Hogging in" can be caused by too great a lip clearance; too little clearance causes riding, and failure to cut
Chips clogging flutes; insufficient coolant	Keep flutes and cutting edges smooth and free of metal build-up; steel-wool polishing often helps; coolant with more lubricating properties may be needed; or use a drill with greater helix—40° or more, rather than standard 28°; chipbreakers may be necessary; when drill bushings are used provide one drill diameter space for chip clearance
Speed too slow; defective center, drill "walks off" center; misalignment; loose collet; drill set too far outside of holder or bushing; backlash in work or spindle; another tool working in conjunction with drill may be at fault	Feed rate must be changed proportionately as speed is changed; decrease feed or increase speed. Defective center can cause deflection, poor entrance, eventual splitting or spalling due to unequal loads on lips. Centers, smoothly polished, correctly and evenly sharpened and solidly positioned seldom cause difficulty. Uneven lips cause drill to "walk" as does misalignment; use spot drill; check point, alignment, collet position; springing, belling of hole can result from too great deflection, or drill length. Insure that work is held rigidly, machine is not worn excessively nor has loose bearings
Variation in Drilled Depth	
Dull drill; drill cutting hard	At first sign of hard cutting, drill should be resharpened, and feed rate adjusted to support speed if dulling is consistent
Worn linkage between tool and cam; loose collet; end play in spindle lead, improper drill clearance; part variation	Insure that toolholders, machine, fixtures are in first-class condition. Variations may be intermittent or progressive; when progressive over a run, one or more components may be merely slipping or gradually loosening; when greatly variable, extreme wear is indicated
Oversized Holes	
Wrong size drill; improper grind; bent drill; lip chipped; poor alignment; loose spindle; floating holder defective; damaged center; land clamping	Check drill size; try smaller size; poorly ground drill can drill holes a size larger or more; unequal point angle or varying length of cutting edges can be determined by measurement; check straightness of drill, chipped lips, alignment, defective or dirty holders, play in spindle; improper center can cause drill to wander
Eccentricity of Holes	
Drill with larger included angle than center; faulty center; bent holder shank; bent stock; improper clamping	Drill may start untrue especially when the included angle is smaller than that of the center; check to see that center is rigid, correctly ground; use center drill; bent holder shank can sometimes be detected by noting whether drill turns out of true when running idle
Undersized Holes	
Size of drill too close to tolerance; dull drill; drill many walk off center; untrue hole may cause binding of gage; worn drill margins	Check hole size in thinner piece of metal; if noticeably tight, change to size larger; check for sharpness—a dull drill will generally drill undersize; inspect center and drill approach to guard against wandering, untrue hole; check margins of drill to determine wear

Screw Machine Work on Aluminum—IV

Trouble Chart for Drilling—*Continued*

Possible Causes	Corrections
Rough Holes	
Improper grind; chipped lip; dedective lands; clogged chips; dullness; lack of coolant; wrong lubricant; too much pressure	Examine for unequal clearance of lips or lip angle as one side cuts deeper, deflecting the drill; improper grind can cause spalling, chipping of lips and margins and consequent roughness of holes; dullness may well result in excessive feed pressure and cause deflection and roughness; inadequate chip disposal may call for thinned web
Tapered Holes	
Bent drill; misalignment; poor center, drill bushing; defective bearings; squeezing of hole	Check for bent drill, especially near base; misalignment can be caused by chips between drill shank and holder, or worn components; check center hole and drill approach, determine if included angles of each are equal; note if there is play in bearings or other components
Chipping of Lip or Cutting Edge	
Excessive lip clearance; too much pressure	Too much clearance of cutting area can cause "hogging in" and chipping of lips; excessive clearance weakens the cutting edges (review drill breakage); too little lip clearance can cause rubbing and failure to cut, with resultant excessive pressures; this causes heat and strain

Trouble Chart for Counterbores

Possible Causes	Corrections
Breakage or Chipping of Edges	
Dullness; wrong size tool; hole eccentric between steps; coolant insufficient	Sharpen carefully (see below), to insure all steps are concentric with each other as well as shanks; diameter of a counterbore, for reaming to size, should be approximately halfway between the mean and top limit of hole size, except for fragile conditions where tool size may be somewhat smaller; check for concentricity of each diameter in a stepped hole, examine combination or step drill for fault
Chatter, Poor Finish	
Resharpening needed; wrong clearances; margins too wide; rough tool; chip packing; metal buildup; lack of rigidity; fault of previous operation	Check grinds; for cutting solid stock, flutes should be set at about a 23° helix for counterbores under 1 in. Flues are often straight-milled at an angle of about 10° to tool axis, and 5° to the radius for a positive rake. Lip angles approaching 60° will generally serve, depending upon shoulder angle desired for the part. Web thickness of tools for drilling into solid stock should be thinned to reduce pressure needed for penetration, seldom exceeding 1/16 in. at the point. Speeds will approach 700 fpm and feeds can parallel those for a similar sized drill; for wider steps or flat bottoming, feeds down to nearly 0.002 ipr may be required. Clearances back of each lip run about 12° for the free-machining alloys and 15° for others
Tolerances not Maintained	
Dullness, wrong angles; machine bushings, ways and spindles loose; previous operations off tolerance; cut too shallow or feed too light	Sharpen as recommended above. All loose bearing members should be corrected. Determine and remedy faulty previous operations. Increase feeds (note typicals above); see that all elements in operation are aligned, rigid. Coolants, properly applied in ample amounts, often help

Screw Machine Work on Aluminum—V

By M Bryan Baker, Kaiser Aluminum & Chemical Corp, Oakland, Calif

Trouble Chart for Cutoff Tools

Possible Causes	Corrections
	Breakage
Tool too thin for stock; rake excessive; point too weak from excessive angle	Width depends upon size and shape of stock, depth of cutoff and type of cutoff. For cutting off 1 in. round rod, representative tool widths are: 3/64 in. for depth of cut below 0.125 in.; 1/16 in. for 0.125 in. to 0.1875 in.; ⅛ in. for 0.313 in. to 0.500 in.
	Length Tolerance Not Held
Tool too thin; point angle wrong; tool dull; side clearance wrong; feed too heavy; tool contacting hex corners; toolslide loose; end play in spindles or work spindle head; pick-up attachment damaged; previous operation causes piece to push back	Determine if deflection is caused by tool being too thin for cut; measure point angle and side clearance; resharpen tool; (note angle recommendations below). Feeds for round stock generally run from about 0.0035 ipr for tools up to 1/16 in. wide, to 0.005 ipr for those 3/16 in. to ¼ in.; feeds for square and hexagonal stock commonly run about 0.0005 in. less. Check for contact at hex corners. Determine if slides, spindles, tool holders have play. Check pick-up attachment for wear or damage. Measure part following preceding operation
	Tool Chatter
Tool too thin for depth of cut or stock size, too much rake; dullness; off center; too much play in spindles; breakdown of hex stock incomplete	Note tool-width recommendations above. Check for rake of 0° to 3° for 2011, 2017, and 2024 and 5° to 10° for most other alloys. Sharpen with front angle of up to 23°, according to machinability of alloy, a front clearance of 8° to 12° and side clearance of from 0° to 4°. Gage for off-center difficulty
	Rough Finish
Dull tool; improperly set tool; point angle too blunt; side clearance too small; rake angle insufficient; tool "hogging in", off center; worn bearings; defective holder; damaged toolslide; heavy feed; coolant insufficient; chips interfering with coolant	Where no forming of the bar or cutoff piece is necessary, other than parting, the thin, straight-type blade is usually employed. An angular cutoff is used when coning or pointing is required. Where chamfering or a similar operation is needed, the circular cutoff tool can be shaped accordingly, and it holds the same contour through repeated sharpenings. The thin-blade type has a square cutting edge at right angles to the centerline of the tool point; this cutting edge may be beveled to about 15° to prevent burrs on the cutoff. Often it is advisable to provide hook for additional rake and chip curling, when cutting softer or more gummy alloys. Holder, toolslide must be rigid. Slacking off feed at end of cut may eliminate burring

Trouble Chart for Tapping

Possible Causes	Corrections
	Breakage
Dull tap, dragging on cut, improper land, wrong cutting angle, wrong number of flutes, wrong type of tap; percentage of thread too high, 75% maximum, 65% preferred	Sharpen to recommended angles (about 10° hook and 10° to 20° rake angle); when radial thread relief is insufficient, dragging is encountered, check angles, relief; select appropriate number of flutes for hole diameter—2 flutes for holes under ¼ in.; 3 flutes for those ¼ to ½ in., 4 flutes for holes over ½ in.

Screw Machine Work on Aluminum—VI

Trouble Chart for Tapping—*Continued*

Possible Causes	Corrections
	Breakage
Drilled hole too small; hole out of line; inadequate clearance at bottom of hole; tap lead-angle too short	Check size of hole and alignment; if hole is too small, tap may show reaming action. Out-of-line hole will show deep and shallow threads. Provide generous clearance at bottom of hole for all except bottoming taps
Chip clogging; rough tap flutes, poor grinding and polishing; lack of lubricant; speed too slow; work-hardened hole	If alloy is stringy, it may be necessary to use a fluteless or spiral-flute tap. Grinding marks on a tap or roughness can cause clogging; lubricants are essential. Speed in small diameters is more important than for larger ones. If torque is excessive, a 2- or 3-flute tap should be used. If surface of hole has been work-hardened by using too fine a feed or a dull drill, the tap will probably break or the thread will show a poor profile
Tap pushing or dragging on cut	Try more relief, change to more flutes; try different lubricant; grind more hook and rake angle; where threads are torn and rough, fewer (but larger) flutes may be the solution
Miscellaneous causes	Loose bushings or spindles can cause breakage, as can loose threading clutches; tap holder may be wrong type or worn. Speed may be too fast for feed or vice versa. Polished or plated flutes may be necessary if sticking persists. Out-of-round holes produce threads only on two sides
	Rough Threads
Rough or worn tap, dragging on entry; build-up on cutting edges; flutes packing with chips, flute size insufficient; poor alignment; hole too small; eccentric hole; insufficient coolant	If tap is cutting too small as well as rough, more rake often helps. Check chamfer on tap lead—hook and rake angles should include the first full tooth. If excessive heat is causing build-up, provide coolant-type lubricant; try fewer flutes, but larger ones. If hole size is too small, tap has excessive stock to remove, resulting in torn, rough work. Eccentric hole results in much the same condition as too small a hole. If work hardening has occurred in drilling, reaming may be necessary; reaming will also reclaim eccentric holes
	Variation in Pitch Diameter
Worn tap, defective tap, oversize or undersize tap, too much land; insufficient clearance, pushing or dragging; rough first thread; oversized hole	Resharpening without change in form often helps, but added rake is probably better. An undersize tap is more likely to cause pitch-diameter variation; too large a tap causes excessive torque. Relief should be ground on taps with too much land, or a tap with a greater number of flutes substituted. More radial thread relief can alleviate shaving thread on backing out. Insufficient clearance for chamfer can cause tap to "ride", especially where chip buildup has occurred. Lubricant, correct sharpening, with relief on the high side, and correct tap and hole size will generally relieve pushing or dragging of tap
	Tapered Threads
Threads shaved by tap retraction; heat caused by excessive land, poor grind, tapered hole, misalignment	Grind more radial thread relief, polish lands, use more efficient lubricant; change to more flutes if breakage will not be a problem (2- and 3-flute taps are stronger than 4-flute); grind to recommendations; check ID of hole at several points; gage for straightness

Screw Machine Work on Aluminum — VII

By M Bryan Baker, Kaiser Aluminum & Chemical Corp, Oakland, Calif

Trouble Chart for Threading

Possible Causes	Corrections
Damaged Chasers	
Dull chasers; chaser touching shoulder; loose fits at corner, alignment; loose front plate; high feed; poorly designed cams; too small a die for the job	Sharpen to recommended angles (see below). Check alignment with correct-size blank; close-tolerance blank diameter must have slightly chamfered edge for proper starting. Obtain rigidity in components, tighten front plate. Speed should be adjusted. Provide ample clearance for chip movement. Initial damage to chasers is sometimes due to faulty cams; especially check entry; try larger die if available or smaller-size stock when possible
Rough Threads	
Dull chaser, poor grind, wrong clearance or chamfer angles; defective diehead; misalignment; chasers hitting shoulder, improper position in diehead; improper feed; chip packing; coolant inadequate; chamfer undersize; blank too large; speed or feed too high	Resharpen at first signs of dulling. Circular chasers should have a face angle of 20° to 25°, clearance of 10°, a top rake of 25° and a chamfer angle of 20° to 22°. Tangent chasers have a top rake of 30°, lip hook of 5° to 10° and a chamfer angle to 20° to 22°. Starting ends of lands can be chamfered to about 30° for the first 3 or 4 teeth, which do most of the cutting. An angle of 8° to 10° should be ground on the face to give side rake. Check for proper positioning, shoulder contact. Provide adequate coolant-lubricant to avoid heating, buildup or chip packing. Provide more chamfer on part to be threaded, try smaller blank. Blank size will depend upon accuracy needed (see "Pitch Variation"); try speeds under 200 sfpm, especially if swing holder is used (see "Tapered" Threads"). Feeds include about 0.0005 ipr for ⅜ in. root dia of blank at 14 threads per in.
Tapered Threads	
Defective chasers, dieheads or start; alignment; feed variation; bad camming; bent diehead shank; chasers loose in diehead; improper speed	Check sharpening, clearances, positioning and start of cut, alignment and feed. Although dies are generally self-leading, it is sometimes advisable to control the longitudinal movement. For 9 through 14 pitch, 200-250 sfpm is nominal, 14 to 18 requires 325 sfpm and 18 or finer, 475 sfpm
Pitch Variation	
Foreign matter in diehead; worn loose locking pin or binding screw; blank size too large or too small; threads off lead	Use air hose and/or brush and cleaner to clean diehead. Check for play in locking pin and binding screw. Experiment with different blank sizes. When thread rolling, there are three classes of blank size —for ¼ to ½ in., blanks are 0.002 to 0.0025 in. larger than pitch diameter; for ½ to 1 in., blanks are 0.0025 to 0.003 in. larger; those from ¼ to ½ in., with comparatively free fit, can vary from 0.002 to 0.003 undersize; above ½ in., 0.003 to 0.005 in. undersize; those smaller than ¼ in., from 0.001 to 0.0015 under pitch diameter. Start chamfer below root of chaser teeth, averaging approximately 22° for straight cutting by circular tangent chasers. For close-to-shoulder work use shorter chamfer of about 45°. Be sure all chasers are ground to begin cutting at same time

Screw Machine Work on Aluminum – VIII

Trouble Chart for Skiving

Possible Causes	Corrections
	Breakage of Tool Edge
Angle too sharp; metal build-up; improper cooling of cutting edge; poor tool support	Grind skive tool to recommended angle for work (end cutting edge angle 10° to 35°, front relief about 15°); provide coolant that has lubricating quality; check toolholder and tool overhang
	Breakage of Work
Feed too great; approach angle too shallow; support of work poor; work extends too far from collet; tool digs in; skive tool not set properly	Check feed, (depends on approach angle, support); reduce approach speed; increase approach angle of skive tool; check tool approach angle; examine cross-slide for looseness
	Chatter and Poor Finish
Work improperly supported, tool dull; rake and approach angle incorrect; edge build-up; stringy chips between cutter and work; machine loose, excessive skive tool overhang	Support work and tool for greater rigidity; sharpen cutter. Increase rake and/or reduce approach angle (see below). Arrange for proper chip breaking. Check clearances between work and cutter; examine lubricant. Be sure machine is tight, tool is held firmly and that slides for other operations are tight. Avoid excessive overhang
	Work Undersize, Oversize or Eccentric
Approach angle wrong, lack of support; non-rigid tool; defective tool; machine spindle and slides loose; wrong speed, feed	Most approach angles vary from 15° to 20° except where pressure is unlikely to cause bending, as in the case of large diameters; here smaller approach angles are used. Check angles, sharpness of cutter. Make certain machine spindle bearings, as well as slide-bearing surfaces, are snug. Recheck machine feeds, speeds. Usual practice is 700 to 1100 sfpm for round, 450 to 550 sfpm for hexagonal and 300 to 350 for square work. For alloys with poor machinability, these figures should be reduced approximately 250 to 550 sfpm for round, 100 to 150 sfpm for hexagonal, and about 50 sfpm for square work. For unsupported stock, ratio of length to smallest turned diameter is 8 to 10 to 1. For stock under 3/8 in., feed will average 0.002 ipr with a 0.150 in. depth of cut; with 0.004 in. feed, depth of cut will approach 0.125 in.

Blanks for Dovetail Forming Tools

These blanks are listed according to group number, and makes of machines on which they can be used.

Group Number	Type of Machine	Maximum Capacity*
1	No. 00 Brown & Sharpe	3/8
	No. 19 Brown & Sharpe	3/8
	Index "0"	7/16
	3/8 Cleveland	5/8
2	3/8 Gridley	3/8
	1/2 Davenport	7/8
	9/16 Acme Gridley	9/16
	No. 0 Brown & Sharpe	5/8
	5/8 Cleveland	3/4
	5/8 × 7/8 Cleveland	1 1/16
	No. 204 New Britain	5/8
	7/8 Greenlee	1
	7/8 × 1 1/4 Cleveland	7/8
3	7/8 Gridley	7/8
	1 Acme Gridley	1
	No. 172 New Britain	1
	No. 2 Brown & Sharpe	1 1/8
	1 1/4 Gridley	1 1/4
	1 1/4 Cleveland	1 1/4
	1 1/4 Cleveland	1 3/8
	1 1/4 × 1 1/2 Cleveland	1 3/4
	1 1/4 × 1 1/2 Cleveland	1 1/2
	1 1/4 Greenlee	1 1/4
	1 3/8 Gridley	1 3/8
	1 1/2 Greenlee	1 1/2
	1 5/8 Gridley	1 5/8
4	1 5/8 Gridley	1 5/8
	1 5/8 Acme Gridley	1 5/8
	No. 206 New Britain	1 5/8
	No. 415 New Britain	1 5/8
	No. 410 New Britain	...
	1 3/4 Gridley	1 3/4
	1 3/4 Greenlee	1 3/4
	No. 4 Brown & Sharpe	1 7/8
	2 Greenlee	2
	2 Gridley	2
	2 Cleveland	2 1/2
	2 × 2 3/4 Cleveland	3 1/4
	2 1/4 Cleveland	2 1/2
	2 1/4 × 2 3/4 Cleveland	3 1/4
	2 1/4 Gridley	2 1/4
	2 1/4 Greenlee	2 1/4
5	No. 6 Brown & Sharpe	2 3/8
	No. 208 New Britain	2 1/2
	No. 425 New Britain	2 1/2
	2 5/8 Gridley	2 5/8
	2 3/4 × 3 3/4 Cleveland	3 1/4
	2 3/4 × 4 Cleveland	2 3/4
	3 Gridley	3
	3 5/16 Gridley	3 5/16
6	3 1/4 Gridley	3 1/4
	3 1/2 Gridley	3 1/2
	4 Gridley	4
	4 Cleveland	4
	4 1/4 Cleveland	4 1/4
	4 1/2 Cleveland	4 1/2
	4 1/4 Gridley	4 1/4
	4 1/2 Gridley	4 1/2
	5 Gridley	5
	5 1/2 Cleveland	5 1/2
	6 3/4 Cleveland	6 3/4
	7 3/4 Cleveland	7 3/4

All dimensions are given in inches.

* The group classification numbers apply to all machine models of the respective makes listed having the maximum capacities indicated.

NOTE: Technical Committee No. 10 in preparing these machine capacities and classifications used all of the information available. They will welcome additional information leading to the improvement of these classifications.

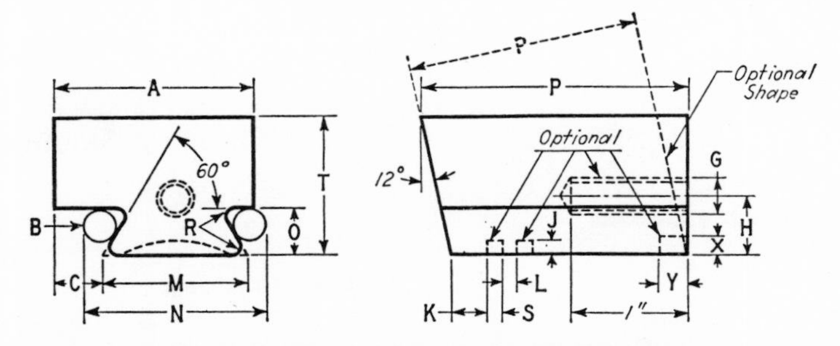

Dimensions for Finished Blanks for Dovetailed Forming Tools

Group No.	Basic Blank Size[1]	Width A		Thickness T		Length	Dovetail			
		Max	Min	Max	Min	P	Width M[2]	Height O	Radius R	To Face C
1	1	1 1/32	1 1/64	29/32	57/64	1 1/2	0.732	9/32	1/32	1/4
	1 1/4	1 9/32	1 17/64	29/32	57/64	1 1/2	0.951	19/64	1/32	19/64
2	1 1/4	1 9/32	1 17/64	29/32	57/64	2	0.951	19/64	1/32	19/64
	1 1/2	1 17/32	1 33/64	29/32	57/64	2	0.951	19/64	1/32	19/64
	1 3/4	1 25/32	1 49/64	1 5/32	1 9/64	2	1.250	13/32	1/16	7/16
	2 3/4	2 25/32	2 49/64	1 5/32	1 9/64	2 7/16	2.000	33/64	1/16	1/2
3	1 3/4	1 25/32	1 49/64	1 5/32	1 9/64	2 7/16	1.250	13/32	1/16	7/16
	2 1/4	2 9/32	2 17/64	1 5/32	1 9/64	2 7/16	1.250	13/32	1/16	7/16
	2 3/4	2 25/32	2 49/64	1 5/32	1 9/64	2 7/16	1.250	13/32	1/16	7/16
4	2 3/4	2 25/32	2 49/64	2 17/32	2 33/64	2 5/8	1.614	35/64	1/16	1/2
	2 3/4	2 25/32	2 49/64	2 17/32	2 33/64	2 5/8	1.882	35/64	1/16	1/2
	3	3 1/32	3 1/64	2 17/32	2 33/64	2 5/8	1.882	35/64	1/16	1/2
5	2 3/4	2 25/32	2 49/64	1 9/32	1 17/64	2 7/16	2.000	33/64	1/16	1/2
	2 3/4	2 25/32	2 49/64	3 1/32	3 1/64	3	2.000	33/64	1/16	1/2
	3	3 1/32	3 1/64	3 1/32	3 1/64	3	2.000	33/64	1/16	1/2
	3 1/4	3 9/32	3 17/64	3 1/32	3 1/64	3	2.000	33/64	1/16	1/2
6	3 1/4	3 9/32	3 17/64	3 17/32	3 33/64	4	2.238	35/64	1/16	9/16
	3 1/2	3 17/32	3 33/64	3 17/32	3 33/64	4	2.883	43/64	1/16	5/8
	4	4 1/32	4 1/64	3 17/32	3 33/64	4	2.883	43/64	1/16	5/8
	4 1/2	4 17/32	4 33/64	3 17/32	3 33/64	4	2.883	43/64	1/16	5/8

Group No.	Basic Blank Size[1]	Across Plugs	Plug Diam	Adjusting Holes		Slots					Notch for Hook Bolt Adj. Nut	
		N[3]	B	G	H[4]	Number	J[4]	K	L	S[5]	Width Y	Depth X
1	1	0.834	5/32	5/16–18	3/8	2	1/8	5/16	1/8	1/8	1/8	1/8
	1 1/4	1.035	5/32	5/16–18	1/2	2	1/8	5/16	1/8	1/8	3/16	3/16
2	1 1/4	1.035	5/32	5/16–18	1/2	3	1/8	11/32	1/8	1/8	3/16	3/16
	1 1/2	1.035	5/32	5/16–18	1/2	...	..	....	...	...	3/16	3/16
	1 3/4	1.464	1/4	5/16–18	1/2	3	1/8	11/32	1/8	1/8	1/4	1/4
	2 3/4	2.771	1/2	7/16–14	3/8	3	1/8	7/16	5/32	5/32	5/16	5/16
3	1 3/4	1.464	1/4	5/16–18	1/2	3	1/8	7/16	5/32	5/32	1/4	1/4
	2 1/4	1.464	1/4	5/16–18	1/2	...	..	....	...	...	1/4	1/4
	2 3/4	1.464	1/4	5/16–18	1/4	...	..	....	...	..	1/4	1/4
4	2 3/4	2.349	1/2	5/16–18	1/2	...	..	....	...	...	1/4	1/4
	2 3/4	2.617	1/2	7/16–14	3/8	...	..	....	...	...	1/4	1/4
	3	2.617	1/2	7/16–14	3/8	...	..	....	...	...	1/4	1/4
5	2 3/4	2.771	1/2	7/16–14	3/8	...	..	....	...	...	5/16	5/16
	2 3/4	2.771	1/2	7/16–14	3/8	...	..	....	...	...	5/16	5/16
	3	2.771	1/2	7/16–14	3/8	...	..	....	...	...	5/16	5/16
	3 1/4	2.771	1/2	7/16–14	3/8	...	..	....	...	...	5/16	5/16
6	3 1/4	2.973	1/2	7/16–14	3/8	...	..	....	...	...	5/16	5/16
	3 1/2	3.815	5/8	7/16–14	3/8	...	..	....	...	...	5/16	5/16
	4	3.815	5/8	7/16–14	3/8	...	..	....	...	...	5/16	5/16
	4 1/2	3.815	5/8	7/16–14	3/8	...	..	....	...	...	5/16	5/16

All dimensions are given in inches.

[1] Blanks made of high-speed steel shall be stamped H.S.
[2] Dovetail width M is to the theoretical sharp corners.
[3] The tolerance for width of dovetail N measured across plugs B is +0.003 or −0.003.
[4] Methods of adjustment are optional.
[5] Tolerance for width of adjusting slot S is +0 003, −0.000.

Chapter 11

Brake and Press Bends

Tonnage for air bends, 1

Capacity ratings of press brakes are based on air bends in which dies do not strike solidly on the metal. All pressure is used in forming—none in coining or squeezing

Air-bend dies produce a bend with an inside radius approximately 15% or 5/32 of the die opening.

This means that less than an 8-times die opening must be used if a smaller radius is desired—requiring higher tonnage.

Press ratings are based on die openings of 8 times the material thickness up to about ⅝ in. plate. Die openings up to 10 or 12 times are used for forming heavier thicknesses of plate.

If the die opening is too large, an excess amount of metal is drawn into the die, causing a curve to form in the metal each side of the point radius.

If the metal is formed over a die opening less than 8 times the plate thickness, there is danger of fracturing the metal in the heavier thicknesses, unless a small amount of preheat is applied.

Effective width of die opening

When the punch radius is equal to or less than the material thickness, the effective width of die opening to use with the tonnage table, page 99, is die width W.

When the punch radius is greater than the plate thickness, the effective width of die opening to use with the tonnage table is 2 times X shown on sketch.

Pressure per foot

Check tonnage required from table to be sure it is within the capacity of the machine, making allowance for coining and drawing forces required on other than air-bend dies.

Bending pressure is proportional to the ultimate strength of the material for the same thickness and die opening.

By Roy F Dehn,
Director of engineering,
Cleveland Crane & Engineering Co,
Wickliffe, Ohio

The inside radius of a bend is approximately 5/32 of the die opening and is about the same for varying thicknesses of material bent on the same die set.

Heavier thicknesses of plate contain higher carbon content in order to maintain full ultimate strength. This results in more bending fractures which can be reduced by 10 or 12 times die opening or the use of special flanging steel.

High tensile steel plates are usually formed over 10 to 12 times die opening.

The manufacturers of special steels usually recommend the radius of die opening to use with their materials.

Bends across grain will show less breakage than when bent in line with the grain of the plate, especially in the thicker plates.

It helps to avoid cracks by rounding the edges of thick plates at each end of the bend, on the outside of the bend.

Approximate spring back:

Low carbon steel, 1° to 2°
0.40 to .50 carbon steel, 3° to 4°
Spring steel annealed, 10° to 15°

The same size of press brake, as formerly used for bends of 8 times plate thickness in mild steel, is suitable for 12-times bends in the popular low-alloy steels.

If less flange width is required, a smaller die opening must be considered, and this will affect the tonnage rating needed.

Forming practice

Material bent on too wide a die opening may not come square. Re-hitting with dies set closer in trying to square up the bend frequently overloads the press. The forming of channel or offset bends may require more than six times the load needed for a single right angle bend in the same material.

Effective width of die opening

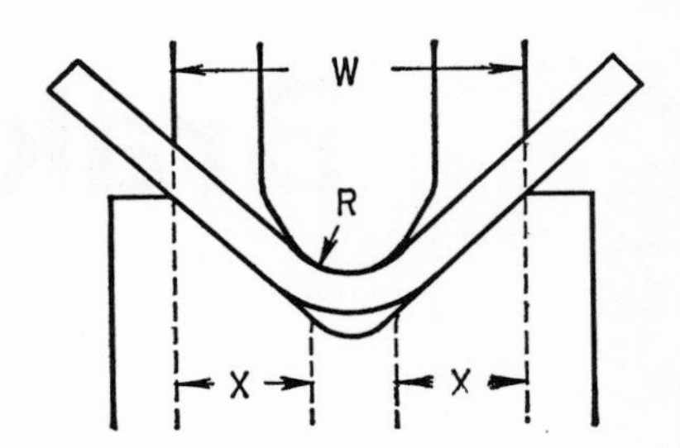

To adjust a mechanical press brake ram and bed parallel under load follow these three steps:

1. With the eccentrics on bottom center, adjust right hand pitman or screw (or pitman with drive motor) 0.015 in. or 0.025 in. above left-hand end.
2. Run both screws down together until left hand end bottoms in die and stalls adjusting motor.
3. Release adjusting clutch on cross shaft and run right hand screw down until it stalls.

Ram and bed are then parallel under pressure and it is only necessary to back up adjustment for the thickness of the material.

Another method is to use a short test piece under each end of the press, adjusting until equal results are obtained on both ends. Use wide enough test pieces so that the unit pressure will not be high enough to indent the dies. Another method is to start with a shallow bend and run the adjustment down between strokes, until the desired angle is formed. However, one must check for equal angles at both ends of the bend.

Dies should preferably be of a closed height so that the adjusting screws project about one to three inches.

If a load is put on one end of the press so that it is substantially performed by one pitman, it should be limited to half the press capacity to avoid overloads.

Tonnage for air bends, 2

Tons pressure per foot for 90° air bends in mild steel

Thickness of metal		Width of vee die opening – Inches																									
Inches	Gauge	1/4	3/8	1/2	5/8	3/4	1	1 1/4	1 1/2	2	2 1/2	3	4	5	6	7	8	9	10	11	12	13	14	16	18	20	22
.036	20	3	1.7	1.1																							
.048	18	5	**3**	2.2	1.7	1.3																					
.060	16	9	5.5	**3.8**	2.8	2.2	1.5																				
.075	14		9	6.3	**4.7**	3.5	2.5	1.8																			
.105	12			13	10	8	5.6	4.1	3.2																		
.135	10				18	14	**9.5**	7	5.5	4																	
3/16						25	19	14	**11**	7	5.5	4															
1/4							38	29	22	**15**	11	8.5	6														
5/16								48	38	26	**19**	15	10	7.5													
3/8									60	41	30	**24**	16	12	9												
1/2										80	60	46	**32**	23	18	15	10										
5/8												78	55	**40**	30	24	20	15									
3/4													85	63	**48**	40	32	25	21								
7/8														95	74	58	**48**	43	35	30							
1															105	85	70	**64**	56	53	40	36					
1 1/8																118	95	82	**73**	68	62	55					
1 1/4																	126	109	94	86	**80**	74	64				
1 3/8																		138	119	106	93	**84**	78	69			
1 1/2																			148	132	118	104	**96**	82	75		
1 5/8																					141	130	120	**101**	86	80	
1 3/4																						155	144	123	**105**	92	87
2																							188	171	148	**129**	116

Underlined tonnages are for minimum vee die openings recommended

This chart for mild steel of 55,000 to 65,000 psi

Tons pressure per foot for mild steel 90° air bend

Tonnage vs stroke of press brakes

This chart can be used to check the capacity of press brakes, because it shows tonnage available at different points in the stroke

Data are based on a die opening width W, and are correct for the usual proportions of width of die opening to material thickness.

Distance A is punch travel required to make the bend, and equals 40% of die width W. Full tonnage to make the bend is required at 0.7 A above bottom, where W, or width of die opening = 8 or more. Also, under these conditions, $0.7\ A = 0.28\ W$.

Problem:

What length of 3/8 in. mild steel plate can be bent on a 320-ton brake with a 5-in. stroke?

Solution:

A 3-in. die opening would normally be used. Therefore the height above bottom for full tonnage $= 3 \times 0.28 = 0.84$ in.

Percentage of stroke above bottom stroke $= 0.084 \div 5 = 16.8\%$.

Enter the chart at 16.8% of stroke above bottom stroke and draw a dash line to the stroke-capacity curve.

Drop down a dash line and read the tonnage available at 16.8% of stroke equals 1.3 times full capacity, or $1.3 \times 320 = 416$ tons.

From the chart "Tonnage for air bends," (**AM**—March 28, '66, p99) find that the pressure to bend 3/8-in. plate in a 3-in. wide die equals 24 tons per foot.

Then, $416 \div 24 = 17$ ft, or maximum length of air bend that can be made on the 320-ton brake, using 3/8-in. mild steel plate.

This value would have to be adjusted upward or downward for other materials.

If alloy-steel plate is to be bent on extra-wide die openings, consult the press-brake manufacturer with regard to the limiting effect of the available flywheel energy.

This chart may be used also for work to be done on mechanical press brakes rated with a bottom stroke capacity equal to 150% of mid-stroke capacity. ■

By Roy F Dehn
Director of engineering
Cleveland Crane & Engineering Co
Wickliffe, Ohio

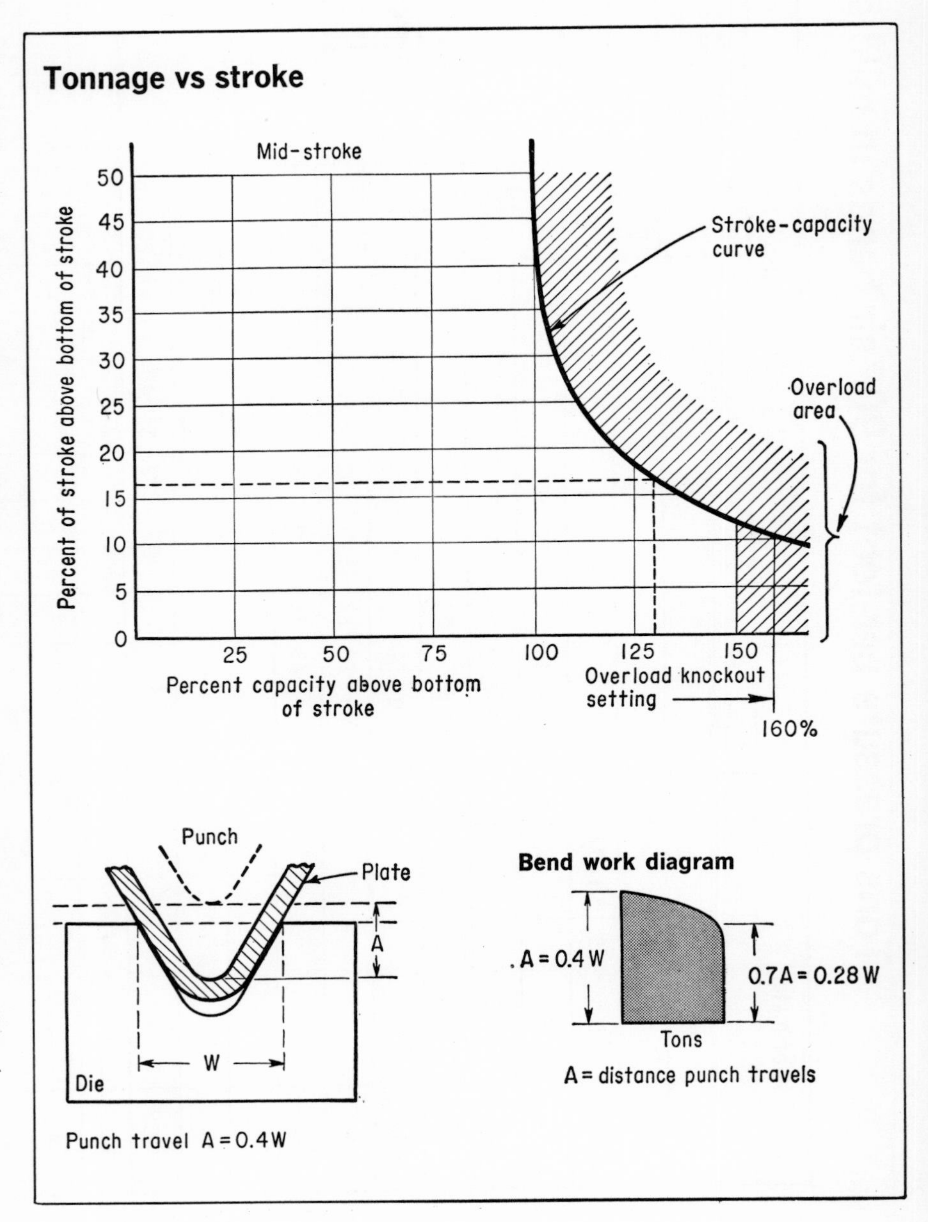

Tonnage chart for various bend angles

The chart published earlier (**AM**—March 28, '66, p99) gives the tonnage per foot to produce 90° air bends in various plate thicknesses and using various die openings. In many cases, however, material must be bent to less than a 90° bend angle, and then the accompanying chart provides a means of estimating the tonnage in relation to that required for a 90° air bend.

By Roy F Dehn
Director of engineering
Cleveland Crane & Engineering Co,
Wickliffe, Ohio

Example: What is the percentage of tonnage for a 90° bend that is required to bend plate to an inside bend angle of 175°?

According to the tabulation, the percentage is 50%. Now cross check this by using the curve AEP.

Solution:

Follow the 175° inside bend angle to the right until the dash-line extension cuts curve AEP. Drop down to the scale for 100% air bend tonnage, and read that 50% of that tonnage is required. If the full 90° bend in 2 in. plate requires 171 tons per foot, a 175° bend will require 50% of it, or 85 tons per foot. ■

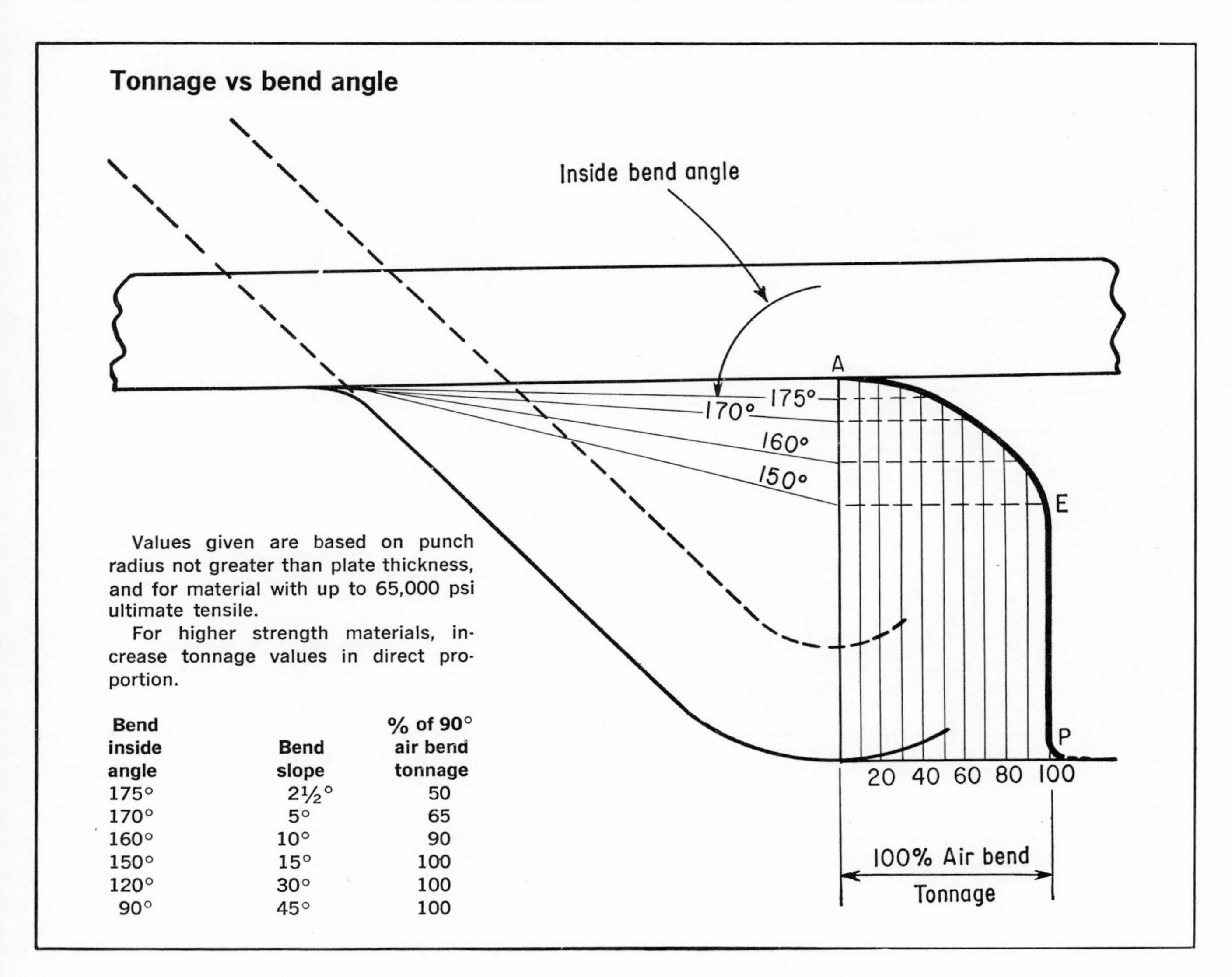

Bend inside angle	Bend slope	% of 90° air bend tonnage
175°	2½°	50
170°	5°	65
160°	10°	90
150°	15°	100
120°	30°	100
90°	45°	100

PRESS TOOLS FOR BENDING...I

DON R KING, tool engineering consultant

Selection of a suitable method is the initial step in designing press tools for parts that require bending. Often, for a given shape, there are several possible methods. To select the one best suited to your job, we give here schematic drawings of press tools to serve as a reference guide. These are classified according to standard arrangements to produce basic bend configurations. Notes under each sketch give the advantages and disadvantages of the particular design.

In most cases, the bend is shown as accomplished at the last station of a progressive die, in order to indicate the relationship of cut off. Of course, the construction may apply to intermediate stations, where the bend includes only part of the strip, or is turned parallel to the direction of feed. Dimensions "x" on certain sketches are likely to be critical in respect to part dimensions. They should be checked for limitations of die wall thickness or space.

RIGHT-ANGLE BENDS

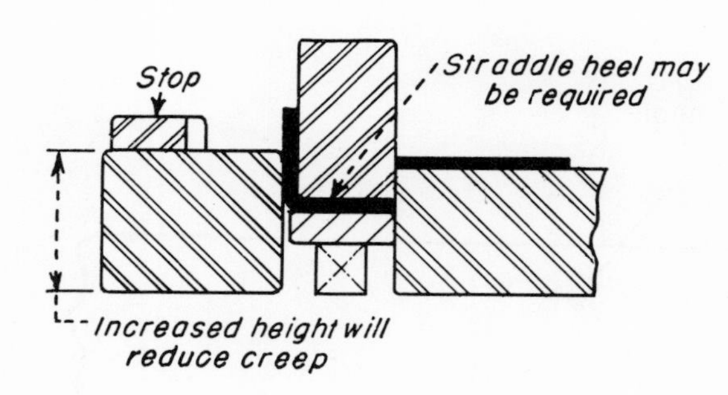

No. 1
- Good location and alignment
- Inclined ejection possible
- Slight tendency for part to creep
- No scrap waste

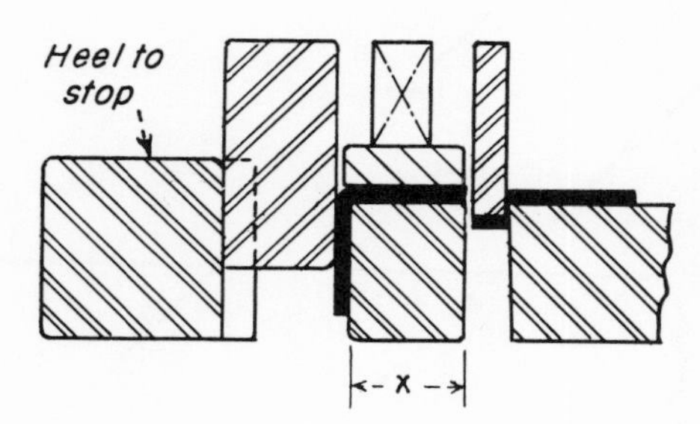

No. 2
- Good location and alignment
- Inclined ejection preferred
- Scrap slug wasted
- No creep if other punches are engaged
- Large spring space needed in punch holder

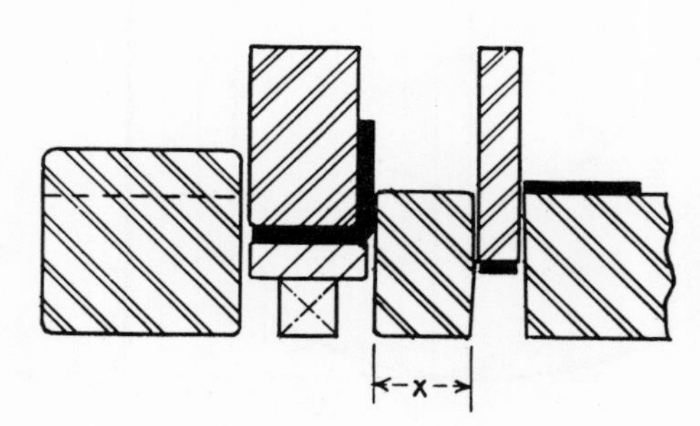

No. 3
- Alignment may depend on stock fit in stripper
- Push-through ejection is possible
- Some tendency to creep
- Scrap slug wasted
- Long cut-off punch required

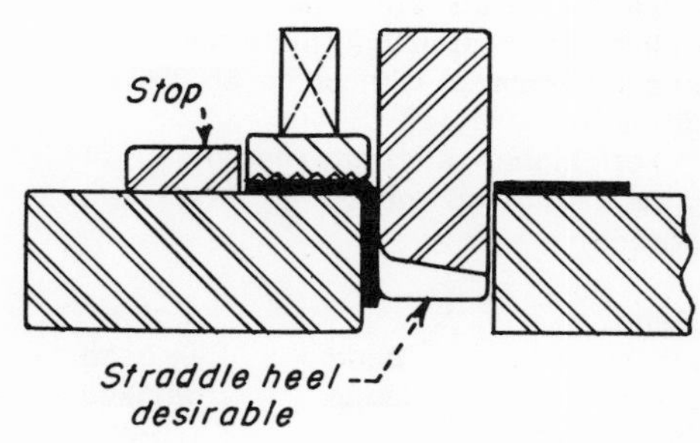

No. 4
- Inclined ejection required
- Tendency to creep
- No scrap waste
- Punch sharpening more difficult than other designs
- Large spring space needed in punch holder

PRESS TOOLS FOR BENDING...II

RIGHT-ANGLE BENDS ... continued

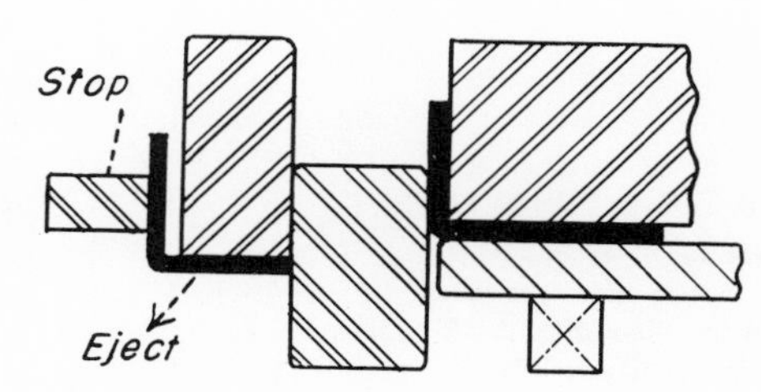

No. 5
Requires inverted pierce and notch operations
No creep if other punches are engaged
No scrap loss
Good ejection

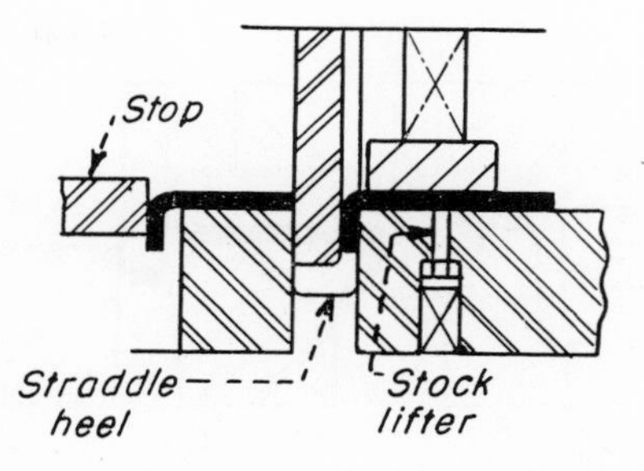

No. 6
For bends with short legs only
No creep if other punches are engaged
No scrap loss
More difficult to resharpen
Large spring space needed in punch holder

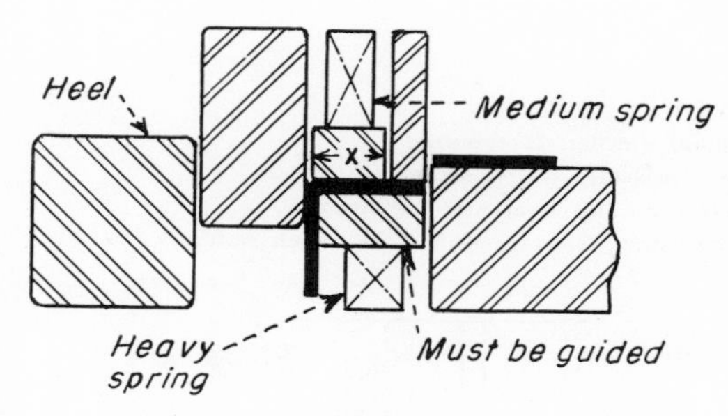

No. 7
More complicated and costly than other designs
Eliminates scrap slug, when forming downward is necessary
No creep occurs

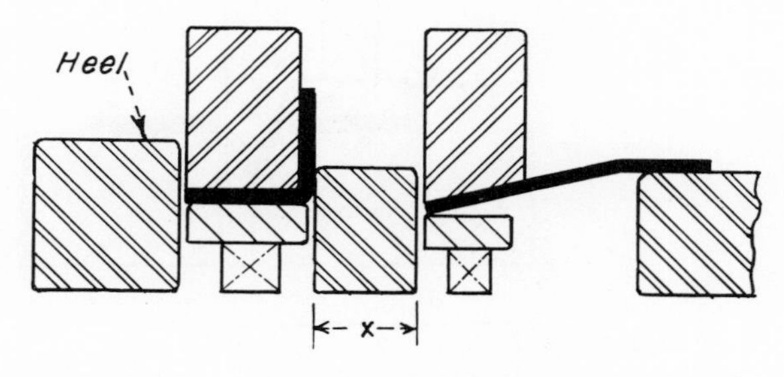

No. 8
Limited to thin material and short forming travel
Eliminates scrap slug, when inside form-up is necessary

ACUTE-ANGLE BENDS

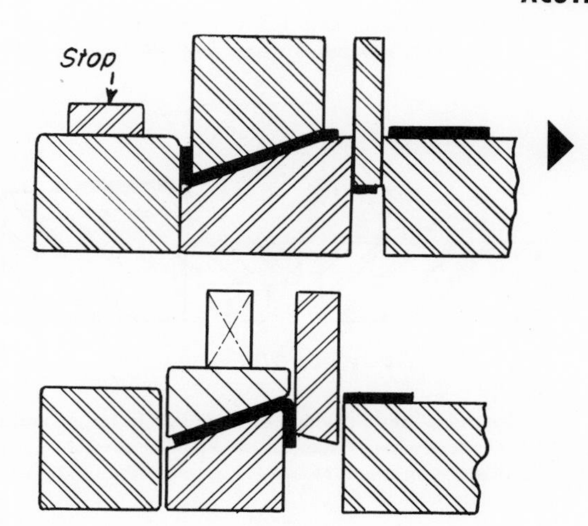

No. 9
Not suited to parts of all proportions
Resharpening die may cause difficulty
Scrap slug is wasted
Distortion of stock and creep are possible

No. 10
Suitable only for moderately acute angles
No scrap waste
Some distortion of stock is likely

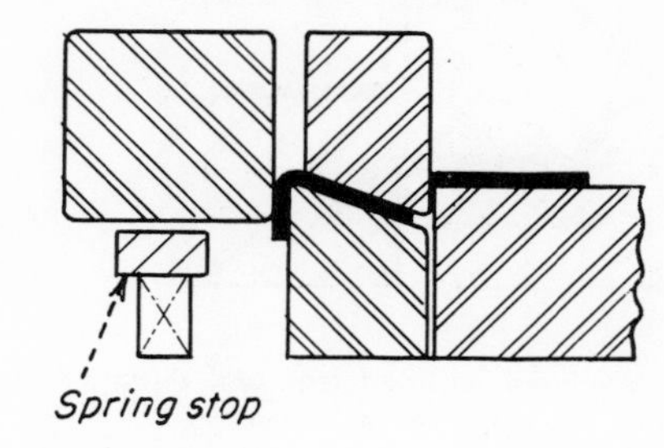

No. 11
No scrap waste
Not suited to parts of all proportions
Resharpening may cause some difficulty
Distortion of stock and creep are likely

ACUTE-ANGLE BENDS ... continued

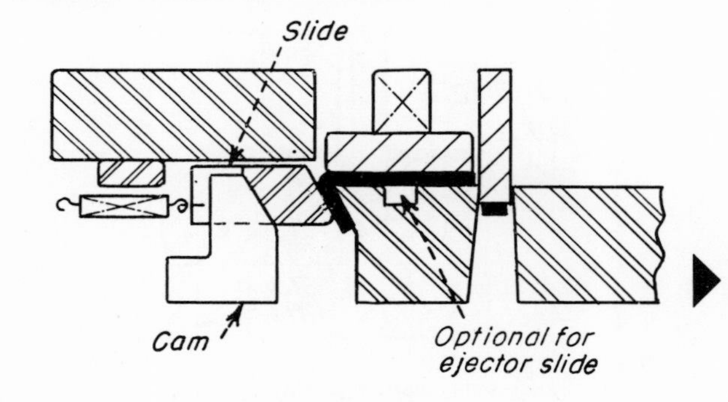

No. 12
- Widest adaptability
- Good-quality bends are produced
- More costly than other dies
- Inversion of design is possible

OBTUSE-ANGLE BENDS

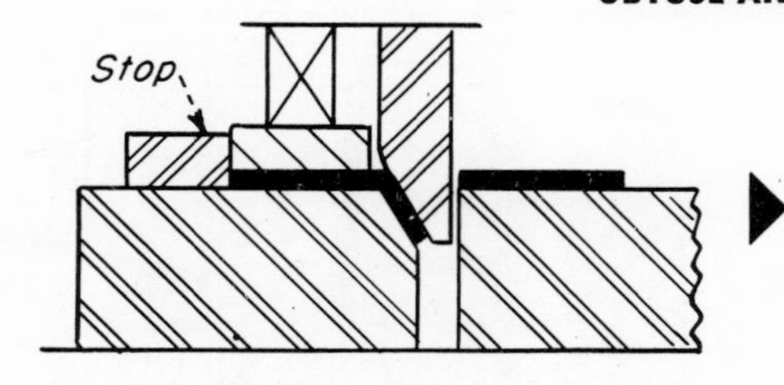

No. 13
- Inclined ejection is required
- Some difficulty in resharpening
- Special backup heel may be needed
- Large spring space required in punch holder

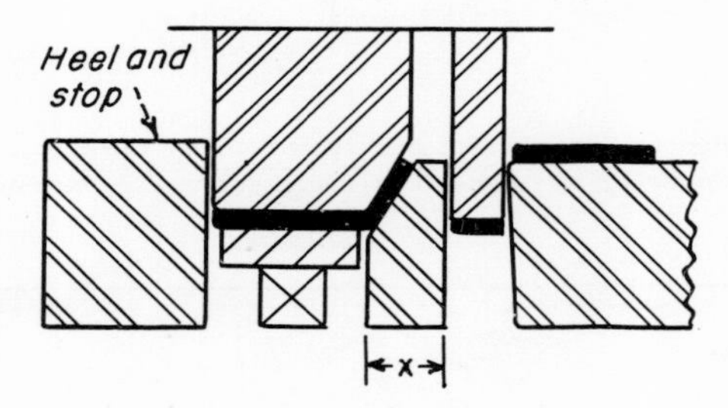

No. 14
- Inclined ejection is desirable
- Scrap slug wasted
- Some difficulty in resharpening

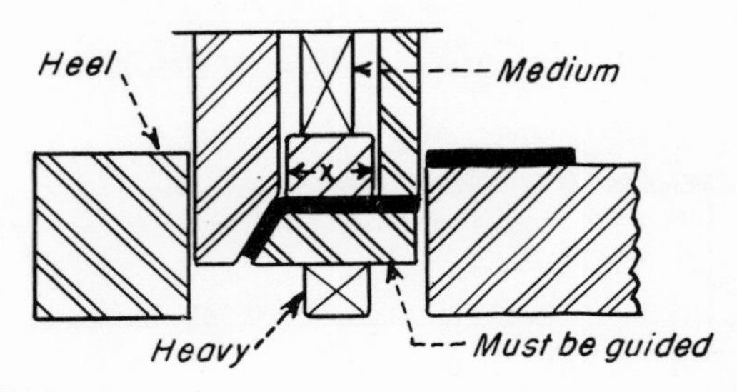

No. 15
- More complicated and costly than other designs
- Scrap slug is eliminated when form down is necessary

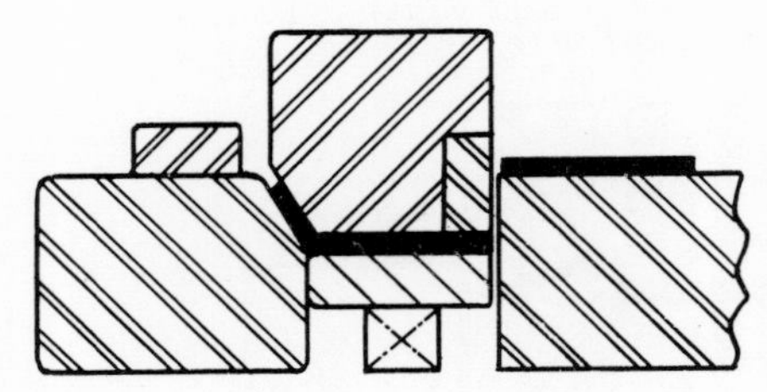

No. 16
- Good-quality bends produced on short legs and sharp corners
- Stock distortion occurs on long legs or when angle is close to 90°

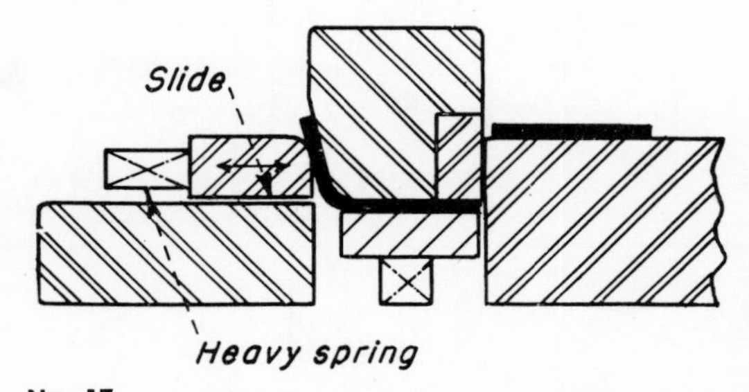

No. 17
- Desirable design when bend radius is large, because follow wiper minimizes stock out of control
- Usually unsuited to angles greater than 120°

PRESS TOOLS FOR BENDING...IV

DON R KING, tool engineering consultant

OBTUSE-ANGLE BENDS — Continued

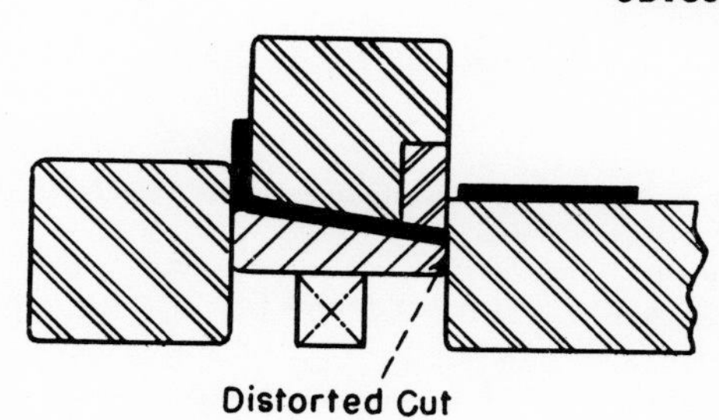

No. 18
Good design only when angle is close to 90°
Bends of good quality produced regardless of leg length or radius

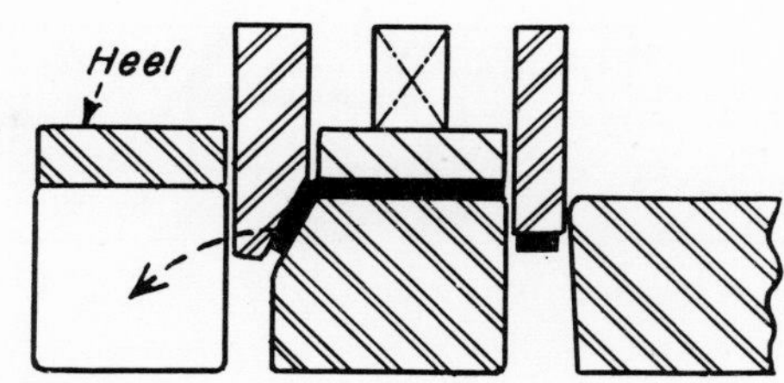

No. 19
Good-quality bends when legs are short
Push-through ejection is possible
Scrap slug is wasted
Large spring space needed in punch holder

CHANNELS

No. 20
Good location and alignment of bends
Inclined ejection is desirable
Scrap slug is wasted
Long cutoff punch is required

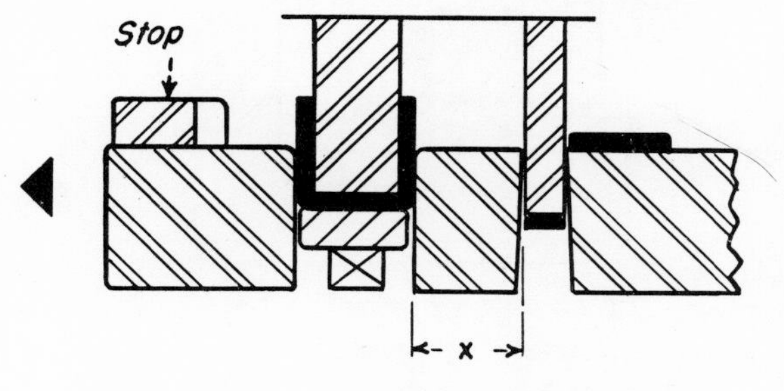

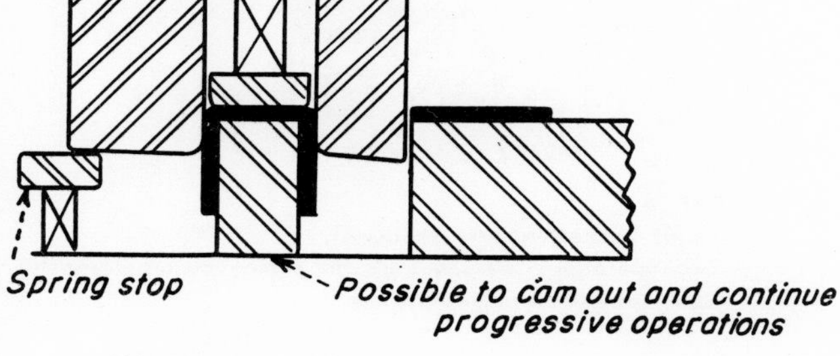

No. 21
Good design for cross-transfer operation
Inclined ejection is desirable
Large spring space needed in punch holder

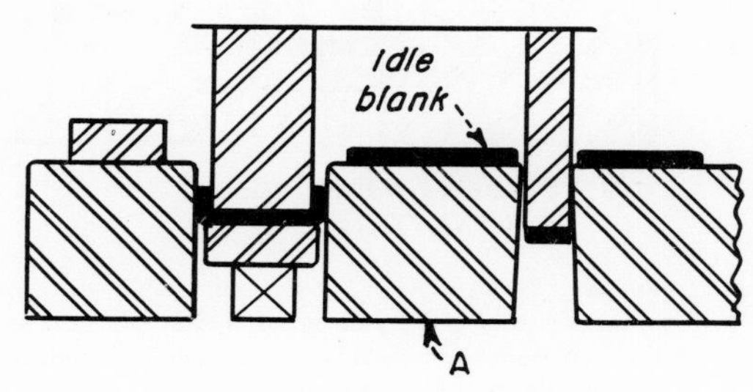

No. 22
Same notes as for No. 20 above
Introducing an idle station avoids thin die wall

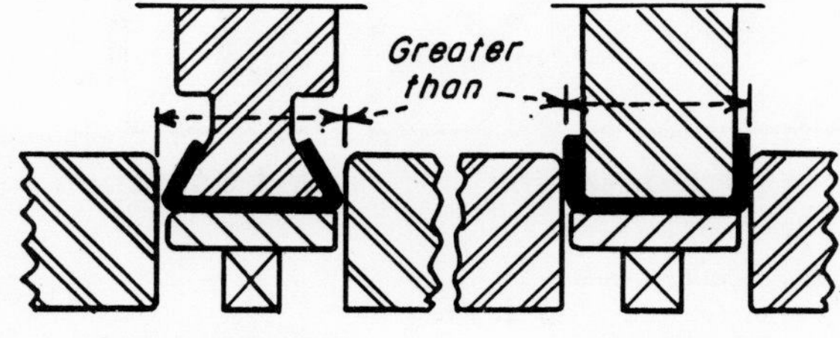

ACUTE CHANNELS

No. 23
Good quality bends are produced
Special ejection requirements must be met, but inversion may help
Cross-transfer or cut-and-carry progressive operation may be employed
Closely fitted guides or nest are required on 2nd operation

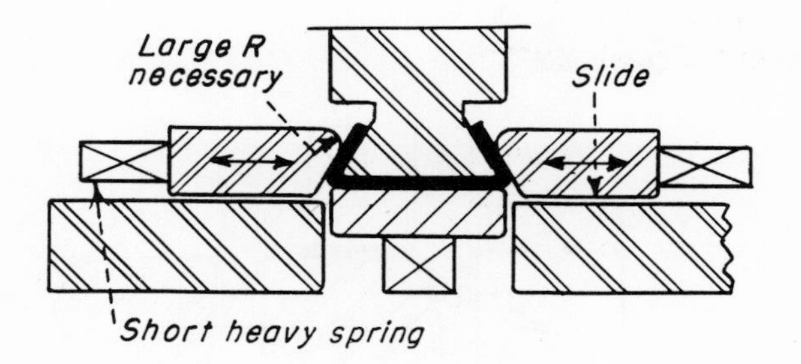

ACUTE CHANNELS — continued

No. 24

Good quality bends produced; even 90° bends in springy material
Not used for heavy stock or extreme acute angles
Special ejection means are required

WINGED CHANNELS

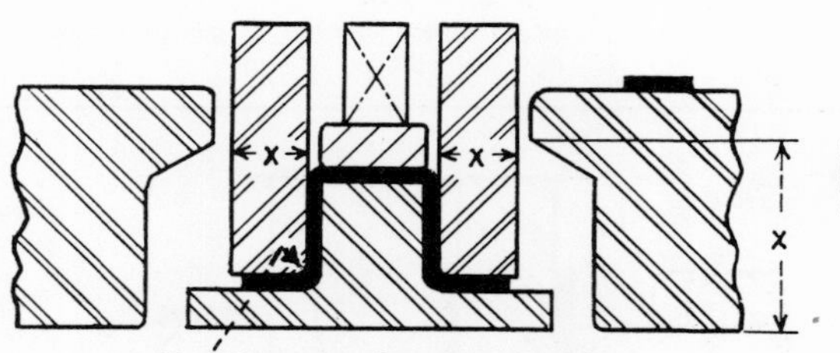

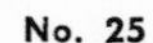

No. 25

Limited to small parts
Fair-quality bends produced, but wings will not be square unless spanked
Large spring space may be required in punch holder

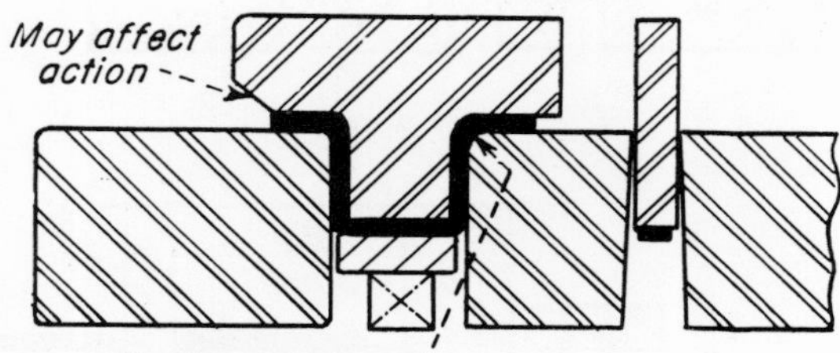

No. 26

Bends usually of poor quality, but die cost is low
Distortion remains from "slip forming" unless straightened by spanking
Not suited to parts of all proportions

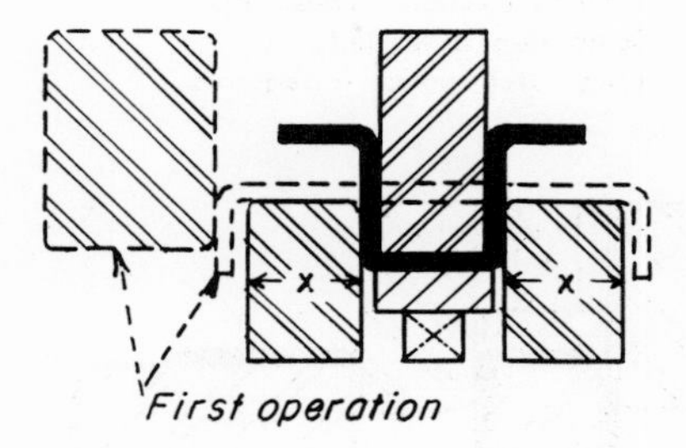

No. 27

Bends of best quality produced
Two operations are required; these may be cross-transfer or cut-and-carry progressive

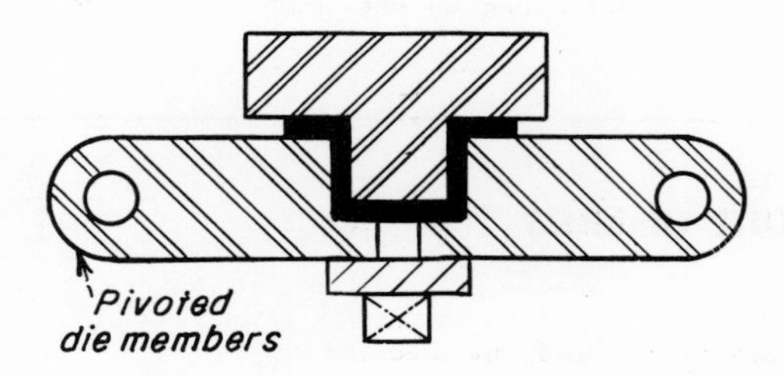

No. 28

Bends of fair quality
Do not use for heavy stock
Die more difficult to construct, but useful for odd angles

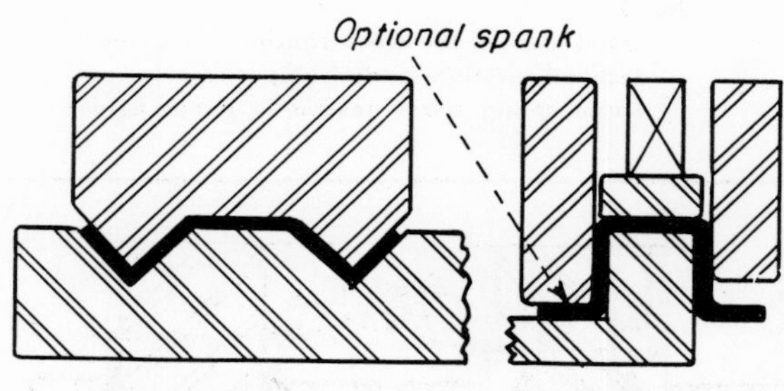

No. 29

Fair quality bends
Useful for odd wing angles
Inversion of design is possible
Die may be cross-transfer or cut-and-carry progressive type

PRESS TOOLS FOR BENDING...VI

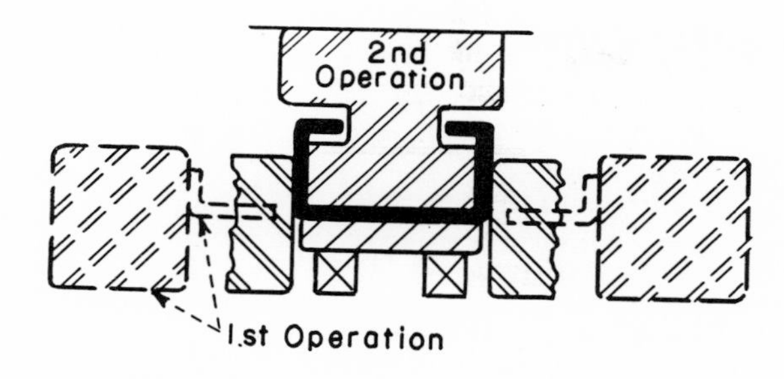

RETURN FLANGES

No. 30
- Two operations are required, but method produces good-quality bends
- Special ejection means required, but inversion may aid ejection
- Piece may be cross transferred or made in cut-and-carry progressive die

Z-BENDS AND OFFSETS

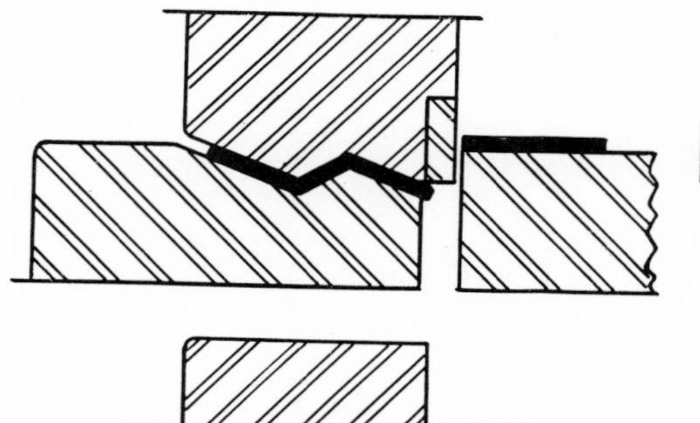

No. 31
- Good quality bends on small offset
- Low die cost

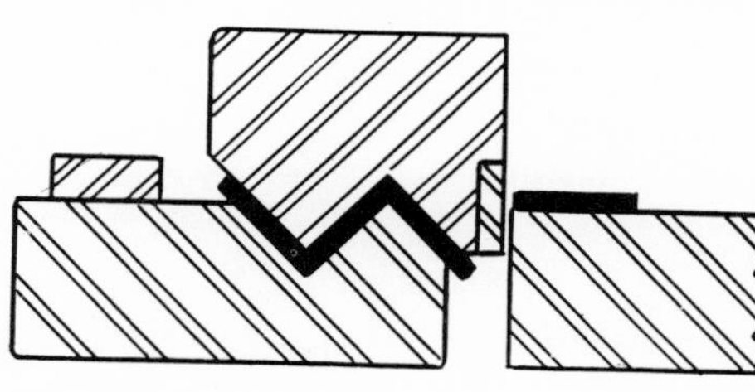

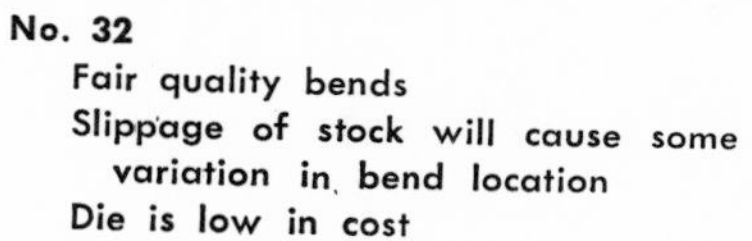

No. 32
- Fair quality bends
- Slippage of stock will cause some variation in bend location
- Die is low in cost

Medium

Heavy

No. 33
- Good quality bends produced
- Slight tendency to creep is minimized by balancing spring pressures
- Large spring space required
- Pads must be guided

END HOOKS

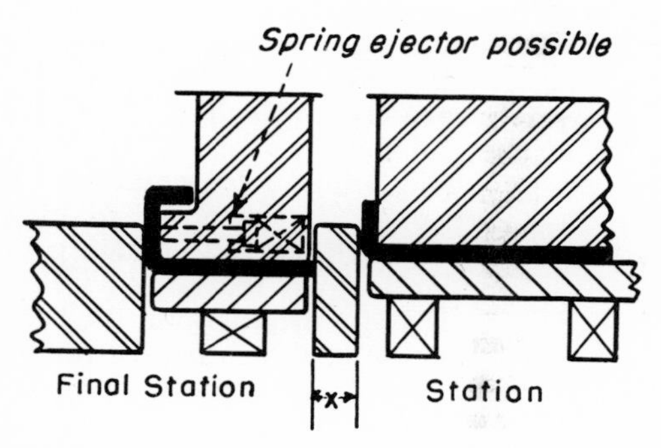

No. 34
- Good quality bends
- Special ejection requirements
- Inverted piercing and notching required

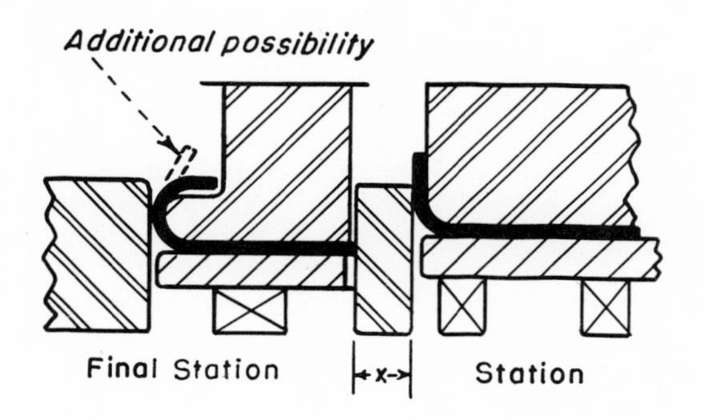

No. 35
- Notes same as at left, except that slight distortion of bend radius may occur

Wing bending methods

Tangent and stretch bending methods and folding techniques are opening up new economies in making sheet-metal products by wrap-around instead of multi-panel construction

Wing-type and stretch-bending equipment are used in the metal working industry for the production folding, tangent bending, and stretch bending of preformed sheet stock, bars, tubes, structurals, and extrusions with sections like those shown in Fig. 1. The greatest proportion of these jobs is done on the tangent bender due to the capabilities of the machine, its relatively high production rate, and the quality of work attainable.

Material—Use of wing-type and stretch bending machinery involves processing parts made from various grades of these materials:

(1) Low carbon steel
(2) Stainless steel
(3) Drawing-quality steel
(4) Copper and brass
(5) Aluminum
(6) Magnesium (in heated dies)

Material properly selected and prepared for bending operations meets these conditions:

(1) Sufficient elongation
(2) Preform section design suitable for proper dies and insert support during bending.
(3) More complicated preform designs can be supported by mandrels, or mechanically actuated die inserts during bending.
(4) A preformed section properly dimensioned for the tangent bending process is illustrated in Fig. 2. In this typical section, as dimensioned, it is understood that mill tolerance of +0.005 in., −0.003 in. metal thickness (MT) is acceptable.

The bending methods illustrated are normally used for making up to 90° metal folds or radius bends in flat or preformed metal.

Metal folding is a wing-type bending method by which comparatively sharp bends are made in flat or preformed sheet metal to produce cabinet shells and parts with sheer or well-defined corners. Where preformed material is processed through a metal-folding operation, corner notching is utilized so that no side flange upsetting or stretching is done, as in the case of tangent bending.

By Edward P Schneider,
Metalforming specialist
Taylor-Winfield Corp, Warren, Ohio

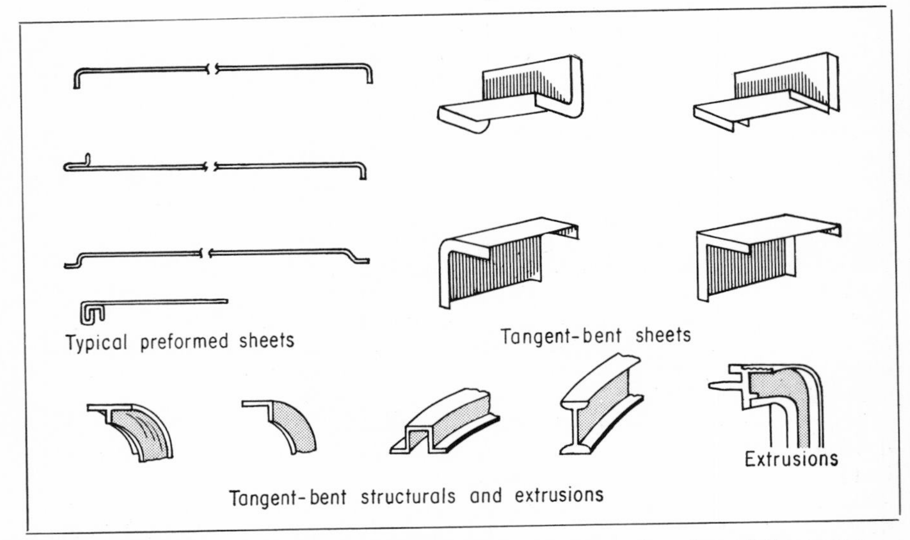

1. These preformed sheets and tangent-bent parts are only a few of the many shapes that lend themselves to wrap-around construction of products

Tangent bending is a metal-forming process in which flat or preformed material is precision-bent to a specified radius. This type of bending is accomplished by a controlled rocking movement of a straight block-type 'rocker die' around a properly positioned 'radius die.' The bending takes place in small, progressive increments, always at the tangent point of rocker die application. At this tangent point the material is confined so tightly between the die elements that each stage of progressive metal movement causes continuous flow into a wrinkle-free bend.

The rocker die component in a properly designed tangent bending die does not merely slide around the bend to set the metal down; its movement with respect to the material being processed is positively controlled by mechanical means.

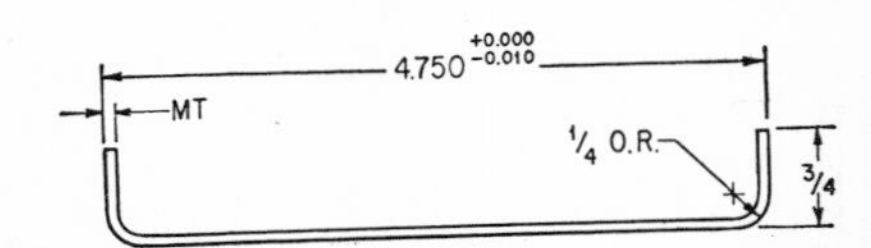

2. Tangent-bent parts should be dimensioned to the outside of the form

The bending methods illustrated are normally used for making up to approximately 100° bends in flat or preformed metal. Variations in design of a product and the bending machine result in bends up to 180° for the wing-type bending methods shown in Figs. 5 and 6 on special applications.

The pivot point in each of the four bending methods shown corresponds to the swing center of the bender wing in a properly designed and efficiently maintained setup.

For either of the fold-action bending methods in Figs. 3 and 4, the pivot point is exactly in line with the top surfaces of clamp and wing dies, and is at a distance of MT (or slightly less) from the nose of the radius die.

Right angle bend—fold action

This wing-type bending, or folding method, Fig. 3, is usually employed in leaf-type sheet metal bending brakes or plain wing type folding machines where a 'sharp' corner is desired. Although the nose on the radius die may be machined to get a small inside radius in a fold, actual production will check between MT and 2 MT inside radius.

For 'sharp' folds in sheet metal having inside radius = MT (or less) it is always advisable to consider coining them on a press brake.

Radius bend—fold action

Technically this is the same action, Fig. 4, as utilized in Fig. 3, differing mainly in that a greater uniformity of forming from end to end of the bend is possible. This method of fold-action radius bending requires a much greater overbending allowance as well as a means of checking the unwinding of the bend when the bender wing swings down.

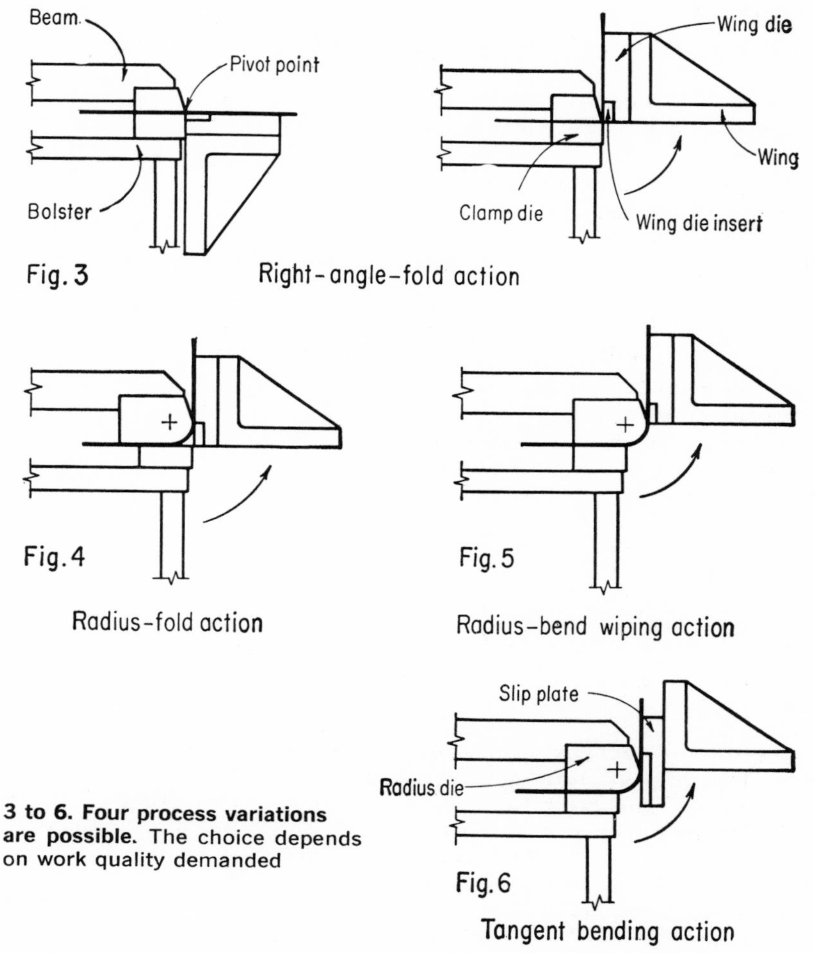

Radius bend—wipe action

In this type of wing-action radius bending, Fig. 5, the pivot point, or wing swing center, corresponds exactly with the center of the radius on the nose end of the radius die.

The wing die pushes the part metal against the radius die and skids on the outer surface of the bend area as the bender wing swings up. This action 'wipes' a bend in the part metal. Overbend requirements are less and more positively predicted in this method of radius bending of flat sheet metal than in bending methods described under Figs. 4 and 6. The wiping pressure is readily varied by shimming or spring loading the wing die to give desired setting effect to the bend as it is made.

3 to 6. Four process variations are possible. The choice depends on work quality demanded

Radius bend—tangent bending action

In this type of wing-action radius bending, the pivot point, or wing swing center corresponds exactly with the center of the radius on the nose end of the die.

Die space distances are at standard 3-place decimal dimensions, plus 0 000 in., minus 0.002 in., for:

(1) Bolster to pivot point, vertical direction.

(2) Pivot point vertically to beam in clamping position

(3) Pivot point horizontally away from wing to beam

In this method of wing-type bending, the rocker die applies canti-lever-beam type bending forces to the material in progressive increments, always tangent to the radius die during the completion of a bend.

Normally there is no skidding of a rocker die on part material as a tangent bend is made or as bender wing returns. This feature makes the process ideal for bending precoated or pre-finished stock.

SPRING-BACK CONTROL...I

DON R KING, tool engineering consultant

The problem of springback occurs in almost every bending operation. Even with dead-soft materials, springback may become serious where accuracy of the bend is required.

The pressure-pad type of wiped bend, Fig 1, is usually considered superior, but it is not always the complete answer. A point often overlooked is the rather high pad pressure required. This may be difficult to obtain unless a cushion is available in the press.

Standard formulas will give the approximate holding pressure, but in all cases the die should be tried out in "slow motion". If at any point in the bending stroke, a separation can be detected between the stock and the punch, more spring is needed. A relief angle on the punch, as shown, and a tight wiper setting are of some help.

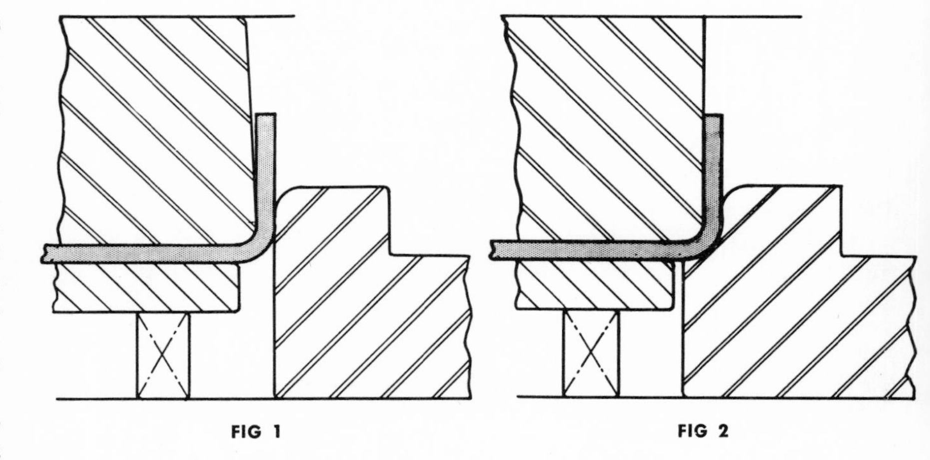

FIG 1 FIG 2

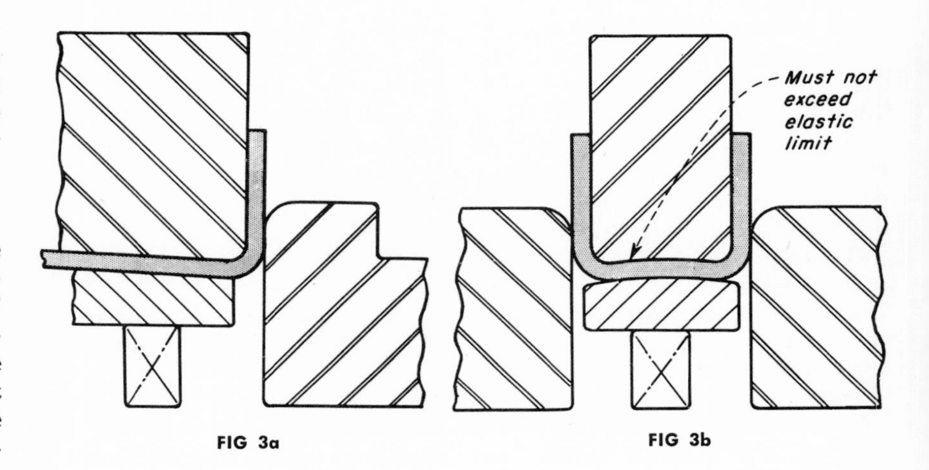

FIG 3a FIG 3b

"SPANKING" THE BEND

Another means of giving a more definite "set" to the material is illustrated in Fig 2. Here "spanking" of the bend area occurs at the bottom of the stroke. Careful stroke adjustment is required to prevent abnormally high pressures. Some improvement may be gained by altering the spanking radius to reduce the area of contact, but this partially defeats the purpose, and does not set the entire bend area.

One of the most satisfactory control methods is shown in Fig 3*a*. Here the pad and punch are constructed on an angle to compensate for the amount of springback. Uniformity of bends is good and not affected by the press stroke. As applied to channels, Fig 3*b*, this method is limited to combinations of channel width and material thickness that will not be permanently overformed by the compensation.

The usual explanation offered for the superiority of the pressure-pad, wiped-type of bend is that the material is ironed or stretched. It seems doubtful that the friction involved can produce stretching, and ironing mostly takes place after the bend is complete. A more likely explanation is that the wiper block or the entire die yields, and after passing tangency of the bend, re-

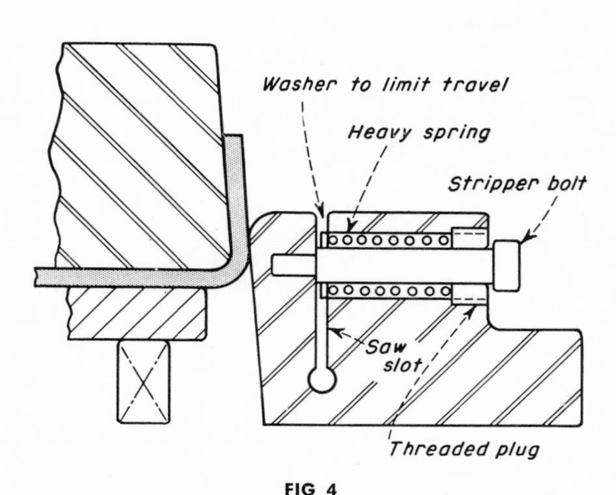

FIG 4

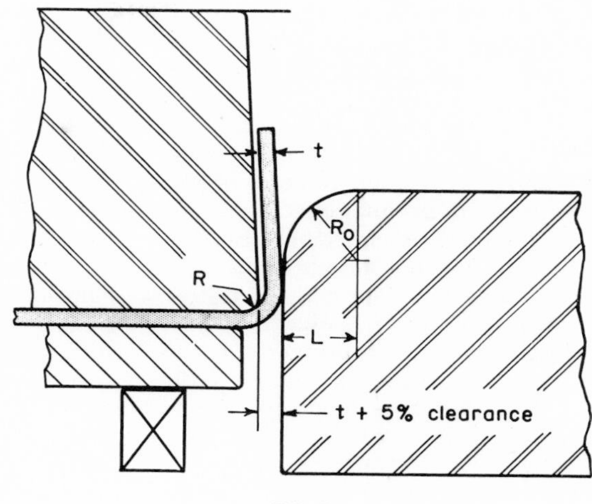

FIG 5

covery takes place and produces a small amount of overbending.

Extension of this reasoning led to the construction shown in Fig 4. Controlled movement of the wiper is introduced by means of springs and stops. The amount of movement required is very small. A few thousandths past tangency will correct a considerable angle of spring-back, and the simple spring-hinged wiper will be satisfactory in most cases. On light material, auxiliary springs may not be required.

Until now, we have considered only cases where the wiper is closely fitted, with stock clearance only, and have ignored the wiper-nose radius, as it has little effect under this condition. In Fig 5, a small clearance is provided between wiper and punch, in addition to stock thickness, and the wiper radius (actually length L) plays an important part in controlling spring back. When length L is in proper proportion, a small amount of "air bending" takes place at some point in the bending cycle. This is added to the formed bend, resulting in an overbend sufficient to counteract springback. The amount of overbend increases with an increase in L. It is also affected by the clearance, but it is easier to maintain this at about 5% of stock thickness and alter the nose radius. Exact dimensions will usually have to be determined by experiment. For a starting point, Table 1 shows the approximate dimension that will produce a slight overbend.

When the bend radius is large in proportion to stock thickness, the wiper radius becomes rather large. If space is limited, the nose contour shown in Fig 6 may be substituted. With this arrangement, L may be about 60% of the dimension for a plain radius.

table 1 . . LENGTH OF L TO OVERBEND

BEND RADIUS	LENGTH L
R = Sharp	4 t
R = ½ t	6 t
R = t	10 t
R = 2 t	15 t

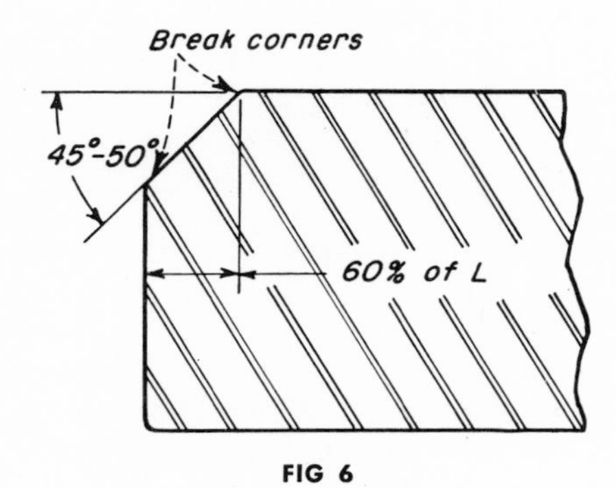

FIG 6

Control of U-Shaped Bends—I

By Bernard K H Bao, chief tool engineer, Cribben & Sexton Co, Chicago 12, Illinois

Long U-shaped pieces are often troublesome to make when close tolerances are required across the bends. Various corner-setting means are used in die design to compensate for spring back. And usually strip-stock tolerances are required in production.

The method described in this article, in the author's opinion, is the most efficient way to control spring back when the inside bending radius is less than three times the stock thickness. Stock with a close thickness tolerance is not required. Regular sheet-stock tolerance is satisfactory in piece-part production. It is also possible to overbend a piece 2° to 3° when required.

The set corner on the die block is ground with a radius $R + t$ and is stopped at $\frac{1}{2}t$, as shown in Fig. 1. For soft and ¼ hard CRS or aluminum, the punch is made square (without back taper) for a 90° bend. Back taper is allowed on the punch side when overbend is required. Clearance between punch, die and pad is made to maximum stock thickness ($t_{max.}$) In calculating the dimension h, use minimum stock thickness. A nomograph, page 157 simplifies the calculation of h.

A regular rubber stamp, as shown in Fig. 2, may be used to show detailed dimensions of the set corner. Dimensions are filled in after all calculations are made.

Usually it is difficult to hold a close-tolerance bend in one bending operation for a part as shown in Fig. 3. For example, a piece with a large radius (say 8 to 10 times stock thickness) on one side and a smaller bending radius (say less than 3 times stock thickness) on the other side. By using the set corner described, it is possible to bend this part in one operation. The punch face and pad are ground at an angle to compensate for the spring back of the longer leg, Fig. 4. For the shorter leg the punch side is ground with a back taper to over-bend the shorter leg for the amount needed.

Punch radius R' for the longer leg usually is smaller than R on the finished part. It may be calculated* once the angle of spring back is determined by tryout.

*Refer to BAO Stamping Calculator sold by BAO Slide Co, P.O. Box 7902, Chicago 80, Ill.

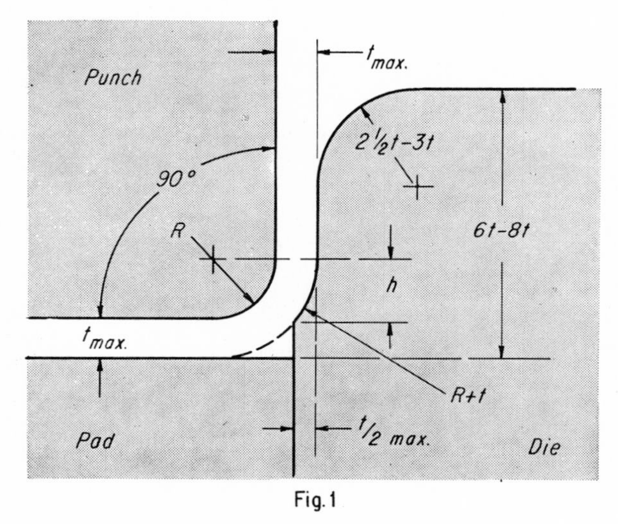

Fig. 1

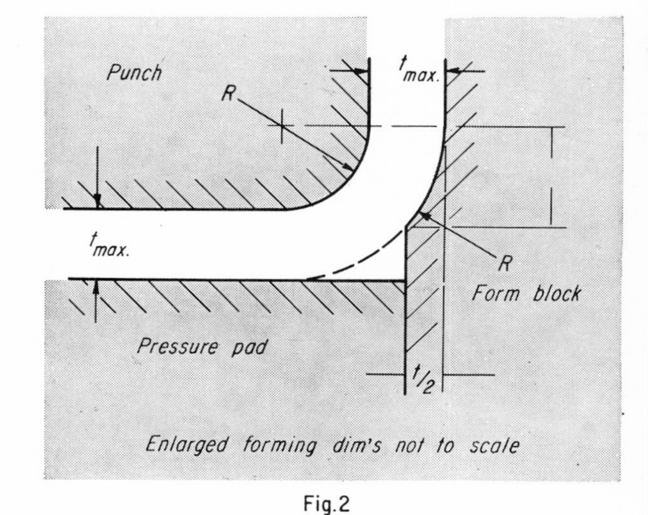

Fig. 2

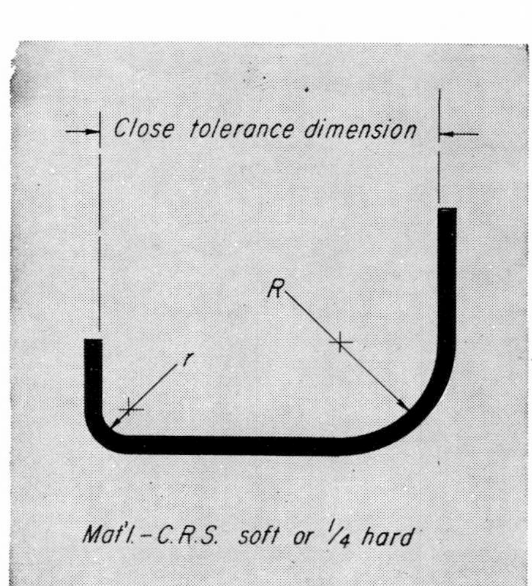

Fig. 3

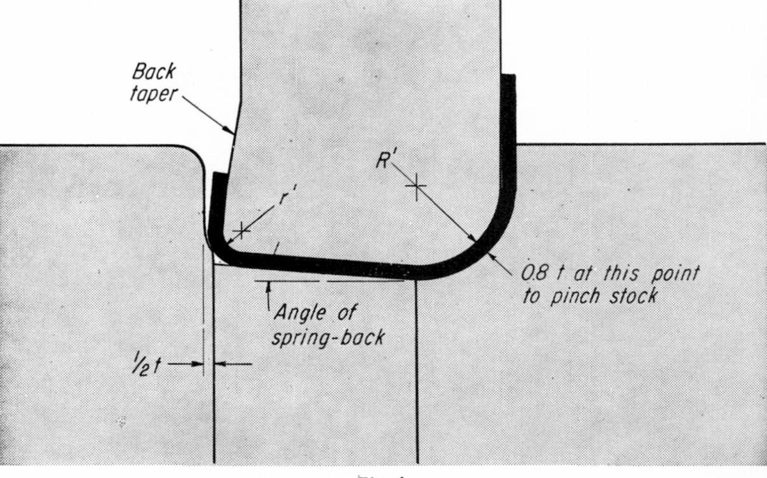

Fig. 4

Control of U-Shaped Bends — II

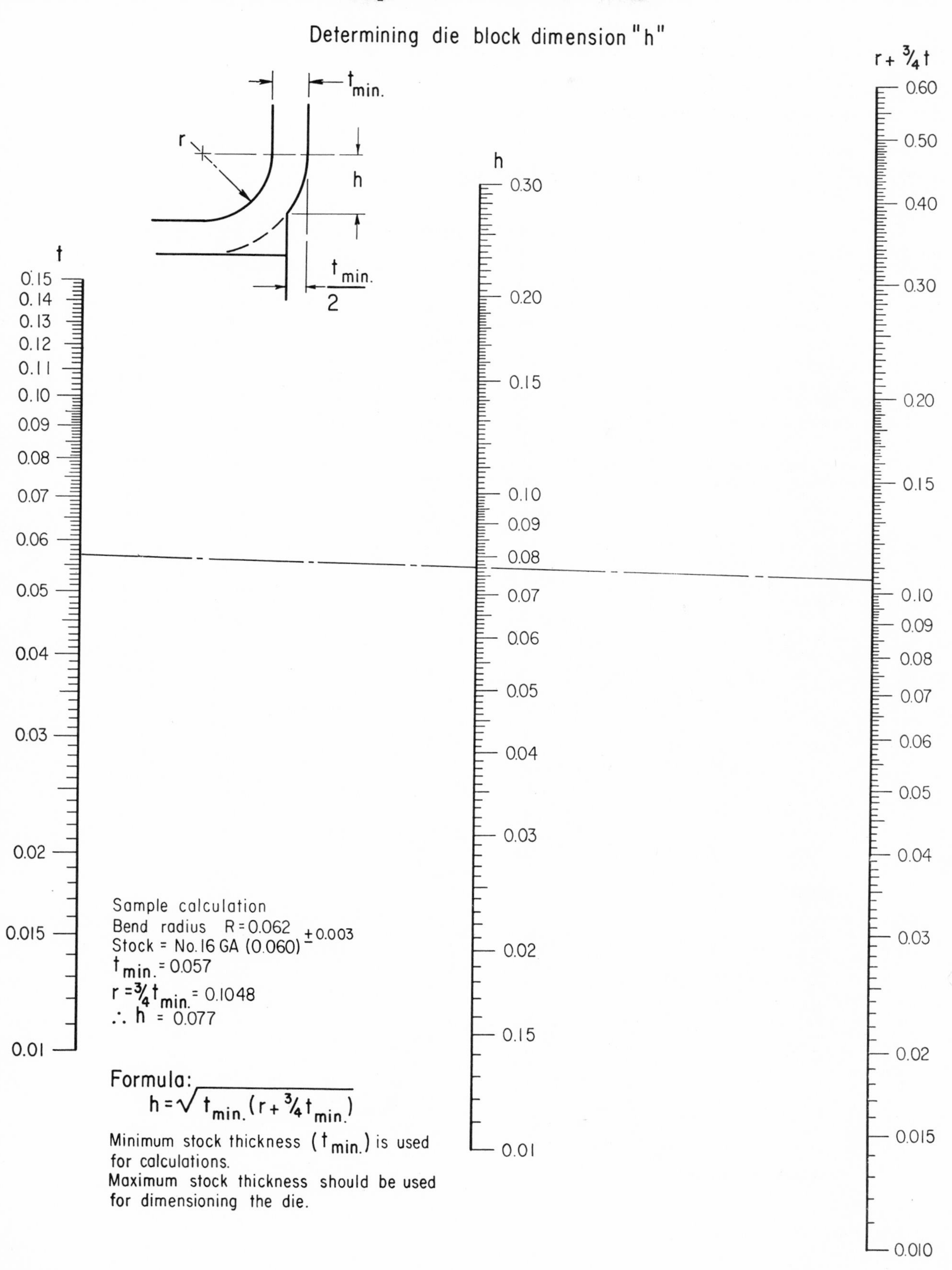

Bend radii for aluminum sheet and plate

Here are approximate bend radii for 90° cold bends in aluminum sheet and plate of various tempers, and also minimum bend radii for painted aluminum sheet

Approximate Bend Radii① for 90-degree Cold Bend

ALLOY	TEMPER	RADII FOR VARIOUS THICKNESSES EXPRESSED IN TERMS OF THICKNESS "t"							
		1/64 in.	1/32 in.	1/16 in.	1/8 in.	3/16 in.	1/4 in.	3/8 in.	1/2 in.
1100	-O	0	0	0	0	0	0	0	1t-2t
	-H12	0	0	0	0	0-1t	0-1t	0-1t	1t-3t
	-H14	0	0	0	0	0-1t	0-1t	0-1t	2t-3t
	-H16	0	0	0-1t	½t-1½t	1t-2t	1½t-3t	2½t-3½t	3t-4t
	-H18	0-1t	½t-1½t	1t-2t	1½t-3t	2t-4t	2t-4t	3t-5t	3t-6t
Alclad 2014	-O	0	0	0	0	0-1t	0-1t	1½t-3t	3t-5t
	-T3	1t-2t	1½t-3t	2t-4t	3t-5t	4t-6t	4t-6t	5t-7t	5½t-8t
	-T4	1t-2t	1½t-3t	2t-4t	3t-5t	4t-6t	4t-6t	5t-7t	5½t-8t
	-T6	2t-4t	3t-5t	3t-5t	4t-6t	5t-7t	6t-10t	7t-10t	8t-11t
2024	-O②	0	0	0	0	0-1t	0-1t	1½t-3t	3t-5t
	-T3②③	1½t-3t	2t-4t	3t-5t	4t-6t	4t-6t	5t-7t	6t-8t	6t-9t
	-T36②	2t-4t	3t-5t	4t-6t	5t-7t	5t-7t	6t-10t	7t-10t	8t-11t
	-T4②	1½t-3t	2t-4t	3t-5t	4t-6t	4t-6t	5t-7t	6t-8t	6t-9t
	-T81	3½t-5t	4½t-6t	5t-7t	6½t-8t	7t-9t	8t-10t	9t-11t	9t-12t
	-T86	4t-5½t	5t-7t	6t-8t	7t-10t	8t-11t	9t-11t	10-t13t	10t-13t
3003	-O	0	0	0	0	0	0	0	1t-2t
	-H12	0	0	0	0	0-1t	0-1t	0-1t	1t-3t
	-H14	0	0	0	0-1t	0-1t	½t-1½t	1t-2½t	1½t-3t
	-H16	0-1t	0-1t	½t-1½t	1t-2t	1½t-3t	2t-4t	2½t-4t	3t-5t
	-H18	½t-1½t	1t-2t	1½t-3t	2t-4t	3t-5t	4t-6t	4t-7t	5t-8t
3004	-O	0	0	0	0	0-1t	0-1t	½t-1½t	1t-2t
	-H32	0	0	0	0-1t	0-1t	½t-1½t	1t-2t	1½t-2½t
	-H34	0	0	0-1t	½t-1½	1t-2t	1½t-3t	2t-3t	2½t-3½t
	-H36	0-1t	½t-1½t	1t-2t	1½t-3t	2t-4t	2t-4t	2½t-5t	3t-5½t
	-H38	½t-1½t	1t-2t	1½t-3t	2t-4t	3t-5t	4t-6t	4t-7t	5t-8t
5005	-O	0	0	0	0	0	0	0	1t-2t
	-H12	0	0	0	0	0-1t	0-1t	0-1t	1t-3t
	-H14	0	0	0	0-1t	0-1t	½t-1½t	1t-2½t	1½t-3t
	-H16	0-1t	0-1t	½t-1½t	1t-2t	1½t-3t	2t-4t	2½t-4t	3t-5t
	-H18	½t-1½t	1t-2t	1½t-3t	2t-4t	3t-5t	4t-6t	4t-7t	5t-8t
	-H32	0	0	0	0	0-1t	0-1t	0-1t	1t-3t
	-H34	0	0	0	0-1t	0-1t	½t-1½t	1t-2½t	1½t-3t
	-H36	0-1t	0-1t	½t-1½t	1t-2t	1½t-3t	2t-4t	2½t-4t	3t-5t
	-H38	½t-1½t	1t-2t	1½t-3t	2t-4t	3t-5t	4t-6t	4t-7t	5t-8t
5050	-O	0	0	0	0	0	0	—	—
	-H32	0	0	0	0	0-1t	½t-1½t	2t-3½t	2½t-4t
	-H34	0	0	0	0-1t	½t-1½t	1t-2t	2t-4t	3t-4t
	-H36	0-1t	0-1t	½t-1½t	1t-2t	1½t-3t	2t-4t	2½t-4t	3t-5t
	-H38	½t-1½t	1t-2t	1½t-3t	2t-4t	3t-5t	4t-6t	4t-7t	5t-8t
5052	-O	0	0	0	0	0-1t	0-1t	½t-1½t	1t-2t
	-H32	0	0	0	0-1t	0-1t	½t-1½t	1t-2t	1½t-2½t
	-H34	0	0	0-1t	½t-1½t	1t-2t	1½t-3t	2t-3t	2½t-3½t
	-H36	0-1t	½t-1½t	1t-2t	1½t-3t	2t-4t	2t-4t	2½t-5t	3t-5½t
	-H38	½t-1½t	1t-2t	1½t-3t	2t-4t	3t-5t	4t-6t	4t-7t	5t-8t
5083	-O	—	—	0-½t	0-1t	0-1t	½t-1½t	1½t-2t	1½t-2½t
5086	-O	0	0	0	0-1t	0-1t	½t-1t	½t-1t	½t-1½t
	-H32	0-½t	0-1t	½t-1½t	1t-2t	1½t-2t	1½t-2½t	2t-2½t	2½t-3t
	-H34	0-1t	½t-1½t	1t-1½t	1½t-2½t	2t-3t	2t-3t	2½t-3½t	3t-4t
	-H36	—	—	—	2t-3½t	2½t-4t	3t-4½t	3t-5t	3½t-5½t
	-H112	0-½t	0-½t	½t-1t	½t-1½t	1t-1½t	1t-2t	1t-2t	1½t-2½t

Courtesy: The Aluminum Association, New York, NY

Bend radii for aluminum [continued]

ALLOY	TEMPER	RADII FOR VARIOUS THICKNESSES EXPRESSED IN TERMS OF THICKNESS "t"							
		1/64 in.	1/32 in.	1/16 in.	1/8 in.	3/16 in.	1/4 in.	3/8 in.	1/2 in.
5154	-O	0	0	0	0-1t	0-1t	½t-1½t	½t-1½t	1t-2t
	-H32	0	0	0-1t	½t-1½t	1t-2t	1½t-3t	2t-4t	2½t-5t
	-H34	0-1t	0-1t	½t-1½t	1t-2t	1½t-3t	2t-4t	2½t-4½t	3t-5t
	-H36	0-1t	½t-1½t	1t-2t	1½t-3t	2t-4t	2t-4t	2½t-5t	3t-6t
	-H38	1t-2t	1½t-3t	2t-4t	3t-5t	4t-6t	4t-6t	5t-8t	5t-8t
	-H112	—	—	—	—	—	1½t-3t	2½t-4t	3t-5t
5456	-O	—	—	—	0-1t	½t-1t	½t-1t	½t-1½t	½t-2t
	-H321	—	—	—	2t-3t	3t-4t	3t-4t	3t-4t	3t-4t
	-H323	—	—	1t-2t	1½t-3t	1½t-3½t	2t-4t	—	—
	-H343	—	—	1t-2t	1½t-3t	2t-4t	2½t-4½t	—	—
5457	-O	0	0	0	0	0	0	0	1t-2t
	-H38	½t-1½t	1t-2t	1½t-3t	2t-4t	3t-5t	4t-6t	4t-7t	5t-8t
6061	-O	0	0	0	0	0-1t	0-1t	½t-2t	1t-2½t
	-T4②	0-1t	0-1t	½t-1½t	1t-2t	1½t-3t	2t-4t	2½t-4t	3t-5t
	-T6②	0-1t	½t-1½t	1t-2t	1½t-3t	2t-4t	3t-4t	3½t-5½t	4t-6t
7075	-O	0	0	0-1t	½t-1½t	1t-2t	1½t-3t	2½t-4t	3t-5t
	-T6②	2t-4t	3t-5t	4t-6t	5t-7t	5t-7t	6t-10t	7t-11t	7t-12t
7079	-O	0	0	0-1t	½t-1½t	1t-2t	1½t-3t	2½t-4t	3t-5t
	-T6②	2t-4t	3t-5t	4t-6t	5t-7t	5t-7t	6t-10t	7t-11t	7t-12t
7178	-O	0	0	0-1t	½t-1½t	1t-2t	1½t-3t	2½t-4t	3t-5t
	-T6②	2t-4t	3t-5t	4t-6t	5t-7t	5t-7t	6t-10t	7t-11t	7t-12t

①Minimum permissible radius over which sheet or plate may be bent varies with nature of forming operation, type of forming equipment, and design and condition of tools. Minimum working radius for a given material or hardest alloy and temper for a given radius can be ascertained only by actual trial under contemplated conditions of fabrication.

②Alclad sheet can be bent over slightly smaller radii than the corresponding tempers of the uncoated alloy.

③Immediately after quenching, this alloy can be formed over appreciably smaller radii.

Recommended minimum bend radii for painted aluminum sheet in terms of metal thickness T

The table at right shows minimum bend radii for 90° bends in sheet with alkyd or acrylic films (recommended for moderate forming) and for 180° bends in sheet with vinyl films which are recommended for severe forming.

Minimum radius over which painted sheet may be bent varies with the nature of the forming operation, type of forming equipment, and design and condition of the tools. Minimum radius for a specific material, or hardest alloy and temper for a specific radius can be closely determined only by actual trial under contemplated conditions of fabrication.

BASE ALLOY	TEMPER BEFORE FILM APPLICATION	THICKNESS OF BASE SHEET—Inches					
		0.016	0.025	0.032	0.040	0.050	0.064
1100	—0	1T	1T	1T	1T	1T	1T
	—H12	1T	1T	1T	1T	1T	1T
	—H14	1T	1T	1T	1T	1T	1T
	—H16	1T	1T	1T	1T	2T	3T
	—H18	2T	2T	3T	3T	4T	5T
3003	—0	1T	1T	1T	1T	1T	1T
	—H12	1T	1T	1T	1T	1T	1T
	—H14	1T	1T	1T	1T	1T	1T
	—H16	1T	1T	2T	3T	3T	4T
	—H18	2T	3T	4T	5T	6T	7T
5005	—0	1T	1T	1T	1T	1T	1T
	—H32	1T	1T	1T	1T	1T	1T
	—H34	1T	1T	1T	1T	1T	1T
	—H36	1T	1T	2T	3T	3T	4T
	—H38	2T	3T	4T	5T	6T	7T
5050	—0	1T	1T	1T	1T	1T	1T
	—H32	1T	1T	1T	1T	1T	1T
	—H34	1T	1T	1T	2T	2T	3T
	—H36	2T	2T	3T	3T	4T	5T
	—H38	2T	3T	4T	5T	6T	7T
5052	—0	1T	1T	1T	1T	1T	1T
	—H32	1T	1T	1T	1T	1T	1T
	—H34	1T	1T	1T	2T	2T	3T
	—H36	2T	3T	3T	3T	4T	5T
	—H38	2T	3T	4T	5T	6T	7T

Limits for bends in aerospace alloys

Critical bend limits in degrees are given for 11 alloys. For limits below these values, you can interpolate to find the minimum radius

Bending tests were carried out by air bending; that is, the workpiece was supported at its outer edges only and the length of the ram stroke determined the bend angle α. Brake forming limits were extended to the 1500 F tooling limitation in order to find optimum temperatures .

Use of tables:

1. Compare the required bend angle with the tabulated critical bend angle.
2. If this angle is less than critical, interpolate R/T values and α below critical, to determine the required R/T ratio.
3. Multiply R/T by T, to determine the minimum bend radius R.

Example 1: Material—2024–T3 Al in 0.020 in. gage to be bent longitudinally to 130° at room temperature.

Required $\alpha = 130°$ is greater than critical bend angle of 111°. Therefore the required bend bend limit becomes the critical bend limit, or 111°, and $R/T = 2.02$. And R min $= R/T \times T = 2.02 \times 0.020 = 0.040$ in.

Example 2: Material—A –286, $T = 0.040$ in., required $\alpha = 52°$. From table, critical angle $= 128°$. By interpolation, $R/T = 0.079$ when $\alpha = 52°$. R min $= R/T \times T = 0.079 \times 0.040 = 0.0032$ in.

Extracted from supplemental handbook for "Final Report on Advanced Theoretical Formability Manufacturing Technology." This report was prepared under the direction of William W Wood, chief of manufacturing research and development, Aeronautics Division, Ling-Temco-Vought, Inc, Dallas, for the U S Air Force.

Bend limits at room and elevated temperatures

Material	Temp	Crit. Bend	Bend Limits	Bending limits R/T — For various angles α below critical						
	F	α	R/T	30	45	60	75	90	105	120
2024-T3	70	111	2.02	0.50	0.90	1.38	1.70	1.90	2.0	2.0
	600	122	1.07	0.17	0.28	0.45	0.75	1.00	1.15	1.18
Ti-8-1-1	70	104	2.60	0.80	1.30	1.90	2.30	2.55	2.6	2.6
	1500	128	0.60	0.05	0.07	0.13	0.23	0.43	0.57	0.60
TZM Moly	70	121	1.12	0.30	0.55	0.81	0.98	1.10	1.12	1.12
	500	118	1.40	0.10	0.20	0.25	0.30	0.50	1.10	1.40
Cb-752	70	121	1.20	0.38	0.60	0.85	1.05	1.18	1.20	1.20
	500	115	2.00	0.15	0.20	0.40	0.75	1.30	1.80	2.00
Ti-13-11-3	1000	126	0.75	0.08	0.10	0.15	0.23	0.38	0.60	0.73
	1200	129	0.48	0.03	0.05	0.07	0.14	0.26	0.38	0.47
PH 15-7 Mo	1500	126	0.65	0.05	0.08	0.12	0.20	0.38	0.55	0.63
AM-350	1000	123	1.03	0.12	0.20	0.33	0.48	0.68	0.90	1.03
	1500	131	0.35	0.02	0.03	0.04	0.07	0.15	0.28	0.35
A-286	1500	128	0.60	0.04	0.06	0.10	0.20	0.38	0.52	0.59
Rene 41	1000	125	0.86	0.09	0.15	0.25	0.39	0.58	0.80	0.87
L-605	1000	127	0.67	0.04	0.07	0.12	0.20	0.33	0.52	0.65
Tungsten	500	103	2.70	2.50	2.70	2.70	2.70	2.70	2.70	2.70
	1000	107	2.00	1.30	2.25	2.40	2.40	2.40	2.40	2.40

Note: Grain direction—longitudinal in all cases

Test conditions for checking bend limits

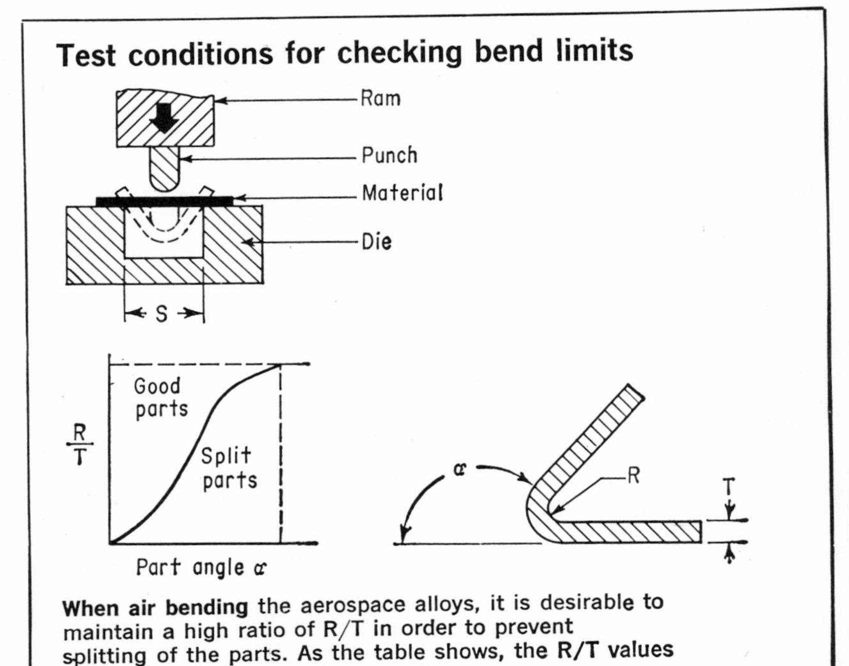

When air bending the aerospace alloys, it is desirable to maintain a high ratio of R/T in order to prevent splitting of the parts. As the table shows, the R/T values must increase with the bend angle

Bending jigs

Twelve bending jigs, or fixtures, offer the shop man a variety of means to make simple or complex bends in low-production parts

There are many advantages to using bending jigs to form wire or strip, instead of employing special press tools. Low-volume production and prototype work are examples of situations where bending jigs, or fixtures, are employed. These jigs are used because:

1. They allow complicated forming in one setting of the workpiece.
2. They are self-contained.
3. Such jigs can be installed quickly in any convenient place.
4. Usually they are hand-powered; accident hazards are low.
5. Parts produced on these jigs are square, uniform, flat on the straight portions, free from twists.
6. Building costs are often much less than for corresponding bending or forming dies.
7. The production rate can be high.

As will be seen in the sketches, bending jigs may be used satisfactorily whenever the traditional methods may present troubles and difficulties because of any of the following reasons:

1. Lot size too small; press tool amortization unsatisfactory.

By Federico Strasser

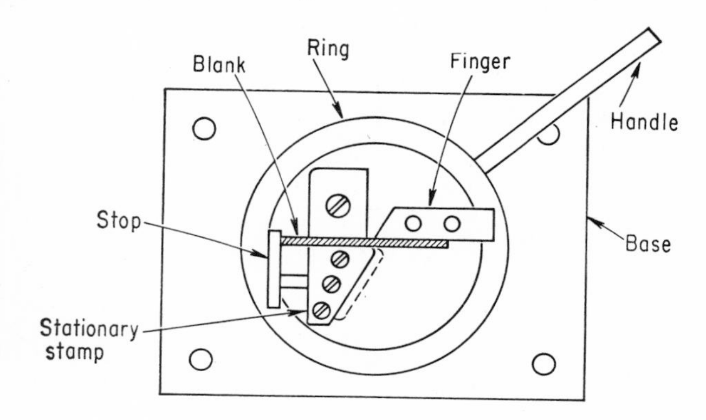

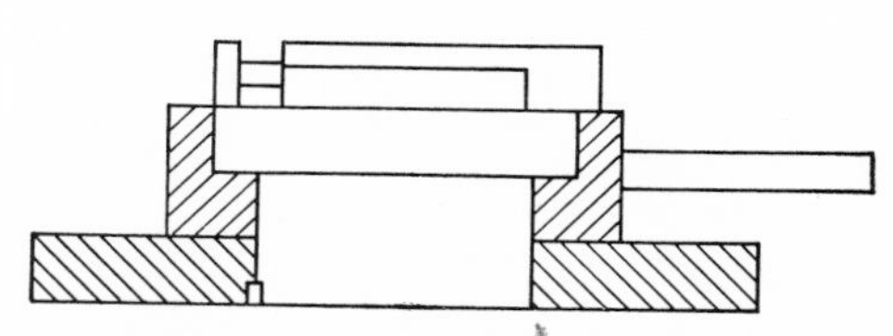

Universal jig accepts strip or round or square bars; bends to any angle. Only the stamp needs changing

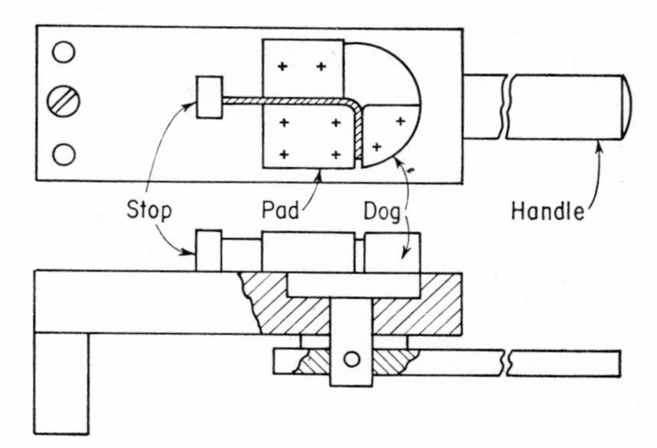

Small parts are formed in this jig. The rotating dog, or stamp, is guided by the jig body

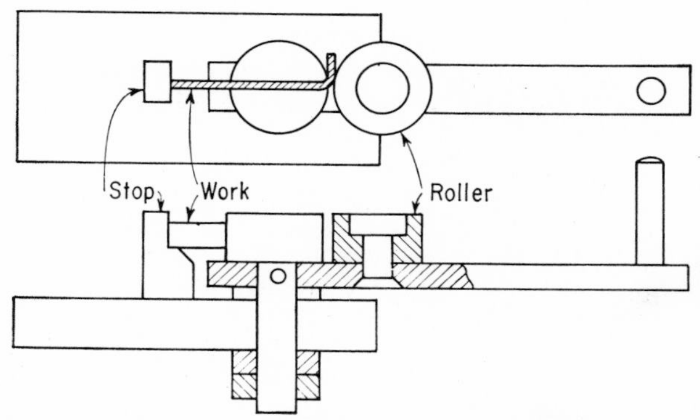

Parts with unequal legs are smoothly bent by roller, which avoids relative movement of part and tool

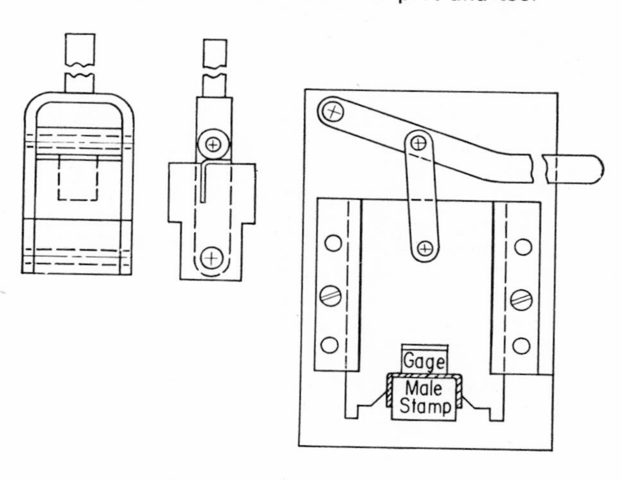

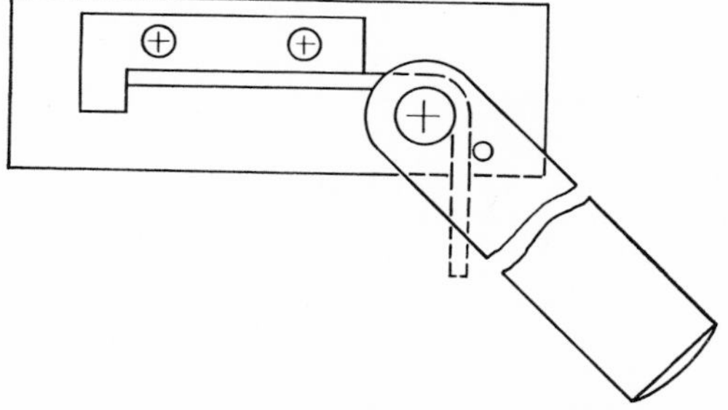

Lever (upper left) has long radius. Part shape (upper right) is in slide. Wire (below) is bent

Bending jigs [Continued]

2. Delivery dates too short; not enough time for the building of proper press tools.
3. Workpieces too small or too cumbersome for easy handling in press-tools.

The chief limitation of bending jigs derives from the fact that they are usually hand-powered. Therefore, they cannot be employed when force requirements are high.

Basic components of jigs

The chief members, or components, of bending jigs are:

1. A base for fastening or clamping to the workbench.
2. Stops or nests for the flat, or pre-formed, blanks. There are usually two stops in every bending jig: one stationary and one movable.
3. A rotating spindle or shaft, which carries out the actual forming operation. Its position may be vertical or horizontal, according to working conditions; it may be held in one bearing or in two bearings. Sometimes there are two (or more) spindles in a jig. Quite often the spindle is replaced by a slide.
4. A handle for moving the spindle or slide.
5. Sometimes clamping devices are provided to hold the blanks in the nests.

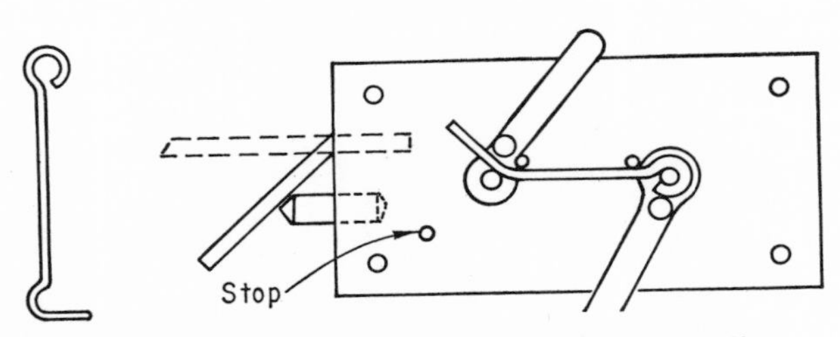

Hook is formed by bending ends to pointed pin, then rolling the preform by use of the two handles

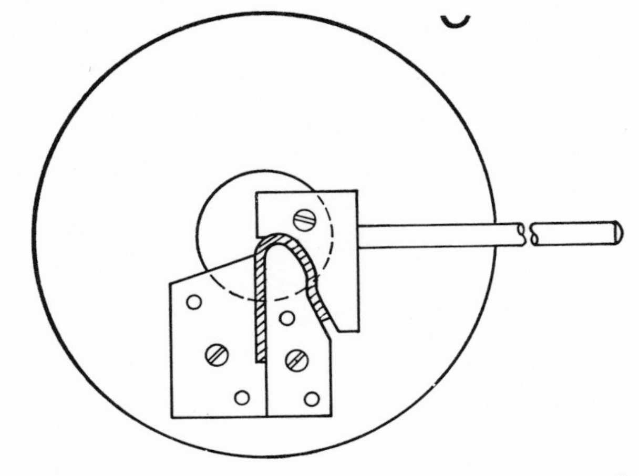

Compound bends are produced with one movement of the rotating tool member. Form blocks are easily changed

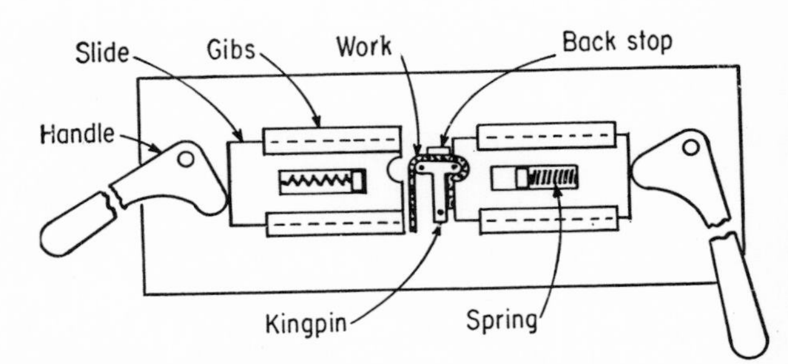

Two double bends are produced on a preformed, or U-shaped, workpiece. Springs return the slides

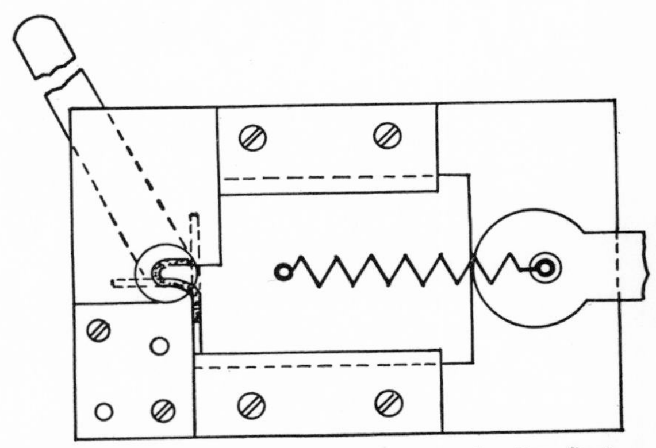

An eccentric drives the slide left to make the first form, and then the handle completes compound bend

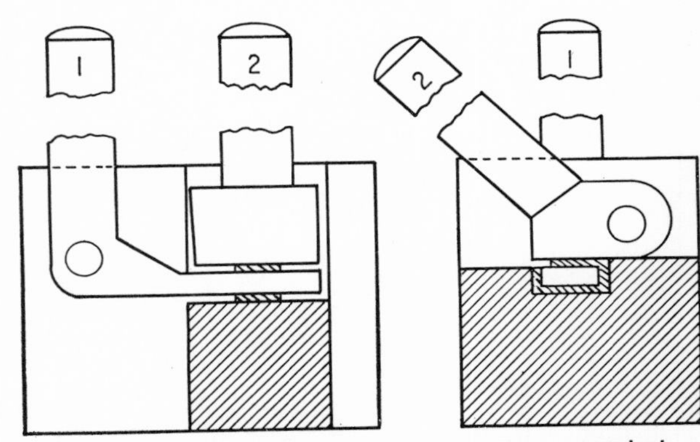

Handle 1 forms a U shape, then acts as a mandrel, while handle 2 completes a rectangular box

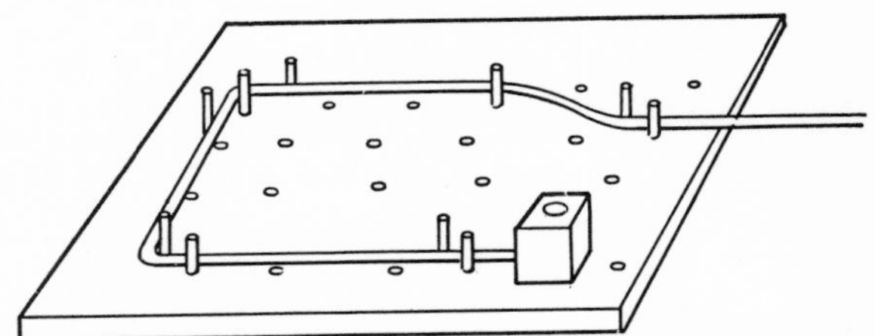

Pegboard jig with movable stops allows making wire forms by fitting and bending the stock by hand

Chapter 12

Punching, Blanking, Forming and Drawing

Punch and die clearances, 1

Punch and die clearances, die life, pieces per grind are controversial topics among stampers. Here are some definitive guidelines on how to avoid error

Two methods of measurement of stamping die performance are in general use and both have built-in errors. One is based on the number of parts produced per press run, the other on the number of parts produced per sharpening grind.

For accurate performance comparisons, the measurement system must be based on burr heights related to specific part quality levels. Five classes of cut-edge characteristics are recommended for defining part quality levels. By limiting productivity measurements in terms of burr height, the individual effect of the many contributing factors can be properly evaluated.

By Harding R Hugo
Product manager
Punchrite Components
Danly Machine Corp, Chicago

Part features

The desired dimensions on the periphery of a blank or hole are called the 'feature' sizes. There is an established relationship between feature sizes and the punching components. The hole feature size establishes the punch size, and punch-to-die clearance is added to the die cavity. Conversely, the blank feature size establishes the die cavity size, and clearance is subtracted to determine the punch size. These generally accepted rules are subject to variations.

Punch size

The punch size is not necessarily the same as the desired hole size—it may produce a smaller hole or a larger hole, depending upon the clearance between punch and die, the type of stripper used, the condition of the cutting edge, and the type of material being punched. Conventional per-side clearances of 5% of the stock thickness for ferrous materials and 3-4% of the stock thickness for nonferrous materials usually result in producing a part hole approximately 0.0005 in. smaller than the punch. Doubling these conventional values results in holes that are equal to or about 0.0005 in. larger than the punch which produced them.

A spring-loaded stripper restricts movement of the top surface of the material and will generally produce a part hole size different from that produced with a channel-type stripper.

Dull cutting edges require deeper penetration to effect fracture than sharp edges and therefore have a minor effect on the resultant hole size. The material being punched is worked

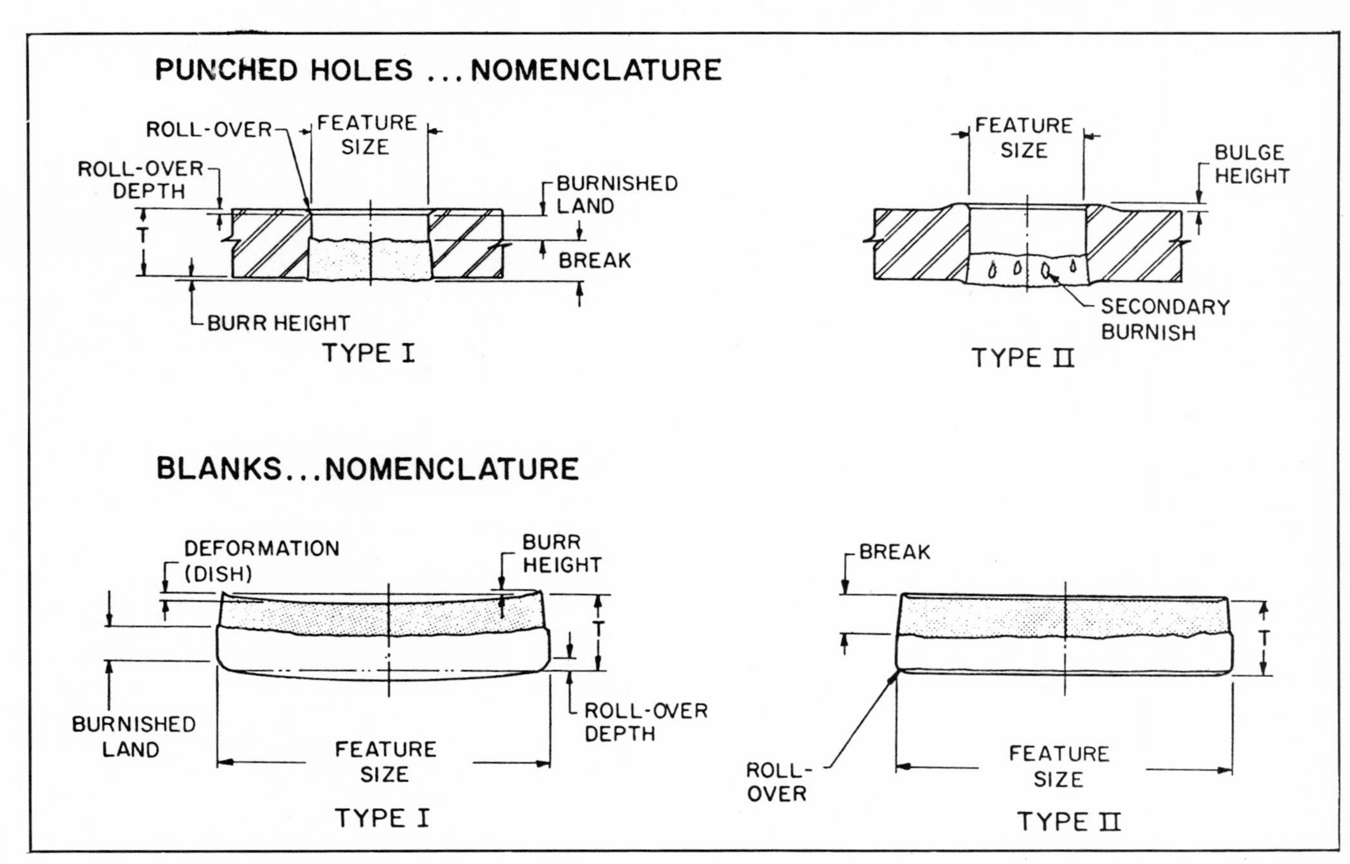

Punch and die clearances [Continued]

harder and this work-hardening effect influences hole size.

The proportion of blank size to stock thickness may be large enough to introduce an oil-can effect. Unless the part is supported in the cavity, the blank may be larger than the cavity due to deformation.

Cut-edge nomenclature

The American Metal Stamping Association has prepared a series of 'Design Guide Sheets,' intended to provide a basic understanding of stamped parts and to define practical limits for general-purpose applications. AMSA terminology is used in this report.

Definitions are (See Figs. 1, 2):

Roll-over is the rounded edge caused by plastic flow under a compression load. Some clearance is required between the punch and die to effect fracture. The resulting torque causes the material to roll as the punch initially penetrates the stock, and the underside of the stock is displaced into the die cavity.

Bulging or lateral displacement is generally combined with the roll-over when the punch-to-die clearance is insufficient and the punched material is soft or ductile. This edge characteristic occurs in Type II punched holes.

Roll-over depth is the distance from the surface where the rolling begins to the intersection with the burnished land.

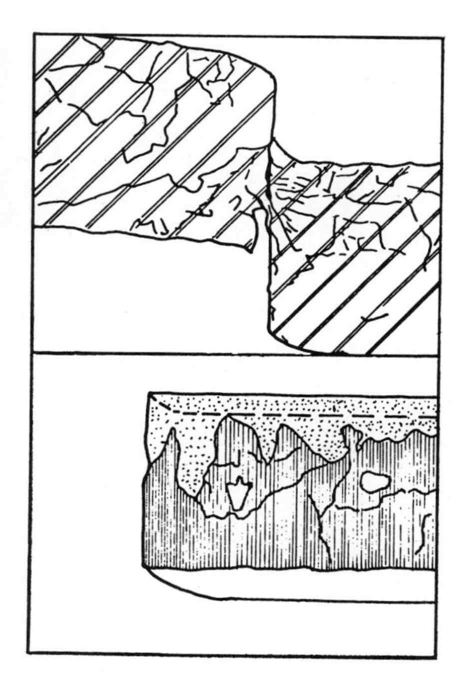

1. Burr formation (above) at 75x. Cut-edge characteristics are shown below

Burnished land in the hole or on the blank edge is the relatively straight portion of the edge produced by the plastic deformation of the material being punched. The length of the land is the distance from the intersection with the roll-over to the point where fracture or separation occurs. Burnished land may also be referred to as the useful length of the feature size or as the sheared length. For general-purpose holes, the burnished land length is said to be one-third of the stock thickness.

Angular break is the edge portion that results from sudden separation due to fracture. The texture of the break area is dull and appears to be pock-marked. When insufficient punch-to-die clearance is utilized, small burnished areas will become apparent, as illustrated in the Type II punched hole.

Magnitude of the angular break is very small. For example, a total clearance value of 0.003 in. for 0.030-in.-thick stock in a general-purpose hole results in a break of only 0.0015 in. in 0.020 in. length. Unfortunately the angle is so small that it can only be indicated by exaggerating it.

Burr height is the distance from the top of the burr to the surface of the material from which it projects. The burr on the hole edge results from the fracture generated by the cutting edge of the die cavity. The burr on the slug or blank is caused by the fracture at the punch cutting edge.

These characteristics are affected by the mechanical properties of the material being punched, the punch-to-die-clearance, the cutting-edge conditions, the type of stripper, and the ratio of feature size to stock thickness. Fig. 1 illustrates the effect of worn cutting edges on the punch. It is evident that a considerable amount of smearing took place because of the plastic metal that accumulated below the corner radius on the punch.

Recommended specifications

To reduce the cost of tooling and stamping production, several basic recommendations are presented:

- Feature size should be measured only within the burnished land.
- Shape deviations within the feature size limits should be permissible.
- The length of the burnished land should not be specified unless the part function will be impaired. If it is necessary to specify this dimension, only the minimum value shall be given.
- Burr height is to be specified only as a maximum, and the largest value consistant with part function shall be specified.

Deflection is the most harmful factor of all because it almost always results in punch breakage. Deflection may be caused by misalignment of the punch to the die, by unbalanced loads relative to the center of pressure, and by misfeeds. Careful consideration of each of these factors is absolutely essential to their control.

Stripping factors

Component wear can only be caused by frictional forces. The punch wears at two to three times the rate of the die cavity. The basic explanation is that the punch goes through the stock twice while the part or slug goes through the cavity only once. Stripping friction generates approximately twice the wear rate of the shearing friction. For example, the burnished land is produced by plastic deformation of the material being punched. When the stripping cycle begins, the burnished land has become work-hardened and results in an increase of the coefficient of friction.

Until very recently, there was not enough knowledge relative to the factors that influence the force of friction. As a consequence, performance improvement was limited to a change in the punch material. With a fixed value of friction, the only course of action was to select a new tool steel with greater abrasion resistance. Unfortunately, the shock resistance of tool steel is reduced as the abrasion resistance is increased. As a consequence, it was not unusual to exchange the component failure from premature wear to that of breakage.

Punch-to-die clearance offers the greatest potential for increasing stamping die performance. Since larger than conventional clearance values generally result in slug-pulling problems, there must be a change in slug size.

The problem of slug-pulling can be resolved in several ways. Ejector pins installed in the perforating punch will strip the slug. Other methods include an air blast through the punch, a rubber insert in the punch, or a vacuum bushing placed below the cavity when the punch point is small. ■

Punch and die clearances, 2

Most stampers think of punching operations as those where the punch is smaller than the die cavity. This is not always so with respect to burr height

There are three conditions of punch-to-die clearance: negative, neutral, and positive.

Discussion of these conditions must be predicated on burr height. Most objections to increased or optimum clearances are the result of the mistaken notion that burr height will become objectionable.

To substantiate the contention that increased burr height results only from tight clearances (rather than from increased values), it is necessary only to examine a situation where both kinds of clearance can exist simultaneously —a misaligned punch and die cavity.

Study of the resultant hole in the work will reveal that the length of the burnished land is not uniform. Further, the burr height is not uniform around the periphery of the punched hole. Correlating these two factors, it will be discerned that the highest burr is in the same relative position as the longest burnished land. If the hole is studied in relation to the punch and die, it can be seen that the tightest side produced both the longest burnished land and the highest burr.

Negative clearance

As Fig. 2 indicates, the punch is larger than die cavity. The hole produced with this clearance is a half-hole or 'squirt.' The punch only penetrates a portion of the stock thickness and fracture will not occur.

When the load exceeds the elastic limit, the material becomes plastic or easily deformed. In this state, the material displaced by the punch is forced to flow into the die cavity. The material that projects below the bottom surface of the stock may be called a compressive burr,' because compression was the origin of its displacement.

By Harding R Hugo, product manager
Punchrite Components
Danly Machine Corp, Chicago

Remember that the punching load for this type of operation is generally a maximum.

Neutral clearance

Many die builders pride themselves on their ability to produce die assemblies with supposedly no clearance between the mating components. Line-to-line or metal-to-metal fits are possible, although they are not likely—especially if punch penetration into the die cavity is desired.

The punching load must be large to cause plastic deformation of the material. The burr will be relatively high because it consists of a compressive element and a tensile element, Fig. 3. Like the burr caused by plastic deformation in the case of negative clearance, the burr in the neutral clearance condition will exhibit some of the same characteristics. The material in this case has failed in tension and the residual string-ends will also project beyond the bottom of the stock.

The stripping load will be substantial because of the long burnished land and the heavy lateral displacement of material around punch periphery.

Positive clearance

This clearance condition, Fig. 4, results when the punch is smaller than the die cavity. Depending upon the space between the punch and die, the compressive squirt may or may not be present. If compressive displacement is present, it will be less than in the first two clearance conditions.

Conventional clearances are positive clearances in that the punch is usually smaller than the die. The fact remains, however, that the resulting burr does contain an element of plastic displacement. As the clearance is increased, the amount of plastic squirt will be physically reduced until, with sufficient clearance, the plastic squirt will be entirely eliminated, leaving a burr composed only of the string-ends. This

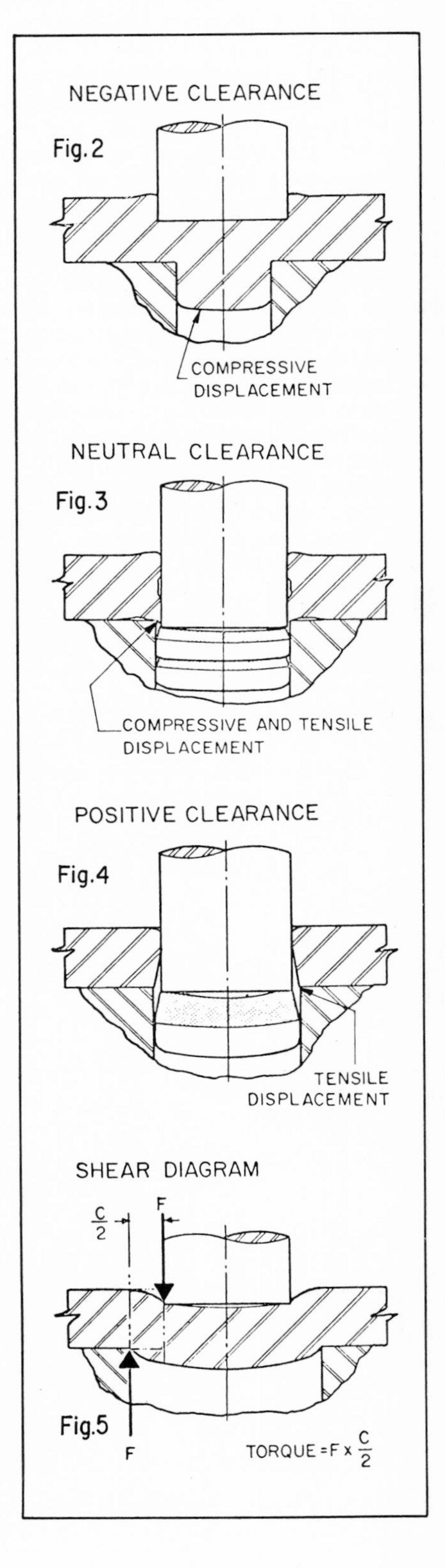

Punch and die clearances [Continued]

tensile burr is the smallest burr that can be obtained when punching any given material.

Burr height depends not only on the clearances used, but also on the cutting edge condition and the properties of the material being punched. Dull punches or die cavities tend to smear the metal when it is in the plastic state, causing longer burnished lands and increased burr heights. The correlation between these two factors will be demonstrated. Ductile materials tend to be plastically deformed for a considerable distance before fracture occurs and, as a result, increase the burr height.

Brittle materials, on the other hand, cannot be plastically deformed to any major degree and fracture readily. The burr tends to be small.

Shear diagram

The shear diagram, Fig. 5, illustrates the effect of punch-to-die clearance. By definition, a couple is the system of forces that produces torque. It consists of two equal and opposite forces that are applied parallel to each other and separated by some distance. The resulting torque value can be calculated by the formula:

$$\text{Torque} = F \times \frac{C}{2}$$

where: F = one of the forces

$\frac{C}{2}$ = perpendicular distance between the forces

In the case of punched holes, the perpendicular distance between the forces is the per-side clearance. Obviously, the greater the clearance, the greater the torque generated. Torque is a rolling action and is as destructive to the material being punched as the compressive load.

Fig. 6 shows the reduction in press energy required to punch a hole when the clearance is increased. The solid lines indicate the burnished land length and minimal rollover that results from conventional clearance values. The phantom lines indicate the corresponding characteristics when larger clearances are utilized. By using the hole-edge characteristics to indicate penetration, the load curve can be drawn.

The maximum load must be applied at the point where the rollover meets the burnished land. By drawing a line from the 0-0 reference to this point, the rate of load application will be shown. The load maximum must be applied until fracture occurs (this point is located at the end of the burnished length).

A horizontal line drawn the length of the burnished land represents the duration of the application of the maximum load. At fracture, the slug is separated from the stock and the only load necessary is that required to push the slugs out of the die cavity.

In the case of the conventional clearances, the back line is somewhat sloped with a large sweeping curve at the end. This is intended to indicate the force is not consistent. Overcoming the inertia of the slugs contained in the die cavity requires a larger force than that necessary to keep them in motion; hence, the straight line and the sweeping curve.

The larger clearances, however, do not tend to permit the slugs to remain in the cavity. Instead, each slug is 'shot' out of the cavity by the slug ejector action. This is the reason for the almost vertical nature of the back slope. It will be noted that the depth of the roll-over is greater with increased clearance, resulting in slower rate of load buildup.

The burnished land is also shorter and this reduces the time that the maximum load is applied. Remember, the torque effect of larger clearances is also destructive!

By comparing the load curve 'envelopes,' one can readily see that the conventional clearances have generated a load curve that encompasses a very substantial area. The shaded areas illustrate the greater energy requirements for tight clearances. Obviously, the larger the clearance, the less energy will be required. ■

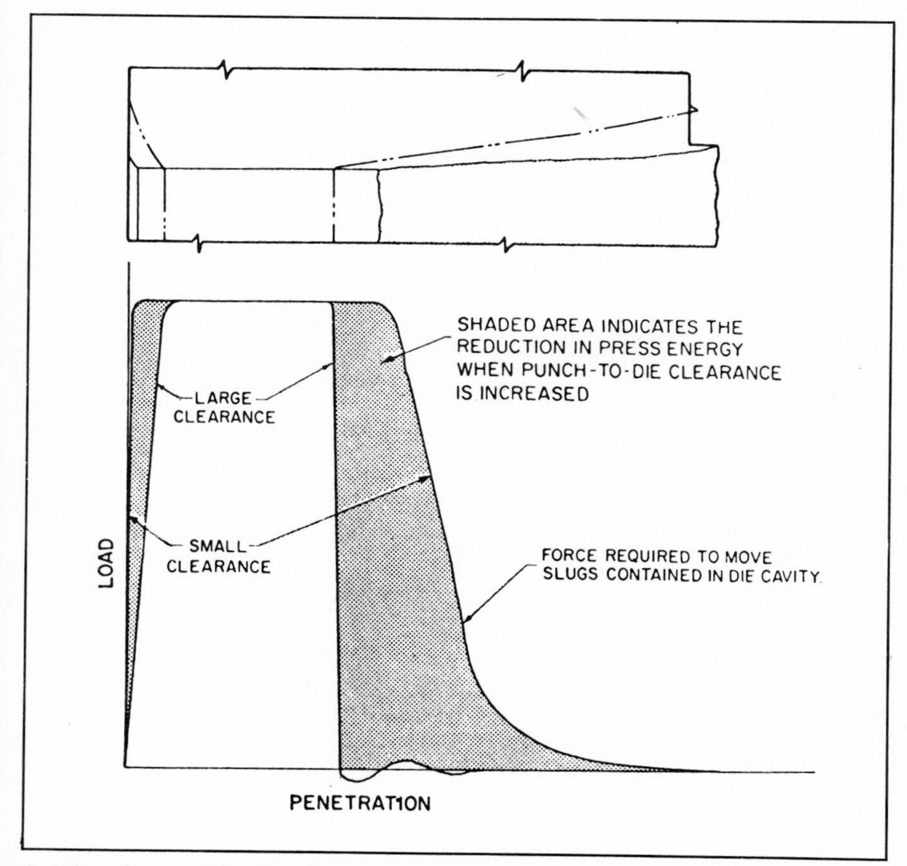

6. When the punch-to-die clearance is large, the slugs are 'shot out' of the die cavity. The shaded areas of the 'load curve' envelopes illustrate the greater energy requirements for tight clearances

Punch and die clearance

Courtesy: W A Whitney Corp, Rockford, Illinois

Punch sizes		Die sizes for material thicknesses			
Frac. & No.	Dec.	Through 14 ga	16 ga −5/32	11 ga −9/32	3/16 −½
1/16	.0625	.0685	.0745		
52	.0635	.0685	.0745		
51	.0670	.0745	.0780		
50	.0700	.0780	.0841		
49	.0730	.0780	.0841		
48	.0760	.0841	.0901		
5/64	.0781	.0841	.0901		
47	.0785	.0841	.0901		
46	.0810	.0870	.0940		
45	.0820	.0901	.0940		
44	.0860	.0940	.0998		
43	.0890	.0940	.1020		
42	.0935	.0998	.1058		
3/32	.0938	.0998	.1058		
41	.0960	.1020	.1090	.1185	
40	.0980	.1058	.1090	.1214	
39	.0995	.1058	.1090	.1214	
38	.1015	.1090	.1154	.1214	
37	.1040	.1090	.1154	.1260	
36	.1065	.1125	.1185	.1260	
7/64	.1094	.1154	.1214	.1310	
35	.1100	.1154	.1214	.1310	
34	.1110	.1185	.1214	.1310	
33	.1130	.1185	.1260	.1370	
32	.1160	.1214	.1260	.1370	
31	.1200	.1260	.1310	.1420	
1/8	.1250	.1310	.1370	.1466	
30	.1285	.1345	.1420	.1526	
29	.1360	.1420	.1466	.1570	
28	.1405	.1466	.1526	.1622	
9/64	.1406	.1466	.1526	.1622	
27	.1440	.1526	.1570	.1682	
26	.1470	.1526	.1570	.1682	
25	.1495	.1570	.1622	.1682	
24	.1520	.1570	.1622	.1750	.1880

Punch sizes		Die sizes for material thicknesses			
Frac. & No.	Dec.	Through 14 ga	16 ga −5/32	11 ga −9/32	3/16 −½
23	.1540	.1622	.1682	.1780	.1930
5/32	.1562	.1622	.1682	.1780	.1930
22	.1570	.1622	.1682	.1780	.1930
21	.1590	.1650	.1720	.1830	.1990
20	.1610	.1682	.1720	.1830	.1990
19	.1660	.1720	.1780	.1880	.2020
18	.1695	.1750	.1830	.1930	.2070
11/64	.1719	.1780	.1830	.1930	.2091
17	.1730	.1780	.1830	.1950	.2091
16	.1770	.1830	.1880	.1990	.2151
15	.1800	.1880	.1930	.2020	.2151
14	.1820	.1880	.1930	.2020	.2190
13	.1850	.1930	.1990	.2070	.2248
3/16	.1875	.1930	.1990	.2091	.2248
12	.1890	.1950	.2020	.2115	.2270
11	.1910	.1990	.2020	.2115	.2270
10	.1935	.1990	.2070	.2151	.2308
9	.1960	.2020	.2091	.2190	.2350
8	.1990	.2070	.2115	.2190	.2350
7	.2010	.2070	.2115	.2248	.2404
13/64	.2031	.2091	.2151	.2248	.2404
6	.2040	.2091	.2151	.2270	.2404
5	.2055	.2115	.2190	.2270	.2404
4	.2090	.2151	.2190	.2308	.2464
3	.2130	.2190	.2248	.2350	.2500
7/32	.2188	.2248	.2308	.2404	.2560
2	.2210	.2270	.2350	.2404	.2560
1	.2280	.2350	.2404	.2500	.2620
15/64	.2344	.2404	.2464	.2560	.2716
1/4	.2500	.2560	.2620	.2716	.2872
17/64	.2656	.2716	.2776	.2872	.3029
9/32	.2812	.2872	.2932	.3029	.3185
19/64	.2969	.3029	.3089	.3185	.3341
5/16	.3125	.3185	.3245	.3341	.3498
21/64	.3281	.3341	.3401	.3498	.3654

This table 'standardizes' punches and dies for short run work on manual duplicating, gaging, and numerically controlled presses. Its purpose is to minimize the number of dies that must be purchased by compromising on the amount of die clearance provided between the punch and die for various stock thicknesses, yet maintain high quality sheet metal parts. The size of the punched hole is determined by the punch size, and the die size only affects the condition of the hole.

The chart shows fractional and number punch sizes in the vertical columns and the material thicknesses across the top. The die size for each material thickness range is based upon 8% to 20% over-all clearance which will produce a Type 3 hole, but in some cases will have slight amount of secondary break as in Type 2.

By examining this table, you will see that one die can be used with more than one punch. Example: 1.0216 can be used with 63/64 in. punch on 3/16 to ½ in. material and with 1 1/64 in. punch on 14 ga stock.

Die clearance is normally expressed as a percentage of the thickness of material being punched.

Excessive clearance —Type 1 hole

Burr

Excessive deformation around the hole. Burr around the bottom of the hole

Insufficient clearance—Type 2 hole

Secondary break

Short tool life. Excessive stripping force causes distortion of the part

Proper clearance —Type 3 hole

Punch and die clearance [continued]

Punch sizes		Die sizes for material thicknesses			
Frac. & No.	Dec.	Through 14 ga	16 ga –5/32	11 ga –9/32	3/16 –½
11/32	.3438	.3498	.3558	.3654	.3810
23/64	.3594	.3654	.3714	.3810	.3966
3/8	.3750	.3810	.3870	.3966	.4122
25/64	.3906	.3966	.4026	.4122	.4279
13/32	.4062	.4122	.4182	.4279	.4435
27/64	.4219	.4279	.4339	.4435	.4591
7/16	.4375	.4435	.4495	.4591	.4748
29/64	.4531	.4591	.4651	.4748	.4904
15/32	.4688	.4748	.4808	.4904	.5060
31/64	.4844	.4904	.4964	.5060	.5216
1/2	.5000	.5060	.5120	.5216	.5372
33/64	.5156	.5216	.5276	.5372	.5529
17/32	.5312	.5372	.5432	.5529	.5685
35/64	.5469	.5529	.5589	.5685	.5841
9/16	.5625	.5685	.5745	.5841	.5998
37/64	.5781	.5841	.5901	.5998	.6154
19/32	.5938	.5998	.6058	.6154	.6310
39/64	.6094	.6154	.6214	.6310	.6466
5/8	.6250	.6310	.6370	.6466	.6622
41/64	.6406	.6466	.6526	.6622	.6779
21/32	.6562	.6622	.6682	.6779	.6935
43/64	.6719	.6779	.6839	.6935	.7091
11/16	.6875	.6935	.6995	.7091	.7248
45/64	.7031	.7091	.7151	.7248	.7404
23/32	.7188	.7248	.7308	.7404	.7560
47/64	.7344	.7404	.7464	.7560	.7716
3/4	.7500	.7560	.7620	.7716	.7872
49/64	.7656	.7716	.7776	.7872	.8029
25/32	.7812	.7872	.7932	.8029	.8185
51/64	.7969	.8029	.8089	.8185	.8341
13/16	.8125	.8185	.8245	.8341	.8498
53/64	.8281	.8341	.8401	.8498	.8654
27/32	.8438	.8498	.8558	.8654	.8810
55/64	.8594	.8654	.8714	.8810	.8966
7/8	.8750	.8810	.8870	.8966	.9122
57/64	.8906	.8966	.9026	.9122	.9279
29/32	.9062	.9122	.9182	.9279	.9435
59/64	.9219	.9279	.9339	.9435	.9591
15/16	.9375	.9435	.9495	.9591	.9748
61/64	.9531	.9591	.9651	.9748	.9904
31/32	.9688	.9748	.9808	.9904	1.0060
63/64	.9844	.9904	.9964	1.0060	1.0216
1	1.0000	1.0060	1.0120	1.0216	1.0372
1-1/64	1.0156	1.0216	1.0276	1.0372	1.0529
1-1/32	1.0312	1.0372	1.0432	1.0529	1.0685
1-3/64	1.0469	1.0529	1.0589	1.0685	1.0841
1-1/16	1.0625	1.0685	1.0745	1.0841	1.0998
1-5/64	1.0781	1.0841	1.0901	1.0998	1.1154
1-3/32	1.0938	1.0998	1.1058	1.1154	1.1310
1-7/64	1.1094	1.1154	1.1214	1.1310	1.1466
1-1/8	1.1250	1.1310	1.1370	1.1466	1.1622
1-9/64	1.1406	1.1466	1.1526	1.1622	1.1780
1-5/32	1.1562	1.1622	1.1682	1.1780	1.1930
1-11/64	1.1719	1.1780	1.1830	1.1930	1.2091
1-3/16	1.1875	1.1930	1.1990	1.2091	1.2248

Punch size		Die sizes for material thicknesses			
Frac. & No.	Dec.	Through 14 ga	16 ga –5/32	11 ga –9/32	3/16 –½
1-13/64	1.2031	1.2091	1.2151	1.2248	1.2404
1-7/32	1.2188	1.2248	1.2308	1.2404	1.2560
1-15/64	1.2344	1.2404	1.2464	1.2560	1.2716
1-1/4	1.2500	1.2560	1.2620	1.2716	1.2872
1-17/64	1.2656	1.2716	1.2776	1.2872	1.3029
1-9/32	1.2812	1.2872	1.2932	1.3029	1.3185
1-19/64	1.2969	1.3029	1.3089	1.3185	1.3341
1-5/16	1.3125	1.3185	1.3245	1.3341	1.3498
1-21/64	1.3281	1.3341	1.3401	1.3498	1.3654
1-11/32	1.3438	1.3498	1.3558	1.3654	1.3810
1-23/64	1.3594	1.3654	1.3714	1.3810	1.3966
1-3/8	1.3750	1.3810	1.3870	1.3966	1.4122
1-25/64	1.3906	1.3966	1.4026	1.4122	1.4279
1-13/32	1.4062	1.4122	1.4182	1.4279	1.4435
1-27/64	1.4219	1.4279	1.4339	1.4435	1.4591
1-7/16	1.4375	1.4435	1.4495	1.4591	1.4748
1-29/64	1.4531	1.4591	1.4651	1.4748	1.4904
1-15/32	1.4688	1.4748	1.4808	1.4904	1.5060
1-31/64	1.4844	1.4904	1.4964	1.5060	1.5216
1-1/2	1.5000	1.5060	1.5120	1.5216	1.5372
1-33/64	1.5156	1.5216	1.5276	1.5372	1.5529
1-17/32	1.5312	1.5372	1.5432	1.5529	1.5685
1-35/64	1.5469	1.5529	1.5589	1.5685	1.5841
1-9/16	1.5625	1.5685	1.5745	1.5841	1.5998
1-37/64	1.5781	1.5841	1.5901	1.5998	1.6154
1-19/32	1.5938	1.5998	1.6058	1.6154	1.6310
1-39/64	1.6094	1.6154	1.6214	1.6310	1.6466
1-5/8	1.6250	1.6310	1.6370	1.6466	1.6622
1-41/64	1.6406	1.6466	1.6526	1.6622	1.6779
1-21/32	1.6562	1.6622	1.6682	1.6779	1.6935
1-43/64	1.6719	1.6779	1.6839	1.6935	1.7091
1-11/16	1.6875	1.6935	1.6995	1.7091	1.7248
1-45/64	1.7031	1.7091	1.7151	1.7248	1.7404
1-23/32	1.7188	1.7248	1.7308	1.7404	1.7560
1-47/64	1.7344	1.7404	1.7464	1.7560	1.7716
1-3/4	1.7500	1.7560	1.7620	1.7716	1.7872
1-49/64	1.7656	1.7716	1.7776	1.7872	1.8029
1-25/32	1.7812	1.7872	1.7932	1.8029	1.8185
1-51/64	1.7969	1.8029	1.8089	1.8185	1.8341
1-13/16	1.8125	1.8185	1.8245	1.8341	1.8498
1-53/64	1.8281	1.8341	1.8401	1.8498	1.8654
1-27/32	1.8438	1.8498	1.8558	1.8654	1.8810
1-55/64	1.8594	1.8654	1.8714	1.8810	1.8966
1-7/8	1.8750	1.8810	1.8870	1.8966	1.9122
1-57/64	1.8906	1.8966	1.9026	1.9122	1.9279
1-29/32	1.9062	1.9122	1.9182	1.9279	1.9435
1-59/64	1.9219	1.9279	1.9339	1.9435	1.9591
1-15/16	1.9375	1.9435	1.9495	1.9591	1.9748
1-61/64	1.9531	1.9591	1.9651	1.9748	1.9904
1-31/32	1.9688	1.9748	1.9808	1.9904	2.0060
1-63/64	1.9844	1.9904	1.9964	2.0060	2.0216
2	2.0000	2.0060	2.0120	2.0216	2.0372

Blanking Pressure Requirements

By BERNARD PACKER, tool engineer, E F Hauserman Co, Cleveland

This table is designed to minimize mathematical computation in determining the pressure necessary to blank various materials. The table is based on the equation

$S = 50{,}000\ P \times t$ (for cold rolled steel)

where S = blanking pressure, lb
P = hole perimeter, in.
t = metal thickness, in.

For materials other than cold rolled steel the following multiplication factors are used:

Stainless steel	1.4
Monel metal	1.4
Aluminum	0.4
Brass and copper	0.8

PROBLEM: What is the blanking pressure for a 2.5-in.-dia hole in 14-gage stainless steel?

SOLUTION: Perimeter of 2.5 in. hole = 7.854 in.
Blanking pressure for 7-in. perimeter = 26,250 lb
Blanking pressure for 0.85-in. perimeter = 3188 lb

Total = 29,438 lb

This is for hot-rolled or cold-rolled steel. We must multiply by 1.4 to get the blanking pressure for stainless: 29,438 × 1.4 = 41,213 lb.

Hole Perimeter, in.	Gage Number of Stock: 10	12	14	16	18	20	21	22	23	24
Equivalent Thickness, in.	0.135	0.105	0.075	0.060	0.048	0.036	0.033	0.030	0.027	0.024
	BLANKING PRESSURE, lb									
0.10	675	525	375	300	240	180	165	150	135	120
0.15	1013	788	563	450	360	270	248	225	203	180
0.20	1350	1050	750	600	480	360	330	300	270	240
0.25	1688	1313	938	750	600	450	413	375	338	300
0.30	2025	1575	1125	900	720	540	495	450	405	360
0.35	2363	1838	1313	1050	840	630	578	525	473	420
0.40	2700	2100	1500	1200	960	720	660	600	540	480
0.45	3038	2363	1688	1350	1080	810	743	675	608	540
0.50	3375	2625	1875	1500	1200	900	825	750	675	600
0.55	3713	2888	2063	1650	1320	990	908	825	743	660
0.60	4050	3150	2250	1800	1440	1080	990	900	810	720
0.65	4388	3413	2438	1950	1560	1170	1073	975	878	780
0.70	4725	3675	2625	2100	1680	1260	1155	1050	945	840
0.75	5063	3938	2813	2250	1800	1350	1238	1125	1013	900
0.80	5400	4200	3000	2400	1920	1440	1320	1200	1080	960
0.85	5738	4463	3188	2550	2040	1530	1403	1275	1148	1020
0.90	6075	4725	3375	2700	2160	1620	1485	1350	1215	1080
0.95	6413	4988	3563	2850	2280	1710	1568	1425	1283	1140
1	6750	5250	3750	3000	2400	1800	1650	1500	1350	1200
2	13,500	10,500	7500	6000	4800	3600	3300	3000	2700	2400
3	20,250	15,750	11,250	9000	7200	5400	4950	4500	4050	3600
4	27,000	21,000	15,000	12,000	9600	7200	6600	6000	5400	4800
5	33,750	26,250	18,750	15,000	12,000	9000	8250	7500	6750	6000
6	40,500	31,500	22,500	18,000	14,400	10,800	9900	9000	8100	7200
7	47,250	36,750	26,250	21,000	16,800	12,600	11,550	10,500	9450	8400
8	54,000	42,000	30,000	24,000	19,200	14,400	13,200	12,000	10,800	9600
9	60,750	47,250	33,750	27,000	21,600	16,200	14,850	13,500	12,150	10,800
10	67,500	52,500	37,500	30,000	24,000	18,000	16,500	15,000	13,500	12,000
11	74,250	57,750	41,250	33,000	26,400	19,800	18,150	16,500	14,850	13,200
12	81,000	63,000	45,000	36,000	28,800	21,600	19,800	18,000	16,200	14,400
13	87,750	68,250	48,750	39,000	31,200	23,400	21,450	19,500	17,550	15,600
14	94,500	73,500	52,500	42,000	33,600	25,200	23,100	21,000	18,900	16,800
15	101,250	78,750	56,250	45,000	36,000	27,000	24,750	22,500	20,250	18,000

Note: Add columns corresponding to total perimeter being blanked to obtain blanking pressure in pounds

Tonnage for blanking and piercing

Press tonnages for short-run Kirksite blanking and piercing dies can be calculated by use of a formula or by multiplication factors related to material thickness and condition

Two options are provided in respect to determining the tonnage required to blank and pierce aluminum and plastic details. You can use the tonnage-multiplication factor table on this page or the nomograph on page 172. In the first case, the shear length is multiplied by the tonnage factor listed, whereas the nomograph supplies the answer by finding the shear length and then joining the appropriate scales.

The two problems shown in the illustrations are solved by using the tonnage factors. In both cases the material is 0.050 in. aluminum sheet 2014-T3. In the case of the rectangular part, the perimeter by scale measurement is 14.50 in., and the perimeters of the several holes equal 4.95 in. Therefore, the tonnage required equals the total perimeter (19.45) times the tonnage factor from table (0.92) or 17.89 tons precisely, or 18 tons rounded off.

In the second example, the perimeter measures 11 in. and the circumferences of the seven holes add up to 3.78 in. Therefore, 14.78 times the tonnage factor of 0.92 establishes the tonnage at 13.59, or 14 tons rounded off.

If the nomograph is used, assume that the shear length of a 0.040 in. 7075-T6 aluminum alloy part measures 14 in. To use the nomograph, start by joining 14 in. on scale I by line A to 0.040 in. on scale IV. At the intersection of line A with scale II, draw line B to scale V at shear strength of 49,000 psi for the alloy involved. Line B intersects scale III at a tonnage of 14 tons.

When using the nomograph, remember that the shear length is the total perimeter of the part to be blanked plus the circumferences of all holes. The perimeter may easily be measured by using a full-size print. And to get hole perimeters, use a table of circumferences of common hole sizes.

By Murray M. Shtulman
Mechanical Systems Engineering
Grumman Aerospace Corp.
Bethpage, N.Y.

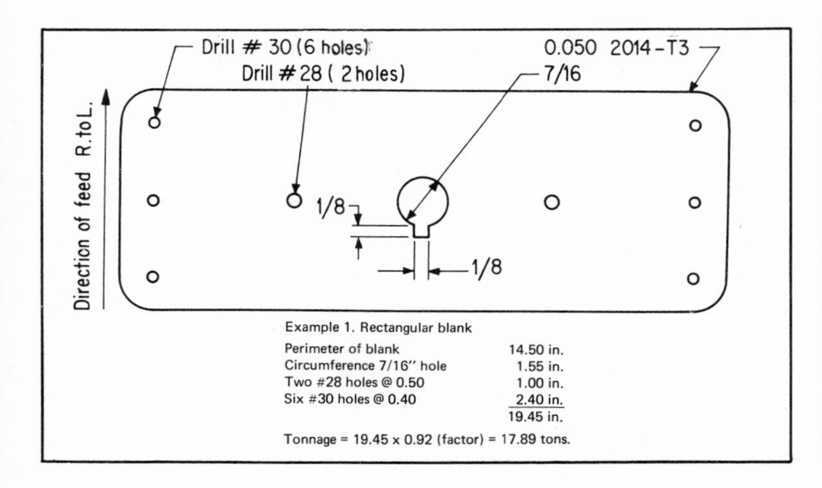

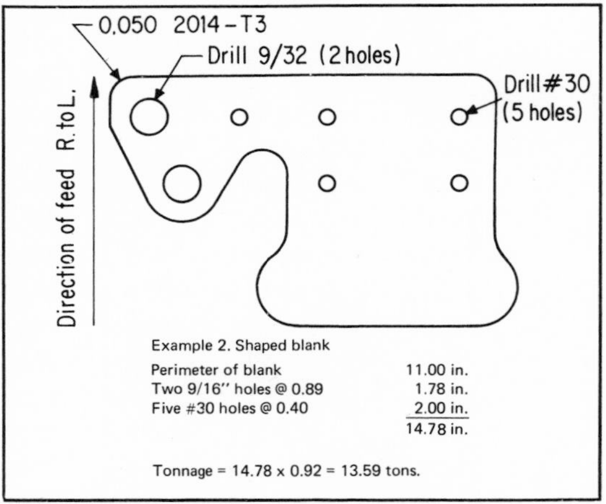

Tonnage multiplication factors

Thickness / Material	2014-0 2024-0 5052-0 6061-0 7075-0	2014-T3	2024-T4 2014-T6	7075-T6	6061-T4	2017-T4	3003-H14	ABS	Polycarb	Acrylic FRP
.016	.14	.30	.34	.39	.19	.30	.11	.04	.08	.36
.020	.18	.37	.42	.49	.24	.38	.14	.05	.10	.45
.024	.22	.45	.51	.59	.29	.46	.17	.06	.12	.54
.032	.29	.59	.67	.78	.38	.61	.22	.08	.16	.72
.040	.36	.74	.84	.98	.48	.76	.28	.10	.20	.90
.050	.45	.92	1.05	1.22	.59	.95	.35	.125	.250	1.125
.063	.58	1.18	1.34	1.57	.77	1.21	.45	.158	.316	1.418

Tonnage for blanking and piercing (Cont.)

Information on plastics has been added to the nomograph and to the table of tonnage multipliers, because the means for ascertaining blanking pressures for these non-metallics are often highly inaccurate. In fact, cases have been known where the assumed tonnage was 85% larger than required.

These data were worked out for a revision of the Grumman standards sheets for manufacturing engineering procedures—and specifically, in this case, for the procurement of Kirksite dies. However, the data are generally useful. ■

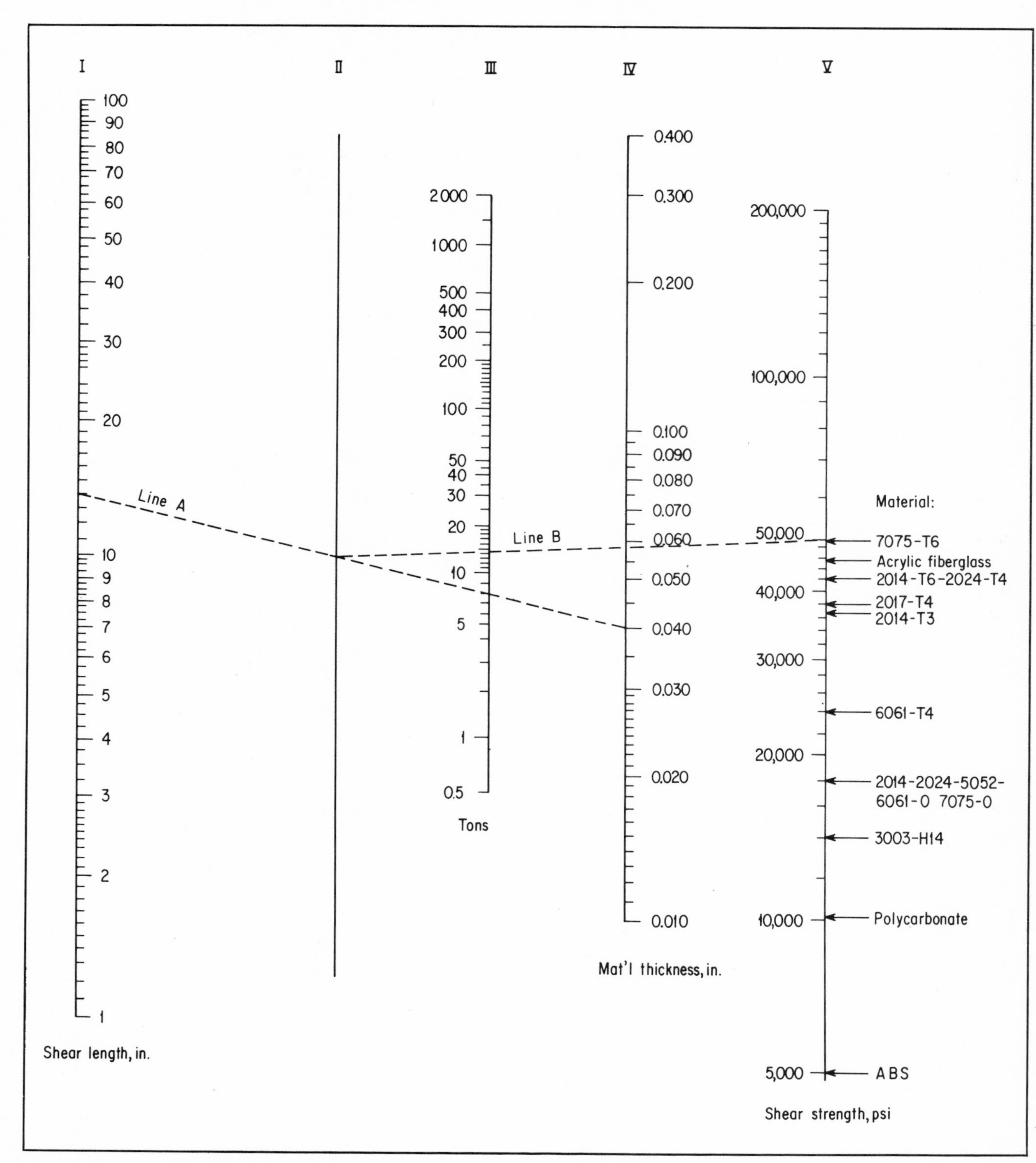

Pressures to Pierce Sheet Steel...I

BY E. V. SARGEANT, ASST. CHIEF TOOL DESIGNER, GENERAL MOTORS OF CANADA, LTD., OSHAWA

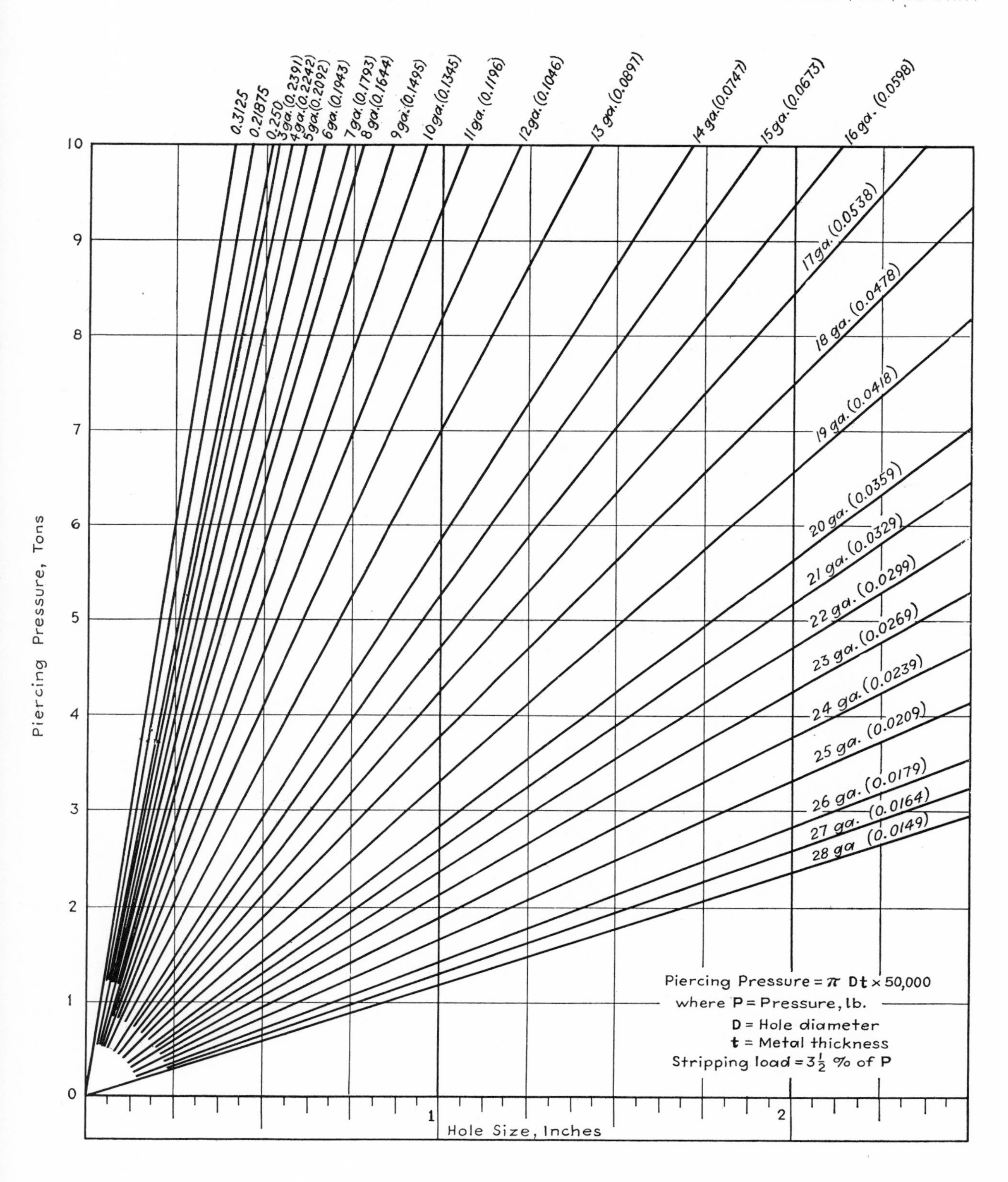

Pressures to Pierce Sheet Steel...II

BY E. V. SARGEANT, ASST. CHIEF TOOL DESIGNER, GENERAL MOTORS OF CANADA, LTD., OSHAWA

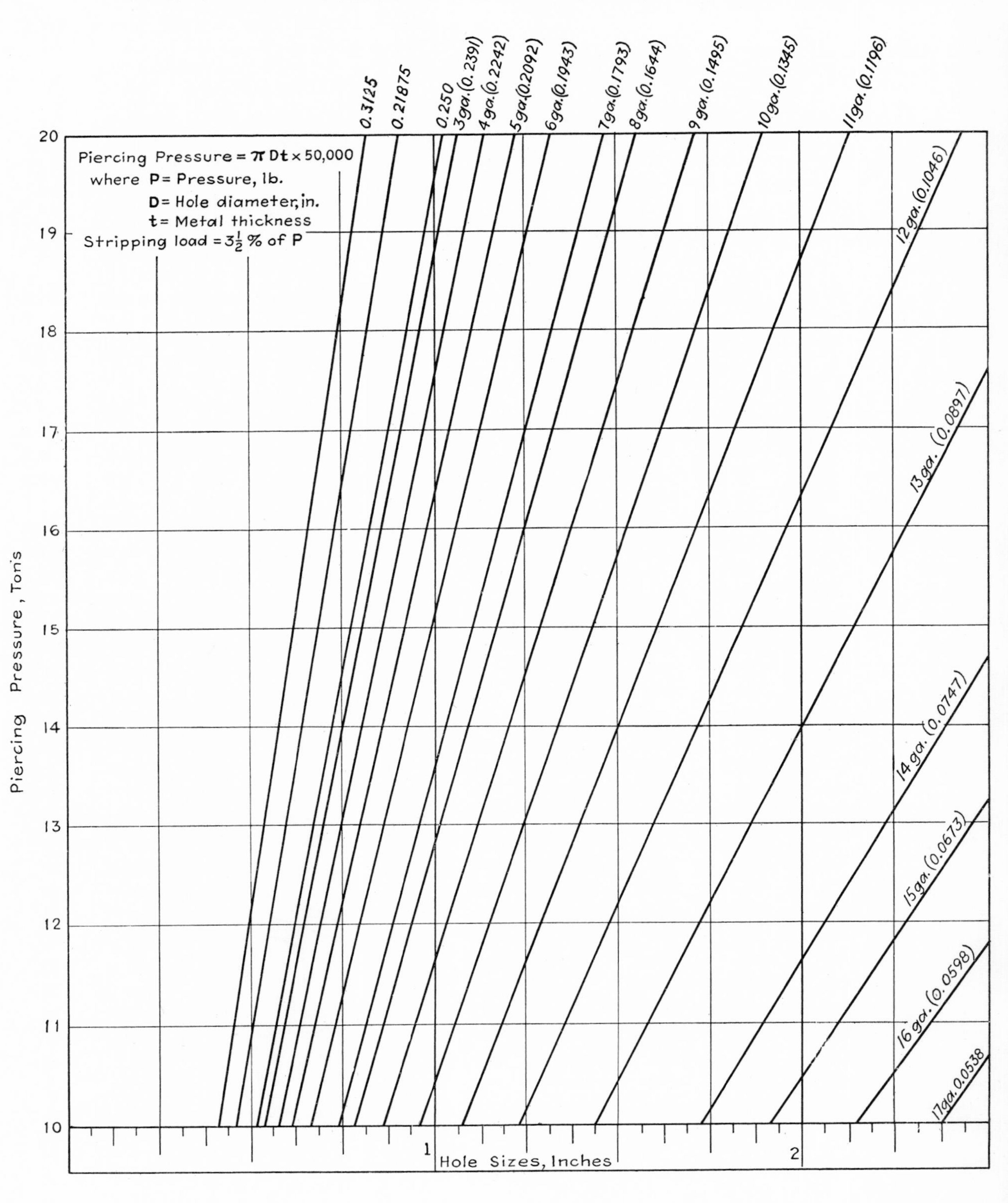

Stripping Pressure Requirements

By BERNARD PACKER, tool engineer,
E F Hauserman Co, Cleveland

This table is based on the relationship: stripping pressure equals 0.02 times the blanking pressure.

PROOF OF TABLE:

Perimeter of 2.5 in. hole = 7.854 in. Then stripping pressure equals:

For 7.0 in. perimeter: 525 lb.
For 0.85 in. perimeter: 63.8 lb.

588.8 lb.

For stainless steel, multiply by 1.4:
Then 589 x 1.4 = 824.6 lb.

Hole Perimeter, in.	Gage Number of Stock									
	10	12	14	16	18	20	21	22	23	24
	Equivalent Thickness, in.									
	0.135	0.105	0.075	0.060	0.048	0.036	0.033	0.030	0.027	0.024
	STRIPPING PRESSURE, lb									
0.10	13.5	10.5	7.5	6.0	4.8	3.6	3.3	3.0	2.7	2.4
0.15	20.3	15.8	11.3	9.0	7.2	5.4	5.0	4.5	4.1	3.6
0.20	27.0	21.0	15.0	12.0	9.6	7.2	6.6	6.0	5.4	4.8
0.25	33.8	26.3	18.8	15.0	12.0	9.0	8.3	7.5	6.8	6.0
0.30	40.5	31.5	22.5	18.0	14.4	10.8	9.9	9.0	8.1	7.2
0.35	47.3	36.8	26.3	21.0	16.8	12.6	11.6	10.5	9.5	8.4
0.40	54.0	42.0	30.0	24.0	19.2	14.4	13.2	12.0	10.8	9.6
0.45	60.8	47.3	33.8	27.0	21.6	16.2	14.9	13.5	12.2	10.8
0.50	67.5	52.5	37.5	30.0	24.0	18.0	16.5	15.0	13.5	12.0
0.55	74.3	57.8	41.3	33.0	26.4	19.8	18.2	16.5	14.9	13.2
0.60	81.0	63.0	45.0	36.0	28.8	21.6	19.8	18.0	16.2	14.4
0.65	87.8	68.3	48.8	39.0	31.2	23.4	21.5	19.5	17.6	15.6
0.70	94.5	73.50	52.5	42.0	33.6	25.2	23.1	21.0	18.9	16.8
0.75	101.3	78.8	56.3	45.0	36.0	27.0	24.8	22.5	20.3	18.0
0.80	108	84	60	48.0	38.4	28.8	26.4	24.0	21.6	19.2
0.85	114.8	89.3	63.8	51.0	40.8	30.6	28.1	25.5	23.0	20.4
0.90	121.5	94.5	67.5	54.0	43.2	32.4	29.7	27.0	24.3	21.6
0.95	128.3	99.8	71.3	57.0	45.6	34.2	31.4	28.5	25.7	22.8
1	135	105	75	60	48	36	33	30	27	24
2	270	210	150	120	96	72	66	60	54	48
3	405	315	225	180	144	108	99	90	81	72
4	540	420	300	240	192	144	132	120	108	96
5	675	525	375	300	240	180	165	150	135	120
6	810	630	450	360	288	216	198	180	162	144
7	945	735	525	420	336	252	231	210	189	168
8	1080	840	600	480	384	288	264	240	216	192
9	1215	945	675	540	432	324	297	270	243	216
10	1350	1050	750	600	480	360	330	300	270	240
11	1485	1155	825	660	528	396	363	330	297	264
12	1620	1260	900	720	576	432	396	360	324	288
13	1755	1365	975	780	624	468	429	390	351	312
14	1890	1470	1050	840	672	504	462	420	378	336
15	2025	1575	1125	900	720	540	495	450	405	360

Note: Add columns corresponding to total perimeter being stripped to obtain stripping pressure in pounds

Clearance for Cupping Dies

By **Ferenc J Kuchta,** mechanical engineer, J Wiss & Sons Co, Newark, N J

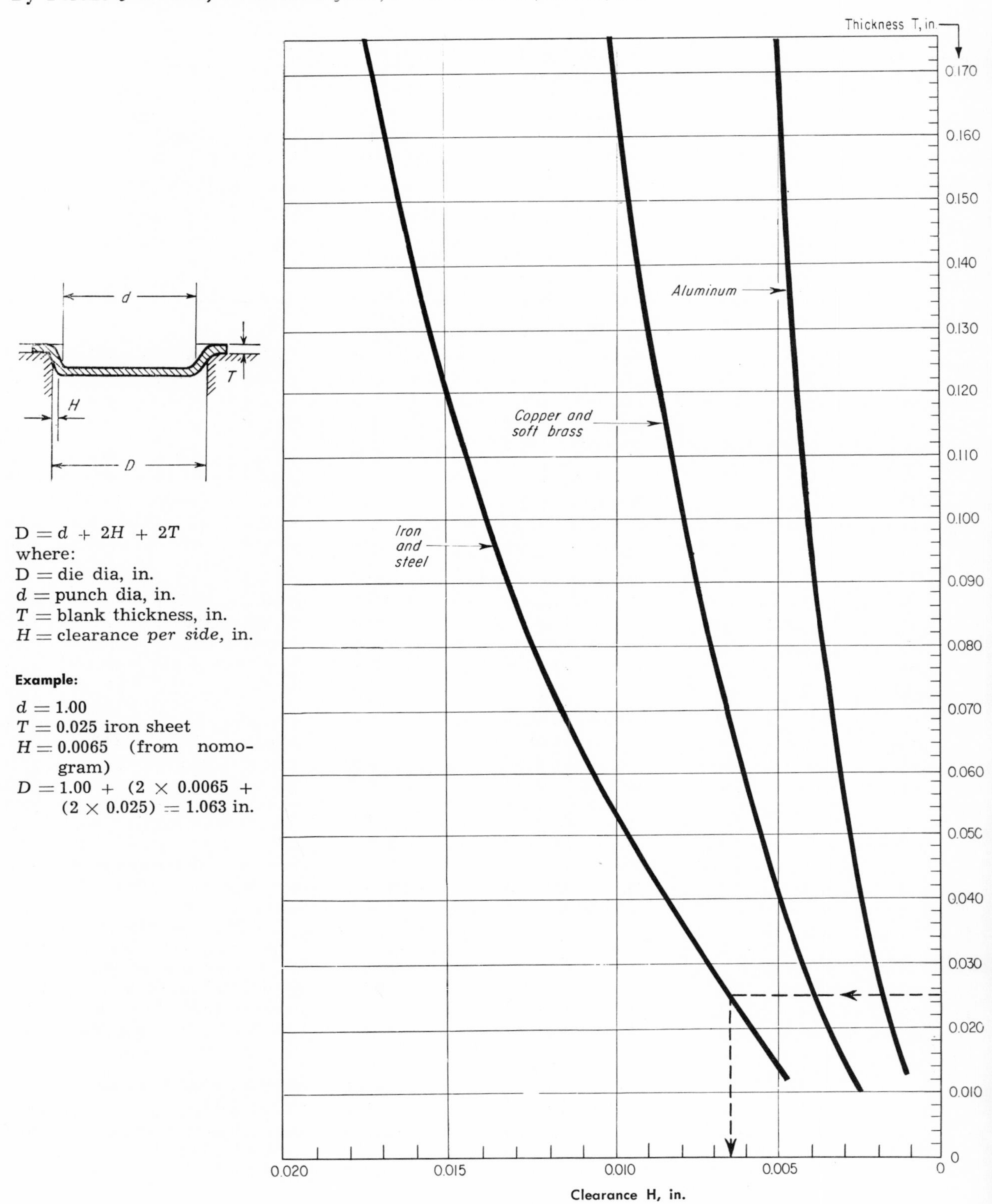

$D = d + 2H + 2T$

where:

D = die dia, in.

d = punch dia, in.

T = blank thickness, in.

H = clearance *per side,* in.

Example:

$d = 1.00$

$T = 0.025$ iron sheet

$H = 0.0065$ (from nomogram)

$D = 1.00 + (2 \times 0.0065 + (2 \times 0.025) = 1.063$ in.

One-Stroke Rim Holes—I

By David M Bauer, Norristown, Pa

Rim holes are applied to sheet-metal parts for a wide variety of purposes:

- To provide metal for a sufficient number of tapped threads, to hold a fastener securely.
- To act as a tubular rivet to secure thin springs and flat stock when headed with simple riveting tools.
- To grip solid or spring pins.
- To act as stops for levers.
- To act as locators for mating parts. Two blanks of different thickness can be rim-holed together in one stroke. The two parts are easily separated but fit snugly when put together.
- To provide a durable bearing surface for sliding rods or hinge pins. A rim hole will outlast a punched hole many times before showing appreciable wear.

There are many names for rim holes. They have been called flanged holes, extruded holes, burred holes, and holes with hubs or rims.

Various ways have been developed for making rim holes. The tooling and formulas supplied apply to one-operation tooling for rim holes.

Almost any metal can be rim holed, provided its ductility permits extrusion. The inside surface of the rim is burnished to a mirror polish and is held to the exact size of punch without spring-back to a smaller diameter.

The height of the rim depends on metal thickness, punch diameter, and relationship of the punch diameter to the die bore. Rim height will be approximately 2½ to 2¾ times metal thickness.

Limitations on diameters and stock thicknesses are not easily found without experimentation. Factors such as type of material, its thickness and ductility must be considered.

My tools did not go above 0.160 in. ID for rims. So, I made a punch for rim ID of 5/16 in. and a die with three different sized bores of 11/32, 3/8 and 27/64 in.

A batch of rims was extruded in different metals of varied thicknesses with these results:

For luster C R steel 0.035 in. thick, I was able to extrude a rim without cracks and a rim height 5½ times stock thickness. Rim edge was remarkably smooth for such a long draw. Rims of this type are used for joining tubing to sheet metal for brazing, as in refrigeration, automobile radiators and many other similar products.

Not having any soft 1/16 in. C R steel in stock, I used what was rated half hard. Rims with good proportions resulted, but with some roughness and cracking at the rim's outer edge.

Soft aluminum 1/16-in. thick produced "beautiful" rims, using 40% control. Also used a larger die bore than 40% and they came out just as good but with larger radius at rim's base.

Both the 1/16 in. aluminum and CR steel produced rim height 3 times stock thickness.

However, I suggest using a 5/16 in. rim ID for material up to 1/16 in. thick as the largest size. I feel sure it's possible to go to a larger diameter and thicker material.

Tooling Requirements

Tools are comparatively inexpensive to make, require less accuracy than tools for plain punched holes and last longer before requiring sharpening. The average machine shop, with only the usual run of equipment, can successfully make the punches and dies.

Presses without tight-gibbed rams can be used, from arbor presses to power punch presses. Power required for a 5- to 6-rim hole die can be obtained from a hand press.

Slug Control

The piercing tip of the punch normally pops out the slug and it drops through the die opening. However, some metals stick to

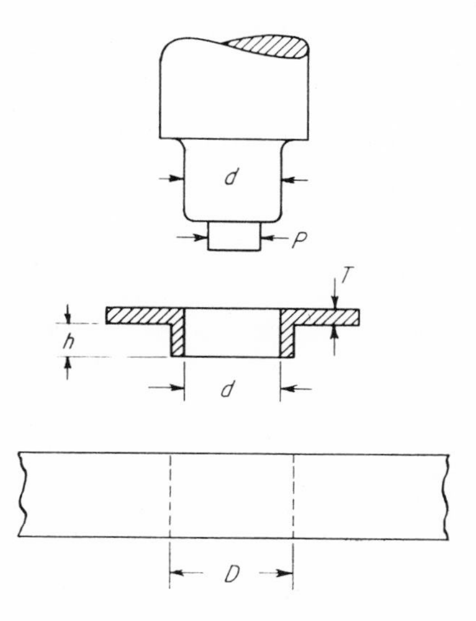

Fig. 1

the punch tip by adhesion. Lubricants often have a tendency to hold some slugs by suction. If slugs are not removed before the next press stroke, the punch will not be damaged but the edges of the rim will be rough and sometimes cracked. Slug removal can be accomplished by an air blast or flat springs that rub across the punch tip on the upstroke of the press.

Pressure pads are not required. The metal surrounding the rim at its base will be flat without wrinkles or crimp marks.

Ejector pins or plates are not used. After the rim has been formed, the punches on the upstroke will withdraw the rims from the die bores and the stripper will shed them from the punches.

Strippers are important. They must be rigid. Stripper holes should be a close fit and locate the punch in the center of the die bore; otherwise the top of the rim will be lopsided and thin walled on one side.

Formulas for Rim Holes

These formulas are based on a rim control percentage that determines rim-wall thickness and

One-Stroke Rim Holes—II

height. The punch has two diameters: piercing and extruding (Fig. 1)

Let D = OD of rim hole (equal to die bore)
d = ID of rim hole = dia extruding section of punch
P = dia piercing section of punch = $d/2$
T = stock thickness
K = rim control percentage (40% for general run of sheet stock used in production)

To find rim diameter, when rim hole is known:

$$D = d + (2T - K \times 2T) = d + 2T\ (1 - K)$$

Example:

Stock thickness T = 0.035; rim control percentage = 40%, and d = 0.145

$$D = 0.145 + 2 \times 0.035\ (1 - 0.40) = 0.145 + (0.07 \times 0.6) = 0.145 + 0.042 = 0.187$$

To find rim hole diameter when outside diameter is known:

$$d = D - 2T\ (1 - K)$$

Example: D = 0.187, T = 0.035, and K = 40%

Then d = 0.187 − 0.042 = 0.145

In both cases, the diameter P of the punch's piercing section is taken as 50% of d or 0.145/2 = 0.073.

By using a rim control percentage of 40%, the metal is thinned 20% per side. A rim with a smooth, thinner and slightly higher wall can be achieved by increasing the rim-control percentage to 50 to 60%. The type of stock used, its hardness, surface finish and lubricant will affect results. The 40% figure gives excellent results for the general run of sheet stock.

Experimental Tools

The tool elements for making rim holes in different materials for testing are shown in Fig. 2. A hand punch press is used so that the action, while making the rim, can be observed. The ¼ in. dia. punch is adequate for stock up to ⅛-in. thick without possibility of breakage.

PUNCH—The section A is formed with the cutting edge of the tool at right angles to the blank axis. The shoulder created will act as a stop for the press ram, leaving the stock flat back of the rim. There should be a slight radius at the shoulder.

The nose of section A, or forming portion, is set approximately at 26° back to produce an angle for later forming a radius. At the junction of sections A and B, the radius should be formed with a small fine-cut file to remove any slight corners. Section A should be polished.

Punches should not be tried experimentally before hardening for they will be scored at the extruding section. However, light scoring can be leveled out with a fine grit stone and oil.

DIE—When only a limited number of experimental rim holes are to be extruded, say about 25, a piece of ¼-in. cold-rolled flat stock can be used.

The die bore is drilled and reamed to size. A slight radius is required at the bore edge. A shop-made abrasive-cloth polishing spindle, used in the drill press, will give an excellent finish to the inside of the hole.

STRIPPER—Most hand presses are furnished with a stripper, but make sure it is tightly secured and rigid to prevent bending when shedding the rim from the punch, or the punch will break off.

LOCATOR—This device fits in the bore, and the punch piercing tip is lowered and located in its center hole. Then the die can be secured to the press bed and the tools will be in alignment.

LENGTH OF PUNCH—The piercing

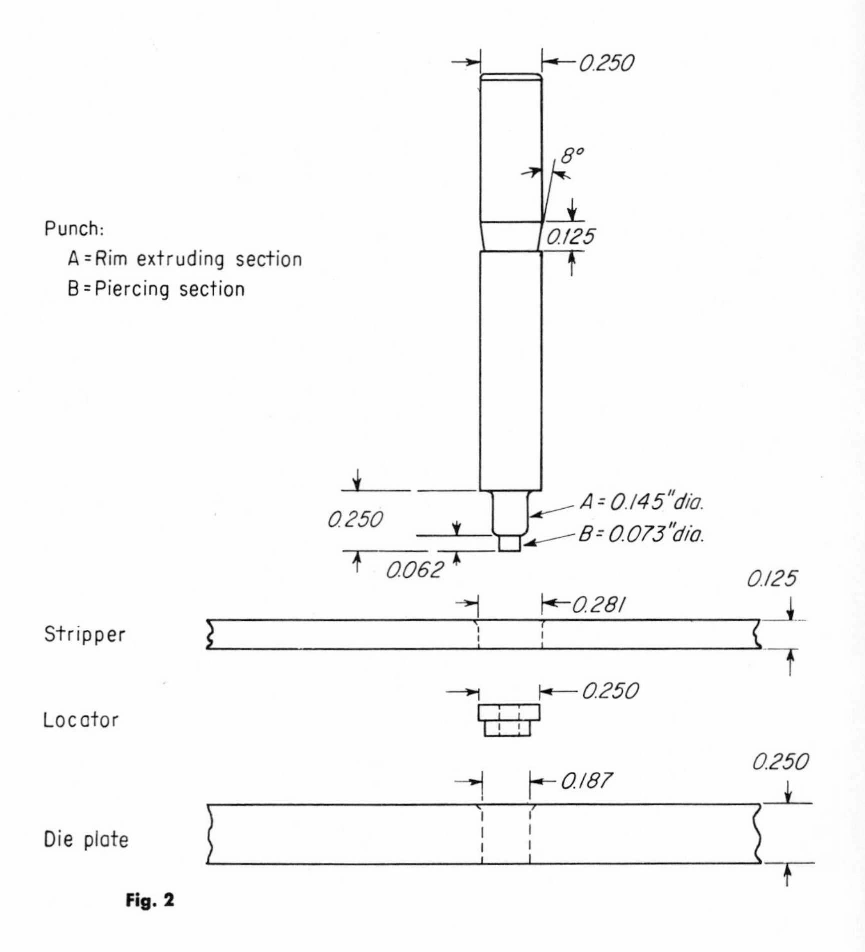

Fig. 2

One-Stroke Rim Holes—III

section should be 0.0625 in. long for stock up to 1/16 in. thick. Rim extruding section must have a length not less than four times stock thickness, for the punch must proceed through the material far enough so the rim diameter will be the same to the other edge.

DIE BORE—The die bore is drilled and reamed to size. A reamer that is a few thousands over or under size will not materially affect the finished rim, unless the outside of the rim is to be used and must be an exact diameter. A slight radius is required at the bore edge.

Effects of Material

Various metals can be rim holed but the temper is important. Brass, copper, aluminum and cold-rolled steel have all been worked successfully. However, avoid use of material that has been skin rolled. This mill operation produces a hard surface skin that will cause cracking and torn or weak rims. If multiple rim holes are required in the part, the stresses setup in skin-rolled or tempered stock will induce distortion or bowing.

The effects of material and temper were determined by rim holing various metals with an experimental die. The results using 40% rim control were:

Test No. 1 — Luster-finished cold-rolled steel, 0.035 in. thick. Height of rim hole, about 2 T. Uneven, rough edge on rim.

Test No. 2—2S-½ H aluminum, bright one side, 0.032 in. thick. Results: *h* less than 2 T; rim hole uneven.

Test No. 3—23S-73 full-hard aluminum, 0.032 in. thick. Rim torn out before it was formed.

Test No. 4—copper, ⅛ to ¼ hard, 0.032 in. thick. Results: *h* about 2 T. Rough, burred rim.

Test No. 5—CR steel, 0.035 in. Good rim hole that headed smoothly when it was used to attach 0.0115 spring stock.

Test No. 6—Luster-finished CR steel, two thicknesses rim holed in one stroke. Upper piece, shed by stripper, had thin sharp rim hole about one-metal thickness in height. Lower part, lifted from die, had rim hole as produced in Test No. 1. Parts nest firmly.

The following tests were made with 60% rim control, and five punches in a row:

Test No. 7—Luster CR steel. Height *h* about 3 T. Thin rim wall, sharp opening. Some of the rims were torn or cracked.

Test No. 8—2S-½ H aluminum. Rims did not tear or crack, but the outer edges were sharp and thin.

5-Hole Die

A die for producing 5 rim holes at once is shown in Fig 3. Desired ID of the rim holes is 0.1425 in.; the material is luster-finished CR steel 0.035 in. thick. Here 60% rim control is applied. In this case:

$$D = 0.1425 + 2 \times 0.035\ (1 - 0.60) = 0.1705$$

$$P = d/2 = 0.1425/2 = 0.071$$

The die was made by use of progress data sheets, because complete information is supplied to the diemaker. In essence, these sheets call for drilling, redrilling, reaming and tapping the seven pieces of stock used for the punch plate, punch holder, stripper, stock guides, die plate and base.

The seven pieces of stock are sawed to size and two adjoining edges are finished square and the corner stamped *S*. Next, the punch holder is blued with the *S*-side up and the coordinate lines for hole centers are scribed, working from the two squared edges. Punch holder, stripper, die plate and base are clamped together and four spring-pin holes are drilled in the corners. Spring pins are pressed in with an arbor press and two leader pin holes are drilled, likewise five undersize holes at the punch locations. Then the spring pins are removed and the holes finished to their final sizes.

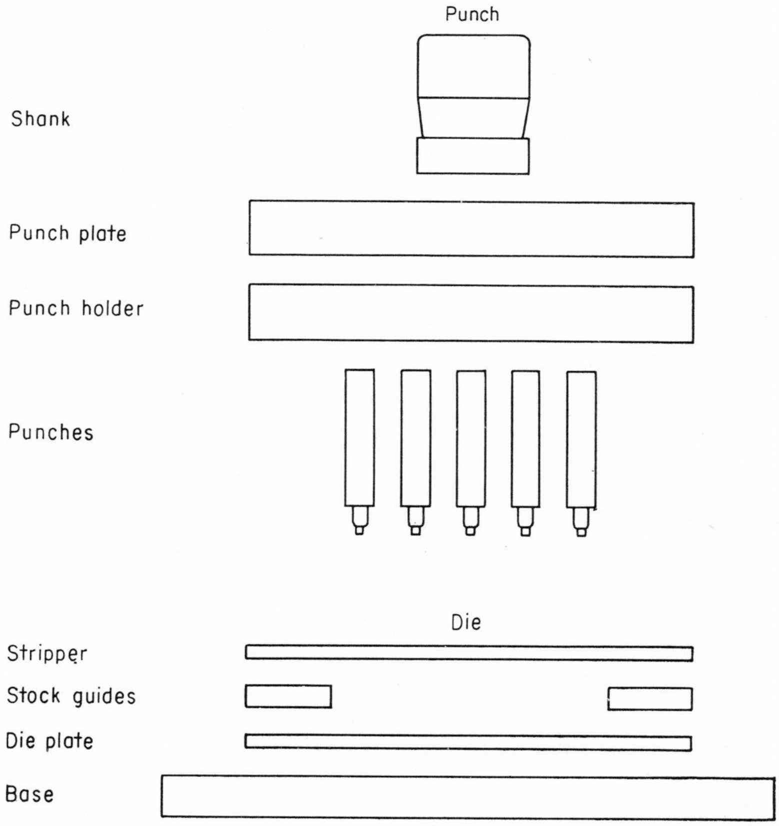

Punch
Shank
Punch plate
Punch holder
Punches
Die
Stripper
Stock guides
Die plate
Base

Fig. 3

Definitions for Forming Operations—I

Source: SAE Aeronautical-Automotive Drawing Standards, March 1960, as published by the Society of Automotive Engineers, 485 Lexington Avenue, New York, N Y

Beading (Sheet Metal)—The forming of a small groove, concave impression or corrugation in metal, usually for the purpose of strengthening the part.

Bend Allowance—The amount of sheet metal required to make a bend over a specific radius to a specified angle.

Counterpunch — A conical-shaped depression around a punched hole so that the installed heads of rivets, bolts and screws may be flush with the surface.

Crimped Edge—A condition produced on sheet metal by turning the edge. It may be used to conceal the raw edge of the metal or to strengthen and stiffen the part.

Datum Line—A base line from which calculations or measurements are taken; in body drawing, inch lines or station lines are commonly used as datum lines.

Die Balance Line—A line at right angles to the direction of pierce or draw, shown on engineering drawings to assist in eliminating back draft.

Die Model—An actual size model of a part from which surface is taken for the manufacture of a die.

Die Plate—A master detailed draft of a die formed part.

Flange—A stiffening edge turned on sheet metal, or that portion farthest from the neutral beam axis, designed to carry the bending stresses.

Flare—A small flange around the circumference of a tube at the end.

Forming—Contour lines of formed parts should be proportioned to flange height to prevent unreasonable elongation of material in relation to physical properties. Excessive stretch requires either extra forming operations, annealing or both. Even when flanges are finally formed under such difficulties, the material has thinned out to a degree requiring a review of stress. With SAE 1010 steel blanks having sheared edges, 25% stretch is considered a safe maximum. See Fig. 1.

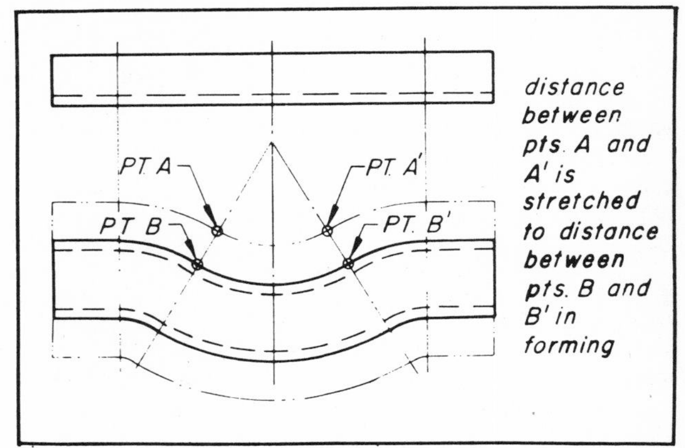

Fig. 1

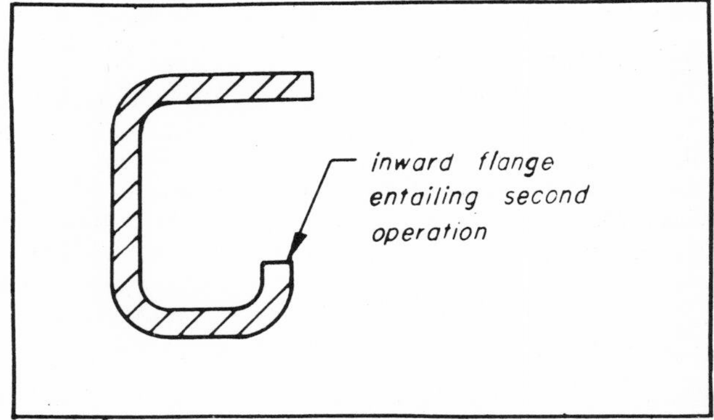

Fig. 2

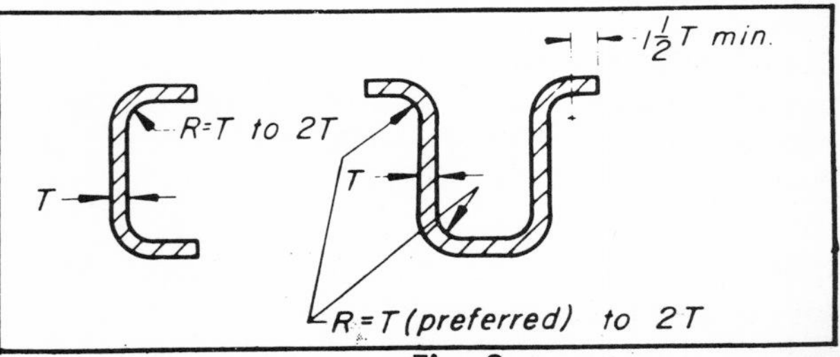

Fig. 3

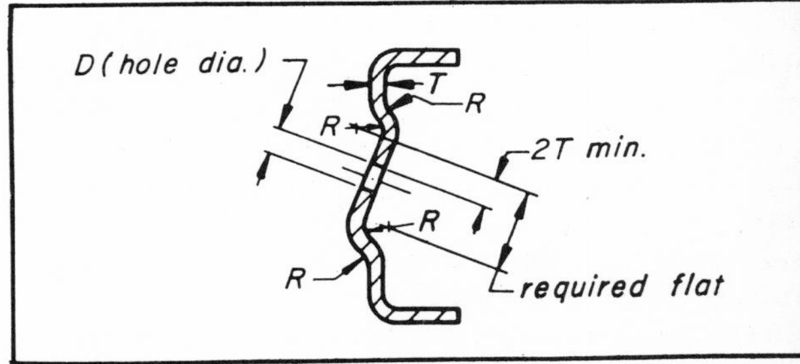

Fig. 4

Flanges formed as shown in Fig. 2 require extra operations and should be avoided.

Inside corner radii of channel and flange forms should be a minimum of metal thickness to give

Definitions for Forming Operations—II

optimum conditions on the finished stampings. See Fig. 3. In cases where panels are formed or embossed into a surface, the radii should be as generous as clearances will allow, but never less than the stock thickness, and the flat panelled area should be as large as possible. Large corner radii prevent metal tearing and breakage. The large flat allows some tolerance on corners of forming tools after they have worn slightly and tends to minimize tool repair cost. It also gives a greater area seat for perforating dies when holes are punched after forming, thus preventing coining action by die buttons. See Fig. 4.

When formed sumps or panels are required, a suitable slope should be allowed to eliminate locking on the punch and consequent risk of cracked stampings. See Fig. 5.

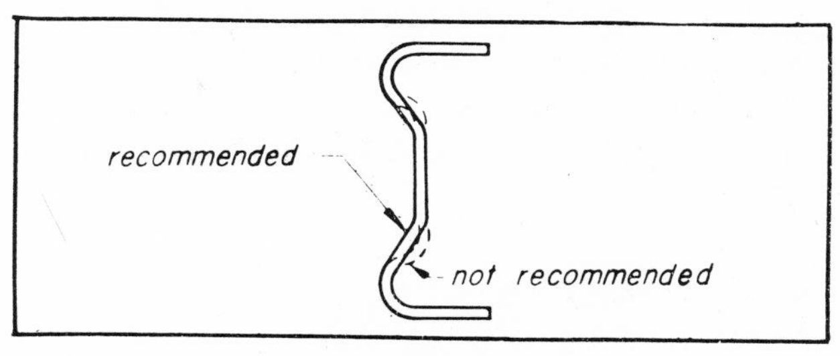

Fig. 5

Ends of bars tend to crack in the corner radii when forming. To relieve this condition radii should be increased if possible. See Fig. 6.

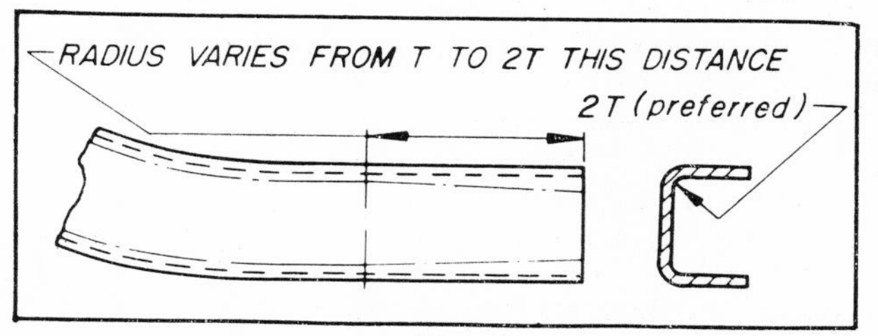

Fig. 6

On any formed part which has a decidedly sharp contour radius the flange should be increased. See Fig. 7.

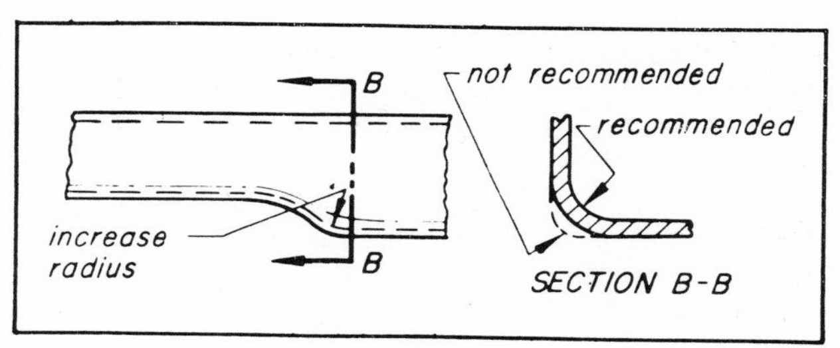

Fig. 7

Forming and Gage Notch—Similar to Gage Hole.

Gage Hole—A hole located in a part to insure proper location during fabricating operations. Also known as "Fool Proof Hole" or "Tooling Hole."

Hem Flange—A flange which is formed through 180° to "double back" on the basic metal surface. It is used to avoid exposed raw edges of metal or to join two panels together.

Holes—In general, holes are perforated in the flat blank whenever possible. This is not possible in extreme contours, nor too close to the bend line of a channel section unless distortion is permissible. Where such can be tolerated, permissible distortion should be specified. As a general rule, the edge of a hole should not be closer than three metal thicknesses from the outside face of web. This must be increased proportionately for large holes and holes in contoured sections. See Fig. 8.

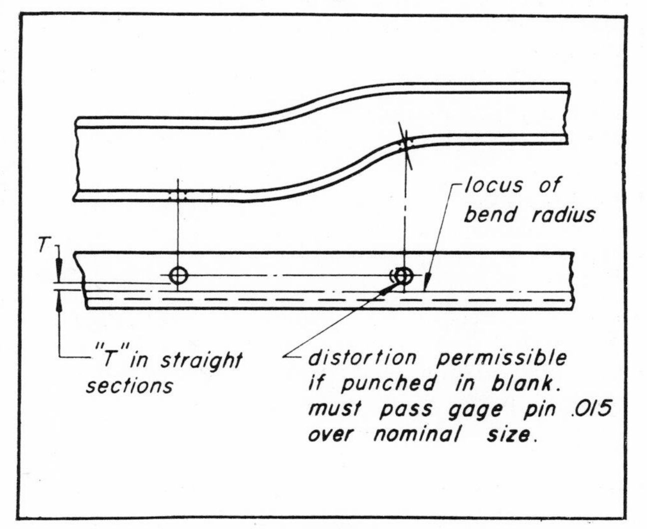

Fig. 8

The conditions in Fig. 9 (a) through (d) show minimum stock requirements around holes used for bolts, clips or rivets. Spacing of groups of pierced holes should be such that minimum stock between holes is no less than shown.

The straight side of an oval hole can be flanged to any height within the limits of the material available; however, this does not apply to the rounded ends of the hole. At these points, allowable elongation must be the controlling factor. A 15° angle is desired on cupped holes. See Fig. 10.

Definitions for Forming Operations—III

Inside of Metal—The surface of metal which is formed by the male part or punch side of the die; the surface of metal to which the die model is made; the surface of metal to which the part drawing is made; the surface to which contour lines are drawn on the metal draft.

Lightening Holes—Holes that are incorporated in parts for the purpose of reducing weight.

Metal Draft—A thin sheet of metal, usually aluminum, properly painted, on which lines are drawn for accurate layout work. Sometimes referred to as the aluminum draft or plate.

Normal Surface of Metal—The general contour of the part excluding any local deformations.

Ogee Curve (OG)—An "S" curve; a reverse curve; a compound curve.

Panel Tipping—The act of redrawing a part in a newly desired position. In body work it usually refers to revolving a part into the die strike position.

Piercing—The process by which any hole or inside portion of a part is formed by means of a pointed tool.

Pinchweld—The spotwelding together of two metal flanges with raw edges pointing in the same direction.

Punching—The process of forming holes out of sheet metal by use of a punch and die.

Rolled Section—A continuous shape such as angles or channels, formed by passing strip stock through a series of progressive rolls.

Shear—The separating of metal by displacing two adjacent sections in opposite directions by means of a cutting tool. This operation includes blanking, perforating, trimming, slitting and parting.

Spring Back (Metal)—The angular amount by which a metal returns toward its former position, after being bent through some specific angle.

Templet—A pattern or guide used for fabricating or checking the shape of parts. A template may be made of any suitable material, but is often made of aluminum because of its rigidity and noncorrosive properties.

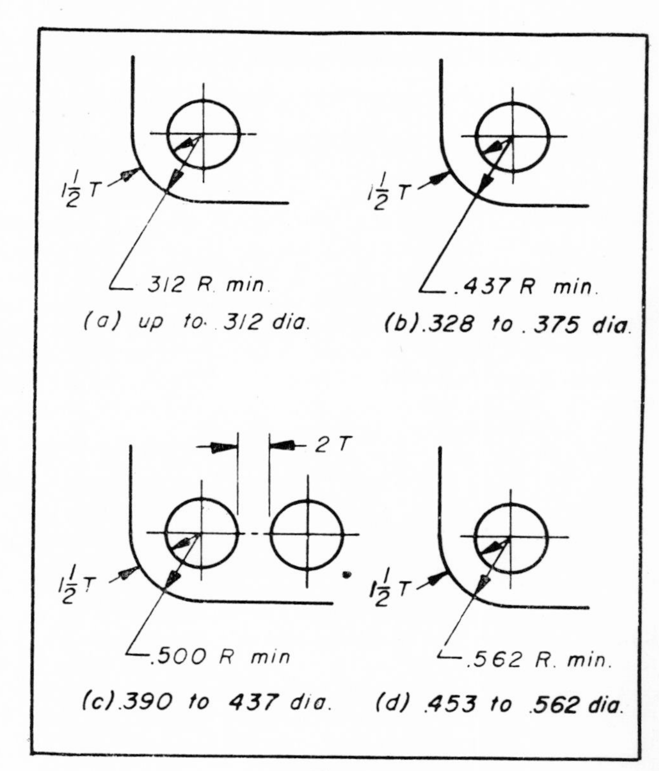

Fig. 9

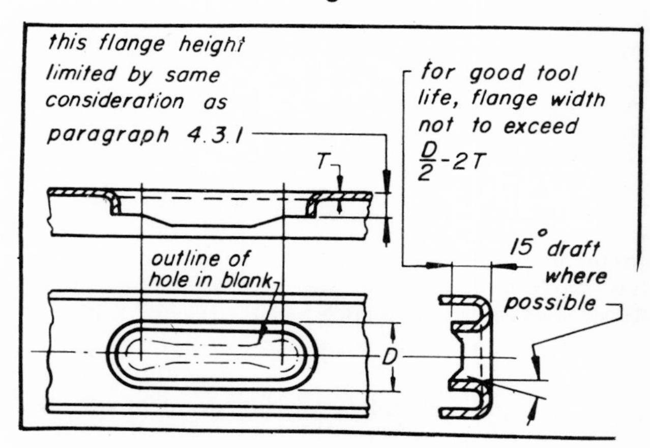

Fig. 10

True Sweep—A true radius; usually refers to a large radius. Also, the wood, plastic or metal templates used as guides in drawing true arcs with radii too large to be drawn with conventional drafting instruments.

Workline (WL)—A term used by draftsmen to indicate the datum or reference from which distances are measured when drawing or checking a view. In theory, this line represents the intersection of two planes of projection which are at right angles to each other.

Deep drawing

To work metals in the plastic range requires some knowledge of press characteristics and behavior of metals when stressed beyond the elastic limit

Deep drawing of metals involves working metals in the plastic range, which lies above the yield point. In this range, the metal flows under tensile or compressive stresses and does not recover its original shape or thickness, or both. Curves in Fig. 1 show the difference in yield point with carbon content for annealed steels, and the characteristics of the respective plastic ranges.

Stresses set up in deep drawing are more severe than in forming, because the stresses occur over a larger portion of the metal. In fact, deep drawing is more of an art than a science.

The actual plastic action of metal during drawing can be depicted by Fig. 2. When the punch enters the die, as at *A*, it tends to pull the metal around the edge of the die, or from horizontal to vertical. Obviously tensile stresses are created in the vertical wall of the shell. As the metal continues to be 'drawn' in this manner, the outside diameter of the blank is 'reduced', creating compressive stresses in the flange, as shown at *B*. The greater the reduction in diameter the greater is the compressive stress in the flange, and also, the greater is the induced tensile stress in the vertical wall of the shell.

Two conclusions may be drawn from this phenomenon. First, the tonnage required depends on the amount of reduction in diameter of the blank; second, there is a point at which the tensile stresses induced in the shell wall will exceed the tensile strength of the metal. Thus the amount of reduction in diameter obtainable is limited.

Apply correction factor

The tonnage required to tear out the bottom of a circular shell is to be obtained by multiplying the cross-sectional area of the shell wall by the ultimate tensile strength of the metal. This is a maximum figure, and should be modified by a correction factor N, dependent on the amount of the blank is reduced. This is usually expressed:

$$\text{Per cent reduction} = \frac{D-D_1}{D}$$

Where

D = Blank diameter
D_1 = Mean diameter of shell

The formula for drawing tonnage for circular shells then becomes:

$$P_d = \pi N D_1 TS$$

Where

P_d = Drawing tonnage
N = Correction factor (Table 1)
D_1 = Mean diameter of shell
T = Wall thickness of shell
S = Tensile strength of metal

Table 1. Correction factors for various reductions

$\frac{D-D_1}{D}$	0.45	0.40	0.35	0.30	0.20
N	1.0	0.9	0.8	0.7	0.6

Abstracted from "Types of Mechanical Presses and Some Considerations Affecting Their Selection," to be published by the McKay Machine Co., Youngstown, Ohio.

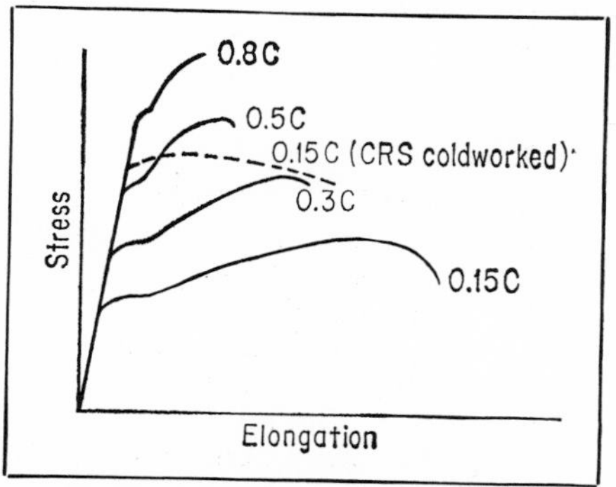

1. Carbon content has a marked effect on the yield point of steel

Note: the tonnage obtained from this formula is only the drawing tonnage. To this tonnage must be added the blankholder tonnage, if a cushion is used, and if any other loads are present, and the total must be converted to press tonnage.

The drawing tonnage for other than circular shells may be determined using the principles discussed, modified by judgment and experience. As an example, take the case of a rectangular draw, which is frequently required.

Here, the metal is not drawn all around the circumference. The metal in the corners is drawn, but that along the four sides is bent over the die radius. Therefore, the tonnage required is only a part of that to draw a circular product with similar circumferential area. For a rectangular shape with corner radius not less than 1/3 to ¼ the depth of the draw, and punch or die radius greater than 10 to 15 times metal thickness, apply a fac-

Deep drawing [continued]

tor C. Values of C range from 0.75 when part periphery is 10 to 12 times the corner radius, to 0.5 where periphery is 35 to 40 times the corner radius.

The formula for draw tonnage for a rectangular shell thus becomes:

$P_d = NCLTS$
where

P_d = Drawing tonnage
N = Correction factor (Table 1, using $(D-D_1) \div D$ for the smallest ratio)
C = Corner radius factor
L = Length of periphery
T = Wall thickness of shell
S = Tensile strength of metal

The limiting reduction in diameter obtainable as noted above (and consequently the limiting depth of draw) depends on a number of factors. A thick blank will not build up compressive stresses as fast as a thinner blank. Consequently greater reductions are obtained with relatively thick material. More generous draw radii and bottom radii allow the metal to flow easier, resulting in greater reductions. Also, increased reductions are obtained by using more ductile metal, and due attention to use of a good drawing lubricant.

The tendency for the edge of the blank to wrinkle is suppressed or eliminated by use of a blankholder ring. Some drawn parts that are relatively thick in proportion to their diameter may be drawn without a blankholder, as the edge of the blank, being thick, is strong enough to resist the compressive stresses without buckling. This type of draw is called a single-action draw. In order to perform such an operation, the blank diameter should be not more than 30 to 40 times the thickness, for reductions $\frac{(D-D')}{D}$ of roughly 30%.

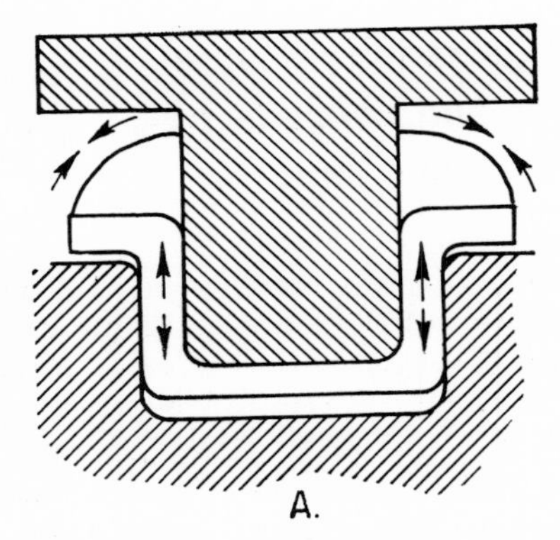

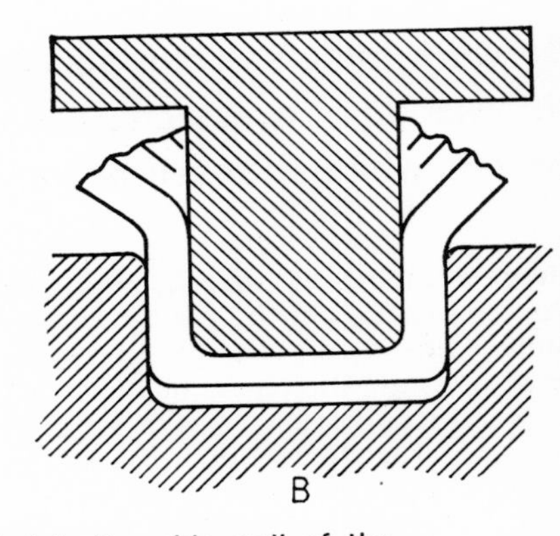

2. In drawing, tensile stresses are created in the side wall of the shell by pulling the metal over the die radius, as at A. As the process is carried further, compressive forces are developed in the flange

Thicker blanks may be drawn with greater reductions.

Percent reductions for draws using a blankholder (double-action draws) vary from 30% for blanks 500 times metal thickness to over 45% for blanks 200 times metal thickness

The reduction limit in each succeeding draw is lower due to strain hardening of the metal. Annealing restores ductility and again permits maximum reduction. Percent reduction obtainable with double-action re-draws on thinner stock starts with about 25% and descends in steps of 20%, 16%, 13% and 10%. The single-action re-draw series starts with about 20%. Again, greater reduction may be obtained with heavier stock.

Blankholder tonnage

The following empirical formula is offered to calculate tonnage for round or rectangular shells with straight sides and flat bottoms.

$P_b = \pi KD_t TS$
where

P_b = Blankholder tonnage
K = Stock thickness factor (Table 2)
D_t = Mean diameter of shell
T = Wall thickness of shell
S = Tensile strength of metal

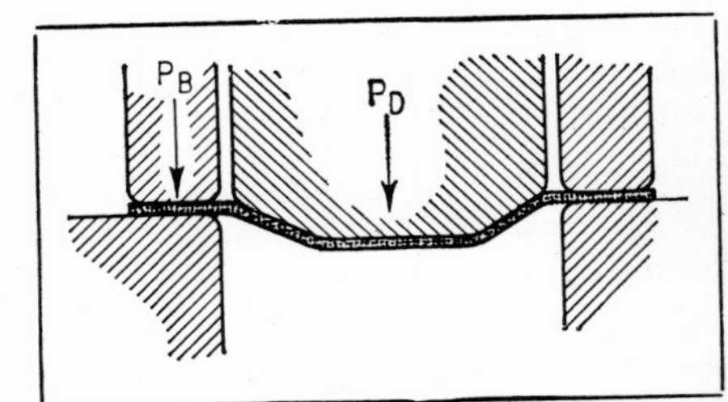

3. If draw beads are not used in the die, blankholder pressure rises

Table 2. Stock thickness factors

Stock thickness	K factor	Stock thickness	K factor
0.020	0.415	0.070	0.117
0.030	0.333	0.080	0.100
0.040	0.250	0.090	0.083
0.050	0.167	0.100	0.067
0.060	0.133	& up	

When drawing shallow shapes, with curved bottoms, such as is sketched in Fig. 3, the blankholder tonnage must be greatly increased in relation to the draw tonnage, because the vertical component P_d of the force necessary to stretch and set the metal is small, whereas the force P_b necessary to hold the blank from slipping is quite large. The use of draw beads will substantially reduce the blankholder pressure required.■

Selection Chart for Deep-Drawing Dies...I

COURTESY: THE BRIDGEPORT BRASS COMPANY

BY USE of the chart on page 186 the die designer can lay out a series of drawing dies so the appropriate reduction in area per draw is achieved for material involved, type of operation and required properties of the shell. This chart can be used for these non-ferrous materials: 66-34 brass, 70-30 (cartridge brass), 85-15 and 80-20. The 95-5 and 90-10 alloys are generally redrawn without thinning the cup wall, because high-copper alloys lack tensile strength to withstand punch pull.

The chart is also applicable to silicon-bronze, except that one-half of the normal 30% reduction must be used for the final operation in order to maintain uniform wall thickness temper.

In this chart, reduction in area does not refer to reduction in total cross-sectional area of the shell, but instead the reduction of the cross-sectional area of the shell wall.

Example: What die sizes are required to produce a flashlight case 1½ in. in dia. by 0.015-in. wall thickness? Calculated blank size is 4½-in. dia. by 0.037 in. thick.

Step 1—Mark the chart at the intersection of the vertical line representing 4½-in. dia. and the horizontal line coinciding with metal thickness of 0.037 in.

Step 2—To find the cupping die size, move left from point 1 to the 44% reduction-in-area curve, and thence downward, so it is found that a 2½-in. cupping die will be satisfactory.

Note: Normally a reduction in area of 40-45% for brass can be assumed for cups, but reductions up to 49% can be used by paying special attention to tooling and equipment. Percentage of reduction on thin-metal cups should be kept near 40% as a general rule.

Step 3—Determine the number of draws. Count the spaces between curves from the size of cup diameter at point 2 and the finish-drawn size at point 6. Twelve spaces are noted. A reduction in area of only about 30% is practical for each redraw operation. The chart shows that three spaces are equivalent to a reduction of 29%. Hence, twelve spaces divided by three spaces yields four draws as necessary to make the shell.

Step 4—To select draw die sizes, connect points 2 and 6 with a smooth curve, not a straight line. The ideal curve would be perpendicular at intersections with the reduction-in-area curves. In this case, the curve as drawn favors this proposition only near points 2 and 3.

Point 3 on the curve (three spaces) is equivalent to a first-draw size of 2⅛-in. dia. by 0.031-in. wall thickness.

Point 4 on the curve (the second three spaces) is equivalent to a second-draw size of 1 13/16 x 0.026 in. wall thickness.

Point 5 on the curve (another three spaces) coincides with a third draw size of 1⅝ in. by 0.019 in. wall thickness.

Point 6 of course represents the final draw size of 1½x0.015 in.

Customarily, a greater amount of reduction is applied at the first and second draws, resulting in say the use of 3½, 3½, 3 and 2 spaces on the curve instead of equal steps of 3 spaces. However, the observance of this nicety in practice should depend on whether the resulting draw-die sizes can be worked out to dimensions of potential further use on subsequent jobs.

With a 30% reduction per draw, a shell about ¾ hard will be produced. Hence, to permit subsequent redrawing it is necessary to anneal and pickle between drawing operations. And if the shell must be less than ¾ hard, use reductions of 1, 2 or 2½ spaces, depending upon requirements. Sometimes, greater hardness (or temper) may be wanted in the finished shell. A higher final reduction will achieve this requirement.

The chart can also be used when a big cup must be folded; that is, redrawn without thinning the wall. From point 2 merely step off the equivalent of 30% reductions, moving horizontally to the left. However, sleeve-type drawing will probably be required. This does not pertain to eyelet or multiple-plunger machines where a 41% reduction in the cupping operation and 15% reductions for succeeding draws, with no intermediate anneals, is common practice.

Other Cases:

1. Shells with a thick bottom and uniform wall—Use up to 45% reduction of blank diameter to cup diameter for material up to ⅛ in. thick. Above ⅛-in. thickness, the type of equipment will influence the reduction.

2. The chart can be used to lay out draw-die sizes so the wall thickness is reduced say 25 to 30% per operation and only about 5% reduction in diameter. In this case, the construction curve will not be perpendicular to the reduction-in-area curves. This is good practice only when working with heavy-gage materials.

3. The chart can be used for work larger than 10-in. diameter by multiplying the readings by 10, because a logarithmic scale is used.

Selection Chart for Deep-Drawing Dies...II

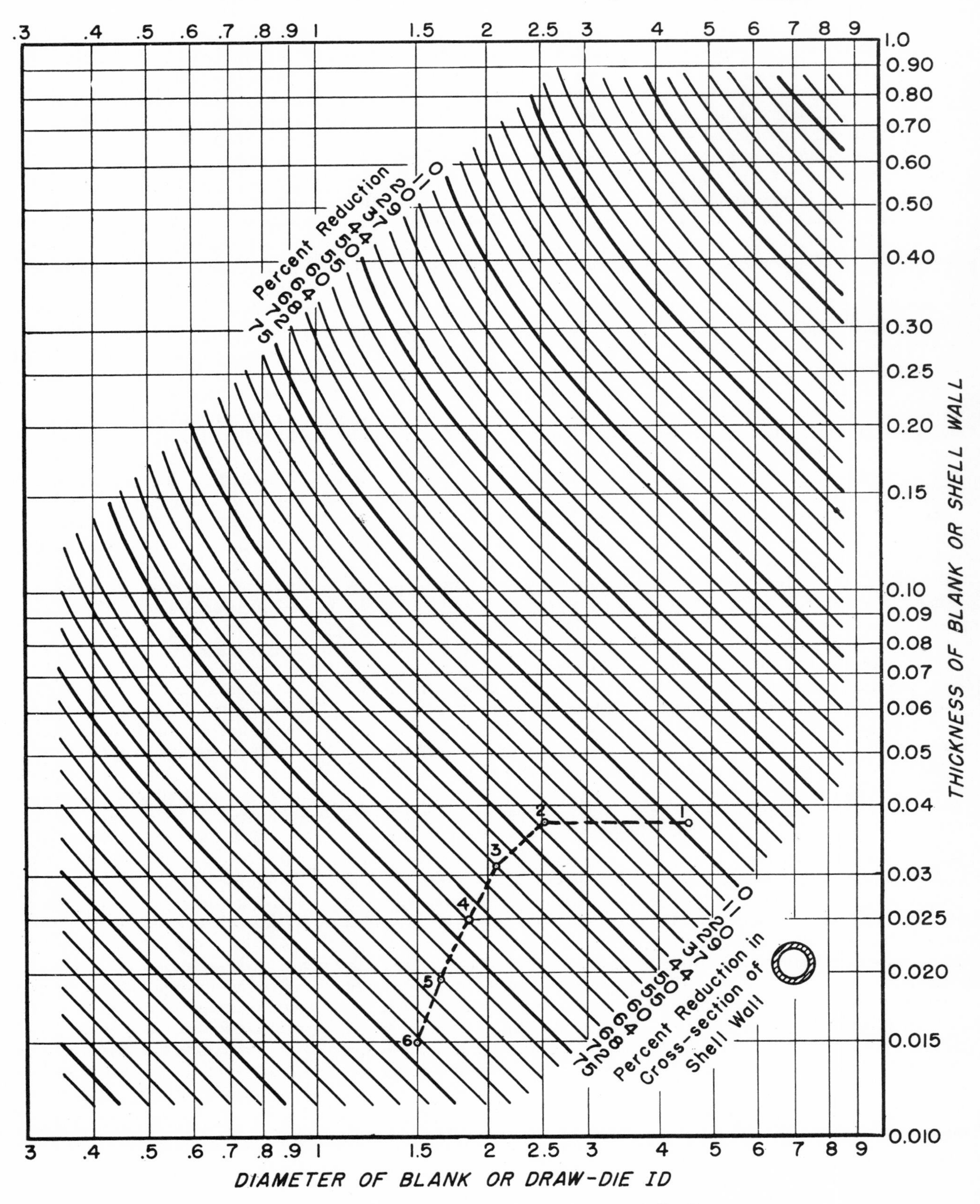

Rules for Drawing Round Shells—I

COURTESY DAYTON ROGERS MANUFACTURING COMPANY

IN metal-stamping work, the expression "drawing" describes an operation in which a flat sheet or metal blank is formed into a shell or cup of fairly uniform thickness. In producing round shells without flange, the factors that govern the possible height of single-operation drawing of a shell are:

1. Ratio of height to diameter of shell
2. Ductility of the material
3. Corner radius

When drawing a round shell (without flange) in one operation, the maximum ratio of height divided by diameter may vary between 1/4 and a possible 3/4, depending on the corner radius and ductility of the material. A generous corner radius helps to secure greater height in one operation, while too small a corner radius may cause the shell to fracture at the radius. Soft, ductile material will permit drawing to a greater height in one operation.

Ductility of materials is measured as percent elongation in 2 in. or percent reduction of area. In general, materials such as deep-drawing steel, annealed sheet steel (SAE 1005 - 1015), dead-soft cold - rolled strip steel (SAE 1010-1020), some stainless steels (304, 410, 420 and 430), soft-temper aluminum (2SO, 3SO and 53SO), soft-temper brass and copper will allow drawing to a maximum height in one operation. Other less ductile materials may require one or more pre-cupping operations; that is, additional drawing dies for gradually reducing the blank to the diameter required, and possibly annealing between operations.

Corner radius of the shell has a considerable effect on the possible height of single-operation drawing. Corner radius should, when possible, be specified at a minimum of four times thickness of material when the height of the shell exceeds one-third of the diameter. When a smaller radius is specified, additional drawing or flattening operations may be

Draw-Reduction Ratios for Round Shells

(Without Flange)

Shell height = inside height of shell
Shell diameter = inside diameter of shell plus one thickness of material

Height-Dia. Ratio	% Reduction of Dia.	Blank-Draw Ratio	Height-Dia. Ratio	% Reduction of Dia.	Blank-Draw Ratio	Height-Dia. Ratio	% Reduction of Dia.	Blank-Draw Ratio
.01	2.0	1.02	.48	41.5	1.70	.95	54.4	2.194
.02	3.8	1.03	.49	41.9	1.72	.96	54.5	2.201
.03	5.5	1.05	.50	42.3	1.73	.97	54.7	2.208
.04	7.1	1.07	.51	42.7	1.745	.98	54.9	2.216
.05	8.7	1.09	.52	43.1	1.758	.99	55.1	2.224
.06	10.2	1.11	.53	43.5	1.770	1.00	55.3	2.231
.07	11.5	1.13	.54	43.8	1.780	1.05	56.2	2.28
.08	13.0	1.15	.55	44.2	1.790	1.10	57.0	2.32
.09	14.3	1.16	.56	44.5	1.800	1.15	57.8	2.37
.10	15.4	1.18	.57	44.8	1.810	1.20	58.5	2.41
.11	16.6	1.20	.58	45.1	1.820	1.25	59.2	2.45
.12	17.8	1.21	.59	45.4	1.830	1.30	59.9	2.49
.13	18.8	1.23	.60	45.8	1.840	1.35	60.5	2.53
.14	19.9	1.24	.61	46.0	1.850	1.40	61.0	2.56
.15	21.0	1.26	.62	46.4	1.862	1.45	61.6	2.60
.16	21.9	1.28	.63	46.7	1.873	1.50	62.2	2.64
.17	22.8	1.29	.64	47.0	1.885	1.55	62.7	2.68
.18	23.7	1.31	.65	47.3	1.898	1.60	63.2	2.72
.19	24.6	1.32	.66	47.6	1.910	1.65	63.7	2.76
.20	25.5	1.34	.67	47.8	1.920	1.70	64.2	2.79
.21	26.3	1.35	.68	48.1	1.930	1.75	64.6	2.83
.22	27.1	1.37	.69	48.4	1.940	1.80	65.1	2.86
.23	27.8	1.38	.70	48.7	1.950	1.85	65.5	2.90
.24	28.6	1.40	.71	49.0	1.960	1.90	65.9	2.93
.25	29.3	1.41	.72	49.2	1.970	1.95	66.3	2.97
.26	30.0	1.43	.73	49.5	1.980	2.00	66.7	3.00
.27	30.7	1.44	.74	49.8	1.990	2.05	67.0	3.03
.28	31.3	1.45	.75	50.0	2.000	2.10	67.4	3.06
.29	31.9	1.47	.76	50.2	2.010	2.15	67.8	3.10
.30	32.6	1.48	.77	50.5	2.020	2.20	68.1	3.13
.31	33.2	1.49	.78	50.7	2.030	2.25	68.4	3.16
.32	33.8	1.51	.79	51.0	2.040	2.30	68.7	3.20
.33	34.4	1.52	.80	51.2	2.050	2.35	69.0	3.23
.34	35.0	1.53	.81	51.4	2.060	2.40	69.3	3.26
.35	35.5	1.55	.82	51.6	2.070	2.45	69.6	3.29
.36	36.0	1.56	.83	51.8	2.080	2.50	69.9	3.32
.37	36.5	1.57	.84	52.1	2.090	2.55	70.2	3.35
.38	37.0	1.58	.85	52.4	2.100	2.60	70.4	3.38
.39	37.5	1.60	.86	52.6	2.110	2.65	70.7	3.41
.40	38.0	1.61	.87	52.8	2.120	2.70	70.9	3.44
.41	38.5	1.62	.88	53.0	2.130	2.75	71.2	3.47
.42	39.0	1.63	.89	53.2	2.140	2.80	71.5	3.50
.43	39.4	1.65	.90	53.4	2.150	2.85	71.7	3.53
.44	39.8	1.66	.91	53.6	2.160	2.90	71.9	3.56
.45	40.2	1.67	.92	53.8	2.170	2.95	72.1	3.58
.46	40.7	1.68	.93	54.0	2.178	3.00	72.2	3.60
.47	41.1	1.69	.94	54.2	2.185			

How to use table: Divide height of shell by diameter; find corresponding ratio in col. 1; Percent reduction is given in col. 2 (use to determine number of reductions required); Blank-draw-ratio is given in col. 3 (blank-draw ratio times shell diameter equals blank diameter approximately).

Rules for Drawing Round Shells—II

required. In no case should the corner radius be less than thickness of material for a one-operation draw.

When the ratio of "height divided by diameter" exceeds ⅝, it will be necessary in most cases to reduce the flat blank to the finished shell by using two or more draw dies of proportionately decreasing diameters. In some cases, one or more annealing operations will be necessary between first and finish draw operations. The necessity of annealing depends to a large extent on the workability of the metal being drawn.

Determination of the number of reductions necessary to draw a shell with ratios (height divided by diameter) greater than ⅝ cannot be done by hard and fast rules. In general, for ductile materials, with generous corner radius in the shell, the requirements are:

1. Height equals ⅝ to 1⅛ times the diameter of the shell—two reductions will be required.
2. Height equals 1⅛ to 2 times the diameter of the shell—three reductions will be required.
3. Height equals 2 to 3 times the diameter of the shell—four reductions will be required.

It may be necessary to anneal the shell when more than two reductions are required. When corner radius is less than four thicknesses of material, add one or two flattening dies and operations, depending on corner radius desired.

For less-ductile materials, it is more difficult to predict the number of operations required. In general, the measure of ductility of the material (percent of elongation or percent reduction of area) will determine the maximum reduction possible in one operation.

Finish of edge depends on the "height divided by diameter" ratio and on the material being drawn. For relatively shallow shells where the "height divided by diameter" ratio is not over 1/3, it is possible to produce an edge within commercial tolerances without requiring finishing operations. That is, the height and uniformity of the edge depends on the size of blank used. For higher shells it is not possible to do this, and one of the following finishing operations will be required:

1. Flange trim and finish draw (Fig. 1).
2. Pinch trim (Fig. 2).
3. Machine trim (Fig. 3).
4. Wedge or "shimmy"-die trim (not suitable for small quantities).

A "Draw Reduction Table" offers a simple means of determining percent of draw reduction and flat-blank diameter. By dividing the inside shell height by the mean shell diameter, a height-diameter ratio is obtained. Find this ratio in Col. I of the table. Directly opposite in Col. II find the percent of reduction of diameter (a measure of the amount of cold-working to be done in drawing the flat blank into a shell). Directly opposite in Col. III, find the Blank-Draw ratio. Multiply the mean shell diameter by this factor to obtain the approximate flat blank diameter.

It must be understood that this table can only be used with round straight-sided shells or cups. Shells or cups with flanges must be investigated by other methods. Care must be used when attempting to predict the number of operations required to produce a flanged shell or cup.

EDGE-TRIMMING METHODS

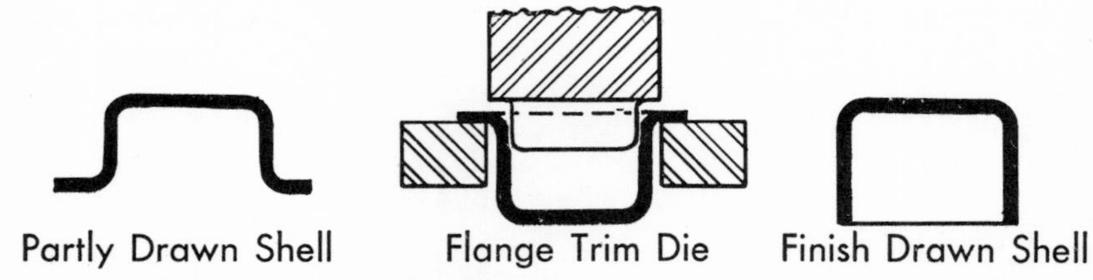

Fig. 1—**FLANGE TRIM AND FINISH DRAW.** This method is satisfactory for most shells, particularly diameters greater than 2 in. Only one additional die—a trimming die—is required.

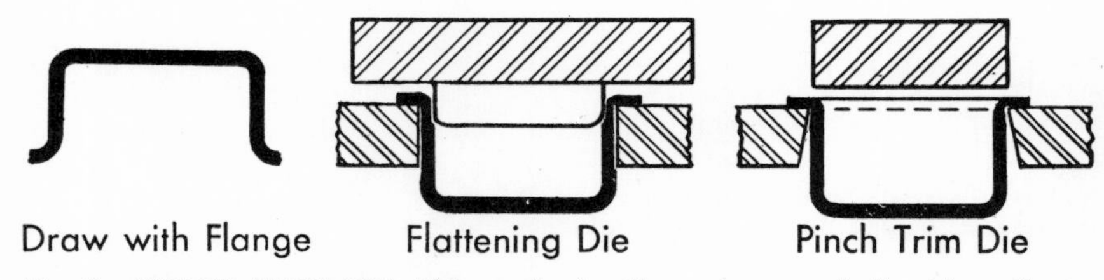

Fig. 2—**PINCH TRIM DIE.** This method will produce a shell with uniform height, but the inside edge is considerably rounded. The flange must be flattened to a sharp corner, which will require one or two dies.

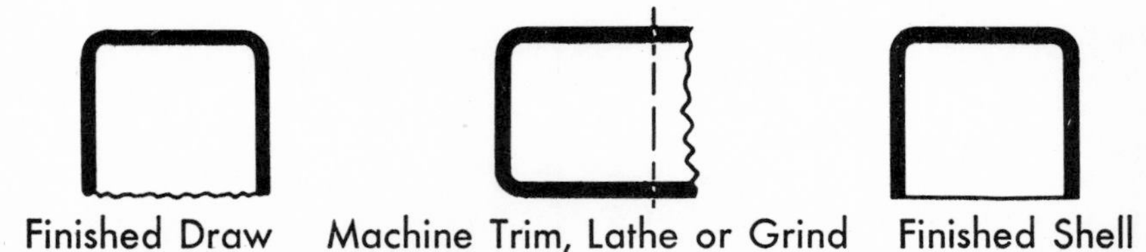

Fig. 3—**MACHINE OR BOX TRIM.** Tooling cost is generally low, but this method is slow. The best-appearing edge is produced, and sometimes this method is the only practical one.

Draw-Die Radius

By **Ferenc Kuchta,** mechanical engineer, J Wiss & Sons Co, Newark, N J

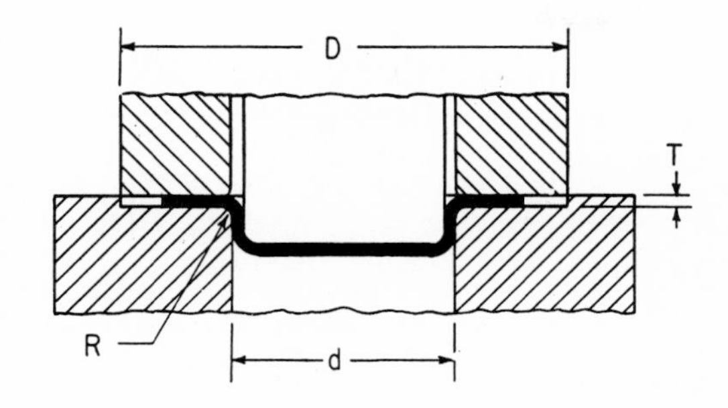

$R = 0.8\sqrt{(D-d)\,T}$

where D = blank diameter, in.

Tolerance for R = ± 0.005 in.

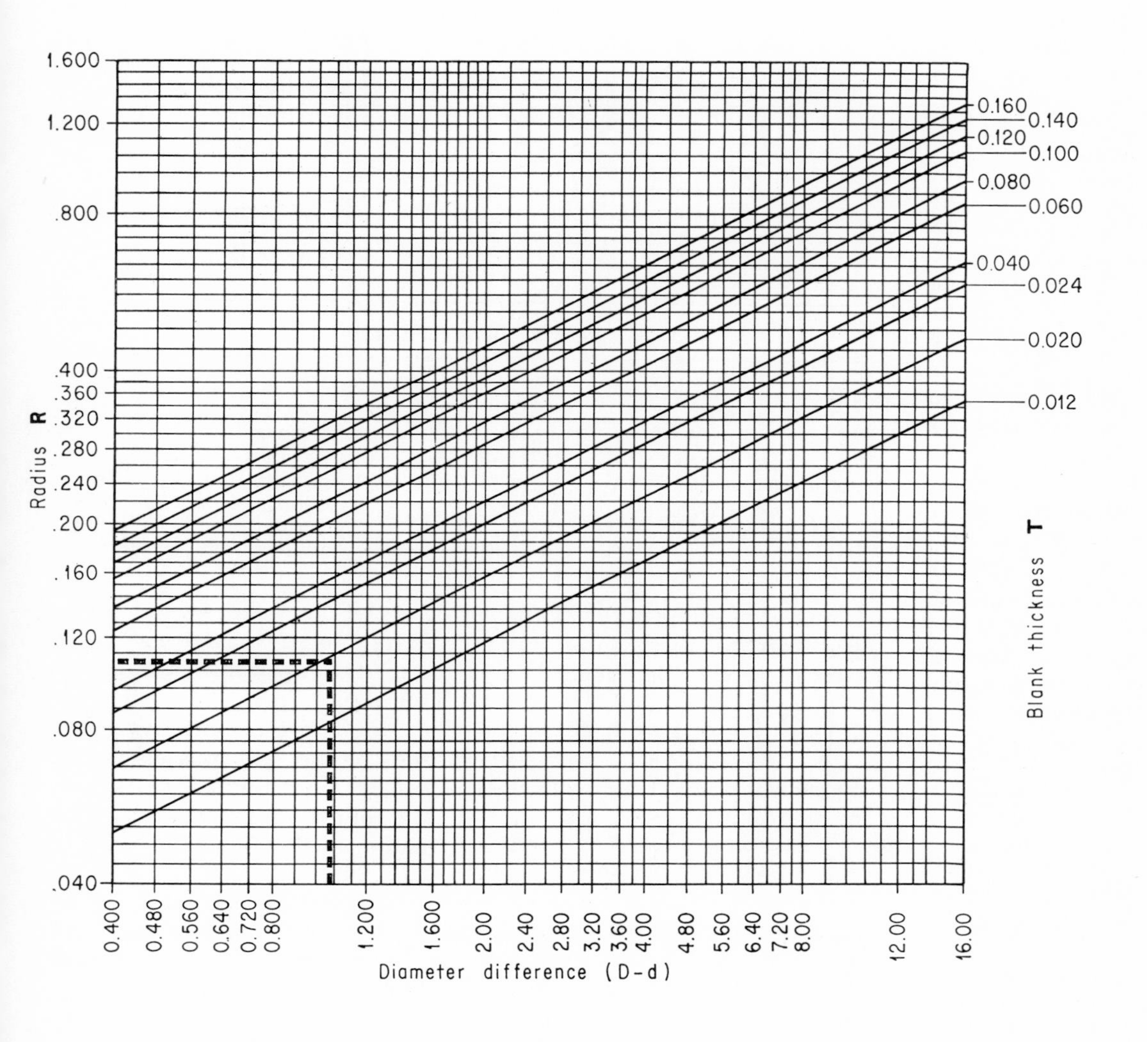

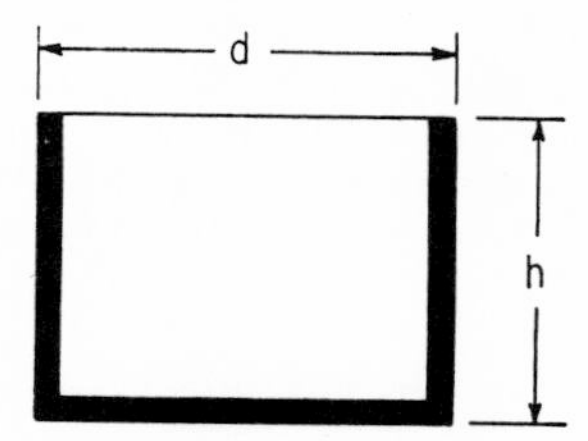

Example:

$d = 1.00$, $h = 0.75$, $T = 0.020$

$D = \sqrt{d^2 + 4dh} = 2.00$

$D-d = 1$

$R = 0.110 \pm 0.005$

Drawing Limit Without Pad

BERNARD K H BAO, chief engineer, R KRASBERG & SONS MFG CO, CHICAGO

Savings in progressive dies can be made if the designer can predict whether a pressure pad is needed at the cupping station. Example: dies for making boot retainers for seals. Most retainers are shallow cups with extruded driving ears at the center. Compound dies are not feasible for drawing shallow shells because the wall of the blanking punch would be too thin.

Progressive dies are a good solution if the parts must be made in one press operation. Blanks can be punched on a pad, pushed back into the strip, and carried to the cupping station. If there is no need of a draw pad, cupping is done by pushing the blank into the draw die and allowing it to drop through the shoe.

Some cups of fairly thick stock relative to blank size can be drawn without a pad if the overhanging portion is thick enough to resist wrinkling. The overhanging flange is under high compressive stresses, which act like the loading on a column that is being squeezed down plastically. If the column is long and thin, it will buckle. Similarly, relatively thin blanks will buckle.

Our problem is to find a relationship whereby we can predict the limits for drawing a cup without a pad. In the sketch, the overhanging portion is $D_B - D_p$. The ratio will be $[D_B - D_p] \div D_p$,* or $\left[\frac{D_B}{D_p} - 1\right]$. Thus, the limits of drawing a cup without pad can be represented by a curve, which is defined by $f\left(\frac{D_B}{t}, \frac{D_B}{D_p}\right)$ or a conjugate hyperbola with these two ratios as coordinates.

Tests were made to confirm the reasoning. Draw-die radii were made three times stock thickness t and punch radii were made to one stock thickness. Steel cups of different ratios of D_B / t and D_B / D_p were drawn until wrinkling started, and the constant determined. The hyperbola was now plotted. Points A, B, D and E fall in the no-pad zone. Point C produced wrinkling marks that were ironed after the cup was all the way in the die.

This curve is tentative. Persons having information on the subject are invited to supply the following information to the author at P O Box 7902, Chicago 80, Illinois:

- Material of cup
- Stock thickness, t
- Draw die radius (3 t preferred)
- Punch radius (one t preferred)
- Punch dia D_p
- Blank dia D_B
- Appearance of cup

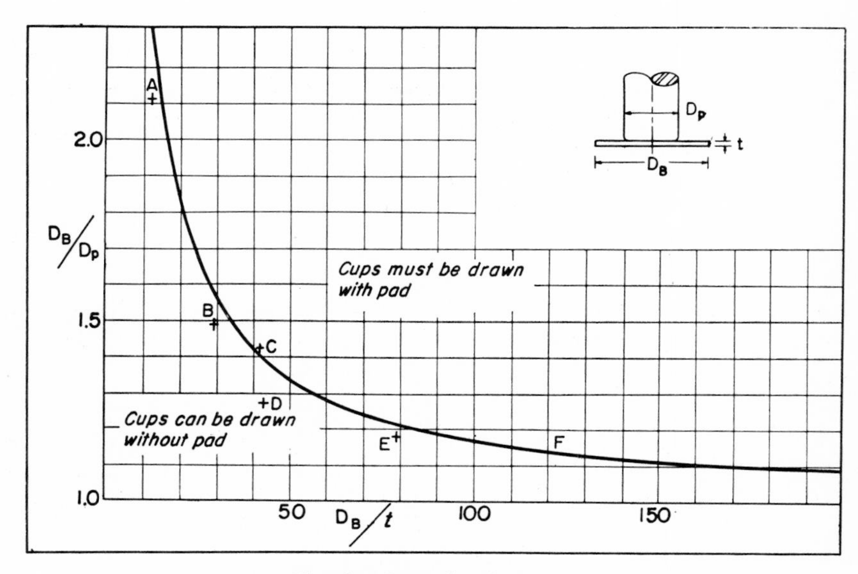

Plotting Data for Curve

Test	D_B/D_p	Constant	D_B/t	Constant
A	1.531/0.719	2.129	1.531/0.125	12.25
B	2.277/1.523	1.495	2.277/0.077	29.57
C	1.280/0.907	1.412	1.280/0.031	41.30
D	0.843/0.660	1.277	0.843/0.020	42.15
E	1.798/1.530	1.183	1.798/0.023	78.17
F	2.517/2.208	1.140	2.517/0.021	119.86

* *Do not confuse this expression with percentage of reduction*

Blank diameters for round shells

By F Caplan, Kaiser Engineers, Oakland, California

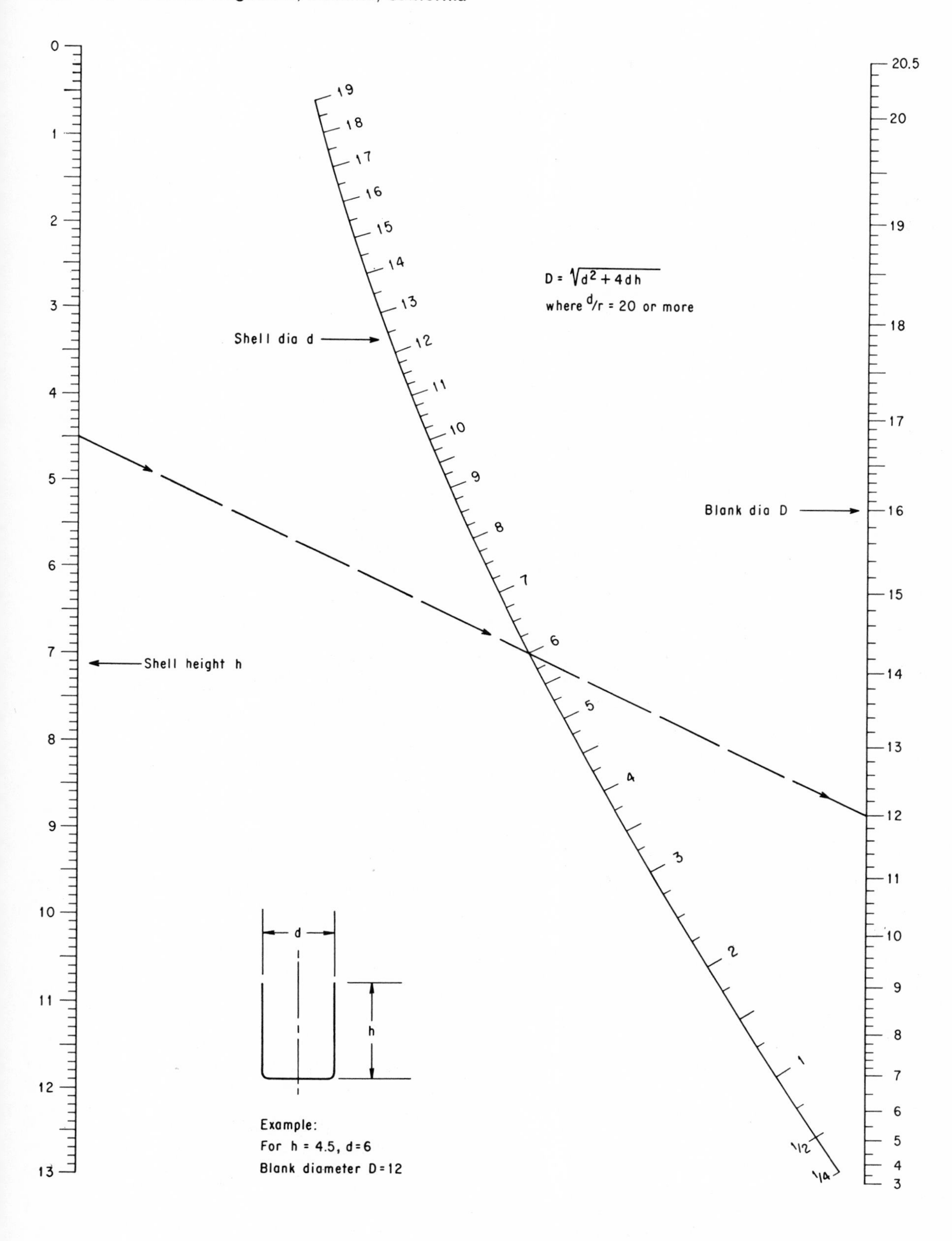

Weight of Coil Stock

By F Caplan, design engineer,
Kaiser Engineers, Oakland, Calif

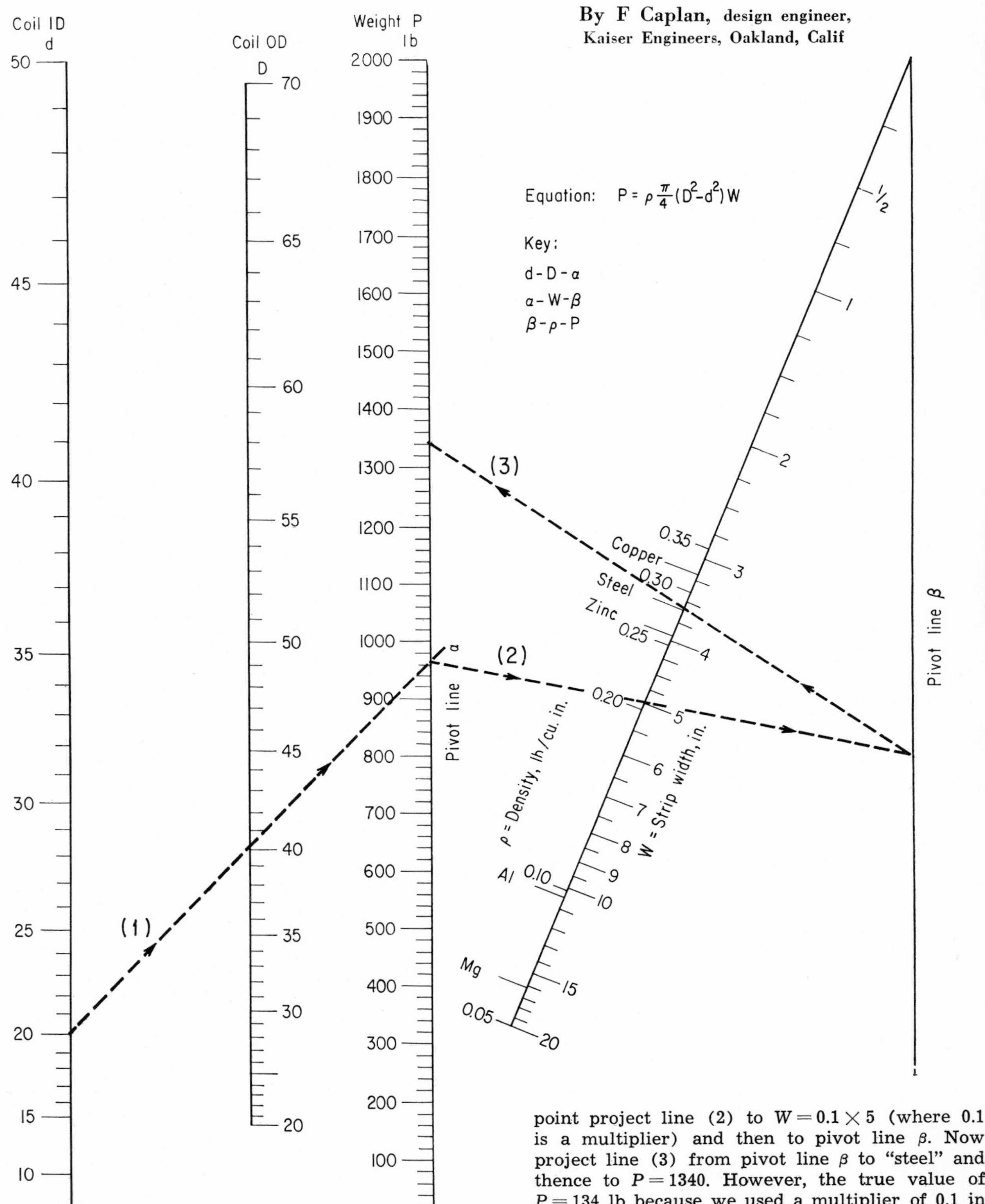

EXAMPLE: What is the weight of a steel coil 40 in. OD, 20 in. ID, and ½ in. wide? Move on construction line (1) from 20 in. ID to 40 in. OD and thence to intersection on pivot line α. From this point project line (2) to $W = 0.1 \times 5$ (where 0.1 is a multiplier) and then to pivot line β. Now project line (3) from pivot line β to "steel" and thence to $P = 1340$. However, the true value of $P = 134$ lb because we used a multiplier of 0.1 in connection with W, to keep within working limits to the nomograph.

NOTE: If scale values of W are multiplied or divided by 10, then scale values of P must likewise be multiplied or divided by 10.

Lubricants for pressworking

Nine types of popular lubricants are reviewed to explain the benefits and short comings of each and the variations that can exist within the grouping

Presswork lubricants have been adopted more on 'trial and error' basis than can be justified in the light of technical information available. These materials, vital to stamping plants, are not chemical mysteries.

Finished products for presswork lubrication may be divided into nine basic categories: chlorinated oils, sulfurized oils, fatty oils and waxes, soap-fat pastes, soap compositions, dry film lubricants, mineral oils, soluble oils, and synthetics. If you understand what can be expected from each type, trial and error testing procedures can be minimized and costly mistakes avoided.

By Leon Salz
manager, lubricant products
Magnus Chemical,
Division Economics Laboratory, Inc,
Garwood, N J

Chlorinated oils

These liquids contain non-volatile chlorinated compounds. They may be purchased as concentrates to be mixed with mineral oil or as ready-to-use blends in various concentrations. The concentrated chlorinated oil is identifiable by its honey-like appearance and a net weight of 525 lb per 55 gal. drum. The blends of chlorinated oil and mineral oil are available in concentrations ranging from 5% to more than 50%.

Concentrated chlorinated oils are widely used for deep drawing, because they prolong die life and minimize scratching on difficult

Rating sheet for pressworking lubricants

Composition	Capability of prolonging die life	Cleanability by Vapor degreasing	Cleanability by Aqueous methods	Rust protection	Non-ferrous metal staining	Remarks
Chlorinated oils						
Concentrate (non-emulsifiable)	1	3	5	5	Usually none	Light color; almost no odor
Concentrate (emulsifiable)	1	3	2	5		
Mineral oil blend (high chlorine content, emulsifiable	1	2	1	5		
Mineral oil blend (low chlorine content, emulsifiable	2	1	1	5		
Sulfurized oils					Non-staining to aluminum, zinc and tin. Active sulfur causes black stain on copper, brass and bronze	Brown color; typical sulfur odor
Concentrate	2*	3	4	3		
Mineral oil blend (high sulfur content)	4*	2	3	3		
Mineral oil blend (low sulfur content)	5*	1	2	3		
Fatty oils and fats					Usually none. Copper alloys develop green stain with compounds having high free fatty acid content	Typical fatty odor. Burns off above 1000°F. leaving no residue
Concentrate	2*	3	3	3		
Mineral oil blend	4*	1	2	3		
Soap-fat compounds						
Non-pigmented	3	4	3	3	Usually tarnishes brass, bronze, zinc and tin	
Pigmented	2	5	4	2		
Liquid lubricating soaps	3	5	1	3	Usually tarnishes brass, bronze, zinc and tin	Foams
'Dry film' soaps	1	5	2	2	Usually tarnishes brass, bronze, zinc and tin	Requires special equipment. Best suited for production work on larger parts
Mineral oil	5	1	2	3	None	Wide range of viscosities
Soluble oils						
General purpose	5	2	1	3	— — —	
Heavy duty	3	2	1	3	May tarnish brass, bronze, zinc and tin	
Vanishing oil	5	—	—	—	None	Evaporates from metal within one hour exposure to air; leaves no residue

*Rating indicated is for ferrous metals. Rating would be upgraded for non-ferrous metals.
Ratings: 1—Best; 2—Good; 3—Average; 4—Poor; 5—Worst

Lubricants for pressworking [Continued]

jobs. Mineral oil blends are used on drawing jobs of lesser severity and for stamping and blanking. Despite these capabilities, chlorinated oils have a major shortcoming—poor rust protection. This frequently limits the applicability of chlorinated oils to non-rusting metals such as stainless steel, aluminum or brass, or to low-carbon steel which will be cleaned within a matter of days.

Considerable improvement has been made in chlorinated oils. A few years ago, expensive vapor degreasing was the only practical way of removing the oil. Now concentrates and blends are available with built-in emulsifiers.

Also, oils having the same chlorine content are now available from SAE 10 to liquids so thick that they barely flow. Such modifications fill the needs of plants that may wish to circulate the oil at full strength to those that find greater efficiency at higher viscosities.

Sulfurized oils

These products are usually blends of sulfur-containing organic chemicals, such as sulfurized fat and mineral oil. The sulfurized additive is generally less than 5%. Such compositions are sold as cutting oils but are used as stamping oils to a greater extent than efficiency warrants. These cutting oils are not effective in prolonging die life for stamped and deep drawn work, but are helpful for cutting, blanking, piercing and related operations where metal shearing rather than stretching is involved. They should be used in preference to chlorinated oils when applicable because of better rust-preventive characteristics.

Fatty oils and fats

These 'glycerides' consist of chemically combined glycerine and fatty acids.

Fatty products have certain merits. When used without mineral oil dilution, they provide reasonably good die life. They give satisfactory rust protection and are usually not difficult to clean.

Fatty oils and fats can be burned off in an annealing operation where the temperature is over 1000 F, eliminating cleaning.

Soap-fat compounds

These water-dispersible greases are of two types: Non-pigmented and pigmented, which contain a spacing agent to mechanically separate the die from the work.

The use of soap-fat pastes for stamping and deepdrawing is common practice. Pigmented pastes provide fairly good die life, better rust protection than most alternatives and are inexpensive, considering that they are mixed with water and/or mineral oil. In addition, pigmented materials often eliminate metal tearing when drawing stock of low ductility. Big problem with soap-fat pastes is their removal. This step is difficult by vapor degreasing or aqueous cleaning.

Liquid soaps

Such compositions, properly compounded, offer several benefits over soluble oils and non-pigmented soap-fat pastes:

- Instant solubility in water
- Exceptional slipperiness
- Water dispersions that wet all varieties of metal surfaces completely, maintaining an unbroken lubricating film (by comparison, other water dispersible products shrink into droplets)
- Easier blowing of stampings to catch bucket
- Good rust protection
- Self-cleaning in hot water without special detergents

Products of this type are excellent for stamping low-carbon steel and for stamping and drawing of copper and brass. On the other hand, they are not cleanable by vapor degreasing, cannot be used in vigorous circulating systems (foam), and will tarnish brass (as most other water soluble products will also do).

'Dry film' soaps

Drawn parts to be coated by dry film soaps are immersed in a 10-25% solution at 180-200 F for 3 minutes and then oven or air dried. The resultant film is thin, dry and plastic and gives much better lubrication than corresponding aqueous solutions. In addition to excellent deep drawing characteristics, there is less mess at the press, drawn parts are easy to handle and cleaning is comparatively easy. Disadvantages include the possible need for special dipping and drying equipment and the adverse effect of high humidity.

Dry film soaps are impractical and needless for drawing small parts. Their principal use is for large-area draws such as refrigerator doors, bath tubs, or gas cylinders.

Mineral oils

These liquids have no extreme-pressure or oiliness additives and are limited to blanking, forming or light duty stamping. The best that can be said for mineral oil and motor oil is that they are cheap.

Soluble oils

Such compositions are comparable to straight mineral oil in effectiveness. But a heavy-duty soluble oil can be of greater benefit for press lubrication because it contains extreme-pressure chemicals. These products provide good rust protection if ratio of soluble oil to water does not exceed 1-10.

Synthetics

A new development in organic chemical technology has produced a product called 'vanishing oil.' This colorless and odorless liquid, which is bringing large savings to its users by providing the slip needed for light-duty work, disappears within one hour after use, completely eliminating the need for cleaning. The product has the lubricating value of mineral oil and is especially suited for blanking, forming, and light-duty stamping of stain-sensitive metals or coated metals such as tin- and zinc-coated steel, aluminum, copper, brass, terneplate, pre-painted steel and plastic-coated steel. The product is approved for use on a metal that will package food. ■

Chapter 13

Forging and Upsetting

Closed-die forgings, 1

Successful building of forging dies and the necessary jigs and fixtures for machining purposes must rest upon a sound knowledge of how forgings are designed

Design of parts for forging, like design for most processes, requires that a number of considerations be taken into account. Some of these are sufficiently well understood to be generalized; others are known to a lesser degree and can be treated only empirically. In the discussion that follows, principles as well as details of forging design are given.

Parting lines

Location of the parting line is a critical factor in forging design. The parting line is the plane in which the two halves of the forging die meet and in which flash is formed. But the parting line is not necessarily in the forging plane—the plane perpendicular to die motion.

An obvious basic requirement is the need to remove the finished forging from the die. Normally, only shapes that are narrow when viewed from the forging plane are forgeable. Undercut shapes can be produced only in dies composed of more than two parts, such as are used on horizontal forging machines and multi-ram presses. Undercut shapes have been produced on hammers, using split bottom dies, but the process often is cumbersome and costly.

Choice of optimum parting-line position is influenced by many interrelated factors often setting opposing requirements. The parting line should be in one plane. This requirement can be fulfilled in most cases, Fig. 1a. However, the shape of the forging may be such as to make a broken parting line more economical or even essential, Fig. 1b. Any portion of the parting line that is inclined to the forging plane gives rise to horizontal components of the forging force which tend to displace the die halves. These forces are balanced when the forging is symmetrical, Fig. 1b, but must be accommodated by a counterlock in the die block, Fig. 1c, for asymmetrical parts. The forging may be tilted to balance the horizontal forces, usually at the expense of a complex parting-line shape and increased die-making costs, Fig. 1d. If feasible, forging two identical components in a mirror-image position, Fig. 1e, is preferable.

Position of the parting line helps determine the depth that steel must flow to fill the impressions. A parting line that divides the forging into two equal halves not only insures minimum depth but also minimum excess material, Fig. 2a. The shape of the part may force placing of the parting line in an unfavorable position, Fig. 2b, and it may be more economical to forge such a shape in the horizontal position, Fig. 2c. In the latter case, however, steel flow into the flash results in a discontinuous structure, Fig. 2d, and the

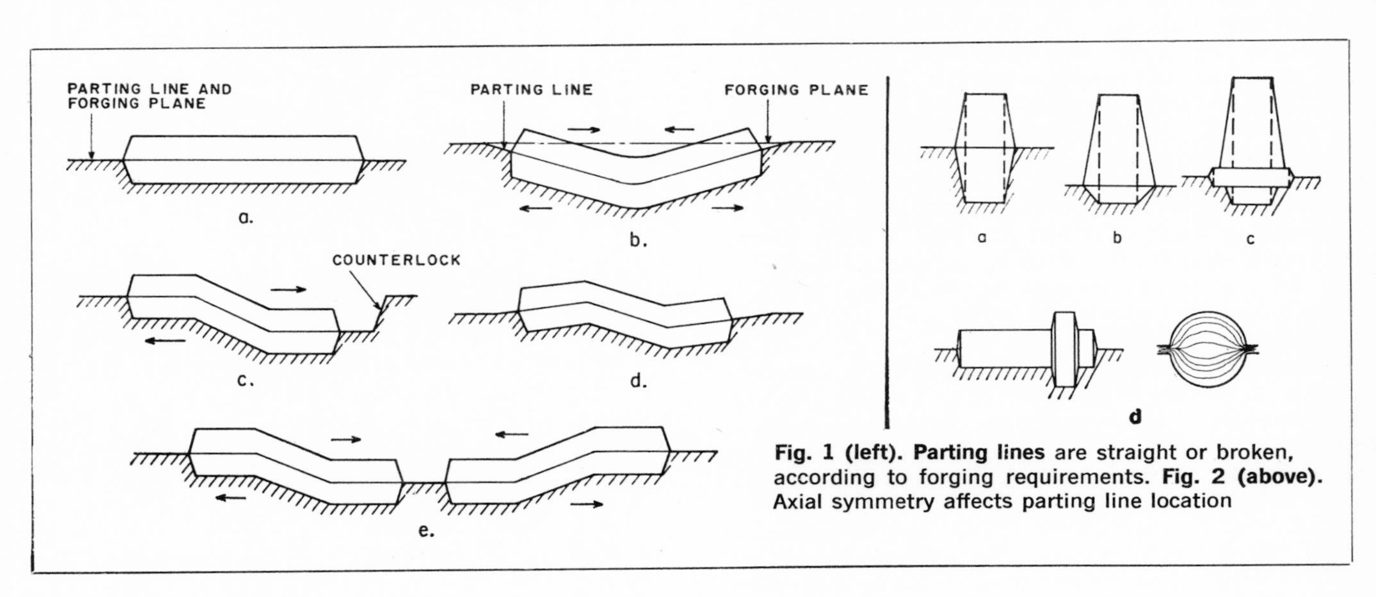

Fig. 1 (left). Parting lines are straight or broken, according to forging requirements. **Fig. 2 (above).** Axial symmetry affects parting line location

Closed-die forgings [Continued]

fibers are cut through when the flash is removed, possibly leading to inferior fatigue properties. Shapes of this kind may be better suited for extrusion, upsetting or roll forging.

Similar conditions apply to hollow components. Parting at the base not only results in weak areas in corners but also increases material consumption, Fig. 3a. Grain structure is uninterrupted when the parting line is placed near the rim, Fig. 3b. This has the further advantage that the wall can be forged with a uniform thickness and the draft eliminated in a subsequent drawing operation, thus obviating the need for machining. Parting for webs and ribs is shown in Figs. 3c and 3d.

Die costs could be much less if the parting line were placed at the top edge of the forging. This practice is permissible in forging certain light metals, but can cause cracking of steel forgings. Flash extruded at the edge of the forging cools more rapidly than the bulk of the material and imposes stresses high enough to deform the hot forging plastically. As a result, tensile stresses are set up on cooling near the flash which cause cracking when the flash is removed.

The forging should be of minimum weight consistent with subsequent machining. A parting line at the edge of the part, Fig. 4a, results in lowest die cost. However, cracking can occur after trimming, and all surfaces must have substantial draft which usually has to be removed by machining. A broken parting line, Fig. 4b, requires an extensive counterlock to take the side thrust developed in forging. The solution also is poor because the machining required is the same as in Fig. 4a. A tilted forging, Fig. 4c, requires a more expensive die but avoids the need for counterlock and eliminates draft from the end surfaces, thus reducing machining costs.

To facilitate subsequent machining, it is desirable that surfaces to be held in chucks or jigs be parallel without side draft. This condition may be difficult to satisfy in many instances, and subsequent operations, such as drawing through a die or close trimming that removes some of the draft, may be satisfactory solutions. If chucking on a conical surface is unavoidable, the parting line should be located to produce as large a reference surface as possible to increase stability in machining.

Preforging of hollows is always advisable because it saves machining time and material. The parting line should be chosen accordingly. This principle may be difficult to follow with forgings of undercut shape, Fig. 5. Parting through the axis of the body permits the forging of the undercut shape, Fig. 5a, but leaves the center solid. Also the radial flash may be undesirable if the part is to sustain fatigue loads. Forged in an upright position. the undercut must be machined afterwards, Fig. 5b, or a split die must be used, Fig. 5c.

For this reason such components often are best produced on horizontal forging machines, where the bore can be preforged without loss of material, Fig. 5d.

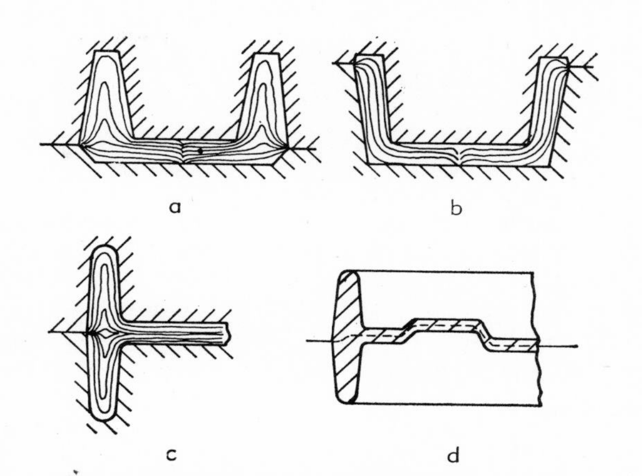

Fig. 3. Parting lines for hollow bodies either follow ribs or are placed to avoid weakness

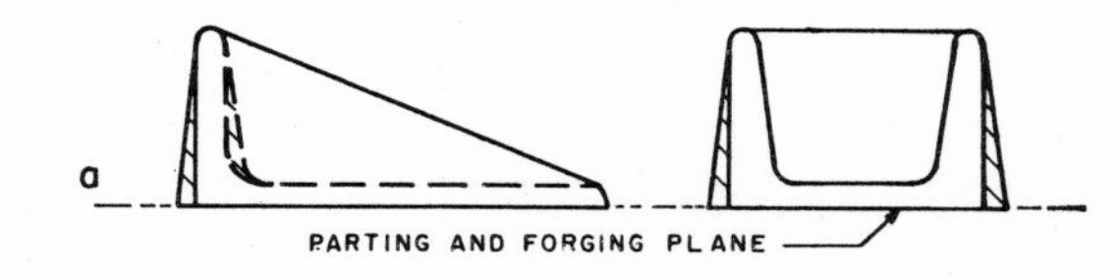

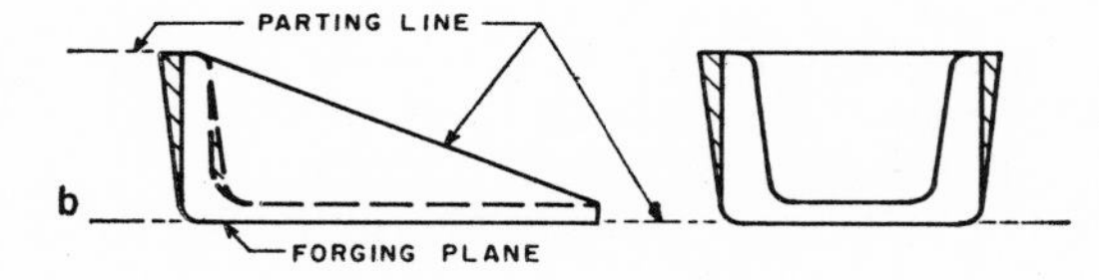

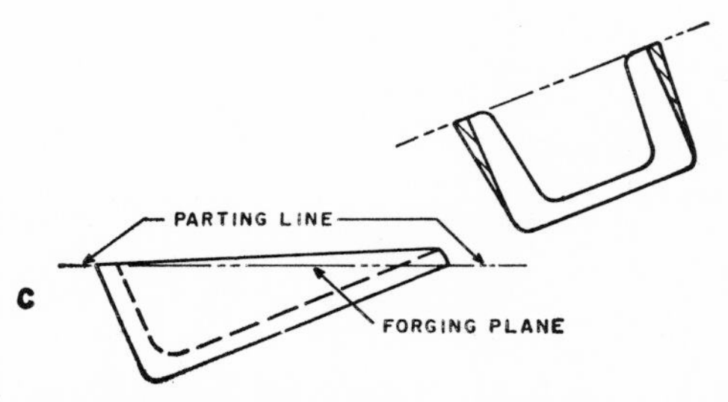

Fig. 4. Shaded areas show excess material that is machined because of parting line location

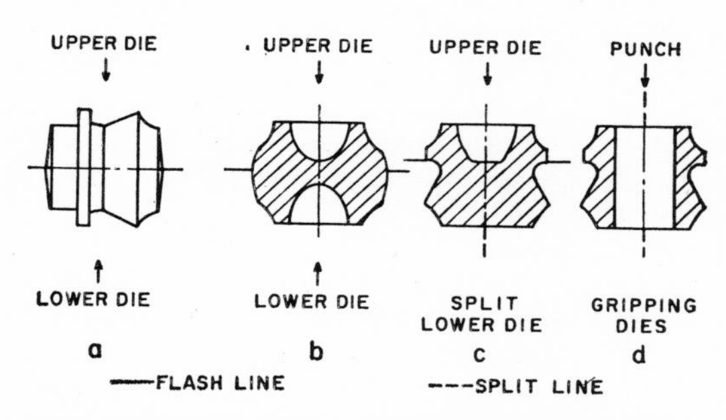

Fig. 5. Location of the parting line for undercut forgings will affect choice of forging method

Closed-die forgings, 2

Necessary draft on impression forgings is a function of several factors. American practice favors depth of cavity as the decisive dimension

Draft is the angle applied to the side walls of a forging, inside and outside, so the part can be removed from the forging dies. The amount of offset that results from applying draft to a forging is known as die-draft equivalent.

In certain instances, draft also adds metal. In most cases, however, draft is wasted material that does not add to the forged component's mechanical utility. So you should make every effort to reduce draft.

As a forging cools, it shrinks, and a gap forms between the outside surfaces of the forging and the die cavity. Therefore, outside draft angles can be smaller than inside draft angles where the steel shrinks onto bosses of the die.

The exact amount of draft necessary is a function of many factors. The deeper the die cavity, the greater the draft required. Impression depth often is used as the sole guiding factor in determining draft (Table I). Commercial standards are used in routine production. Special standards also may be used, but usually at higher cost because more preforming is necessary.

Opinions about optimum draft angles vary. American practice regards cavity depth as the deciding factor. The system of determining draft, based on the combined effect of height and width (Fig. 6) is claimed to lead to considerably less excess material.

Draft angles may have to be increased if the material is difficult to form or if the rate of deformation is slow because of greater cooling during forming. On the other hand, mechanical ejectors permit lower draft angles—generally 2-4° less than on comparable hammer forgings.

You should also keep draft angles low if friction between die and workpiece is high. Otherwise, the increased frictional forces may prevent complete die filling. Draft is occasionaly increased in the top half of hammer dies, so that the forging will not stick in the top die between successive blows.

Spherical shapes

Spherical shapes can be forged without side draft. Horizontally-forged cylinders need draft only on their end faces. Three solutions can be used: Tapered end, rounded end, and conical end (Fig. 7). Of these variations, you should use the conical end when possible because the costs are less.

Draft angles are, of course, very much a function of production method. If preforging is carried to an advanced stage, draft angles can be much reduced, since they no longer are needed to aid metal flow.

Small radii are undesirable for many reasons. They

Courtesy of the American Iron & Steel Institute, New York

Table I. Draft recommendations for impression-die steel forgings

Draft height or depth (in.)	Commercial standards Draft (deg.)	Tolerance* (deg.)	Special standards Draft (deg.)	Tolerance* (deg.)
		Outside draft		
¼ to ½	...	...	3	2
¾ to 1	5	3	...	...
½ to 1	...	...	5	2
1 to 3	7	3	5	3
Over 3	7	4	7	3
		Inside draft		
¼ to 1	7	3	5	3
Over 1	10	3	10	3

*Plus values given; minus tolerance is zero.

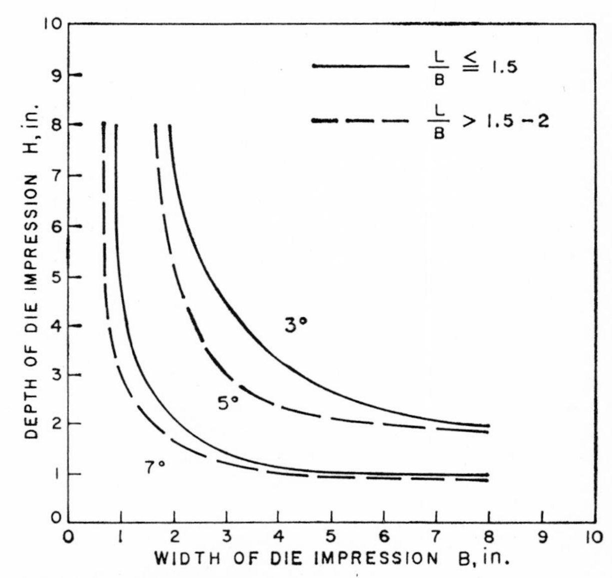

Fig. 6. Recommended draft angles are based on part dimensions. Length (L) refers to length of feature in part, not to over-all length

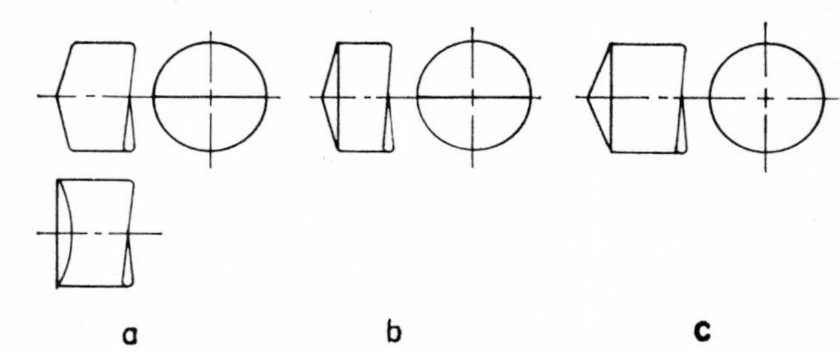

Fig. 7. Various draft configurations are found on the ends of cylindrical forgings

can lead to forging defects (cold shuts), and they often considerably shorten die life. Sharp corners or edges on dies heat up during forging and cool rapidly between operations. Thus, they are subject to thermal fatigue, and are easily eroded by the flow of hot steel. In extreme cases, sharp radii may lead to incomplete filling. Minimum corner and fillet radii normally observed for steel forging are shown in Fig. 8. In more difficult-to-forge alloys, like stainless steel, ¼-in. radii are considered the absolute minimum.

Ribs and bosses are difficult to fill, and radii normally are increased (Fig. 9) to aid metal flow. For more economical die making, you should use the same radius in all parts of the forging if possible.

Adequate radii are mandatory not only because of their effect on forging, but also for optimum service of the part. Small fillet radii cause stress concentrations, whether the component is subjected to bending (Fig. 10) or torsion (Fig. 11).

Large fillet radii are undesirable if a smaller radius subsequently is machined, because the forged fiber is cut through and the part weakened.

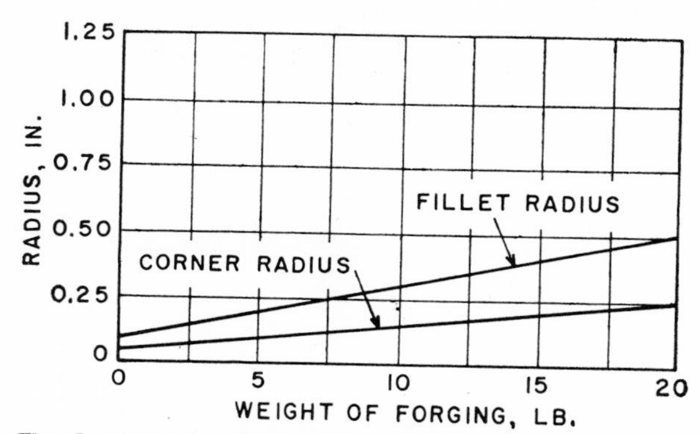

Fig. 8. Minimum fillet and corner radii for steel forgings are based on weight of the forging

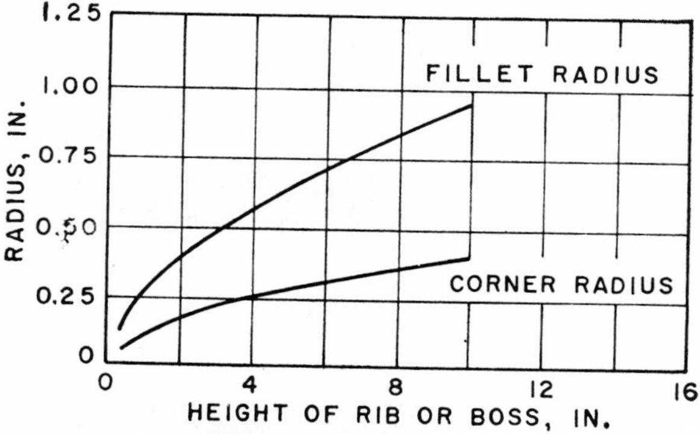

Fig. 9. Definite fillet and corner radii are required for ribs and bosses

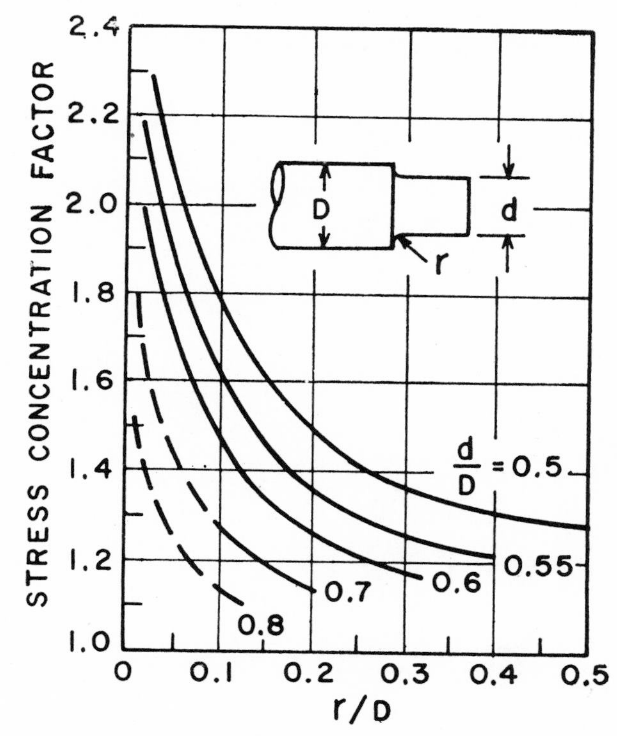

Fig. 10. Stress-concentration factor for shafts in bending shows disadvantage of small fillet radii

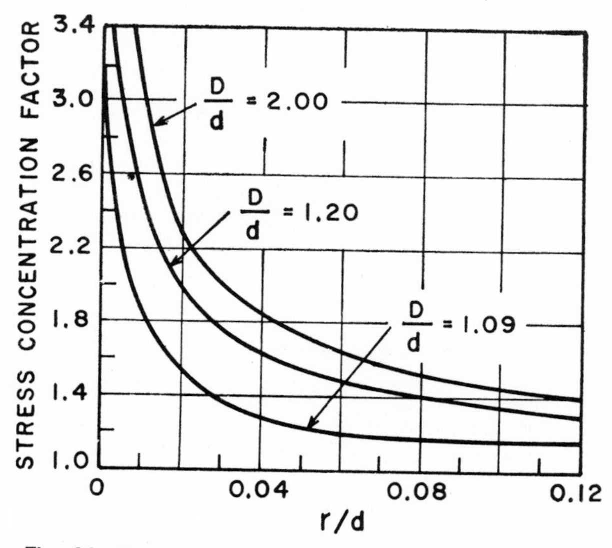

Fig. 11. Stress-concentration factor for shafts in torsion is increased by small fillet radii

Upset forgings, 1

Here are some basic facts extracted from the major published works (both domestic and foreign) on the subject of upset forging

Upsetting, or forging on the horizontal forging machine can be, from the technological point of view, impression die forging, small forging or true flashless (closed die) forging. In its simplest form, it produces a head at the end of a rod, but the construction of the machine permits the forging of a wide variety of shapes.

The workpiece, usually a rod of round cross-section, is gripped in two die halves and the portion extending beyond the grip dies is deformed by the heading tool. The technique is essentially similar to press forging but with the important difference of split gripping dies. This allows the production of undercut shapes and, if the gripping dies are moved with sufficient force they can be used to perform auxiliary operations.

The operational sequence is illustrated in Fig. 1. The hot end of the bar is placed into the stationary grip tool against a stop, at A, the moving grip tool closes, and the stop retracts, B, whereupon the heading tool begins compression, C, and completes it at the end of its stroke, D.

Forging of solid bodies

The maximum length that can be upset in a single stroke is set by the instability, or buckling, of the unsupported rod. From experience, the unsupported length should be not more than three times the diameter of the rod, Fig. 2A. This rule applies whether the stock overhangs the face of the grip dies, or part of it is formed in cavities of grip dies or heading tool as at B.

Extracted from a booklet 'The Principles of Forging Design' sponsored by the Committee of Hot-Rolled and Cold-Finished Bar Producers, the American Iron & Steel Institute, New York

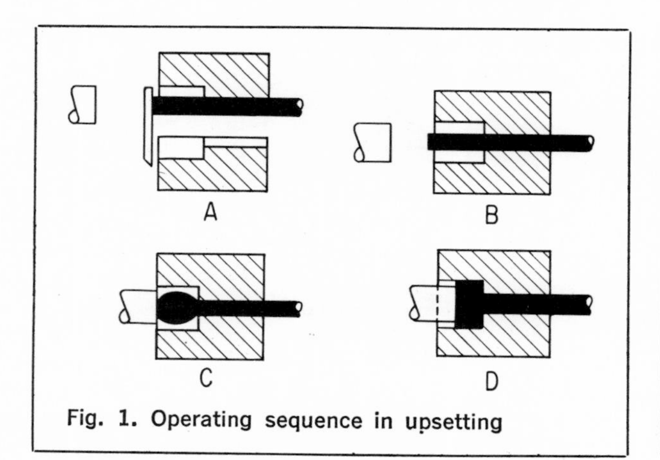

Fig. 1. Operating sequence in upsetting

The permissible length is further reduced if the end of the bar is sheared not exactly perpendicular to its axis, or if an impression is to be formed in the surface. There is no agreement on the exact values; data contained in Table 1 should be considered only as examples.

Greater amounts of material can be gathered

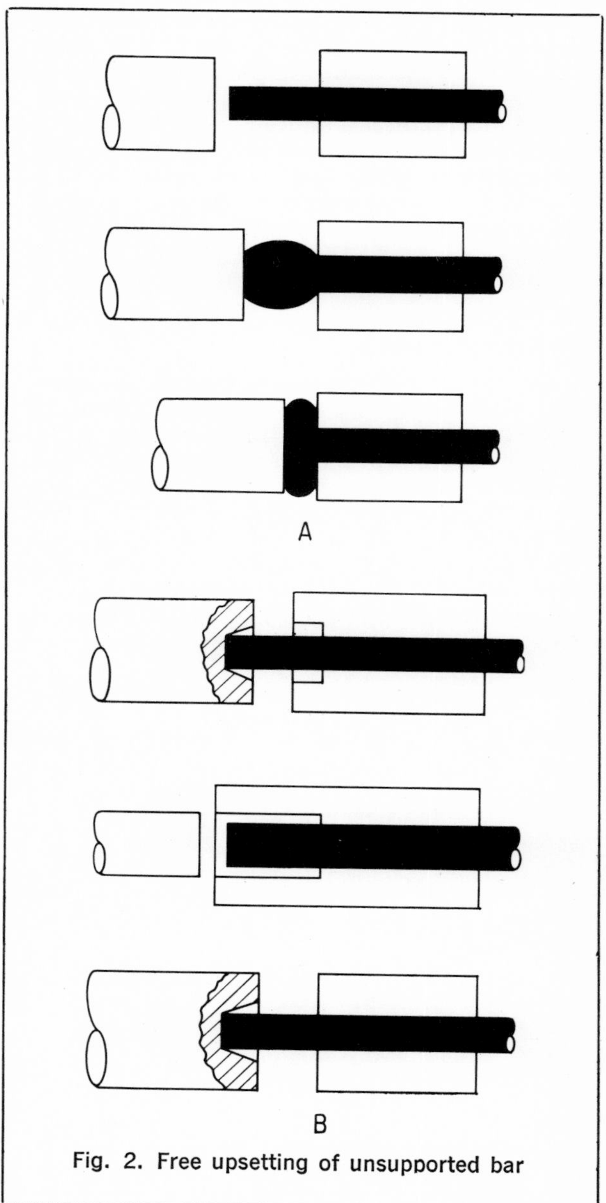

Fig. 2. Free upsetting of unsupported bar

Upset forgings, 2

up only if some support is given the rod against buckling. Thus, in a die where the diameter of the hole is not more than 1½ d, the length upset can, in principle, be infinitely long, as the slight initial buckling is eliminated in further compression without leading to forging defects. Again, the limitation applies whether the metal is gathered in a cavity of the heading tools, of the grip die, or both, and whether the shape is cylindrical or conical.

Recent research has suggested further limitations. Multiple buckling of the bar causes friction on the die surfaces, increases the force required for complete filling of the cavity, and may force some metal into the gap between heading tool and die, thus forming an objectionable flash. The limits of faultless upsetting are shown in Fig. 3 for a given set of experimental conditions. The limiting length/diameter ratios will be larger for rods of greater diameter and can be estimated on the basis of Table 1. If the shape of the forging requires a greater accumulation of metal in the head, several subsequent operations may be necessary with a consequent increase in forging and die costs. From experience, multiple conical upsetting has been found most economical and gives the best material flow.

Stock gathered anywhere

The process is not limited to forging heads at the end of the rod. Material can be gathered at any part of its length, but in this instance the die set will be somewhat more complicated as sliding dies must be inserted in the grip die frames, Fig. 4. The same limitations apply as in upsetting the end of the piece, and care must be taken not to exceed the permissible unsupported length. Otherwise a one-sided flash may be formed when the kinked rod is caught between the faces of the sliding and stationary grip tools. ■

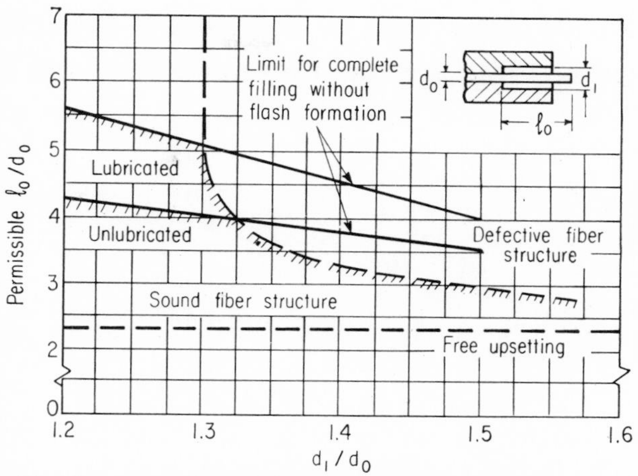

Fig. 3. Permissible upsetting ratio when forging 0.4 in. stock with 0.012 in. clearance between heading tool and die, as determined by H Meyer, Werkstattstechnik und Machinenbau, Vol 50, 1960

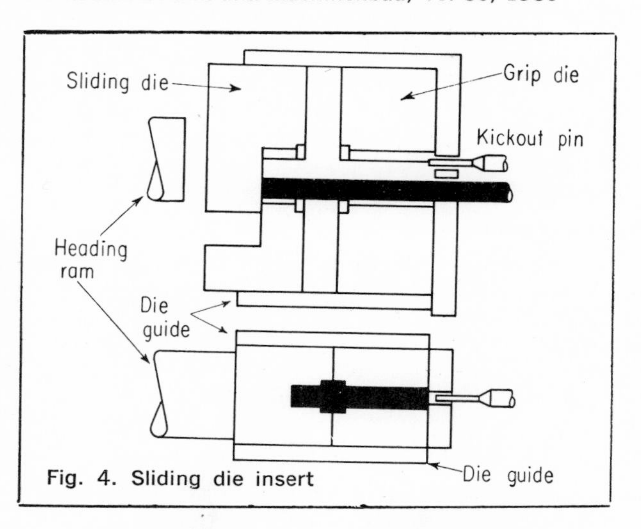

Fig. 4. Sliding die insert

Table 1. Permissible stock length in the free upsetting of bars [1] [2]

Condition of the bar end	Angle of cut	Shape of the heading tool end face	Permissible length, expressed as multiple of given bar diameter: < 1 in.	2 in.	4 in.
Sawed or sheared (even)	≤ 1°	Hollow	2.4		
		Flat	2.2	2.5	3.0
		With preforming punch	1.8	2.0	2.0
	≷ 3°	Flat	1.8	2.0	2.5
		With preforming punch	1.25	1.5	1.5
Uneven	≤ 1°	Flat	1.8		

(1) Metals Handbook, 8th Ed., Vol. 1, ASM, Metals Park, Ohio, 1961
(2) Bruchanow, A. N., and Rebelski, A. W.: Gesenkschmieden und Warmpressen (Closed Die Forging and Hot Pressing—in German), Verlag Technik, Berlin, 1955

Upset forgings, 3

Here are basic data on draft angles for upset forgings from US and German practice, to round out data

Draft angles are required on all surfaces that are withdrawn from a confined cavity and on all surfaces from which a tool is removed. There is little agreement on the actual values of draft. Simple standards, such as in Table 1, are often preferred. The system can be elaborated with some savings in material consumption. An example of this approach is given in Tables 2 to 4. In these tables, length *L*, diameter *D*, and height *H* always refer to the relevant dimensions of that section of the forging which is under consideration and not to the over-all dimensions of the whole forging. Examples are marked on the sketch.

Corner and fillet radii

Corner radii marked *r* contribute little if anything to ease the flow of metal and are, correspondingly, chosen as small as convenient. Usually they are equal to the value of machining allowance. Greater fillet radii marked *R* facilitate metal flow and can be taken as 1/5 of the relevant height (*H*) or depth (*L*). Similarly, generous radii promote the development of sound metal flow in punching operations and radii like those marked *R* should be chosen as large as possible, consistent with need for keeping excess materials within economical limits.

The procedure described above yields different radii for various sections of the forgings. Die making will be more economical if the nearest round number is used and if radii are unified for sections made with the same milling tool.

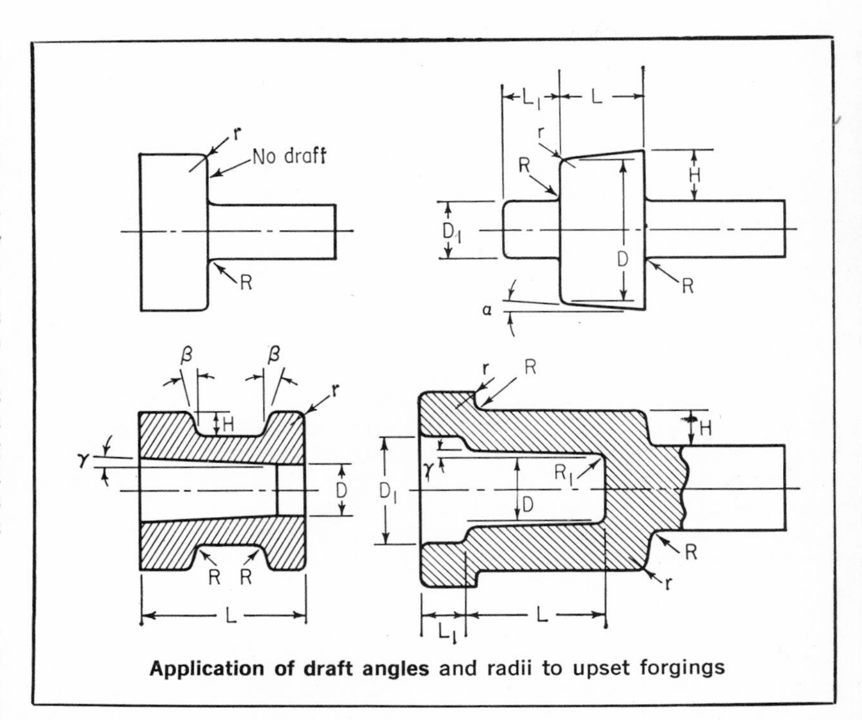

Application of draft angles and radii to upset forgings

General and detailed values for draft angles

Table 1. Draft angles of hot upset steel forgings[1]

Location	Draft Angle: Commercial standard	Draft Angle: Close standard
Outside	3° +2° −1°	3° ±1°
Inside holes and depressions	5° +2° −1°	5° ±1°

Table 2. Draft angles of parts formed in cavities of the heading tool[2]

Height/diameter ratio (H/D)	Angle α
<0.3	0
0.3-1.3	15′
1.3-2.3	30′
2.3-3.3	45′
3.3-4.3	1°

Table 3. Draft angles of parts formed in cavities of the gripping tools[2]

Height (H), in.	Angle (β)
<½	1°
½-1	3°
1-2	5°
2-3	7°
>3	10°

Table 4. Draft angles of cavities formed by the heading tool[2]

Height/diameter Ratio (H/D)	Angle γ
<0.5	0
0.5-2.5	30′
2.5-4.5	1°
4.5-6.5	1°30′
6.5-8.5	2°

[1] Convair Design Manual, Sec 3, Forging, 1955

[2] Bruchanow, A. N., and Rebelski, A. W.: Gesenkschmieden und Warmpressen (Closed Die Forging and Hot Pressing—in German), Verlag Technik, Berlin, 1955

Extracted from 'The Principles of Forging Design' published by the Committee of Hot-Rolled and Cold-Finished Bar Producers, the American Iron & Steel Institute, New York

Lubricants for extrusion, 1

Selection of lubricants for hot or cold extrusion of a particular material are based on factors that may not be compatible. Tables give selections in three groups

In both hot and cold extrusion operations, adequate lubrication is necessary in order to reduce tool wear, reduce friction and power requirements, obtain adequate surface finish and dimensional tolerance, and maintain uniformity of mechanical properties. In addition, in the hot extrusion of high-temperature metals, the lubricant must act as a thermal insulator to reduce cooling of the billet and undue heating of the tool materials.

The inherent reason for lubrication is the complete elimination, if possible, of intermetallic contact and adhesion between deforming tool and deformed metal. Reduction of interfacial friction is of secondary significance. The lubricant thus must possess the necessary physical and chemical properties to achieve this objective.

In hot extrusion, if the lubricant is liquid at the operating temperature, it must have a low surface tension and good wetting characteristics in order to wet the surface uniformly. It must, however, have adequate viscosity so as not to run off or be squeezed out prematurely at the start of the operation. If applied initially as a solid or as a solid in suspension, it must adhere tenaciously to the workpiece material.

For high-temperature operation, the lubricant must, of course, resist thermal breakdown and also act as a protective coating to minimize oxidation during handling and processing. In cold extrusion, particularly backward extrusion, conditions are severe and require that the lubricant film be able to follow the development of large new surfaces without rupture.

In addition, the lubricant should be chemically compatible with both the workpiece and tool materials, should not produce unwanted corrosion under operating conditions, and should not leave residual staining. It should be non-toxic and non-hazardous, relatively inexpensive, and easy to apply and remove.

By A.T. Male, supervisor
Air Force Extrusion Facility
Astronuclear Laboratory
Westinghouse Electric Corp.
Pittsburgh, Pa.

Table 1. Glass compositions and useful temperature ranges

Corning No.	Glass type	Approximate composition, %	Temperature, C
9772	Borate	Principally B_2O_3	870
8871	Potash lead	35 SiO_2, 7.2 K_2O, 58 PbO	870-1090
0010	Potash-soda-lead	63 SiO_2, 7.6 Na_2O, 6 K_2O, 0.3 CaO, 3.6 MgO, 21 PbO, 1 Al_2O_3	1090-1430
7052	Boro-silicate	70 SiO_2, 0.5 K_2O, 1.2 PbO, 28 B_2O_3, 1.1 Al_2O_3	1260-1730
1720	Alumino-silicate	57 SiO_2, 1 Na_2O, 5.5 CaO, 12 MgO, 4 B_2O_3, 20.5 Al_2O_3	1650
7740	Boro-silicate	81 SiO_2, 4 Na_2O, 0.5 K_2O, 13 B_2O_3, 2 Al_2O_3	1540-2100
7810	Silica	96 SiO_2, 2.9 B_2O_3, 0.4 Al_2O_3	1930-2040
7900	Silica	Greater than 96 SiO_2	2200

All the required features may not be obtained in one particular lubricant material, and it is sometimes necessary to use a combination to achieve the desired effect. For instance, in hot extrusion it may be necessary to coat the billet with one lubricant material to prevent oxidation during preheating and then apply a second lubricant just prior to processing in order to obtain the desired friction level. In cold extrusion, it is particularly common to use a chemical coating on the stock material that not only acts as a parting agent but also provides an excellent surface onto which the more fluid lubricants are able to bond to achieve the best lubrication system

Lubricant types

Lubricants can be classified into three categories: liquids, greases and waxes, and solids. In addition, because of their extreme commercial importance, two other more specialized categories can be considered: glasses, and intermediate coatings.

Liquid lubricants—These are usually oils of vegetable or mineral origin. Vegetable oils such as rapeseed or cottonseed oil and petroleum oils of medium fraction are used largely as die lubricants in hot extrusion and are applied by brushing or swabbing. For cold extrusion, these lubricants are normally applied by dipping or spraying.

Greases and waxes—Most industrial greases are composed of sodium, calcium or lithium soap, and a petroleum oil plus filler materials. Some greases have conventional solid lubricant powders such as graphite, MoS_2, ZnO or talc added to impart increased boundary-lubrication properties. They are most often applied by swabbing or spreading. Waxes are soft solids, such as beeswax or lanolin. Additions of graphite or MoS_2 are sometimes made. Waxes are applied by swabbing or spreading, or by dipping the workpiece into molten wax.

Solid lubricants—Non-metallic material used for lubricants are largely of the layer-lattice type such as graphite, molybdenum disulphide, talc, boron nitride and zinc oxide. More recently, polymeric materials, such as PTFE, are being used increasingly. By using a colloidal suspension in a volatile carrier, they are applied by either spraying or dipping the workpiece. These materials are also used in an oil carrier. Other substances used as solid lubricants are soft metals such as lead, tin, zinc, aluminum, copper, and sil-

Lubricants for extrusion [Continued]

ver, which can be applied as a machined shell, a hot dipped coating, a flame sprayed coating, or by electrodeposition. Such coatings are used for both hot and cold extrusion.

Glasses—These are used for lubrication in hot extrusion and range from a simple boron oxide to the soda-lime silicate, borosilicate, and phosphate types. Typical glass compositions and useful temperature ranges[34] are given in Table 1.

Application of glass

Glass lubricants can be applied to extrusion billets by spraying or brushing prior to heating, or by rolling the hot billet down an inclined tray containing powdered glass, glass fiber, or glass cloth. They can also be applied as a bonded glass insert between the billet and the extrusion die. The following information is of assistance in preparing glass mixtures for spray or brush application.[47]

Spray mixtures

1. Dry mix 200 ml of −100 mesh glass (except 7900 glass, which should be −325 mesh) with ¼ to ½ ml of sodium alginate.
2. Add glass and alginate to 150 ml of distilled water (16 to 32C); mix slowly and thoroughly.
3. Spray immediately before glass settles.

Brush mixtures

1. Heat 375 ml of distilled water to 38C.
2. Add 1 gr of Carbopol 934 to heated water and mix thoroughly.
3. Add NaOH solution (7 to 12 pellets or 1 gr NaOH in 59 ml of distilled water) to Carbopol 934 solution to neutralize to pH7.
4. Add 400 to 700 ml of −100 mesh glass to neutralized solution.
5. This mixture may be stored.

Glass may be removed after extrusion either by quenching in water and followed by pickling in hydrofluoric acid, or by mechanical means.

References

1. Monograph on Metalworking Lubrication, prepared by personnel at Illinois Institute of Technology Research Institute on Contract No. AF33(615)-5342 and to be published in book form.
2. W. Gmohling. *Z. Metallkunde, 55,* 567, 1964.
3. E. Faust. *Industrie-Anzeiger, 87* 135, 1965.
4. E. Elliott. *Metal Ind., 101,* 62, 1962
5. H. Ll.D. Pugh. *Metal Treatment, 27,* 189, & 231, 1960.
6. C. Belvedere. *Alluminio, 35,* 163, 1966.
7. G.H. Kitchen. *Lub. Engrg., 23,* 181, 1967.
8. C.E. Pearson and R.N. Parkins. "The Extrusion of Metals", Chapman & Hall, Ltd., London, 1960.
9. Anon. *Iron Age, 180,* 70, 1957.
10. H.L. Shaw, F.W. Boulger and C.H. Lorig. "Development of Die Lubricants for Forging and Extruding Ferrous and Nonferrous Materials", Summary Report, A.F. 33(600)-26272, Oct. 1955.
11. H.L. Shaw, F.W. Boulger and C.H. Lorig. "A Study on Possible Methods for Improving Forging and Extruding Processes for Ferrous and Nonferrous Materials"; Part V—Study on Lubricants for Hot Extrusion Operations. 1st Eng. Summary Rep. A.F. 33(600)-26272. June 1957.
12. J. Williams. *Met. Revs., 3,* 1, 1958.
13. W.W. Beaver. Inst. Metals Monogr. Rep. Ser. No. 28, "The Metallurgy of Beryllium", London, 781, 1963.
14. J.M. Siergiej. Symposium on Newer Structural Materials for Aerospace Vehicles. ASTM. Special Tech. Pub. No. 379, 106, 1965.
15. W.J. Wright and J.M. Silver. Inst. Metals Monogr. Rep. Ser. No. 28, "The Metallurgy of Beryllium", London, 734, 1963.
16. G. Griedman, ASTME tech. paper No. MF67-204, 1967.
17. Northrop Corp. "Program for the Development of Extruded Beryllium Shapes, Interim Rep." No. 5, AF33(600)-36931, 1959.
18. Anon. *Steel, 145,* 102, 1959.
19. A.W. Dana, C.C. McBride, F.J. Ander and W.J. Barnett. "Columbium and Columbium Alloy Extrusion", Rep. AF 33(600)-40700, June, 1960.
20. R.W. Buckman and J.L. Godshall. "Development of High Strength Columbium and Tantalum Alloy Tubing", Tech. Rep. WANL-PR(N)-004, 1964.
21. D.R. Carnahan and V. DePierre. "The Primary Working of Refractory Metals", Tech. Rep. AFML-TR-64-387. 1964.
22. E.L.H. Bastian. "Metalworking Lubricants", McGraw Hill, New York 1951.
23. C.S. Cook & J.W. Spretnak. "The Effect of Specimen Size and Interface Lubrication on the Compressive Flow Stress of Copper and Steel. Tech. Rep. AFML-TR-67-280, 1967.
24. M.T. Watkins, K. Ashcroft, and J. McKenzie. Proceedings of Conference on Technology of Engineering Manufacture, Inst. Mech. Eng., London, 155, 1958.
25. R.W. Tombaugh, R.C. Green, and J.H. Gehrke. "Development of Optimum Methods for the Primary Working of Refractory Metals", Tech. Rep. WADD-TR-60-418, 1961.
26. R.C. Haverstraw. Final Report on High Temperature Extrusion Lubricants, Tech. Rep. ML-TDR-64-256, 1964.
27. D.E. Strohecker, T.G. Byrer, A.F. Gerds, J.H. Gehoke and F.W. Boulger, "Deformation Processing of Nickel-base and Cobalt-base alloys." Tech. Rep. NASA TM X-53439, 1966.
28. E. Tanakea, S. Semoto and Y. Suzuki. Sci. Rep. Res. Inst. Tohoku Univ., *17A*(4), 193, 1965.
29. H. Fischer, *Sheet Metal Ind., 30,* 447, 1953.
30. D. James. *Sheet Metal Ind., 43,* 193, 1966.
31. M. Kunogi. J. Sci. Res. Inst. (Japan), *50,* 215, 1956.
32. C.V. Sciullo. *Automatic Machining, 26,* 63, 1965.
33. A.M. Cooper. *Prod. Eng.,* 40, 283, 1961.
34. P.S. Duletsky and V. DePierre. "Development of Optimum Methods for Primary Working of Refractory Metals", Tech. Rep. WADD-TR-60-418, 1962.
35. A.M. Sabroff and P.D. Frost. Trans. Amer. Soc. Lub. Eng., *3,* 61, 1960.
36. A.M. Sabroff, R.A. Sannincandio and P.D. Frost. Amer. Soc. Tool Eng., 1, 1958.
37. D.R. Carnahan and D.S. Michlin. "Primary Working of Refractory Metals", Int. Tech. Prog. Rep. No. 2, 1966.
38. A.F. Gerds, D.E. Strohecker, T.G. Byrer and F.W. Boulger. "Deformation Processing of Titanium and its Alloys", Tech. Rep. NASA TM X-53438, 1966.
39. A.M. Sabroff and P.D. Frost. *Modern Metals, 13,* (5), 50, and 13. (6), 52, 1957.
40. F.R. Lorenz, W.B. Haynes and E.S. Foster, Trans. AIME, *206,* 1076, 1956.
41. J.P. Pardoe. *Metal Ind. 100,* 426 and 446, 1962.
42. C.E. Lacy and C.J. Beck. Trans A.S.M., *48,* 579, 1956.
43. S. Samanta. *Jernkontorets Ann., 149,* 712, 1965.
44. M.G. Cockcroft. "Lubrication and Lubricants," ed-Braithwaite, Elsevier, 1967.
45. J. Thenenet and J. Buffet. *Rev. Met., 59,* 553, 1962.
46. J. Halapatz. U.S. Atomic Energy Comm., WAPD-RM-159, 1953.
47. I. Perlmutter and V. DePierre. *Metal Prog., 84,* 90, 1963.

Lubricants for extrusion, 2

Successful use of a particular lubricant depends on die angle, extrusion ratio and speed. If glass is specified, refer to the previous installment and to Table 1

Intermediate coatings are applied to the billet either for assistance in lubricant adherence or for protection of the billet against atmospheric or lubricant contamination during processing. Occasionally the intermediate coating can also act as the lubricant. For cold extrusion, the most common intermediate coatings are the phosphate, oxalate, or fluoride-phosphate coatings, which alone give high friction but in conjunction with conventional lubricants give a very efficient lubrication system for severe conditions. Such coatings are applied from chemical solution. In hot extrusion, canning of critical aerospace materials for the prevention of contamination is very common. Canning materials used are copper, steel, nickel, and molybdenum.

Successful lubrication with a particular lubricant system depends to a large extent upon the particular processing parameters such as die angle, extrusion ratio, and speed. Therefore, where alternative lubricants are given in Table 2, the reader is advised to refer to the original publication for precise details of the applicable process. ■

By A.T. Male, supervisor
Air Force Extrusion Facility
Astronuclear Laboratory
Westinghouse Electric Corp.
Pittsburgh, Pa.

Table 2. Guide to lubricants for hot and cold extrusion

Material	Temperature	Lubrication technique	Reference
Aluminum and aluminum alloys	Cold	Do not require conversion coatings. Zinc stearate, zinc stearate plus ethyl palmitate or lanolin, or zinc arachinate are widely used industrially	2, 3, 4
		Any highly viscous or solid lubricant, such as sulphonated tallow, lanolin, graphite, MoS_2, MoS_2 plus lanolin, MoS_2 plus commercial floor wax	5, 6, 7
	300-550 C	Often extruded without lubricant	8
		Die spray of colloidal graphite plus caustic soda	9
		Tetrafluoroethylene resin on the billet and organic grease plus fine flake graphite on the die and container	10, 11
Beryllium and beryllium alloys	400-1100 C	Normally extruded uncanned below 650 C, using MoS_2 or graphite lubricants	12, 13, 14
		Glass is not recommended. Can in mild steel plus Cu electrodeposit plus other lubricants on die and container, such as oil or grease plus asbestos, or bentonite-base grease plus graphite	14, 15, 16 17, 18
Cobalt and cobalt alloys	Hot	Lubrication as for nickel and nickel base alloys	
Columbium and columbium alloys	Hot	It is possible to extrude uncanned billets below 1200 C, using grease, graphite or glass. Can in mild steel, copper or molybdenum and use a glass lubricant	19, 20, 21
Copper and copper alloys	Cold	Do not require conversion coatings.	22
		Zinc stearate, lanolin, waxes, sulphonated tallow and fats containing graphite or MoS are widely used industrially.	
		Can also use graphite grease, teflon or lead	23
	600-900 C	Normally extruded dry, or with a die lubricant of graphite in oil or grease	8
Lead and lead alloys	Cold	Often extruded without lubricant	
		Do not require conversion coatings.	
		Plain or sulphonated tallow, hydrogenated cottonseed oil, zinc stearate soap, wax and fatty acid have all been successfully used	22, 24
	100-250 C	Often extruded without lubricant, although in the extrusion of tube the container and mandrel are normally lubricated with tallow	8

Lubricants for extrusion [Continued]

Table 2. Guide to lubricants for hot and cold extrusion [continued]

Magnesium and magnesium alloys	Cold	Often extruded dry; sulphonated tallow is sometimes used	24
	250-500 C	Often extruded without lubricant	
		Also extruded with billet coated with tetrafluorethylene resin	10
Molybdenum and molybdenum alloys	870-2200 C	Billet coatings of glass are widely used together with tool coatings of proprietary compounds of graphite or MoS_2 in grease, oil or water	21, 25
		Use of a mixture of magnesium meta-borate and graphite in water has also been found successful	26
Nickel and	Cold	Chemical oxalate coating used together with a soap	1
nickel alloys	650-1300 C	Both canned and uncanned billets are extruded using glass lubricants	25, 27
		A mixture of magnesium metal-borate and graphite in water has also been found successful. Lubricants containing sulphides should be avoided	26
Steels-carbon and low alloy	20-400 C	General industrial practice is to phosphate coat the billet and lubricate by the application of soap	22
		High viscosity, high sulphur oils, tallow, palm oil, rapeseed oil and stearic acid are also good.	28, 29
		MoS_2 in water, petroleum jelly or resin bonded has been used	30
	900-1250 C	Conventional practice is to use glass-coated billets and for tool lubricant mixtures containing graphite	1
Steels-stainless	20-400 C	Conventional practice is to oxalate coat the billet and lubricate with aluminum or zinc stearate	31
		Copper-plated billets and conventional lubricants for copper are also used.	32, 33
		High viscosity, high sulphur oils, tallow, stearic acid or MoS_2	28
	900-1250 C	Lubrication as for carbon and low alloy steels	
Tantalum and tantalum alloys	950-2050 C	Extruded uncanned or canned with mild steel or molybdenum and using glass coating	21, 25, 34
Titanium and titanium alloys	Cold	Fluoride-phosphate chemical coating on billet plus gum resin lubricant containing graphite	35, 36
	650-1100 C	Extruded uncanned or canned with mild steel or copper and using glass coating on billet and a grease-base graphitic lubricant on tooling	25, 37, 38 39
Tin and tin alloys	Cold and hot	Lubrication as for lead and lead alloys	
Tungsten and tungsten alloys	1600-2300 C	Extruded bare or canned in molybdenum and using glass coatings	21, 25, 34
Uranium and uranium alloys	800-1000 C	Can in copper and lubricate with MoS_2, glass or barium chloride	40, 41
Vanadium and vanadium alloys	1000-1200 C	Can in mild steel and lubricate with a graphite-oil lubricant Also it is possible to lubricate with glass	42
Zinc and	Cold	Often extruded dry; sulphonated tallow is sometimes used	24
zinc alloys	150-300 C	Often extruded without lubricant, although tallow may be used	
Zirconium and zirconium alloys	Cold	It is necessary to chemically fluoride-phosphate coat the billet and lubricate with graphite or MoS_2 in oil or grease	43, 44
	750-900 C	Extruded uncanned or canned in copper or mild steel and lubricated with glass and/or graphite-based lubricant	41, 45, 46

Chapter 14

Arc Welding

Weld Quality Standards—I

By J W Jakobowski, senior metallurgist, Missile and Space Vehicle Dept, General Electric Co, Philadelphia, Penna

How good does a weld have to be? This question is usually one of the first problems to come up, whether you are dealing with missiles or with other, more common welded products.

Our answers to it, summarized on the following two pages, may not provide a perfect solution for everyone. However, even though they were drawn up for missiles and space vehicles, they are sufficiently basic to cover the manufacture of a variety of other products.

Fundamentally, the quality required in any specific weldment depends a good deal on what we expect it to do and, particularly, on the anticipated service loading it will encounter. Therefore, we have established three classes of weldments, based on the stresses that will be applied in service. In general, the classes are the same for steel, stainless steel, aluminum, and magnesium.

Class I welds are highly stressed joints where the weld bead is oriented within 45° of *normal* to the principal stress, and the design factor is 2.0 or less.

Class II welds are moderately stressed joints where the weld bead is oriented within 45° of *parallel* to the principal stress, and the design factor is greater than 2.0.

Class III welds are lightly stressed joints that have a design factor greater than 3.0.

Cast-steel and wrought-steel weldments are further divided into sub-classes depending on the strength level obtained by heat treatment. This applies only to Classes I and II, as follows:

Highly stressed:
- Class IA: 150,000 psi and up
- Class IB: under 150,000 psi

Moderately stressed:
- Class IIA: 150,000 psi and up
- Class IIB: under 150,000 psi.

Fortunately, a good many of the inspection requirements overlap, especially those of different materials within the same class. Therefore, for Table I, we have grouped the categories that have similar requirements.

For example, all Class I weldments are given the same inspection whether the metal is aluminum, magnesium, stainless steel, or steel. Consequently, they are called Group A, and their requirements are listed in the first column.

A major exception is that Group A includes Class IA steel but not Class IB steel. Because of the lower strength level, Class IB weldments are considered no more critical than the general run of Class II welds. They are therefore placed in Group B, along with Class II stainless steel and Class IIA steel weldments.

For the same reason, Class IIB steel weldments have been downgraded, in effect, and are included in Group D with the Class III weldments in steel and stainless steel.

A further consolidation is possible at this point because Class II and Class III weldments are given essentially the same inspection. The only major difference is that Class III weldments are not inspected radiographically. In almost all other respects they are identical, so Group B and Group D are shown in one column.

The same consolidation has been made for aluminum and magnesium weldments of Class II and III. However, these could not be grouped with steel and stainless steel because their limitations are often stated somewhat differently.

The Table covers three major types of non-destructive inspection:

Visual Inspection. In addition to the specific requirements noted in Table I, visual inspection carries the general requirement that the weldments exhibit good standards of workmanship. The bead sizes that are indicative of good workmanship are shown in Tables II and III.

Magnetic Particle and Penetrant Inspection act as a backstop for radiographic inspection which sometimes misses thin, flat cracks. If such a crack in a steel weldment is open to the surface, it will be detected easily by the magnetic-particle technique. The penetrant method performs the same service, of course, for aluminum, magnesium, and the non-magnetic stainless steels.

Radiographic Inspection is essential for determining the internal quality of weldments. It is a well-known fact that porosity up to approximately 7% will not materially affect the mechanical properties of a weldment. Conversely, it is equally well known that the most serious defects in welds are cracks and lack of fusion. Of the two, cracks appear to be the more dangerous because of the very sharp radius which produces high stress concentration. All these factors have been taken into consideration in establishing the requirements shown in Table I.

Sampling. As the Class I weldments (including IA and IB) are the most critical, they are given a 100% inspection. The other classes are inspected according to the sampling plans below.

Repairs. Defects found during inspection can be corrected by completely removing the defect and rewelding as originally specified. Of course, the repair has to meet the same requirements specified for the original weld. Defects should not be removed and repaired more than two times.

SAMPLING PLAN

Sample size:	Lot size: Class II	Class III
1	1	1
2	2	2
3	3-4	3-8
4	5-6	9-12
5	7-8	13-23
6	9-11	24-35
7	12-15	36-49
8	16-20	49 and up
9	21-26	
10	27-36	
11	37-51	
12	52-82	
13	83-162	
14	163-971	
15	972 and up	

Weld Quality Standards—II

TABLE I INSPECTION REQUIREMENTS

	Steel Class IA Stainless Steel Class I Aluminum & Magnesium Class I	Group B Steel Class IB & IIA Stainless Steel Class II	Group D Steel Class IIB & III Stainless Steel Class III Aluminum & Magnesium Class II & III
VISUAL INSPECTION	See Tables II & III for bead size requirements		
Undercut	5% reduction in T, max (T = minimum parent-metal thickness)		
Contour		must be smoothly rounded	NS (NS = Not Specified)
Dist. from end of weld		20% of bead length but not less than 1.0 in.	NS
Length, each		0.25 in. max	0.20 in. max
Length, cumulative		Group B: 10%; Group D: 20% of weld length, max	NS
Distance between undercuts		0.06 in. min	3.0 in. min
Depth		10% of T, but not to exceed 0.025 in.	25% of T, max
Scratch or burn marks	5% reduction in T, max	NS	25% of T, max
Cracks	None accepted	NS	NS
Crater or depression below level of either parent metal surface	None accepted, face or root bead	NS	NS
End-of-bead crater	None accepted	20% of T, but not to exceed 0.025 in.	0.10 in. max in last 10% of weld bead
Underfill (depression of face or root bead below level of either parent metal surface)	None accepted		
Depth		0.02 in. max	25% of T max
Length, each		1.5 in. max	1.0 in. max
Length, cumulative		3.0 in. max in any 12 in. of weld	3.0 in. max in any 12 in. of weld
Reinforcement of face of root bead	0.050 in. max	Group B: 0.06 in. max Group D: 0.08 in. max	NS
Drop-through beyond root	0.050 in. max	0.12 in. max	NS
Overlap beyond fusion limits, edge or toe	None accepted	NS	NS
Included slag exposed by machining after welding	None accepted	None accepted	NS
Burn-through	None accepted	0.12 in. max	NS
Shrinkage on side opposite weld (lap and fillet welds):			
Length, each	NS	0.31 in. max	NS
Length, cumulative	NS	2 areas max in any 12 in.	NS
Distance between shrinkage areas	NS	3 in. min	NS
Porous areas	NS	NS	0.10 in. max dimension
Fillet weld throat concavity or convexity	NS	NS	20% of theoretical throat, max
MAGNETIC PARTICLE OR PENETRANT INSPECTION			
Porosity	Steel & SS: None accepted within 1.0 in. or 20% of bead length from either end of weld Al & Mg: 0.05 in. max dimension		0.12 in. max dimension
Cracks open to surface	None accepted	None accepted	None accepted
RADIOGRAPHIC INSPECTION		(Class III excluded)	(Class III excluded)
Cracks	None accepted	None accepted	0.06 in. max length
Lack of fusion	Steel & SS: none accepted. Al & Mg: None accepted on side opposite welding side	None accepted	None accepted on side opposite welding side

	Steel Class IA Stainless Steel Class I Aluminum & Magnesium Class I	Group B Steel Class IB & IIA Stainless Steel Class II	Group D Steel Class IIB & III Stainless Steel Class III Aluminum & Magnesium Class II & III
Incomplete penetration	Steel & SS: None accepted. Al & Mg: None accepted on side opposite welding side.	Group B: None accepted on side opposite welding side. Excess penetration is accepted if utility of part is not decreased. Group D: 10% of joint length max. Depth can be 20% of thickness of butt joint or 20% of theoretical leg of fillet weld.	None accepted on side opposite welding side
Voids and Inclusions with sharp tails	None accepted	None accepted	0.06 in. max length
Voids without sharp tails:			
Largest dimension in a single void	Steel & SS: 0.33T or 0.05 in. Al & Mg: 0.03 in.	Group B: 0.5T or 0.06 in. Group D: 0.5T or 0.09 in.	0.06 in. max dimension
Minimum separation between voids	Steel & SS: Longest dimension of the longer void. Al & Mg: NS		NS
Ditto, based on void size: less than 0.03 in.		0.02 in	NS
0.03 to 0.06 in.		0.03 in.	NS
0.06 to 0.09 in		0.06 in.	NS
Sum of longest dimensions of voids, max:			
in any sq in. or any 4 lineal in.	NS	Group B: 2T or 0.25 in. Group D: 2T or 0.38 in.	NS
in any 0.25 sq in.	NS	Group B: 1.3T or 0.12 in. Group D: 1.3T or 0.18 in.	NS
Porosity	Steel & SS: NS. Al & Mg: Two pin-hole porosities per lineal in.	NS	3 pin-hole porosities per lineal in.
Inclusions without sharp tails:			
Largest dimension in a single inclusion	0.33T or 0.50 in.	Group B: 0.3T or 0.25 in. Group D: 0.3T or 0.38 in.	NS
Minimum separation	Same as for voids	Same as for voids	Same as for voids
Sum of largest dimensions in 8T or 4 lineal in.	NS	Group B: 0.25 in. max. Group D: 0.38 in. max.	NS
Combined voids and incl:			
Sum of largest dimensions in 8T or 4 lineal in.	NS	Group B: 2T or 0.25 in Group D: 0.38 in.	NS
When sizes are between 0.025 & 0.050 in. or 0.33T	Steel & SS: No more than 3 per lineal in. nor more than 5 in any 2 consecutive inches. Al & Mg: NS	NS	NS
Random discontinuities, 0.025 in. or less	Steel & SS: Rejected if more than 5 lie in a straight line (±.045 in.) in any lineal inch.	NS	NS
Combined voids, incl, and porosity	NS	NS	2 per lineal in.
Discontinuities not open to surface		Accepted when entirely in weld reinforcement	NS

II—Bead Sizes of Aluminum & Magnesium Welds

Sheet Size	FILLET WELD Leg Max.	Leg Min.	Face Max.	Face Min.	LAP WELD Face Max.	Face Min.	BUTT WELD Bead Width Max.	Bead Width Min.
0.025	0.16	0.09	0.25	0.16	0.06	0.03	0.16	0.09
0.032	0.16	0.09	0.25	0.16	0.06	0.03	0.16	0.09
0.040	0.19	0.12	0.28	0.19	0.06	0.03	0.19	0.12
0.051	0.28	0.19	0.38	0.25	0.09	0.06	0.19	0.12
0.064	0.28	0.19	0.36	0.25	0.16	0.09	0.22	0.16
0.072	0.28	0.19	0.38	0.25	0.16	0.09	0.22	0.16
0.081	0.28	0.19	0.38	0.25	0.16	0.09	0.25	0.19
0.091	0.28	0.19	0.38	0.25	0.19	0.12	0.25	0.19
0.102	0.28	0.19	0.38	0.25	0.19	0.12	0.28	0.22
0.125	0.28	0.19	0.38	0.25	0.28	0.19	0.34	0.25
0.156	0.31	0.22	0.44	0.28	0.34	0.22	0.41	0.28
0.188	0.31	0.22	0.44	0.28	0.44	0.28	0.44	0.34
0.250	0.38	0.25	0.56	0.38	0.56	0.38	0.72	0.47

III—Bead Sizes of Steel & Stainless Welds

Sheet Thickness	FILLET WELD Leg Max.	Leg Min.	Face Max.	Face Min.	LAP WELD Face Max.	Face Min.	BUTT WELD Bead Width Max.	Bead Width Min.
0.049	0.09	0.06	0.19	0.12	0.09	0.06	0.19	0.12
0.058	0.09	0.06	0.19	0.12	0.09	0.06	0.19	0.12
0.065	0.16	0.09	0.19	0.12	0.16	0.09	0.25	0.16
0.083	0.16	0.09	0.19	0.12	0.16	0.09	0.28	0.19
0.095	0.16	0.09	0.25	0.16	0.19	0.12	0.28	0.19
0.125	0.19	0.12	0.28	0.19	0.28	0.19	0.34	0.22
0.156	0.19	0.12	0.28	0.19	0.34	0.22	0.44	0.28
0.188	0.25	0.16	0.34	0.22	0.44	0.28	0.47	0.31
0.250	0.28	0.19	0.44	0.28	0.56	0.38	0.62	0.41

NOTES: All dimensions are in inches, based on the thinnest piece. Deviations are acceptable provided their aggregate length does not exceed 20% of the bead length and no single deviation exceeds 10%. Bead size shall not be less than the minimum allowed for the next thinner gage.

Arc welding aluminum

While selection of the arc welding method depends largely upon the individual application, factors such as thickness of metal, product design and quantities must be considered

The best welding methods for aluminum are the tungsten-inert-gas arc (Tig) and the metal-inert-gas-arc (Mig) processes. Both use inert gas (argon, helium or mixtures thereof) to keep air away from the arc and molten weld pool, thus eliminating the use of a welding flux.

The Tig process is preferred for welding aluminum sections less than ⅛ inch in thickness. This method can also be used on heavier sections but the Mig process is usually chosen for its higher welding speed and economy.

Factors that affect the welding of aluminum, and the properties of aluminum weldments, include melting point, thermal conductivity, thermal expansion and contraction, oxidation, gas porosity and effects of welding. The physical and chemical properties which affect weldability are compared for aluminum, copper and iron in Table 1.

Weldable commercial aluminum alloys start to melt at 1050 F. There is no color change in aluminum during heating. However, it is possible to know when the aluminum is near its melting point and at welding temperature by watching the weld pool. The Tig weld pool, for example, develops a glossy appearance, and a liquid pool or spot forms under the arc when the metal becomes molten.

Thermal conductivity

Aluminum conducts heat three times faster than iron. This means that for welding equal thicknesses of aluminum and steel, a higher heat input is needed to weld the aluminum. Preheating heavy sections of aluminum is generally helpful in reducing heat loss. When using the Tig process for joining such sections, better welding results.

Contraction and expansion

Aluminum welds decrease about 6% in volume when solidifying from the molten state. The stresses induced by this contraction may cause excessive weld-joint distortion unless correct allowances are made prior to welding.

Thermal expansion of aluminum is approximately twice that of steel and one-third greater than copper. The surrounding surface expands due to the heat of welding. Thermal expansion of the adjacent aluminum may reduce the root opening on butt joints during welding. Then, when the metal cools it contracts. This contraction, coupled with shrinkage of filler metal on cooling, may put the weld in tension and increase cracking susceptibility. Excessive restraint of the component sections during cooling of the weld may also result in weld cracking.

Effects of welding heat

Aluminum alloys are of two types—the work-hardenable alloys such as EC (electrical conductor grade), 1100, 3003, 5052, 5083 and 5086; and the heat-treatable alloys such as 6061, 6062, 6063 and 7039.

Although alloys in the 2000 and 7000 series are also heat-treatable, most of them are not recommended for arc-welded fabrications because weldments are low in ductility. Better properties are obtained with resistance welding methods. A notable exception is alloy 7039, now employed for armor plate, cryogenic and other critical applications. Welding qualities of alloys in the 2000 and 7000 series with either resistance of Mig processes are excellent; bead off, as-welded (Mig) strengths are upward of 48,000 psi; ductility of these welds ranges from 8 to 12% elongation in 2 inches.

Improved mechanical properties are attained in heat-treatable alloys in solution heat treatment at elevated temperatures, above 900 F, followed by a low temperature aging treatment, above 300 F.

Aluminum alloys lose individual characteristics of in-

Table 1. Physical and chemical properties of aluminum, copper and iron which affect weldability

Metal	Chemical Resistance of Oxide Film	Melting Point °C	Sp. Heat cal/°C	Heat to Raise Unit Vol. 1°C cal/cm³	Heat Conductivity cal/cm²/cm/°C/sec	Coefficient of Linear Expansion Per °C
Aluminum	Very resistant	660	0.23	0.62	0.46	24×10^{-6}
Copper	Reduced easily	1080	0.09	0.80	0.93	17×10^{-6}
Iron	Reduced with moderate ease	1540	0.13	1.03	0.14	14×10^{-6}

Note: The above values are approximate and are presented for comparison purposes.

Table 2. Typical as-welded properties of TIG and MIG weld joints in aluminum alloys

Base Material	Filler	Typical Transverse Tensile Properties		
		Yield Strength	Ultimate Strength	% Elongation in 2″
EC	1100	4,800	11,000	50
1100	1100	6,000	13,500	23
2219	2319	26,000	38,000	3.0
-T62	2319	40,000	56,000	7.0
3003	1100	8,000	16,000	14
3004	5154	12,000	27,000	19
5052	5356	14,000	28,000	15
5154	5154	17,000	33,000	15
5454	5554	14,800	33,500	14.5
5454	5356	15,100	36,400	15.5
5083	5356	21,000	40,000	16
5083	5183	21,800	43,500	14
5086	5356	18,000	37,000	17
5086	5183	17,000	39,000	21
6061	4043	17,000	27,000	8.5
6061	5356	17,000	29,000	11
6063	4043	12,000	21,000	10
7039	X5039/5183	32,000	50,000	11

Source: 'Weldor's Training Manual—inert gas process,' Second Edition, published by the Kaiser Aluminum & Chemical Sales, Inc, Kaiser Center, Oakland, California

Arc welding aluminum [Continued]

creased hardness and strength when reheated to high temperatures. Heated above 900 F, the aluminum alloys revert to the annealed condition almost immediately. The degree of loss is a function of time and temperature. As mentioned previously, the weld metal is over 1050 F when deposited. Therefore, welding causes some annealing of the parent metal. With the heat treatable alloys, welding also lowers the ductility of the joint.

Preheating

Preheating is necessary if the mass of the parent metal is such that heat is conducted away from the joint so fast that the welding arc cannot supply the heat required to produce fusion. Insufficient heat results in poor fusion of the weld bead and inadequate melting of the parent metal.

In Tig welding, preheating is necessary when welding heavy plate. For the heat-treatable alloys such as 6061, preheat should be used with the knowledge that too high a temperature or too long a preheat period can decrease the as-welded strength of the joint. Recommended preheat temperatures for various thicknesses of aluminum plate and tube are shown in Table 3.

In Mig welding, preheat is seldom required regardless of plate thickness.

Post-weld treatment

Weldments of precipitation-hardening aluminum alloys can be strengthened by post-weld thermal treatment. This consists of either a full solution heat treatment and artificial aging or artificial aging alone. During solution treatment the alloying elements are dissolved by the heating at an elevated temperature and retained in solid solution by rapid quenching. Artificial aging is then carried out at a lower temperature in which the soluble elements are precipitated from the supersaturated solid solution.

Three basic procedures are used for welding heat-treatable alloys:

(1) Starting with 0-temper material: weld, solution heat treat, artificially age.
(2) Starting with T4-temper material: weld artificially or naturally age.
(3) Starting with T6-temper material: weld, naturally age.

The solution treatment requires heat-treating furnaces and quench tanks. Artificial aging requires only a furnace.

For maximum strength in welds requiring a post-weld thermal treatment, it is necessary to use a filler which will respond to heat treatment. Some commercial fillers which can be heat treated are 2319, 4043, 718, X5180 and X5039.

Stress relieving

Residual stresses created in aluminum alloys by the heat of welding may become excessive, due to the total amount of heat input, thickness of the metal and design of the weldment. In extreme cases, such stresses may cause early failure of the weldment. One common method of modifying residual stresses is by peening (localized working of the metal by hammering) to effect limited distribution of the stresses. Peening usually is not advisable on thin sections, however.

For aluminum-magnesium, non-heat-treatable alloys (5000 series), high residual stresses may be reduced by heating at temperatures below 650 F, which is the temperature for complete annealing. Principal limitation on post-weld heating is size—of weldment and/or heat treating oven.

The aluminum-magnesium alloys (5000 series) can be adequately stress relieved, in most instances by post-weld heating at 450 F for approximately 4 hours. As mentioned above, complete annealing is achieved upon heating these alloys to 650 F.

Joint design—Tig process

On relatively thin materials, 1/16 to 3/16 in. thickness, the square butt joint is usually satisfactory for both processes. For thicker metal, either a single-vee bevel or double-vee bevel may be necessary.

Design varies from square butt for ⅛ in. sheet to 60° included angle joint for ½ in. plate. For tubular sections having a wall thickness greater than ⅛ in., the edges should be beveled from a 60 to 90° included angle and have a 1/16 in. square butt lip. For sheet ⅛ in. thick and up, some root opening is recommended to insure complete penetration.

Joint design—Mig process

The joint design and root openings required for Mig welding are determined by metal thickness and structural requirements as in Tig welding. Aluminum sheet up to ¼ in. thick can be welded manually with complete penetration using a square butt design. Heavier sections can be similarly joined if automatic welding is used. For manual welding, material greater than ¼ in. thick may have a single-vee groove or a double-vee groove. The edge of tubular sections are prepared the same as the edges of plate of corresponding thickness. ■

Table 3. Preheat temperature for welding sheet plate and tubular aluminum sections (butt joints)

Tubular Sections

Outside Diameter Inches	Wall Thickness Inches	Approximate Preheat Degrees F	
		TIG	MIG
1-3	1/8	None	NR*
4-6	1/8	Optional–400	NR
1-3	1/4	None	None
4-6	1/4	400	None

Sheet and Plate

Thickness	Approximate Preheat Degrees F	
	TIG	MIG
1/8-1/4	None	None
1/2	600	None
3/4	NR	None
1	NR	None
2	NR	Optional–500
3	NR	Optional–500

* Not recommended

Note: These preheat temperatures are only for use as a guide. Most weldors prefer to increase the welding current and thereby avoid preheating. Preheating is another operation and increases overall costs. Also, if welding the heat-treatable alloys such as 6061, it should be realized that the temperature and length of preheating time can affect the as-welded strength of the joint. It is seldom necessary to preheat when using the MIG process.

Welding austenitic stainless steels

Accepted welding practices are briefed for quick reference when you are setting up schedules to join these chromium-nickel steels by one of the several available procedures

Among the points to watch are: Use a short arc and, deposit a stringer bead when arc welding. Weave width is normally limited to 2 to 3 times core-wire diameter. Before extinguishing or breaking the arc, reduce the molten pool to the smallest possible size. Remove all slag from a bead before the next adjacent bead is deposited.

When employing the MIG (metal inert gas) welding process, you have a choice of spray-type transfer of the molten filler metal, or the newer technique of short circuiting the arc, when using small diameter wire of 0.030 to 0.045 in. diameter with lower amperages. The latter method produces a smaller, more controlled weld pool, and a narrower zone of heat-affected metal.

Normally the TIG (tungsten inert gas) process is used for relatively thin materials, because the procedure is not economical for work over ¼ in. thickness. With the TIG torch, an arc is established between a nonconsumable tungsten electrode and the workpiece under an atmosphere of helium or argon. The process lends itself to manual welding or to a fully automated setup.

Source: 'Welded Austenitic Chromium-Nickel Stainless Steels—Techniques and Properties'—published by the International Nickel Co, New York.

Properties of filler metals for welding stainless

Base Metal (AISI Type)	Filler Metal or Electrode	Comment
301 302, 304, 305	308	Type 308 filler metal has slightly higher alloy content than most of the base metals shown.
303, 303Se	312	Free-machining grades are crack sensitive and not usually welded.
304L	308L	
309	309	
310	310	
316	316, 16-8-2	The 16-8-2 electrode is used for some high temperature applications to obtain improved resistance to embrittlement.
316L	316L	
317	317	
321	321, 347	Type 321 available as bare filler wire but not as covered electrode.
347	347	

* AWS-ASTM specifications designate bare filler wire or rods by using the prefix "ER." Covered electrodes have the prefix "E." In addition, covered electrodes are available with either lime or titania coatings, designated by the suffix "-15" or "-16" respectively.

Examples:

ER308 = bare filler wire or rod;
E308-15 = electrode with lime-type coating;
E308-16 = electrode with titania-type coating.

Suggested amperages for stainless steel electrodes

Electrode Size, in.	Amperes (a/c or d/c)		Maximum Arc Voltage
	Flat, Horizontal and Overhead	Vertical	
3/64	15–25	15–25	23
1/16	20–40	25–40	24
5/64	30–60	35–55	24
3/32	45–90	45–65	24
1/8	70–120	70–95	25
5/32	100–160	100–125	26
3/16	130–190	130–145	27
1/4	210–300	—	28
5/16	250–400	—	29

Submerged-arc welding requirements

Plate Thickness, in.	Wire Diameter, in.	Amperes	Voltage	Travel Speed, ipm
¼	3/16	525–575	30–32	20–24
⅜	3/16	700–900	33–35	12–16
½	¼	800–1000	33–35	12–15

Welding austenitic stainless steels [continued]

MIG welding currents and speeds (two methods)

Type of Arc	Welding Position	Wire Diameter, in.	Current (d/c r/p), amp	Voltage	Travel Speed, ipm
Shorting	Flat or Horizontal	.030	130–140	22–24	5-15 depending on width of weave
	Vertical or Overhead	.030	110–130	20–22	5–15
Spray	Flat or Horizontal	.062	325–375	25–28	5–15

Current and gas requirements for TIG welding

Work Thickness, in.	Welding Position	Tungsten Diameter, in.	Filler Rod Diameter, in.	Welding Current, amp	Gas Flow, cu ft/hr
1/16	Flat	1/16	1/16	70–105	12
1/16	Vertical or Overhead	1/16	1/16	55–90	12
1/8	Flat	3/32	3/32	100–150	15
1/8	Vertical or Overhead	3/32	3/32	75–130	15
3/16	Flat	1/8	5/32	150–200	20
3/16	Vertical or Overhead	3/32	1/8	90–140	20

Cycle times and currents for spot welding

Sheet Thickness, in.	Electrode Diameter, in.	Electrode Force, lb	Weld Time, cycles	Weld Current, amp	Nugget Diameter, in.	Shear Strength of Weld, lb
.010	3/16	230	3	2000	.065	150
.021	1/4	400	4	4000	0.100	370
.040	3/8	900	6	7800	0.160	1000
.062	1/2	1500	10	11,000	0.220	1950
0.125	3/4	3300	20	18,000	0.300	5000

Weight of weld metal

By Omer W Blodgett, Application Engineering Dept, Lincoln Electric Co, Cleveland

d or r	"t" Dimension 1/16	1/8	3/16	1/4	3/8	1/2	"t" Dimension 1/16	1/8	3/16	1/4	r	Included angle 14°	20°	60°	70°	45°	Leg 10% oversize, normal shop practice: Flat	Convex	Concave
1/16											.021								
1/8	.027	.053	.080	.106	.159	.212					.083	.0065	.0094	.031	.037	.027	.032	.039	.037
3/16	.040	.080	.119	.159	.239	.318	.027				.188	.0147	.021	.069	.084	.060	.072	.087	.083
1/4	.053	.106	.159	.212	.318	.425	.035				.334	.026	.037	.123	.149	.106	.129	.155	.147
5/16	.066	.133	.199	.265	.390	.531	.044				.531	.041	.059	.192	.232	.166	.201	.242	.230
3/8	.080	.159	.239	.318	.478	.637	.053	.106			.750	.059	.084	.276	.334	.239	.289	.349	.331
7/16	.091	.186	.279	.371	.557	.743	.062	.124	.186		1.02	.080	.115	.376	.456	.326	.394	.475	.451
1/2	.106	.212	.318	.425	.637	.849	.071	.142	.212		1.33	.104	.150	.491	.595	.425	.514	.620	.589
9/16	.119	.239	.358	.478	.716	.955	.080	.159	.239			.132	.190	.621	.753	.538	.651	.785	.745
5/8	.133	.265	.398	.531	.796	1.06	.089	.177	.266			.163	.234	.766	.930	.664	.804	.970	.920
11/16	.146	.292	.438	.584	.876	1.17	.097	.195	.292	.389		.197	.283	.927	1.13	.804			
3/4	.159	.318	.478	.637	.955	1.27	.111	.212	.318	.424		.234	.337	1.11	1.34	.956	1.16	1.40	1.32
13/16	.172	.345	.517	.690	1.04	1.38	.114	.230	.345	.460		.275	.396	1.30	1.57	1.12			
7/8	.186	.371	.557	.743	1.11	1.49	.124	.248	.372	.490		.319	.459	1.50	1.82	1.30	1.58	1.90	1.80
15/16	.199	.398	.597	.796	1.19	1.59	.133	.266	.398	.530		.367	.527	1.73	2.07	1.50			
1	.212	.425	.627	.849	1.25	1.70	.142	.282	.418	.566		.417	.599	1.96	2.38	1.70	2.06	2.48	2.36
1-1/16	.226	.451	.677	.902	1.35	1.80	.150	.301	.451	.602		.471	.676	2.22	2.68	1.92			
1-1/8	.239	.478	.716	.955	1.43	1.91	.159	.318	.477	.637		.528	.758	2.48	3.02	2.15	2.60	3.14	2.98
1-3/16	.252	.504	.756	1.01	1.51	2.02	.168	.336	.505	.672		.588	.845	2.77	3.36	2.40			
1-1/4	.265	.531	.796	1.06	1.59	2.12	.177	.354	.531	.706		.651	.936	3.07	3.72	2.66	3.21	3.88	3.68
1-5/16	.279	.557	.836	1.11	1.67	2.23	.186	.372	.557	.743		.718	1.03	3.38	4.10	2.93			
1-3/8	.292	.584	.876	1.17	1.75	2.34	.195	.389	.584	.777		.789	1.13	3.71	4.50	3.21	3.89	4.69	4.45
1-7/16	.305	.610	.915	1.22	1.83	2.44	.203	.407	.610	.814		.836	1.24	4.05	4.91	3.51			
1-1/2	.318	.637	.955	1.27	1.91	2.55	.212	.425	.636	.849		.938	1.35	4.42	5.36	3.82	4.62	5.58	5.30
1-9/16	.332	.664	.995	1.33	1.99	2.65	.221	.442	.664	.884		1.02	1.46	4.79	5.81	4.15			
1-5/8	.345	.690	1.04	1.38	2.07	2.76	.230	.460	.690	.920		1.10	1.58	5.18	6.29	4.49	5.43	6.55	6.22
1-11/16	.358	.716	1.07	1.43	2.15	2.87	.239	.477	.716	.956		1.19	1.71	5.59	6.80	4.84			
1-3/4	.371	.743	1.11	1.49	2.23	2.97	.249	.495	.743	.990		1.28	1.84	6.01	7.29	5.20	6.29	7.59	7.21
1-13/16	.385	.769	1.15	1.54	2.31	3.08	.257	.513	.770	1.03		1.37	1.97	6.45	7.81	5.58			
1-7/8	.390	.796	1.19	1.59	2.39	3.18	.266	.531	.796	1.06		1.47	2.10	6.90	8.36	5.97	7.23	8.72	8.28
1-15/16	.411	.822	1.23	1.65	2.47	3.29	.274	.549	.823	1.10		1.56	2.25	7.36	8.94	6.38			
2	.425	.849	1.27	1.70	2.55	3.40	.283	.566	.849	1.13		1.67	2.40	7.85	9.52	6.80	8.23	9.93	9.43

Three methods can be used: Compute cubic inches of weld metal to fill the joint and multiply by 0.283; break down the joint into geometric areas and use the table above; and lastly, compute the weight through consideration of the welding procedure and nomograph (page 278)

Example 1: To compute weight of weld metal based on joint design. See Fig. 1.

Break the cross-section into geometric areas **A, B, C** and **D,** Fig. 2. From the table get values for each area. Total weight: 2.77 lb of weld metal per ft of joint.

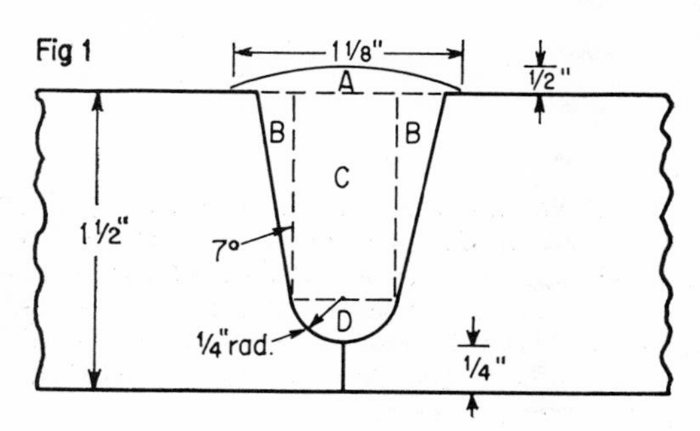

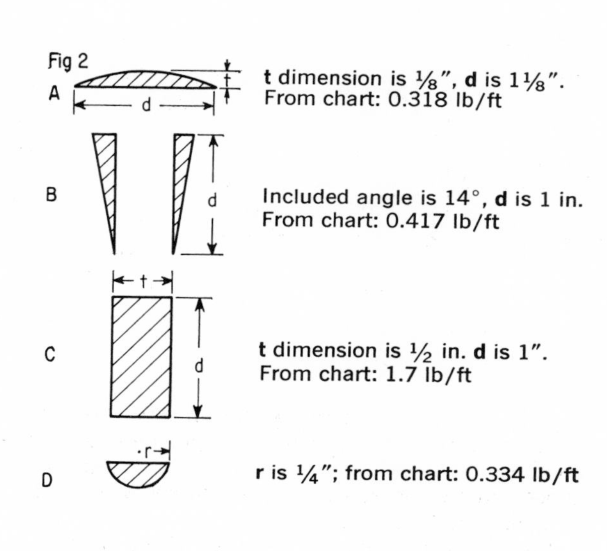

Weight of weld metal (Continued)

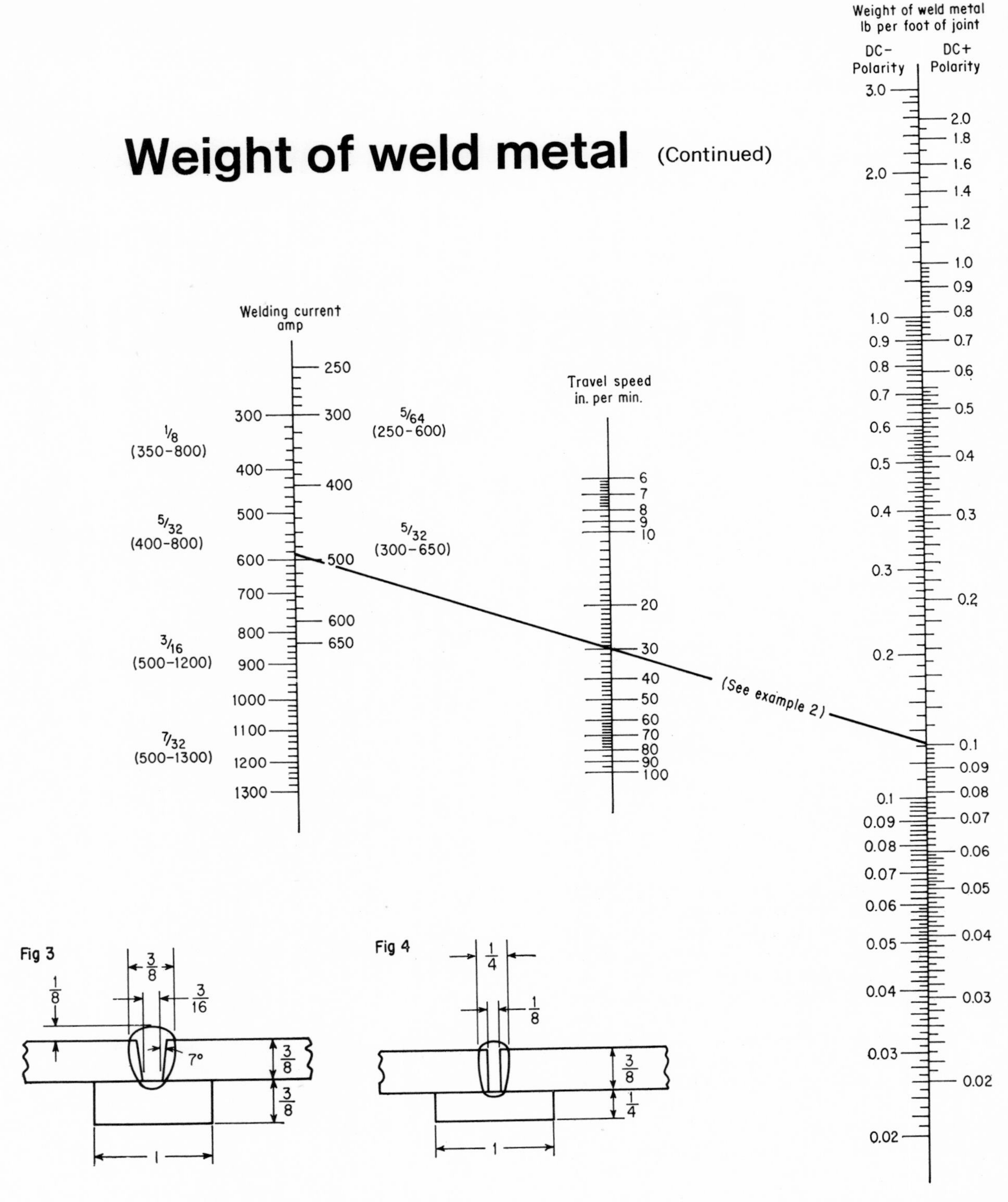

Example 2: To compute weight of weld metal based on welding procedures. Use nomograph. Line up a straight-edge through welding current used (left scale) and travel speed. Where the straight-edge intersects the right scale, read weld metal per foot.

Be sure to use the proper sides of the welding-current and weight-of-weld scales.

The line drawn on the nomograph represents the procedure for 590 amp on 1/8 in. electrode at a travel speed of 30 in. per min. The resultant weight of weld metal is 0.10 lb/ft if DC positive polarity is used, or 0.13 lb/ft for DC negative polarity.

Example 3: For non-standard joints (see Fig. 3), when two variables are known, the third can be found. The variables are: (1) welding current, (2) travel speed, and (3) weight of weld metal. Probably the closest to a standard procedure is shown in Fig. 4. Here: power is DC +; amp = 670; volts = 29; electrode size = 5/32 in., and travel speed = 16 in. per minute.

Slow down the welding speed to provide the fill required. To do this, first determine the weld metal required as outlined with Example 1, or 0.404 lb per ft. Then, employ the nomograph to find that 9 in. per min. should be used to provide adequate fill.

Chapter 15

Resistance Welding

Resistance welding of steel tubes, 1

When production runs are involved, resistance welding can be used to make many different types of joints, using round or square tubing. See sketches and operating data

Thin-wall steel tubing is easily welded tube to tube, tube to sheet, to wire, to bar stock, or to angles by spot or projection welding. In 50% of cases the workpieces need no preparation.

Weld appearance is usually good; sometimes the weld is invisible. The heat-affected zone is

By Max Osborn
Manager, Welding Products Div
Falstrom Co, Passaic, N J

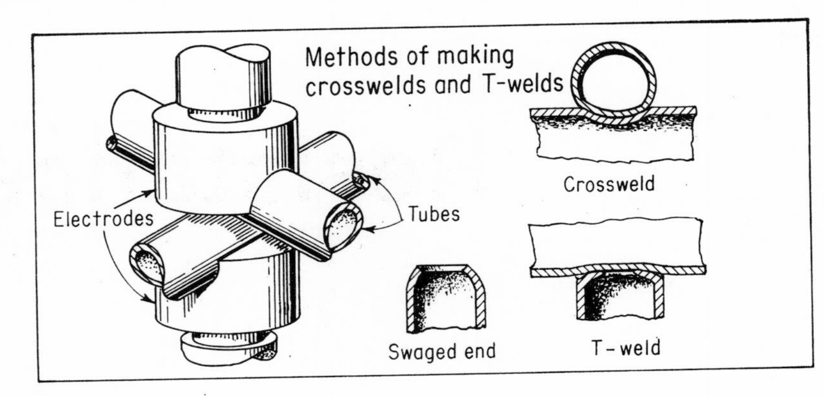

Methods of making crosswelds and T-welds

Cross welding of mild-steel tubing

Welded assembly	Cross-section of weld	Electrode shapes	Remarks
			1. **Tube collapse** is always 10 to 20%. This most used and attractive method requires that wall thickness of both tubes be identical, even if the diameters vary. No tube preparation is required.
			2. **Purpose of 50% collapse** is to achieve an assembly that is 'one tube thick.' Special sequence of welder, fitted with milled electrodes, is: (a) squeeze, (b) weld, (c) apply heat impulses several times, and (4) collapse.
			3. **Both tubes** are preshaped to obtain 50% collapse. Four contact points are produced to provide a strong projection-welded assembly.
			4. **No overlap** occurs in this form of cross welding. Slight distortion of the tube end, at the weld, is unavoidable.
			5. **Contact area** must be reduced by cutting a hole in both tubes approximately the diameter of the internal widths of the square tube. Four welded spots produce a strong joint.

Resistance welding of steel tubes [Continued]

small. Generally speaking, no distortion occurs. Even when chromed tube is used, cleaning of the brown zone eliminates traces of weld heat.

Equipment used

Stationary air-operated equipment is preferred for small, easily handled assemblies, and portable air-operated guns for work that is pre-positioned in large jigs, according to Aro, S. A., Paris, France, which supplied this information.

Tube welding can be easily carried out by standard sequence control panels, except in cases where a 50% tube collapse is needed.

Electrode shapes

Two basic shapes are the V-type and the round milled type. The V type is used when a variety of tube sizes is to be welded. The half--round electrode is always preferable but can be applied to one tube diameter only. It also has the advantage of positioning the tubes more accurately, and eliminates the danger of heating the tube outside of the welding zone, and is used in production.

Minimum clamping force

Tube OD In.	Electrode force lb.	Clamping force lb.
3/8	418	485
1/2	462	572
5/8	462	660
11/16	505	880
3/4	505	925
7/8	505	990
1	528	1055
1 1/8	550	1122
1 3/16	550	1188
1 1/4	550	1300
1 3/8	593	1495

Note: Clamps are made of copper/chrome with an ID about 0.004 to 0.008 in. smaller than the nominal tube OD and a depth of 1 in.

Control settings for cross tube welding

Tube OD in.	Thickness In.	Electrode force In.	Weld time Sec.	Weld power Amp.	Collapse %
3/8	0.036	440	0.25	5000	5
3/8	0.036	440	0.8	9000	15
5/8	0.036	440	0.30	9500	5
5/8	0.036	440	1.2	9500	15
7/8	0.048	485	0.4	12000	5
7/8	0.048	485	1.2	12000	15
1	0.064	528	0.35	14000	5
1	0.064	528	1.15	14000	15
1 3/16	0.064	528	0.35	16000	5
1 3/16	0.064	528	1.10	16000	15
1 3/8	0.080	593	0.40	18000	5
1 3/8	0.080	593	1.50	18000	15

Electrodes

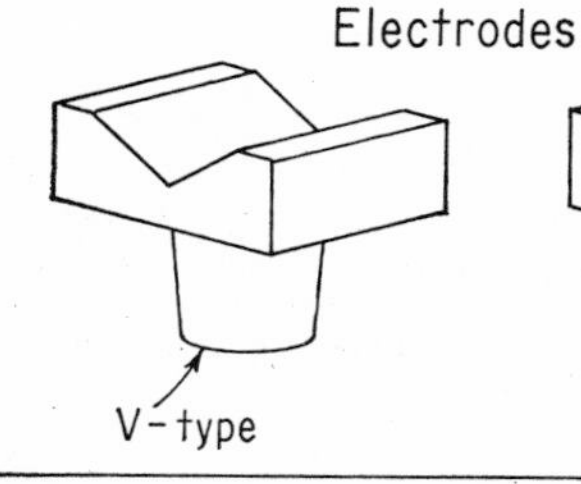

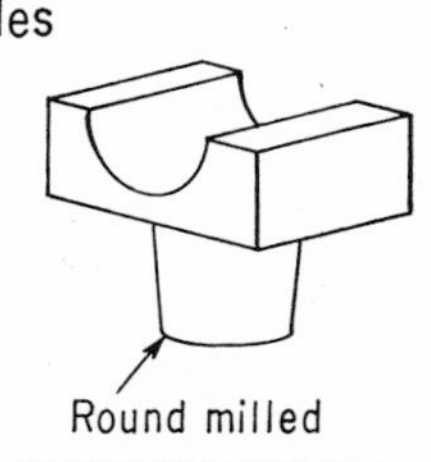

Pneumatic vise for T-welds

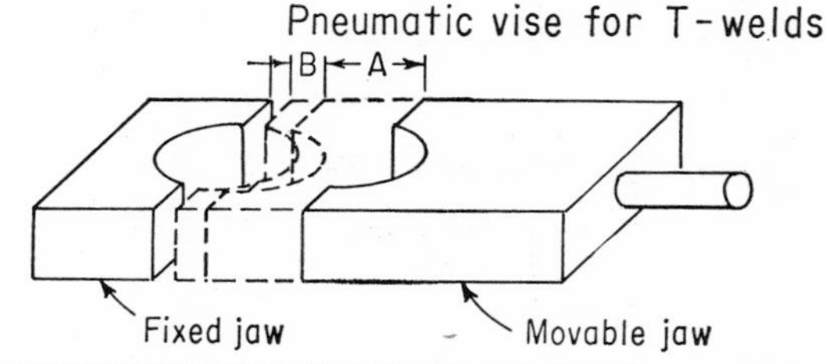

A = Closing stroke
B = Holding stroke

Specialized joints in welded tubing

Welded assembly	Cross-section of weld	Electrode shapes	Remarks
			1. **Welding of rod** or bar to tubing produces excellent results, but indentation of the weld is inevitable unless the rod is of small diameter.
			2. **Back-to-back welding** of bent tubes is required in manufacture of beds and chairs. Shape the electrodes accurately to avoid marks inside the curves.

Resistance welding of steel tubes, 2

T-welds, as well as cross welds (AM—Oct 10 '66, p155) are readily made by unskilled labor in seconds. Success is achieved if you know good weld-preparation procedures

By Max Osborn
Manager, Welding Products Div
Falstrom Co, Passaic, N J

Resistance welding of mild steel tubes can be applied to making good-looking T-joints quickly and economically using unskilled labor. Instead of saddle shaping one of the tubes and silver soldering or brazing the joint, one tube end is swaged and forced into the other tube at welding heat. The penetration that occurs (reduction in length due to entry of one tube into the other during welding) must be taken into account. The loss in length from doming of one end must also be considered.

The nature of the operation is such that weld settings must be reasonably accurate. And they should certainly be constant for production work. The settings will involve weld times of medium length, high welding current, and relatively low weld pressures. The domed tube should collapse slightly into the opposite tube wall.

The two tubes to be welded must have the same wall thickness and preferably they should be of the

T-welding of mild steel tubes

Welded assembly	Cross-section of weld	Electrode shapes	Remarks
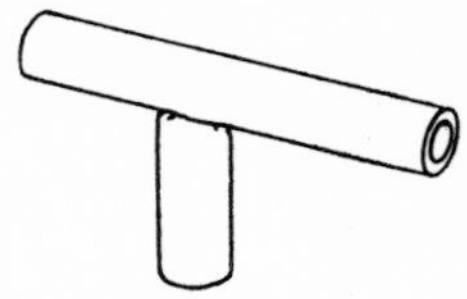	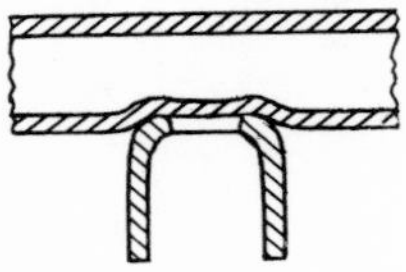	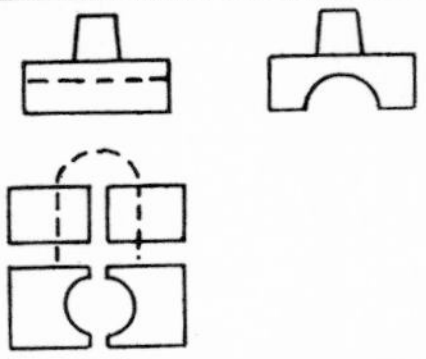	1. **Dome swage** the stud end of one tube until the closure equals one half the diameter. The two tubes must have the same wall thickness.
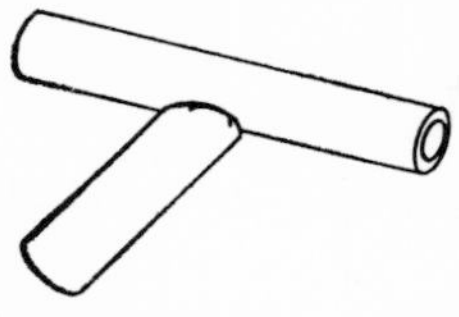	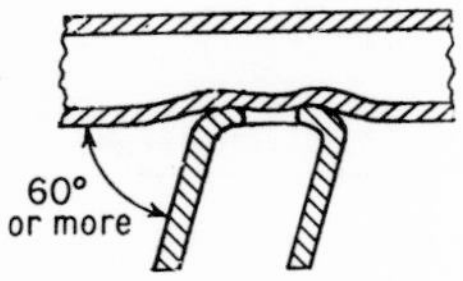 60° or more	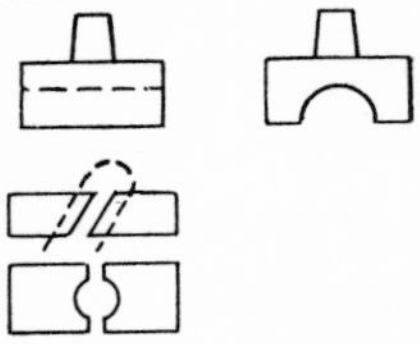	2. **Tube welding** at an angle requires swaging of the open end of one tube, but at an angle to fit against the second tube. The included angle between tubes cannot be less than 60°.
	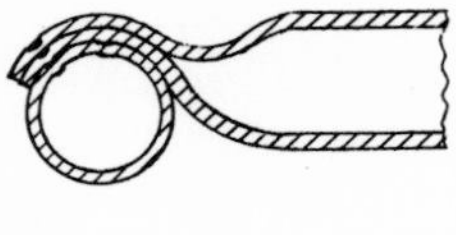	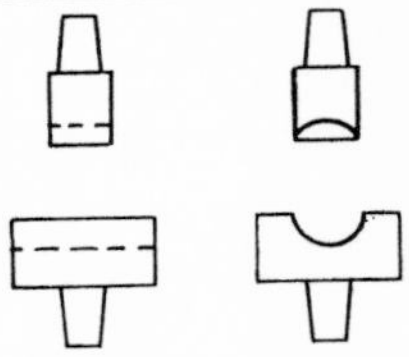	3. **Flatten the end** of one tube and dimple it. Good shear strength is secured but the appearance of the weld is not equal to that of other joints.
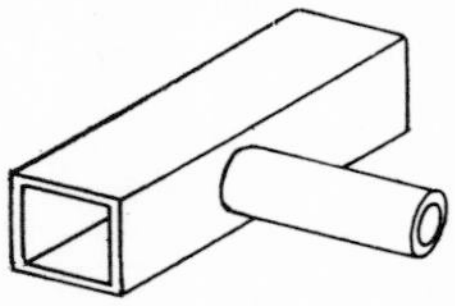	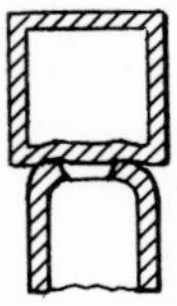	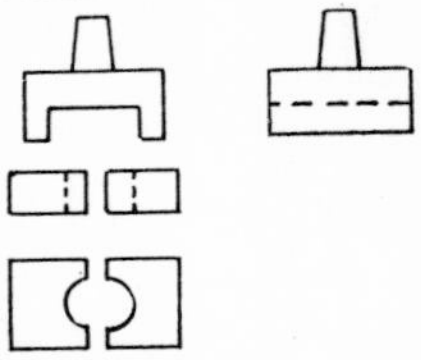	4. **This operation** is similar to welding a tube to a sheet. The round tube must be slightly smaller than the square tube. Both tubes must be of the same wall
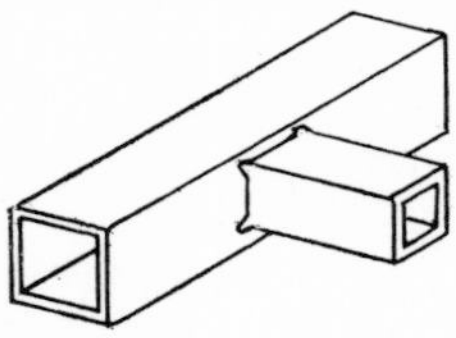	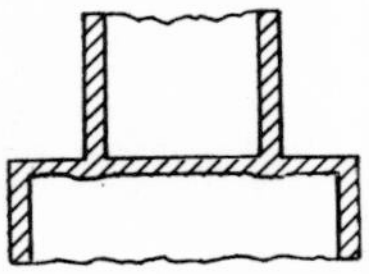	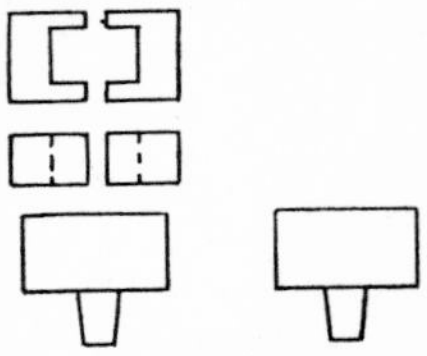	5. **The stub end** of one tube should have the four side walls milled out to a smooth curvature, leaving projections at the corners. There is less chance of burning with this preparation than when the walls are notched square.

Resistance welding of steel tubes [Continued]

Welding of tube to sheet

Welded assembly	Cross-section of weld	Electrode shapes	Remarks
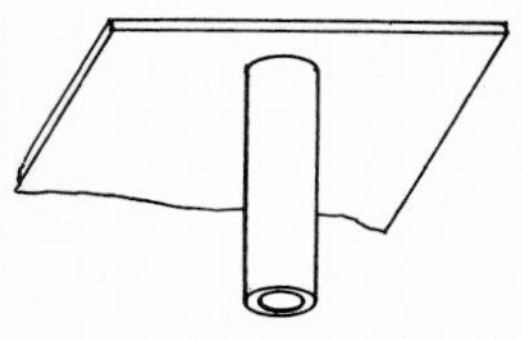	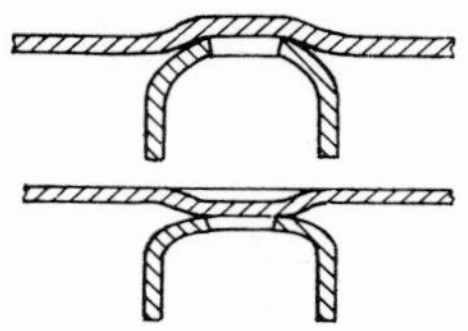	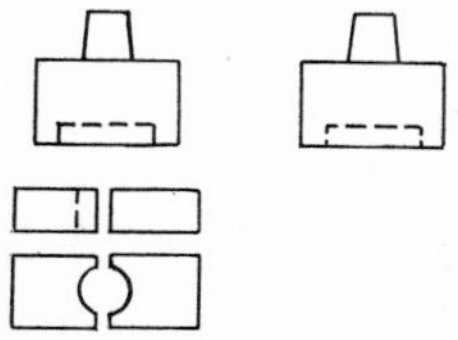	1. **A recessed electrode** in contact with the sheet provides localized heat, so that the swaged dome on the tube can penetrate the sheet. Tube wall and sheet thickness must be the same.
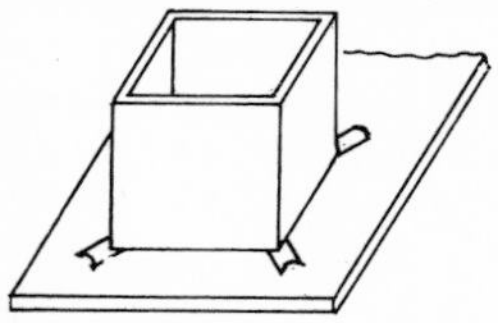	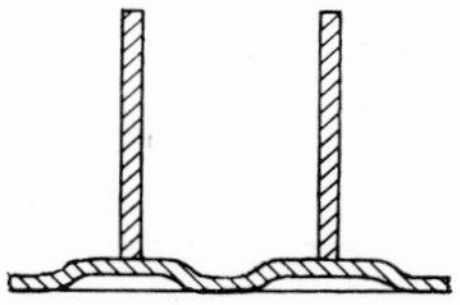	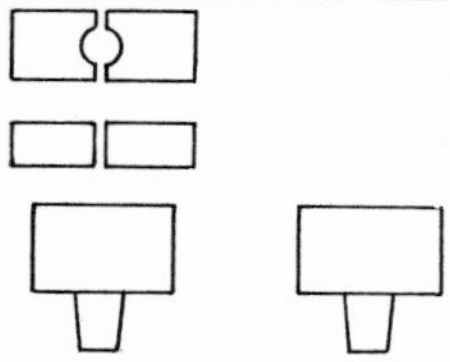	2. **Elongated projections** in the sheet allow welding square or rectangular tube to the flat material.
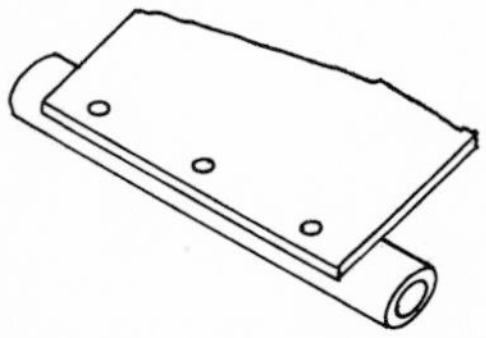	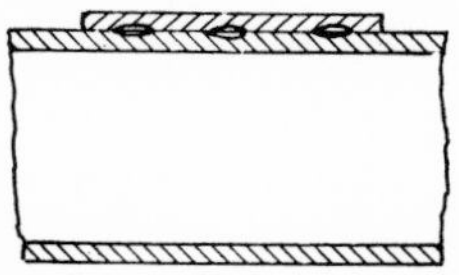	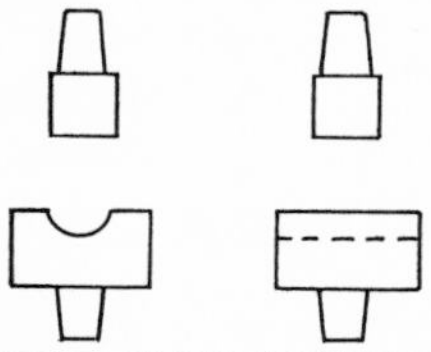	3. **For a good weld,** sheet to tube, the sheet must be thinner than the tube and a V-type electrode must be used.
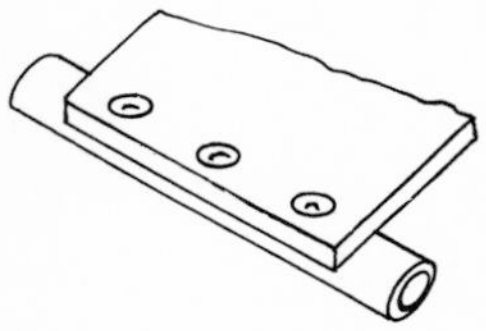	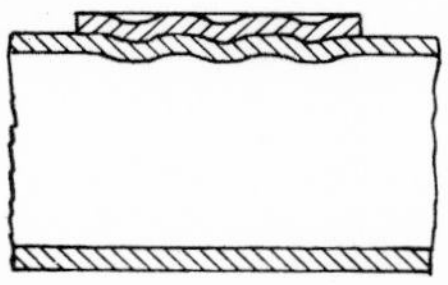	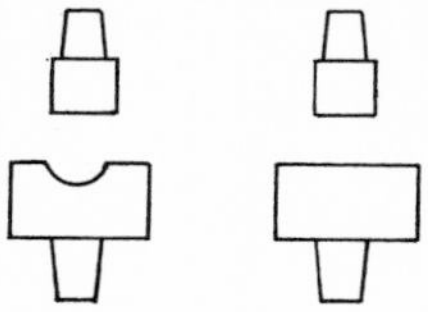	4. **Projection welding** is essential if the sheet is as thick or thicker than the tube. Elongated projections are better than round ones.

same diameter. If necessary, the domed tube can be of a smaller diameter.

To dome swage the stud end of one tube, a swaging tool with a spherical recess is required. After swaging, the opening in the end of the tube should be one-half the tube diameter.

When the domed tube is clamped in the air-operated vise (**AM**—Oct 10 '66, page 157) its end should protrude from the jaws about two-thirds of the tube diameter. The second tube is then positioned and held by a round milled electrode while the joint is made.

Control settings for T-welding of tubes

Tube						
Diameter In.	Thickness In.	Electrode force Lb.	Weld time, Sec.	Weld power, Amp.	Penetration In.
3/8	0.036	418	0.20	11000	0.099
1/2	0.036	462	0.20	12000	0.110
5/8	0.036	462	0.30	13000	0.122
11/16	0.048	505	0.40	13000	0.138
3/4	0.048	505	0.40	14000	0.157
7/8	0.048	505	0.50	14000	0.169
1	0.064	528	0.60	14500	0.197
1 1/8	0.064	550	0.60	15000	0.216
1 3/16	0.064	550	0.80	15000	0.236
1 1/4	0.064	550	1.10	16000	0.248
1 3/8	0.080	593	1.30	19000	0.276

CAP-TYPE RESISTANCE-WELDING ELECTRODES

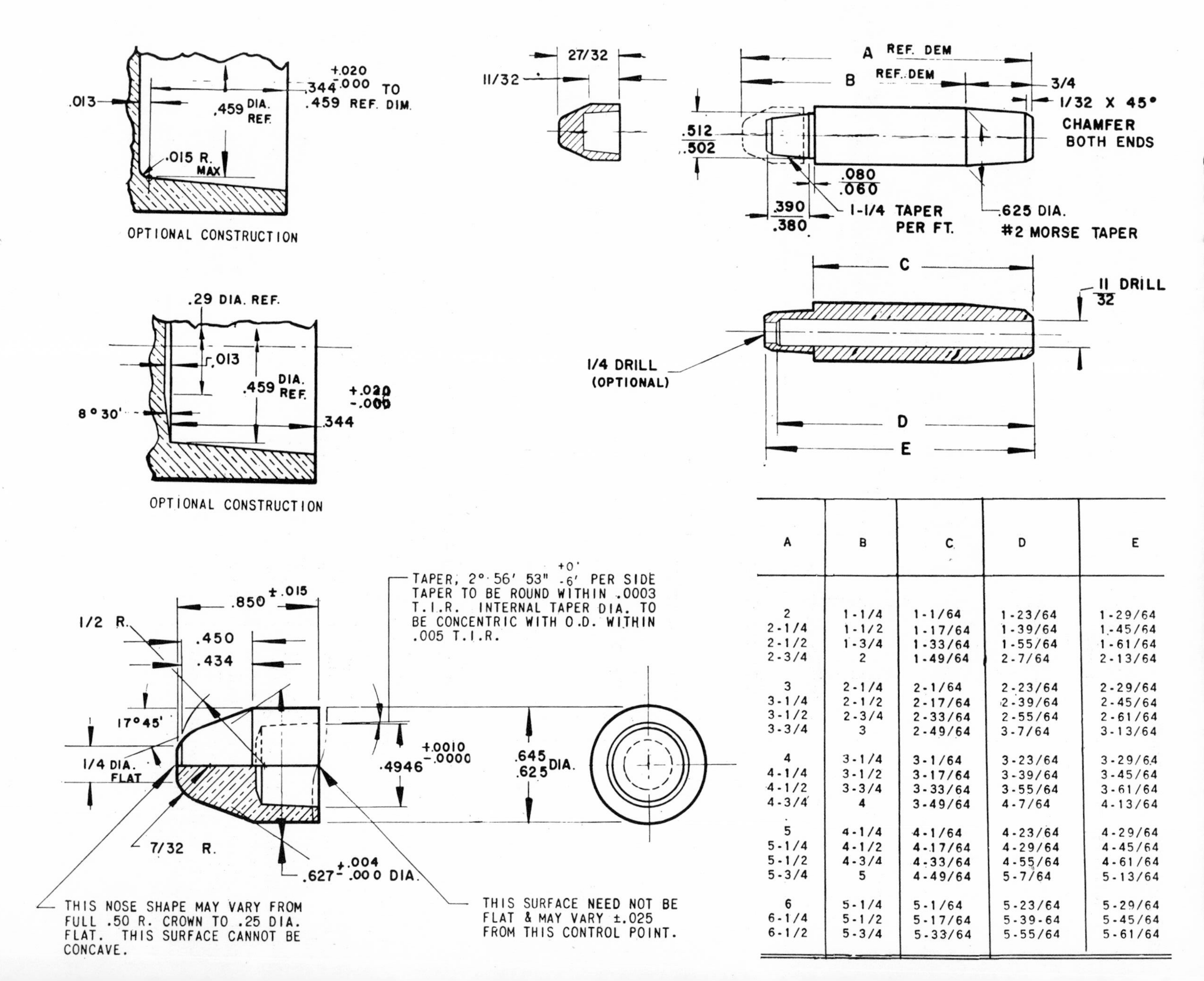

A	B	C	D	E
2	1-1/4	1-1/64	1-23/64	1-29/64
2-1/4	1-1/2	1-17/64	1-39/64	1-45/64
2-1/2	1-3/4	1-33/64	1-55/64	1-61/64
2-3/4	2	1-49/64	2-7/64	2-13/64
3	2-1/4	2-1/64	2-23/64	2-29/64
3-1/4	2-1/2	2-17/64	2-39/64	2-45/64
3-1/2	2-3/4	2-33/64	2-55/64	2-61/64
3-3/4	3	2-49/64	3-7/64	3-13/64
4	3-1/4	3-1/64	3-23/64	3-29/64
4-1/4	3-1/2	3-17/64	3-39/64	3-45/64
4-1/2	3-3/4	3-33/64	3-55/64	3-61/64
4-3/4	4	3-49/64	4-7/64	4-13/64
5	4-1/4	4-1/64	4-23/64	4-29/64
5-1/4	4-1/2	4-17/64	4-29/64	4-45/64
5-1/2	4-3/4	4-33/64	4-55/64	4-61/64
5-3/4	5	4-49/64	5-7/64	5-13/64
6	5-1/4	5-1/64	5-23/64	5-29/64
6-1/4	5-1/2	5-17/64	5-39-64	5-45/64
6-1/2	5-3/4	5-33/64	5-55/64	5-61/64

Chapter 16

Plastics Molding and Machining

Metal inserts for plastic parts

Plastics are increasingly used in automobiles and appliances, and thus a major company compiled these data

Molded parts should be designed around any inserts that are required. This work is done after the type of compound is selected. Inserts are used for two basic reasons:

1. To add strength to the plastic part or to control shrinkage. Sometimes the purpose is to be decorative or to avoid injuries.
2. To provide an attachment means for the conductance of heat or electricity.

Special means of retention are not necessary for inserts not subject to movement with relation to the molding material. Inserts of round bar stock, coarse diamond knurled and grooved, provide the strongest anchorage under torque and tension. A large single groove, as in Fig. 1, is better than two or more grooves and smaller knurled areas. See also examples of inserts in Fig. 3, page 135.

Inserts secured by press or shrink fits

Inserts can be secured in the plastic by a press fit or shrink fit. Both methods rely upon shrinkage of the plastic, which is greatest immediately after removal from the mold. The part should be made so that the required tightness is obtained after the plastic has cooled and shrinkage has occurred. The amount of interference required depends on the size, rigidity and stresses to be encountered in service. The hole to receive the mating part should be round and countersunk. In addition, the mating part should be chamfered and filleted to facilitate proper assembly and to eliminate stress concentration.

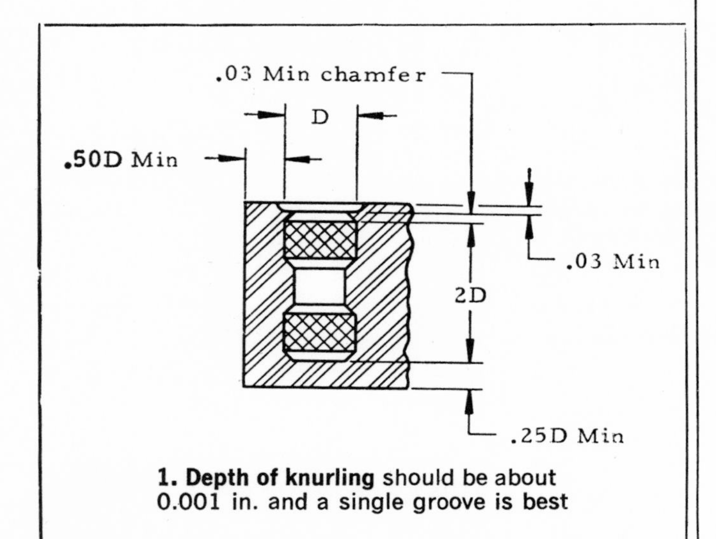

1. Depth of knurling should be about 0.001 in. and a single groove is best

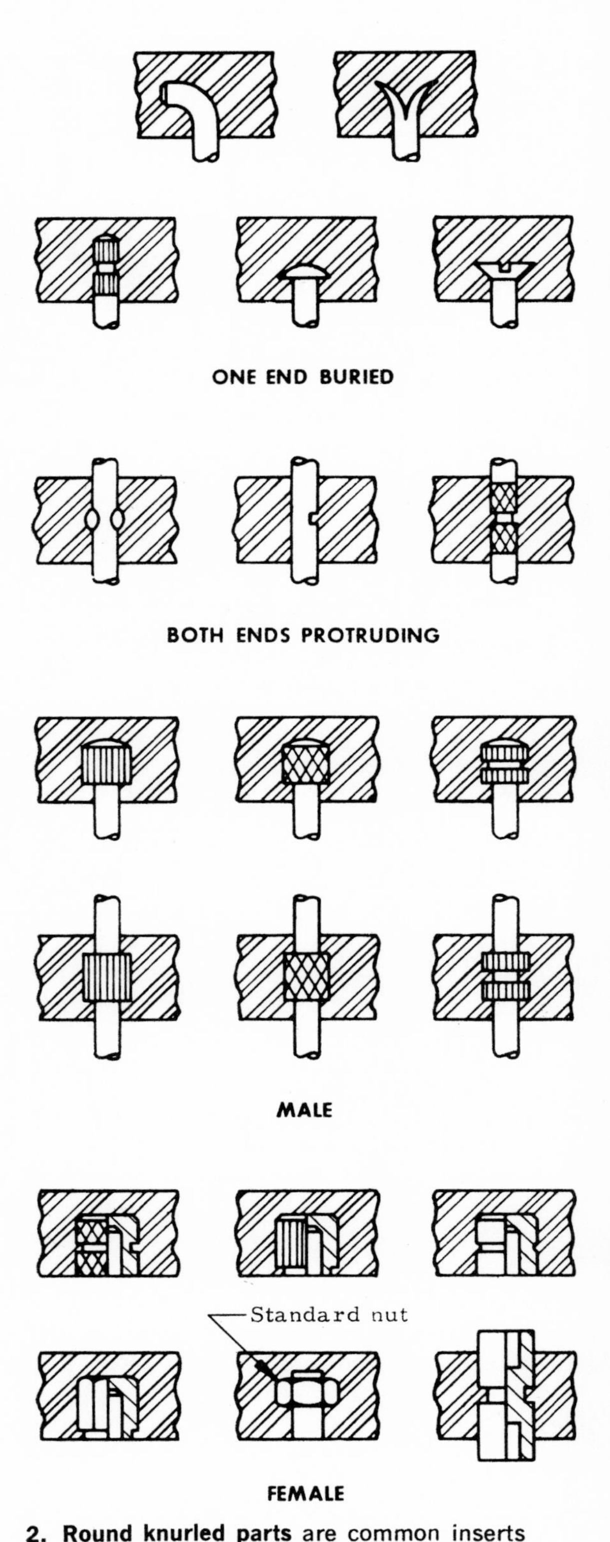

2. Round knurled parts are common inserts

Metal inserts for plastic parts [Continued]

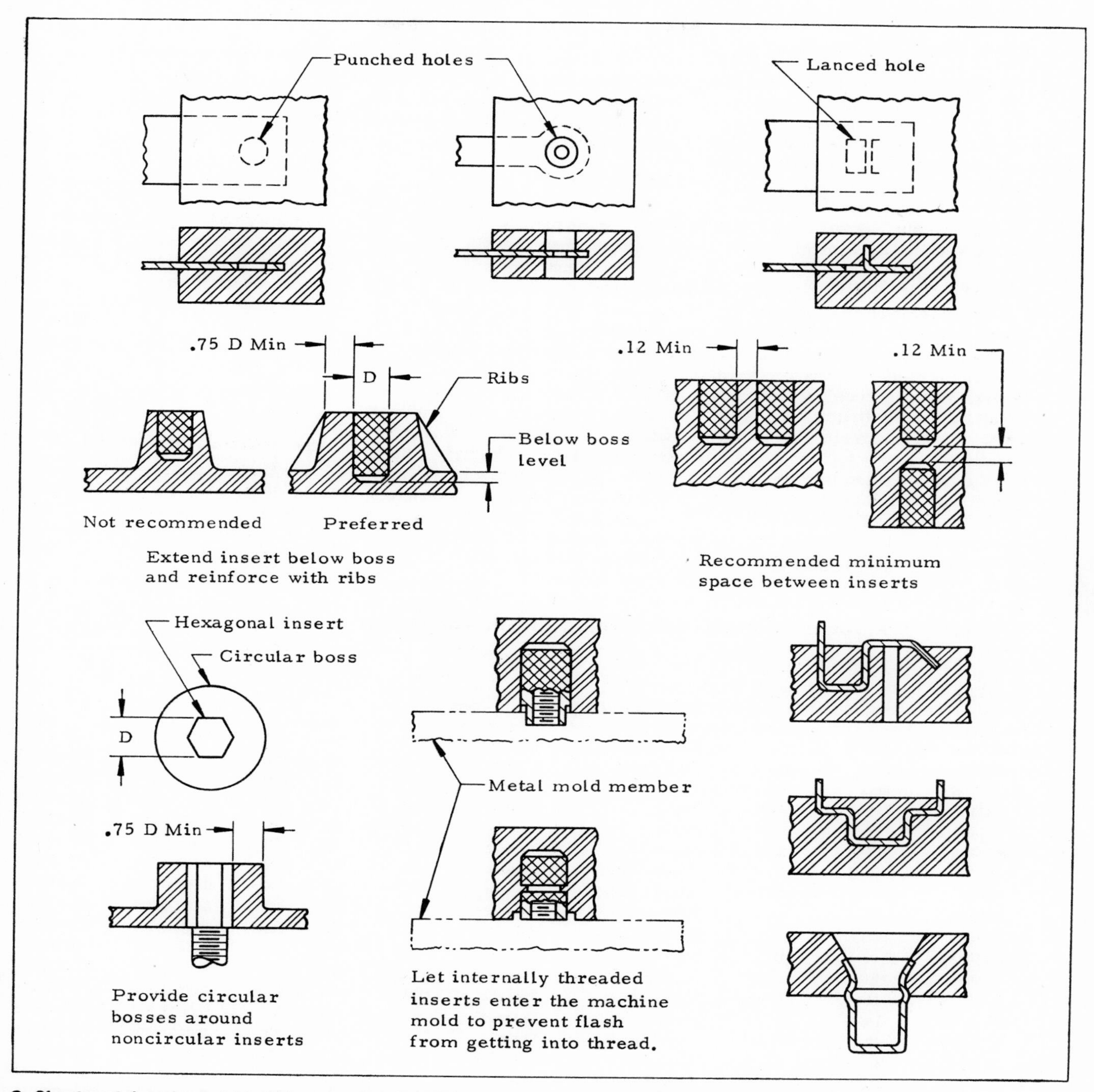

3. Sheet metal and special inserts, as well as knurled inserts, are used to provide connections and closures

Laminated Plastics and Fiber—I

By John T Bishop, production manager, Continental-Diamond Fibre Corp, Newark, Del

Industry is using laminated plastics and fibre to save weight (about 40% of that for aluminum), to improve performance and appearance, and to achieve economies in fabrication processes and assembly. These materials are easily worked on conventional metalworking machines and woodworking equipment.

Nature of materials

The properties of a laminate depend on the resin-filler system used. Resins include: phenolic, Melamine, silicone, epoxy, polyester, fluoro-carbon. Reinforcing materials are cotton fabric, paper, glass fabric, asbestos, and synthetic fibers.

Vulcanized fiber is made up of layers of cellulosic paper, which have been fused by treatment in a zinc-chloride bath.

If the inherent properties of laminated plastics and vulcanized fiber are understood, these materials can be machined as easily as most metals, regardless of form—sheet, tube, rod, or molded.

Shear strength

Laminates have lower densities and generally lower shear strengths than metals. So, they can be machined on relatively light equipment, at high speeds and with low-cost tooling. Often it is advantageous to increase the machine's top speed capability.

Resilience

Both materials are resilient, which means that cutting tools must be kept extremely sharp. Otherwise, size control and finish will suffer. However, tolerances, comparable to those obtainable on metal, cannot be achieved even with the sharpest tools.

In punching and drilling operations, the materials tend to close in. Therefore, tools for these operations must be made over size, according to the material grade and thickness. Hole diameters in punching are affected by stock temperature. The diameter of hole actually produced will depend on whether the material is punched at room temperature or at some higher temperature.

Heating the stock has an important effect on resiliency. Punch diameters vary significantly for material punched cold as compared to stock punched after heating.

Low thermal conductivity

Laminated plastics and vulcanized fiber have low thermal conductivity as compared to most metals. Thus, it is important to make allowance for the fact that

Table 1 — Recommendations for Turning Plastic Laminates

Military Spec	Type	NEMA Grade	Reinforcement	Resin*	Tool Angles ‡ Side cl., deg.	Cutting Speed HSS fpm	Cutting Speed Carbide fpm	Feed Rough	Feed Finish
P-997B	GSG	G-7	Glass Fabric	Silicone	7-10	300	550	0.003-0.005	0.003-0.005
P-997B	GSG	G-6	Glass Fabric	Silicone	7-10	200	550	0.003-0.005	0.003-0.005
P-3115B	PBG	XX	Paper	Phenolic	10-15	400	800	0.005-0.019	0.005-0.019
P-3115B	PBE	XXX	Paper	Phenolic	10-15	400	800	0.005-0.019	0.005-0.019
P-3115B	PBE-P	XXP	Paper	Phenolic	10-15	400	800	0.005-0.019	0.005-0.019
P-3115B	PBE-P	XXXP	Paper	Phenolic	10-15	400	800	0.005-0.019	0.005-0.019
P-8059		A	Asbestos Paper	Phenolic	10-15	400	800	0.005-0.019	0.005-0.019
P-8059		AA	Asbestos Fabric	Phenolic	10-15	400	800	0.005-0.019	0.005-0.019
P-8665A	‡	CF	Cotton Fabric	Phenolic	10-15	400	800	0.005-0.019	0.005-0.019
P-15035	FBG	CE	Cotton Fabric	Phenolic	10-15	400	800	0.005-0.019	0.005-0.019
P-15035B	FBE	LE	Cotton Fabric	Phenolic	10-15	400	800	0.005-0.019	0.005-0.019
P-15035B	FBM	C	Cotton Fabric	Phenolic	10-15	400	800	0.005-0.019	0.005-0.019
P-15035B	FBI	L	Cotton Fabric	Phenolic	10-15	400	800	0.005-0.019	0.005-0.019
P-15936B	GMG	G-5	Glass Fabric	Melamine	7-10	200	550	0.009-0.015	0.009-0.015
P-15047B	NPG	N-1	Nylon Fabric	Phenolic	10-15	400	800	0.005-0.009	0.005-0.009
P-17721C	GMM	G-8	Glass Mat	Melamine	7-10	200	550	0.009-0.015	0.009-0.015
P-18177A	GEE	G-10	Glass Fabric	Epoxy	7-10	200	550	0.009-0.015	0.009-0.015
P-18177	GEB	G-11	Glass Fabric	Epoxy	7-10	200	550	0.009-0.015	0.009-0.015
P-18324A	B		Cotton Fabric	Phenolic	10-15	400	800	0.005-0.009	0.005-0.009
P-19161	GTE		Glass Fabric	Fluorocarbon	7-10	200	550	0.003-0.005	0.003-0.005

Note: If you run into trouble machining glass fabric, leave 0.030 on diameter for finish cut. Then your feed can vary from 0.004 to 0.015 ipr.

† Side rake — 0°, end clearance, 5 — 7°, for all laminates

‡ Postforming type

Laminated Plastics and Fiber—II

heat cannot be dissipated as quickly in machining these materials. In high-speed fabrication, either the material or the tool, or both, may be burned by the heat caused by friction. The cutting tool should be removed occasionally, permitting it and the material to cool. In some operations, tool damage and rejected pieces can be prevented by the use of lubricants or cold air jets. Carbide-tipped tools should be used wherever practical, because they retain a sharp edge at high temperatures.

Material structure

Because they are constructed of layers, laminated plastics and vulcanized fiber are weakest in the direction at right angles to their face. Where design will permit, drilling, tapping or broaching should be done at right angles to the laminations. Otherwise, care should be taken to prevent splitting, as by use of confining jigs or vises. In gear cutting, back the blanks with supporting collars to prevent delamination at runout.

Turning

Honed single-point tools are always advisable. Air-cooling is usually preferable, because liquids may be difficult to remove. Skived forming tools are used, but speeds should be increased on the softer grades.

Higher speeds are possible on thermosetting plastics than on thermoplastic types. Using high-speed tools, most laminates can be turned at 500-800 surface fpm. Table 1 gives detailed recommendations for tooling speeds for rough and finish cuts in turning laminated plastics. Carbide tooling permits double the best surface speed and is especially useful on asbestos or glass-base laminates such as NEMA grades G-5, G-6, G-7, G-10, and G-11. Cutting speed for carbide tools on glass-base laminates should be around 550 fpm.

Successful machining of vulcanized fiber depends on two factors: (1) a sharp tool and (2) plenty of clearance—at least 30°. Because it is extremely hard and tough, vulcanized fiber may crowd against the back of the tool, generating heat and dulling tools.

For light cuts and close work, diamond-cutting tools are best. For high-quantity production, high-speed steel, tungsten carbide or carbide-tipped tooling are used. Coarse cuts at speeds, feeds, and clearances about the same as those used on brass are best for this material.

Milling and planing

Milling and planing are relatively simple operations with laminated plastics and vulcanized fiber and need only normal care, sharp tools, fast chip removal, and quick heat dissipation. Dry cutting is the usual practice.

To prevent the cutter from lifting the laminate layers, climb milling is recommended. Surface speeds may be as high as 1000 fpm, with feed rates up to 20 ipm with carbide cutters. High-speed steel cutters require somewhat lower speeds and feeds. Cutters should have a 10° negative rake, with good clearance. Milling procedures are similar to those for brass.

Planing

Laminated plastics and vulcanized fiber may be planed or shaped at 70-90 sfpm with carbide tools that have up to 10° negative rake and about 20° clearance. For glass-base grades of laminated plastics, such as NEMA grade G-5, G-6, G-7, G-10, and G11 the speed should be 40-50 fpm with a feed of 0.010 to 0.015 in., a side clearance of 30°, a cutting angle of 36°, a cutting angle of 36°, and a shaving angle of 45°. Coolants are not necessary. However, an efficient pneumatic dust and chip-collecting system is necessary to prevent excess deposits of glass dust, which can cause dermatitis and other health hazards.

Forming

Use 0.002-0.003 in. feed when tolerance is less than 0.006 in. on any given dimension.

Use 0.003-0.006 in. feed when tolerance is more than 0.006 in. on any given dimension, providing the wall or diameter left is strong enough to support it.

Cutting off

Laminated plastics—Use 0.002-0.005 in. feed when burr is not permitted. If there is a second operation, which will remove cut-off burr, the feed can be increased to 0.005 to 0.010 in., depending on width of cutoff tool.

Vulcanized fiber—Use 0.001 to 0.003 in. feed to keep cut-off burr to a minimum.

On all automatic screw machines, use carbide tooling and run at top spindle speed with exception of tapping; Then, run at 2/3 maximum spindle speed with a 3-1 ratio to withdraw the tap.

Table 2 — Feeds for Automatic Screw-Machine Work

	Diameter	Feed, ipr
Drilling	0 — 0.125	0.001 — 0.003
	0.125 — 0.500	0.003 — 0.015
	Over 0.500	0.015 —
Turning	0 — 0.125	0.002 — 0.006
	0.125 — 0.500	0.006 — 0.015
	Over 0.500	0.015 —

Note: When tolerance is less than 0.006 in., use feed of 0.005 ipr; when tolerance is greater, use feed of 0.010 ipr.

Laminated Plastics and Fiber—III

Hole-machining operations on these materials are not difficult if simple precautions are observed. Notes on drilling, tapping, reaming and broaching are given from production experience with fabricating non-metallic materials

By John T Bishop, production manager,
Continental-Diamond Fiber Corp, Newark, Del

Drilling

A hole drilled in a laminate will shrink or close in. Unless reaming will be done, always drill a hole 0.002-0.004 in. oversize. Automatic feed is preferred because light, uniform pressure is required. Chipping of the lower surface can be controlled by backing the work and reducing feed pressure at break through. HSS drills, with wide, polished flutes make the best tools for drilling laminates. The wide flutes provide maximum chip removal and the polished surfaces prevent drilling dust from accumulating.

An air jet frequently does a better job of cooling than conventional liquid coolants. The latter tend to mix with drilling dust to form a paste, which generally hampers the operation. If air is used, a thin, high-pressure stream gives the best results. But drilling speed must be cut back to about 10,000 rpm max. and 200 to 350 sfpm, depending on grade and thickness.

Carbide drills with polished flutes and slow helix are particularly well suited to drill glass-base laminates. For through-holes in thin sheets, grind the point to a 55° included angle; 90° is satisfactory for drilling in a thick sheet. Cutting speeds of 150 to 300 fpm, with a feed of 0.002 to 0.003 ipr should be used. Lift the drill regularly to keep the hole clean of chips.

Working with vulcanized fiber requires drills with the same wide, highly polished flutes, steep helix, and narrow web as used for laminated plastics.

The highest speed that won't burn the drills or the fiber is recommended. A No. 60 drill at 10,000 rpm or a ¼ in. drill at 2500 rpm are typical drill speeds.

Hole shrinkage is also experienced when drilling vulcanized fibers, and the same remedial actions can be taken. In addition, you should avoid drilling parallel to the fibrous structure. If this must be done, clamp both sides of the work and lift the drill repeatedly to clear the chips.

Tapping

The resiliency of most laminated materials demands that special attention be given to rake. The tools should have up to 10° negative rake angle, depending on the thread diameter, pitch, and finish required. A positive rake results in a weak thread.

For threading laminates, the most effective tool is a ground HSS tap with three flutes about 0 0015 to 0.002 oversize. A 75% thread with a Class 2 fit is the maximum.

When cutting glass-base laminates, a 5° negative rake on the front of the land minimizes binding or stripping as the tap is removed. Nitrided, chromium-plated taps do the best work on the glass-base materials. Chamfer the hole to prevent lifting of the laminations by the entering tap.

A 75% thread with Class 2 fit is the maximum obtainable in vulcanized fibers. Negative rake of up to 10° is recommended for thread strength. A positive rake on a tap, die, or single-point tool will cause the tool to dig in, crowd its lead, and cut a weak thread.

Moderate application of light machine oil facilitates tapping in vulcanized fibers. The hole should be 0.002 to 0.004 in. larger than its counterpart in metal, and a tap should be 0.004 in. oversize for best results.

Broaching

Because of the laminated structure, the following specialized procedure should be followed in all broaching operations:

To cut across laminations, back up material with a mild-steel or brass plate. This should form a slide fit with the last cutting tooth of the broach. Such a procedure prevents fraying or breakout.

When cutting parallel to the laminations, use a fixture to apply pressure at right angles to the structure. When broaching laminates, take a cut of 0.001 to 0.005 ipt. When broaching vulcanized fiber, the cut can be from 0.02 to 0.03 ipt. Pitch can range from ¼ to ¾ in. depending on grade and thickness of the material being broached.

Reaming

The laminated structure makes it important to ream across the material wherever possible. As in broaching, a fixture should be used to apply pressure to the structure when reaming parallel to the laminations.

Glass-Fabric-Base Laminates

By H V Howe,
Applications engineer,
Technical Service Dept.,
Continental-Diamond Fibre Corp,
Newark, Delaware

In many applications using laminated plastics, particularly copper-clad laminates for critical military and commercial printed circuitry, glass-fabric reinforces the resin system.

The resultant material is tough and hard, and special precautions and tools should be used to assure good fabrication.

Drilling

Drilling of glass-base laminates is best done with carbide drills having polished flutes and a slow helix. For through holes in thin sheets, the point should be ground to a 55° included angle and for thick sheets the point should be ground to 90°. Cutting speeds of 150 to 300 fpm with feed of 0.002 to 0.003 ipr should be used.

For short-run jobs, holes up to 3/16 in. can be drilled with ordinary high-speed drills, having the cutting lip backed off to a negative rake for drilling brass. Drill speed should be held to 80 fpm with a feed of 0.002 to 0.004 ipr, backing out the drill frequently to keep the hole clear. A blast of air clears away dust and cools the drill. Wear will be severe with steel drills.

Removable bushings

If drill jigs or drilling plates are used, removable bushings should be used. Bushings can be removed after the hole is started, but some people use shoulder bushings to support small drills in a special pressure foot.

The workpiece should be firmly clamped to prevent lifting of the laminations when the drill breaks through. If holes are to be drilled parallel to the laminations, the sheet must be securely clamped at the location of the hole to prevent wedge action of the drill point from splitting the piece. Similar care must be taken in tapping of holes drilled into the edge of the part.

Drilled holes will tend to run 0.001 to 0.003 in. undersize for a properly ground drill. Compensating oversize drills may be used to yield the proper hole size, or the drill may be ground with unequal lips to give the full size hole. Grinding a slight radius at the outer end of the lip will also aid in keeping the hole size true.

Shearing

For shearing, a guillotine-type shear with a close setting is recommended. The shearing of glass epoxy is a difficult operation, and in normal practice it is limited to a 1/16 in. maximum thickness.

Sawing

For sawing, segmented diamond cutting wheels are satisfactory when used at 3,600 rpm.

Some glass-base epoxy panels are also fabricated with an abrasive wheel. Break-resistant wheels are recommended. The wheel is operated at 4365 rpm, and is 14 inches in diameter, ⅛ in. thick with a 1⅛ in. bore.

Routing and milling

For routing and milling, the use of carbide-tipped cutters operated at 3000 to 4000 rpm with a 10° negative rake on the blades with good clearance is recommended.

Punching and blanking

Compound dies will give best results in punching or blanking, by holding back on the laminate during stripping. Carbide dies are preferred for glass-base materials, if the quantity involved permits the extra expense. Holes should be no smaller in diameter than the thickness of the sheet, and should be punched no closer to the edge than three times the sheet thickness.

Very close clearance is required between punch and die, and also on the stripper plate, to prevent lifting the laminations around punched edges. Punch and die should be made oversize by about 3% on the thickness to compensate for shrinkage. Punching of glass-base laminates is best done with the stock unheated. In blanking dies, some shear ground on the punch will decrease power needs, but hole punches should be ground flat with no shear. No back clearance should be given to a hole punch.

The stroke of the press should be sufficient to eject the part. Any chips or dust resulting from punching should be blown free of the die to prevent abrasion or hobbing of die parts.

Glass dust collection

In the dry machining of glass-fabric laminates, a thin stream of air at relatively high pressure should be applied to the area of tool contact, and an efficient pneumatic collector system employed to gather the glass dust. A necked-down throat on the end of a flexible metallic tube works very well, because it allows placing of the dust collector close to the machining area.

Setup chart for standardized molds

Use of automatic molds must be based on good design, adequate setup and good maintenance. Leading molders use carefully designed forms and charts for shop guidance

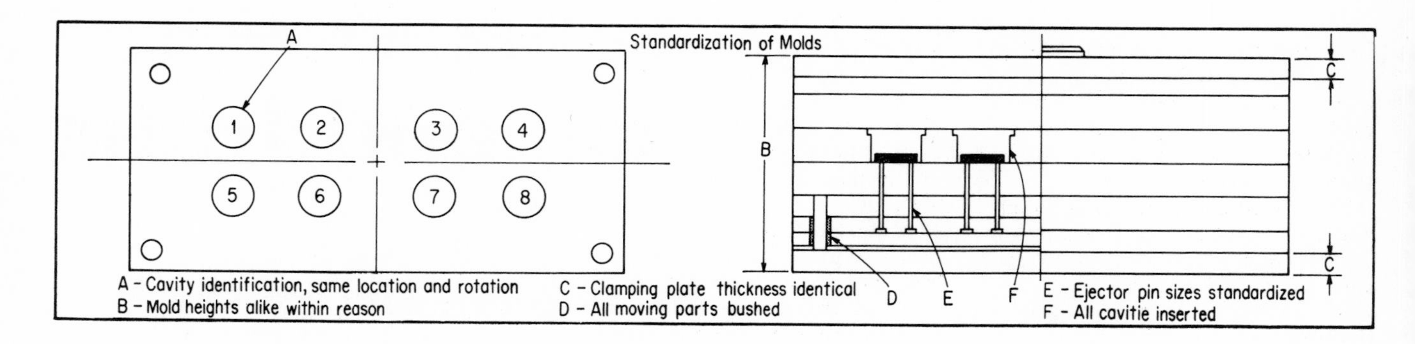

Standardized molds and setup charts aid efficient use of automatic molds (**AM**—Feb 9 '70, p100). This type of mold requires no operator to remove the part and runners (which fall free) but mold design, setup and maintenance are of critical importance. Mold standardization minimizes trouble in the shop and reduces costs. The setup chart leaves nothing to chance and provides an accurate guide to the operator. This chart saves much time getting the mold on cycle and producing consistent parts.

By Harold L Frohriep
Kirsch Co, Sturgis, Mich

Presented before the Regional Technical Conference, Society of Plastics Engineers, Cincinnati, Ohio, Oct 7-8, 1969

Setup chart

MOLD ________ MATERIAL ________

PRESS ________ DATE ________

NOZZLE ________
SCREW SPEED ________
PLASTICIZING DISTANCE ________
CUSHION SETTING ________
NOZZLE TEMP. % ________
ZONE 1 ________
ZONE 2 ________
ZONE 3 ________
INJECTION SPEED ________
OPEN CYCLE TIMER ________
HOLDING TIMER ________
COOLING TIMER ________
CLAMP STROKE ________
EJECTOR BARS ________
CLAMP TONNAGE TURNS ________
WATER, MOLD ________
WATER, CASTING ________
INJECTION PRESSURE ________
HOLDING PRESSURE ________
BACK PRESSURE ________

POWERSTAT SETTING

1. ____	8. ____	15. ____
2. ____	9. ____	16. ____
3. ____	10. ____	17. ____
4. ____	11. ____	18. ____
5. ____	12. ____	19. ____
6. ____	13. ____	20. ____
7. ____	14. ____	21. ____

REMARKS

Chapter 17

Steels and Irons

Cold-finished carbon steel bars

Growing use of steel bars of this quality make it mandatory for the shop man to know the varieties that are now met in production

Cold-finished carbon steel bars are quality steels prepared to exacting requirements. Compared with other materials, their close tolerances in a variety of standard shapes require less production work, result in less waste and scrap loss.

Cold-finished carbon steel bars are produced from hot rolled steel by several different cold finishing processes:

Cold drawing
Turning and polishing
Turning, grinding and polishing
Cold drawing, grinding and polishing
Cold rolling

Cold drawing consists of pulling a hot-rolled bar through a die with a cross section similar to that of the hot rolled bar but dimensionally smaller. The cold-drawn product is usually straightened and cut to length, but coils can be furnished with out straightening.

A reduction or 'draft' generally ranging from 1/64 in. to about ⅛ in. is used in cold drawing, depending on composition, shape and size of the bar to be drawn and the properies desired in the finished product. A 1/16-in. draft is the most common reduction. Carbon steels with a carbon range over 0.55 percent maximum are customarily annealed prior to cold drawing. There are many variations in the process, and some shapes or grades of steel may require several passes and intermediate annealings.

Turned and polished bar is the result of the removal of surface scale and other imperfections formed on hot-rolled round bars. Turning to the desired size is usually done on a lathe. The turned bar is then polished between steel burnishing rolls. The surface is smoother and brighter than that obtained by cold drawing.

Size tolerances are the same as for cold-drawn bars. Turning generally removes decarburization, seams, slivers and other surface imperfections; but the mechanical properties of the resulting bar are substantially the same as those of the hot-rolled steel used.

Turned and polished rounds are relatively free from residual stresses as compared to cold drawn bars in the non-stress relieved condition. Sizes made by turning and polishing commonly range from ¾ to 9-in. in diameter.

Turning, grinding and polishing includes a grinding operation before polishing. Turning and grinding removes decarburization, seams, slivers and other surface imperfections and the resulting bars are relatively free from residual stresses.

This type of cold-finished bar often is preferred when a high degree of size accuracy and surface smoothness are desired. Mechanical properties of the originally hot-rolled bar are essentially unchanged. Sizes commonly range from about ¾ to 9 in. in diameter.

Cold drawing, grinding and polishing produces a high quality surface on small diameter bars while offering mechanical properties similar to those of cold-drawn bars of the same grade.

Cold rolling is often used for flat bars. Hot-rolled bars of any suitable cross section are cleaned of scale, then repeatedly passed through rolls to reduce the cross section and improve surface finish of the bar.

Thermal treatments which may be specified for cold-finished bars include annealing, normalizing, bright annealing spheroidizing, quenching and tempering, and stress relieving.

Steel quality indicates many relative conditions, such as internal soundness and uniformity of chemical composition

Standard-quality bar is produced from special-quality hot-rolled carbon steel bars or special-quality hot-rolled wire rods.

Special-surface bars are produced from steel which has been processed with exacting control through all stages of manufacture, with special surface preparation to minimize the frequency and degree of seams and other surface imperfections.

Cold heading and cold forging quality applies to steel for applications involving severe cold plastic deformation by upsetting, heading or forging. The bars are produced from steel made by closely controlled steelmaking practices.

Cold extrusion quality bars are fully killed fine-grained steel, which also minimizes age hardening. A sound internal structure is necessary and surface requirements are most severe.

The size ranges of sections commonly produced as cold-finished carbon steel bars include:

Rounds, up to 9 in., inclusive
Squares, up to 4 in., inclusive
Hexagons, up to 3⅛ in., inclusive
Flats, ⅛ in. and over in specified thickness, up to 12 in. in specified width

Sizes and sections outside these limits and special bar sections also can be produced.

Following are common size increments for rounds, squares, and hexagons:

To 1 in., inclusive—by 64ths
Over 1 in. to 2 in., inclusive— by 32nds
Over 2 in.—by 16ths

Bars also can be produced in coils in the following sections and sizes:

Rounds: to 1 in., inclusive
Hexagons and squares: to ⅝ in., inclusive
Flats: to 9/16 x ⅝ in. or other sections having cross-sectional areas not more than 0.30 sq. in.

By F E Chepko, Staff Representative
Committee of Hot Rolled & Cold Finished Bar Producers,
American Iron & Steel Institute

Machinability ratings for cold-drawn bars

Cold drawing significantly improves the machinability of steel bars (carbon-steel rounds, squares, and hexagons) by making the chips shear more readily from the work

Cold-drawn material generally produces a better finish than either hot-rolled or hot-rolled annealed stock. Further improvements in finish can be obtained with free-machining steels, which have additions specifically designed to cut tool-workpiece friction, thereby reducing built-up edge and tool wear, and extending tool life by an important amount.

In the accompanying table, machinability for a given material is relative to that of B 1112 (assigned a rating of 100). And this rating is an averaged comparison over a range of machining operations, which might include turning, drilling, tapping, form-tool work, etc. Therefore the machinability rating provides an indication of how tool costs will vary, and serves as a starting point for a detailed investigation on a specific part. Reference must be made to plant experience and case histories when starting to apply these rating data.

Source: 'Cold Finished Steel Bar Handbook,' published by the Committee of Hot Rolled and Cold Finished Bar Producers, American Iron and Steel Institute, New York

Cold work and machinability

It is widely thought that cold working of steel improves machinability regardless of grade, size or shape. This is not always true. As a class, cold-drawn bars are more satisfactory for machining (especially on any kind of automatic machine) because the cold-drawn bars are straight and free from scale and have closer dimensional accuracy. In addition, since the cold-drawn surface is bright and smooth, it can frequently be used without further machining for the maximum diameter of the finished part.

Apart from these advantages, however, certain grades of steel are improved in machinability by cold work, others are affected very little, and some are affected adversely.

Cold work produces a considerable improvement in the machinability of low carbon steels, since the high ductility of these materials in the hot-rolled condition can be lowered markedly without raising strength too much.

In contrast, a steel such as 1144,

AISI No.	Relative Machinability Rating, %, based on B 1112 as 100%	FORMING			TURNING			DRILLING		
		Width of Cut. In.	Surface F.P.M.	Feed In./Rev.	Depth of Cut, In.	Surface F.P.M.	Feed In./Rev.	Size of Hole, In.	Surface F.P.M.	Feed In./Rev.
12 L 14	158	0.500	260	0.0033	0.125	260	0.0093	0.250	145	0.0060
		1.000	240	0.0028	0.250	240	0.0088	0.500	145	0.0066
		1.500	240	0.0027	0.375	235	0.0071	0.750	160	0.0077
		2.000	235	0.0020	0.500	230	0.0060	1.000	160	0.0088
		2.500	230	0.0016				1.250	165	0.0099
B 1113	136	0.500	225	0.0030	0.125	225	0.0085	0.250	125	0.0054
1213	136	1.000	210	0.0025	0.250	210	0.0080	0.500	125	0.0060
		1.500	210	0.0025	0.375	205	0.0065	0.750	140	0.0070
		2.000	205	0.0018	0.500	200	0.0055	1.000	140	0.0080
		2.500	200	0.0015				1.250	145	0.0090
B 1112	100	0.500	165	0.0025	0.125	165	0.0070	0.250	105	0.0045
1119	100	1.000	160	0.0020	0.250	160	0.0065	0.500	105	0.0050
1212	100	1.500	160	0.0018	0.375	155	0.0055	0.750	115	0.0060
		2.000	155	0.0015	0.500	150	0.0045	1.000	115	0.0070
		2.500	150	0.0012				1.250	120	0.0080
B 1111	94	0.500	155	0.0023	0.125	155	0.0066	0.250	99	0.0042
1211	94	1.000	150	0.0019	0.250	150	0.0061	0.500	99	0.0047
		1.500	150	0.0017	0.375	146	0.0052	0.750	108	0.0056
		2.000	146	0.0014	0.500	141	0.0042	1.000	108	0.0066
		2.500	141	0.0011				1.250	113	0.0076
1117	91	0.500	150	0.0022	0.125	150	0.0064	0.250	95	0.0041
1118	91	1.000	145	0.0018	0.250	145	0.0059	0.500	95	0.0045
		1.500	145	0.0016	0.375	141	0.0050	0.750	105	0.0055
		2.000	141	0.0014	0.500	136	0.0041	1.000	105	0.0064
		2.500	136	0.0011				1.250	119	0.0073

Machinability ratings for cold-drawn bars
[continued]

AISI No.	Relative Machinability Rating, %, based on B 1112 as 100%	FORMING			TURNING			DRILLING		
		Width of Cut, In.	Surface F.P.M.	Feed In./Rev.	Depth of Cut, In.	Surface F.P.M.	Feed In./Rev.	Size of Hole, In.	Surface F.P.M.	Feed In./Rev.
1144,	85	0.500	140	0.0021	0.125	140	0.0059	0.250	89	0.0040
Annealed		1.000	136	0.0017	0.250	136	0.0055	0.500	89	0.0045
		1.500	136	0.0015	0.375	132	0.0047	0.750	98	0.0055
		2.000	132	0.0013	0.500	127	0.0040	1.000	98	0.0064
		2.500	127	0.0010				1.250	102	0.0070
1141,	81	0.500	135	0.0020	0.125	135	0.0057	0.250	86	0.0040
Annealed		1.000	130	0.0017	0.250	130	0.0053	0.500	86	0.0045
		1.500	130	0.0015	0.375	127	0.0045	0.750	94	0.0054
		2.000	127	0.0012	0.500	122	0.0037	1.000	94	0.0063
		2.500	122	0.0010				1.250	98	0.0072
1016	78	0.500	130	0.0019	0.125	130	0.0055	0.250	82	0.0038
1018	78	1.000	125	0.0016	0.250	125	0.0051	0.500	82	0.0043
1022	78	1.500	125	0.0014	0.375	121	0.0043	0.750	90	0.0052
		2.000	121	0.0012	0.500	117	0.0035	1.000	90	0.0060
		2.500	117	0.0009				1.250	94	0.0068
1144	76	0.500	125	0.0019	0.125	125	0.0052	0.250	79	0.0037
		1.000	121	0.0015	0.250	121	0.0049	0.500	79	0.0042
		1.500	121	0.0014	0.375	117	0.0041	0.750	87	0.0050
		2.000	117	0.0011	0.500	113	0.0034	1.000	87	0.0058
		2.500	113	0.0009				1.250	91	0.0066
1020	72	0.500	120	0.0018	0.125	120	0.0050	0.250	76	0.0035
1137	72	1.000	115	0.0014	0.250	115	0.0047	0.500	76	0.0040
1045,	72	1.500	115	0.0013	0.375	112	0.0040	0.750	83	0.0047
Annealed		2.000	112	0.0011	0.500	108	0.0032	1.000	83	0.0055
		2.500	108	0.0009				1.250	86	0.0064
1035	70	0.500	115	0.0017	0.125	115	0.0049	0.250	73	0.0034
1141	70	1.000	112	0.0014	0.250	112	0.0045	0.500	73	0.0038
1050,	70	1.500	112	0.0013	0.375	108	0.0038	0.750	80	0.0045
Annealed		2.000	108	0.0011	0.500	105	0.0031	1.000	80	0.0053
		2.500	105	0.0008				1.250	84	0.0062
1040	64	0.500	105	0.0015	0.125	105	0.0044	0.250	67	0.0032
		1.000	101	0.0012	0.250	101	0.0041	0.500	67	0.0035
		1.500	101	0.0011	0.375	98	0.0034	0.750	73	0.0042
		2.000	98	0.0009	0.500	95	0.0028	1.000	73	0.0049
		2.500	95	0.0007				1.250	76	0.0056
1045	57	0.500	95	0.0014	0.125	95	0.0040	0.250	60	0.0028
		1.000	91	0.0012	0.250	91	0.0037	0.500	60	0.0031
		1.500	91	0.0010	0.375	88	0.0031	0.750	65	0.0037
		2.000	88	0.0009	0.500	85	0.0026	1.000	65	0.0044
		2.500	85	0.0007				1.250	68	0.0050
1050	54	0.500	90	0.0014	0.125	90	0.0038	0.250	57	0.0028
		1.000	87	0.0011	0.250	87	0.0035	0.500	57	0.0031
		1.500	87	0.0010	0.375	84	0.0030	0.750	62	0.0037
		2.000	84	0.0008	0.500	81	0.0024	1.000	62	0.0044
		2.500	81	0.0007				1.250	65	0.0050

which has an inherent low ductility because of its higher carbon content, shows little improvement in machining as a result of cold drawing. The increased hardness produced can actually be deleterious, so that it is sometimes helpful to give a stress-relief annealing treatment after cold work in order to reduce hardness somewhat.■

Alloy steel bars, 1

Extracted from the new Steel Products Manual section on alloy steel, semi-finished, hot rolled and cold finished bars, as published by the American Iron & Steel Institute, 150 East 42nd St, New York, N Y

Further improvements in steelmaking practices are reflected in the current standard grades of steel for alloy steel bars. The phosphorus content is now a minimum of 0.035%, instead of 0.040%, and the sulfur content is 0.040% minimum. Moreover, five grades of bar now have limits of 0.030% phosphorus and 0.035% sulfur, when produced by the basic open hearth and basic oxygen processes. These five grades are: axle shaft, bearing quality, cold heading quality, special cold heading quality and rifle barrel or shot quality.

In cold heading, the extra quality now provided means that plasticity is maintained longer and reheating between multiple operations is either eliminated or reduced.

The current list of standard alloy steels, as supplied in bar form, is smaller than formerly. The current grades represent those that have been actually ordered in quantity during the past several years. However, if a user has fabricated parts from some steel that is not on the current list or has been made to a wider spread for sulfur and phosphorus, he can get the ranges and limits that he is familiar with.

Numerical designations of grades

A four-numeral series designates alloy steels specified to chemical composition ranges. Five numerals designate certain types of alloy steels.

The last two digits of the four-numeral series are intended to indicate the approximate middle of the carbon range i.e., 20 represents a range of 0.18 to 0.23 per cent, and the last three digits of the five-numeral series are intended to indicate the approximate mean of the carbon range. It is necessary to interpolate some carbon ranges.

Series Designation	Type and approximate percentages of identifying elements
13xx	Manganese 1.75
40xx	Molybdenum 0.20 or 0.25
41xx	Chromium 0.50, 0.80 or 0.95, Molybdenum 0.12, 0.20 or 0.30
43xx	Nickel 1.83, Chromium 0.50 or 0.80, Molybdenum 0.25
44xx	Molybdenum 0.53
46xx	Nickel 0.85 or 1.83, Molybdenum 0.20 or 0.25
47xx	Nickel 1.05, Chromium 0.45, Molybdenum 0.20 or 0.35
48xx	Nickel 3.50, Molybdenum 0.25
50xx	Chromium 0.40
51xx	Chromium 0.80, 0.88, 0.93, 0.95 or 1.00
5xxxx	Carbon 1.04, Chromium 1.03 or 1.45
61xx	Chromium 0.60 or 0.95, Vanadium 0.13 or 0.15 min.
86xx	Nickel 0.55, Chromium 0.50, Molybdenum 0.20
87xx	Nickel 0.55, Chromium 0.50, Molybdenum 0.25
88xx	Nickel 0.55, Chromium 0.50, Molybdenum 0.35
92xx	Silicon 2.00
50Bxx	Chromium 0.28 or 0.50
51Bxx	Chromium 0.80
81Bxx	Nickel 0.30, Chromium 0.45, Molybdenum 0.12
94Bxx	Nickel 0.45, Chromium 0.40, Molybdenum 0.12

B denotes Boron Steel

Ladle chemical ranges and limits for alloy steels

AISI Number	Chemical Composition Ranges and Limits, per cent								Corresponding SAE Number*
	C	Mn	P Max.	S Max.	Si	Ni	Cr	Mo	
1330	0.28/0.33	1.60/1.90	0.035	0.040	0.20/0.35	—	—	—	1330
1335	0.33/0.38	1.60/1.90	0.035	0.040	0.20/0.35	—	—	—	1335
1340	0.38/0.43	1.60/1.90	0.035	0.040	0.20/0.35	—	—	—	1340
1345	0.43/0.48	1.60/1.90	0.035	0.040	0.20/0.35	—	—	—	1345
4012	0.09/0.14	0.75/1.00	0.035	0.040	0.20/0.35	—	—	0.15/0.25	4012
4023	0.20/0.25	0.70/0.90	0.035	0.040	0.20/0.35	—	—	0.20/0.30	4023
4024	0.20/0.25	0.70/0.90	0.035	0.035/0.050	0.20/0.35	—	—	0.20/0.30	4024
4027	0.25/0.30	0.70/0.90	0.035	0.040	0.20/0.35	—	—	0.20/0.30	4027
4028	0.25/0.30	0.70/0.90	0.035	0.035/0.050	0.20/0.35	—	—	0.20/0.30	4028
4037	0.35/0.40	0.70/0.90	0.035	0.040	0.20/0.35	—	—	0.20/0.30	4037
4047	0.45/0.50	0.70/0.90	0.035	0.040	0.20/0.35	—	—	0.20/0.30	4047

Alloy steel bars, 2

Ladle chemical ranges and limits for alloy steels (continued)

AISI Number	Chemical Composition Ranges and Limits, per cent								Corresponding SAE Number*
	C	Mn	P Max.	S Max.	Si	Ni	Cr	Mo	
4118	0.18/0.23	0.70/0.90	0.035	0.040	0.20/0.35	—	0.40/0.60	0.08/0.15	4118
4130	0.28/0.33	0.40/0.60	0.035	0.040	0.20/0.35	—	0.80/1.10	0.15/0.25	4130
4137	0.35/0.40	0.70/0.90	0.035	0.040	0.20/0.35	—	0.80/1.10	0.15/0.25	4137
4140	0.38/0.43	0.75/1.00	0.035	0.040	0.20/0.35	—	0.80/1.10	0.15/0.25	4140
4142	0.40/0.45	0.75/1.00	0.035	0.040	0.20/0.35	—	0.80/1.10	0.15/0.25	4142
4145	0.43/0.48	0.75/1.00	0.035	0.040	0.20/0.35	—	0.80/1.10	0.15/0.25	4145
4147	0.45/0.50	0.75/1.00	0.035	0.040	0.20/0.35	—	0.80/1.10	0.15/0.25	4147
4150	0.48/0.53	0.75/1.00	0.035	0.040	0.20/0.35	—	0.80/1.10	0.15/0.25	4150
4161	0.56/0.64	0.75/1.00	0.035	0.040	0.20/0.35	—	0.70/0.90	0.25/0.35	4161
4320	0.17/0.22	0.45/0.65	0.035	0.040	0.20/0.35	1.65/2.00	0.40/0.60	0.20/0.30	4320
4340	0.38/0.43	0.60/0.80	0.035	0.040	0.20/0.35	1.65/2.00	0.70/0.90	0.20/0.30	4340
E4340	0.38/0.43	0.65/0.85	0.025	0.025	0.20/0.35	1.65/2.00	0.70/0.90	0.20/0.30	E4340
4419	0.18/0.23	0.45/0.65	0.035	0.040	0.20/0.35	—	—	0.45/0.60	4419
4615	0.13/0.18	0.45/0.65	0.035	0.040	0.20/0.35	1.65/2.00	—	0.20/0.30	4615
4620	0.17/0.22	0.45/0.65	0.035	0.040	0.20/0.35	1.65/2.00	—	0.20/0.30	4620
4621	0.18/0.23	0.70/0.90	0.035	0.040	0.20/0.35	1.65/2.00	—	0.20/0.30	4621
4626	0.24/0.29	0.45/0.65	0.035	0.040	0.20/0.35	0.70/1.00	—	0.15/0.25	4626
4718	0.16/0.21	0.70/0.90	0.035	0.040	0.20/0.35	0.90/1.20	0.35/0.55	0.30/0.40	4718
4720	0.17/0.22	0.50/0.70	0.035	0.040	0.20/0.35	0.90/1.20	0.35/0.55	0.15/0.25	4720
4815	0.13/0.18	0.40/0.60	0.035	0.040	0.20/0.35	3.25/3.75	—	0.20/0.30	4815
4817	0.15/0.20	0.40/0.60	0.035	0.040	0.20/0.35	3.25/3.75	—	0.20/0.30	4817
4820	0.18/0.23	0.50/0.70	0.035	0.040	0.20/0.35	3.25/3.75	—	0.20/0.30	4820
5015	0.12/0.17	0.30/0.50	0.035	0.040	0.20/0.35	—	0.30/0.50	—	5015
5120	0.17/0.22	0.70/0.90	0.035	0.040	0.20/0.35	—	0.70/0.90	—	5120
5130	0.28/0.33	0.70/0.90	0.035	0.040	0.20/0.35	—	0.80/1.10	—	5130
5132	0.30/0.35	0.60/0.80	0.035	0.040	0.20/0.35	—	0.75/1.00	—	5132
5135	0.33/0.38	0.60/0.80	0.035	0.040	0.20/0.35	—	0.80/1.05	—	5135
5140	0.38/0.43	0.70/0.90	0.035	0.040	0.20/0.35	—	0.70/0.90	—	5140
5145	0.43/0.48	0.70/0.90	0.035	0.040	0.20/0.35	—	0.70/0.90	—	5145
5147	0.46/0.51	0.70/0.95	0.035	0.040	0.20/0.35	—	0.85/1.15	—	5147
5150	0.48/0.53	0.70/0.90	0.035	0.040	0.20/0.35	—	0.70/0.90	—	5150
5155	0.51/0.59	0.70/0.90	0.035	0.040	0.20/0.35	—	0.70/0.90	—	5155
5160	0.56/0.64	0.75/1.00	0.035	0.040	0.20/0.35	—	0.70/0.90	—	5160
E51100	0.98/1.10	0.25/0.45	0.025	0.025	0.20/0.35	—	0.90/1.15	—	51100
E52100	0.98/1.10	0.25/0.45	0.025	0.025	0.20/0.35	—	1.30/1.60	—	52100
								V	
6118	0.16/0.21	0.50/0.70	0.035	0.040	0.20/0.35	—	0.50/0.70	0.10/0.15	6118
6150	0.48/0.53	0.70/0.90	0.035	0.040	0.20/0.35	—	0.80/1.10	0.15 Min.	6150
								Mo	
8615	0.13/0.18	0.70/0.90	0.035	0.040	0.20/0.35	0.40/0.70	0.40/0.60	0.15/0.25	8615
8617	0.15/0.20	0.70/0.90	0.035	0.040	0.20/0.35	0.40/0.70	0.40/0.60	0.15/0.25	8617
8620	0.18/0.23	0.70/0.90	0.035	0.040	0.20/0.35	0.40/0.70	0.40/0.60	0.15/0.25	8620
8622	0.20/0.25	0.70/0.90	0.035	0.040	0.20/0.35	0.40/0.70	0.40/0.60	0.15/0.25	8622
8625	0.23/0.28	0.70/0.90	0.035	0.040	0.20/0.35	0.40/0.70	0.40/0.60	0.15/0.25	8625
8627	0.25/0.30	0.70/0.90	0.035	0.040	0.20/0.35	0.40/0.70	0.40/0.60	0.15/0.25	8627
8630	0.28/0.33	0.70/0.90	0.035	0.040	0.20/0.35	0.40/0.70	0.40/0.60	0.15/0.25	8630
8637	0.35/0.40	0.75/1.00	0.035	0.040	0.20/0.35	0.40/0.70	0.40/0.60	0.15/0.25	8637
8640	0.38/0.43	0.75/1.00	0.035	0.040	0.20/0.35	0.40/0.70	0.40/0.60	0.15/0.25	8640
8642	0.40/0.45	0.75/1.00	0.035	0.040	0.20/0.35	0.40/0.70	0.40/0.60	0.15/0.25	8642
8645	0.43/0.48	0.75/1.00	0.035	0.040	0.20/0.35	0.40/0.70	0.40/0.60	0.15/0.25	8645
8655	0.51/0.59	0.75/1.00	0.035	0.040	0.20/0.35	0.40/0.70	0.40/0.60	0.15/0.25	8655
8720	0.18/0.23	0.70/0.90	0.035	0.040	0.20/0.35	0.40/0.70	0.40/0.60	0.20/0.30	8720
8740	0.38/0.43	0.75/1.00	0.035	0.040	0.20/0.35	0.40/0.70	0.40/0.60	0.20/0.30	8740
8822	0.20/0.25	0.75/1.00	0.035	0.040	0.20/0.35	0.40/0.70	0.40/0.60	0.30/0.40	8822
9255	0.51/0.59	0.70/0.95	0.035	0.040	1.80/2.20	—	—	—	9255
9260	0.56/0.64	0.75/1.00	0.035	0.040	1.80/2.20	—	—	—	9260

Defects in barstock

Non-metallic inclusions in steel barstock are not immediate cause for alarm. Parts with stringers can be used safely in most applications

S E Tyson, senior metallurgist
Stainless Steel Dept,
Carpenter Steel Co, Reading, Pa

There are a variety of defects that can occur in barstock as shown in the sketch. Which of these are truly harmful and cause for rejection?

Inclusions

Parts are often discarded needlessly,* at great expense, when non-metallic inclusions are seen on a machined surface. There are two reasons: (1) either the effect of the inclusions is overestimated, or (2) the inclusions are mistaken for cracks.

Make no mistake about it—steel producers generally want, as much as steel users do, metals that are 100% free of non-metallic inclusions. Unfortunately, some non-metallics are inevitable in the final rolled product no matter how carefully the steel is made.

In the first place, the raw material used in steel manufacture contains non-metallic elements. Secondly, all air-melt furnaces are lined with non-metallic refractory materials. And finally, the slags used for metal refinement are non-metallic.

Non-metallics found in molten metal are small, round globules of various substances such as oxides, alumina or silicates. When the molten metal freezes in the ingot, these globules become entrapped and solidify. As the cast ingot is elongated by hot working, the globules or inclusions are also elongated and break up into various lengths called non-metallic stringers.

It is often difficult to distinguish between a stringer and a crack on a machined surface. On turned barstock, the tool will leave drag marks behind surface stringers and, therefore, highlight them. These linear indications are generally small but clearly discernible to the eye.

There is a big difference between the sizes of stringers and cracks. Non-metallic stringers in a good-quality air-melt steel usually are on the order of one thousandth in diameter and from a few thousandths to ½ in. long in length. In vacuum-melt steel, stringers are generally less than ⅛ in. long and the diameter is proportionately smaller.

In contrast, even fine cracks are wider than stringers and several inches long. When cracks are present, they usually run the entire length of a machined piece. A cross-section view generally shows that the cracks run deep from surface to center.

Parts with cracks cannot be used. Parts with some non-metallic stringers *can be used safely in the great majority of cases.*

It should be understood that the discussion of defects up to this point concerns only the internal quality of metals, or that which concerns a producer of parts machined from bar stock. There are seams and laps which are sometimes confused with cracks.

During hot working, longitudinal folds can occur, leading to the formation of seams (perpendicular to surface) and laps angular to (surface). These defects usually measure from a few thousandths to 1/32 in. deep in proportion to bar diameter. Such imperfections are not found on ground bars, only on hot rolled or cold drawn bars.

Cold-drawn shapes usually require no clean-up of surface seams for most applications, such as valve bodies. However, hot drawn or cold drawn bars should be machined in the case of highly stressed parts, because seams might propagate cracks.

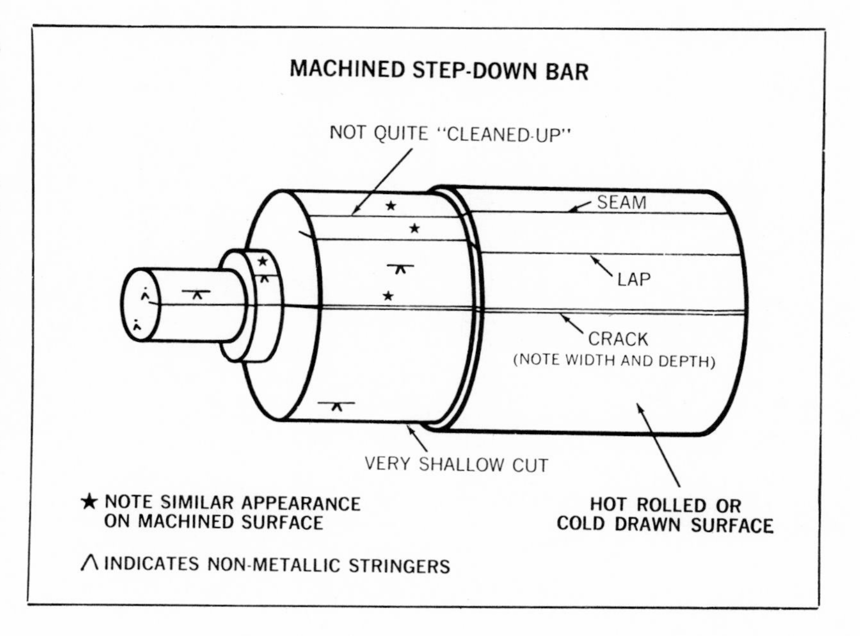

*We recognize that some applications demand the freedom from non-metallics obtainable by vacuum melting methods.

Machining data for stainless steels

Operation	Tool	Parameter	Size		410	416	416 Project '70	No. 5-F	420	420F	430	430F	431	440A
TURNING AND FORMING	HIGH SPEED TOOLS	DEPTH OF CUT	.050	SFM	95	150	165	170	87	100	95	150	75	76
				FEED	.0051	.0058	.0060	.0062	.0053	.0055	.0050	.0058	.0047	.0045
			.250	SFM	86	145	160	165	80	95	86	145	75	76
				FEED	.0040	.0047	.0051	.0051	.0040	.0045	.0038	.0047	.0040	.0040
			.500	SFM	81	135	150	155	74	90	81	135	70	69
				FEED	.0029	.0034	.0045	.0045	.0029	.0037	.0027	.0034	.0030	.0032
	CAST ALLOY TOOLS	DEPTH OF CUT	.050	SFM	143	225	240	250	130	150	143	225	113	114
				FEED	.0051	.0058	.0060	.0062	.0053	.0055	.0050	.0058	.0047	.0045
			.250	SFM	130	217	235	240	120	143	130	217	113	114
				FEED	.0040	.0047	.0051	.0051	.0040	.0045	.0038	.0047	.0040	.0040
			.500	SFM	122	200	225	230	110	135	122	200	105	103
				FEED	.0029	.0034	.0045	.0045	.0029	.0037	.0027	.0034	.0030	.0032
	CARBIDE TOOLS	DEPTH OF CUT	.050	SFM	275	285	295	315	275	285	275	285	250	215
				FEED	0.10	.012	.015	.015	.010	.010	.010	.012	.010	.010
			.250	SFM	200	215	225	245	200	215	200	215	190	175
				FEED	.020	.025	.030	.030	.020	.020	.020	.025	.020	.020
			.500	SFM	175	200	215	230	175	200	175	200	175	160
				FEED	.030	.040	.045	.045	.030	.030	.030	.040	.030	.030
	CUT OFF TOOLS	TOOL WIDTH	1/16"	SFM	90	150	165	170	87	100	90	150	75	77
				FEED	.0011	.0015	.0016	.0019	.0012	.0013	.0011	.0015	.0011	.0011
			1/8"	SFM	97	154	170	175	89	103	97	154	77	78
				FEED	.0013	.0020	.0021	.0025	.0014	.0015	.0013	.0020	.0013	.0013
			3/16"	SFM	100	157	173	178	91	105	100	157	78	79
				FEED	.0015	.0020	.0023	.0027	.0014	.0017	.0015	.0020	.0013	.0013
			1/4"	SFM	100	159	175	180	92	106	100	159	79	81
				FEED	.0019	.0024	.0027	.0030	.0018	.0020	.0019	.0024	.0015	.0015
	FORM TOOLS	TOOL WIDTH	1/2"	SFM	97	154	170	175	89	103	97	154	77	78
				FEED	.0015	.0018	.0021	.0025	.0015	.0017	.0015	.0018	.0013	.0013
			1"	SFM	92	145	160	165	84	97	92	145	73	75
				FEED	.0011	.0014	.0017	.0021	.0012	.0013	.0011	.0014	.0011	.0011
			1½"	SFM	92	145	160	165	84	97	92	145	73	75
				FEED	.0010	.0013	.0015	.0019	.0011	.0011	.0010	.0013	.0010	.0010
			2"	SFM	89	141	155	160	82	94	89	141	70	71
				FEED	.0009	.0011	.0013	.0018	.0010	.0009	.0009	.0011	.0008	.0008
			2½"	SFM	85	135	148	152	78	90	85	135	67	68
				FEED	.0007	.0009	.0012	.0017	.0008	.0008	.0007	.0009	.0007	.0007
DRILLING		DRILL DIAM.	1/16"	SFM	58	91	100	103	53	61	58	91	46	46
				FEED	.0016	.0018	.0020	.0024	.0014	.0017	.0016	.0018	.0014	.0013
			1/8"	SFM	62	97	107	110	56	65	62	97	49	49
				FEED	.0023	.0027	.0030	.0035	.0020	.0025	.0023	.0027	.0020	.0020
			1/4"	SFM	63	100	110	113	58	67	63	100	50	50
				FEED	.0030	.0035	.0036	.0041	.0028	.0032	.0030	.0035	.0028	.0028
			3/8"	SFM	63	100	110	113	58	67	63	100	50	50
				FEED	.0032	.0037	.0040	.0045	.0030	.0035	.0032	.0037	.0030	.0030
			1/2"	SFM	63	100	110	113	58	67	63	100	50	50
				FEED	.0035	.0040	.0043	.0048	.0033	.0037	.0035	.0040	.0033	.0033
			3/4"	SFM	66	105	115	119	61	70	66	105	52	52
				FEED	.0040	.0047	.0050	.0055	.0037	.0042	.0040	.0047	.0037	.0037
			1"	SFM	66	105	115	119	61	70	66	105	52	52
				FEED	.0047	.0055	.0057	.0061	.0045	.0050	.0047	.0055	.0043	.0043
			1¼"	SFM	68	110	120	124	63	73	68	110	55	55
				FEED	.0050	.0064	.0065	.0069	.0049	.0055	.0050	.0064	.0045	.0045
TAPPING		TPI	3-7½	SFM	11	15	15	16	11	11	11	15	6	6
			8-15	SFM	15	22	22	24	15	15	15	22	7	7
			16-24	SFM	19	30	30	32	19	19	19	30	11	11
			OVER 25	SFM	22	35	35	37	22	22	22	35	15	15
THREADING		TPI	3-7½	SFM	15	20	20	22	15	15	15	20	8	8
			8-15	SFM	20	30	30	32	20	20	20	30	10	10
			OVER 16	SFM	25	40	40	42	25	25	25	40	15	15
MILLING				SFM	70-105	100-125	105-135	110-140	35-70	65-95	70-105	110-135	60-80	60-80
				FEED	.003"-.005"	.003"-.006"	.003"-.006"	.003"-.006"	.003"-.005"	.003"-.005"	.003"-.005"	.003"-.006"	.003"-.005"	.003"-.005"
BROACHING				SFM	10-20	15-25	15-30	15-35	8-15	10-20	10-20	15-25	8-15	8-15
				FEED	.001"-.005"	.001"-.005"	.001"-.005"	.001"-.005"	.001"-.005"	.001"-.005"	.001"-.005"	.001"-.005"	.001"-.005"	.001"-.005"
REAMING	UNDER 1/2"		SIZING	SFM	83	132	136	140	76	87	83	132	66	67
				FEED	.0038	.0050	.0052	.0057	.0038	.0042	.0038	.0050	.0030	.0030
			FINISHING	SFM	40	60	60	65	40	40	40	60	35	35
				FEED	.0035	.0035	.0038	.0042	.0035	.0035	.0035	.0035	.0030	.0030
	OVER 1/2"		SIZING	SFM	83	132	136	140	76	87	83	132	66	67
				FEED	.0054	.0072	.0075	.0080	.0054	.0060	.0054	.0072	.0045	.0045
			FINISHING	SFM	40	60	60	65	40	40	40	60	35	35
				FEED	.0035	.0035	.0038	.0042	.0035	.0035	.0035	.0035	.0030	.0030

Machining data for stainless steels [continued]

Operation	Tool	Variable	Size	Unit	440 B	440 C	440 F	443 446	302 304	303 Project '70	304 Project '70	316 Project '70	321 347	No. 20Cb-3
TURNING AND FORMING	HIGH SPEED TOOLS	DEPTH OF CUT	.050	SFM	76	65	76	83	80	115	90	90	80	70
				FEED	.0045	.0043	.0046	.0051	.0051	.0060	.0055	.0055	.0051	.0040
			.250	SFM	76	65	76	80	75	105	85	85	75	65
				FEED	.0040	.0038	.0042	.0040	.0046	.0055	.0050	.0050	.0046	.0030
			.500	SFM	69	60	69	74	70	95	80	80	70	60
				FEED	.0032	.0029	.0032	.0029	.0037	.0043	.0042	.0042	.0037	.0025
	CAST ALLOY TOOLS	DEPTH OF CUT	.050	SFM	114	97	114	125	120	175	135	135	120	110
				FEED	.0045	.0043	.0046	.0051	.0051	.0060	.0055	.0055	.0051	.0040
			.250	SFM	114	97	114	120	110	155	127	127	110	100
				FEED	.0040	.0038	.0042	.0040	.0046	.0055	.0050	.0050	.0046	.0030
			.500	SFM	103	90	103	101	107	140	120	120	107	93
				FEED	.0032	.0029	.0032	.0029	.0037	.0043	.0042	.0042	.0037	.0025
	CARBIDE TOOLS	DEPTH OF CUT	.050	SFM	215	215	215	275	215	285	250	250	200	200
				FEED	.010	.008	.010	.010	.010	.010	.010	.010	.010	.010
			.250	SFM	175	175	175	200	175	215	190	190	165	165
				FEED	.020	.018	.020	.020	.020	.020	.020	.020	.020	.020
			.500	SFM	160	160	160	175	160	200	170	170	150	150
				FEED	.030	.025	.030	.030	.025	.025	.025	.025	.025	.025
	CUT OFF TOOLS	TOOL WIDTH	1/16"	SFM	70	65	77	83	80	110	90	90	80	70
				FEED	.0010	.0010	.0011	.0011	.0013	.0017	.0015	.0015	.0013	.0010
			1/8"	SFM	72	67	78	85	82	113	92	92	82	72
				FEED	.0012	.0012	.0013	.0013	.0016	.0021	.0018	.0018	.0016	.0014
			3/16"	SFM	73	68	79	87	84	115	94	94	84	73
				FEED	.0012	.0012	.0013	.0015	.0016	.0021	.0018	.0018	.0016	.0014
			1/4"	SFM	74	69	81	88	85	117	96	96	85	74
				FEED	.0016	.0016	.0015	.0019	.0019	.0025	.0021	.0021	.0019	.0018
	FORM TOOLS	TOOL WIDTH	1/2"	SFM	72	67	78	85	82	113	92	92	82	72
				FEED	.0012	.0012	.0013	.0015	.0016	.0021	.0018	.0018	.0016	.0012
			1"	SFM	68	63	75	80	78	107	88	88	78	68
				FEED	.0010	.0010	.0011	.0011	.0013	.0017	.0015	.0015	.0013	.0010
			1½"	SFM	68	63	75	80	78	103	88	88	78	68
				FEED	.0009	.0009	.0010	.0010	.0012	.0015	.0013	.0013	.0012	.0010
			2"	SFM	66	61	71	78	75	103	85	85	75	66
				FEED	.0007	.0007	.0008	.0009	.0010	.0013	.0011	.0011	.0010	.0009
			2½"	SFM	63	58	68	75	72	100	81	81	72	63
				FEED	.0006	.0006	.0007	.0007	.0008	.0010	.0009	.0009	.0008	.0008
DRILLING		DRILL DIAM.	1/16"	SFM	42	40	46	50	49	67	55	55	49	42
				FEED	.0012	.0012	.0013	.0016	.0013	.0020	.0014	.0014	.0013	.0010
			1/8"	SFM	45	42	49	54	52	71	59	59	52	45
				FEED	.0020	.0020	.0020	.0023	.0020	.0030	.0023	.0023	.0020	.0018
			1/4"	SFM	47	43	50	55	54	73	60	60	54	47
				FEED	.0026	.0026	.0028	.0030	.0033	.0040	.0035	.0035	.0033	.0030
			3/8"	SFM	47	43	50	55	54	73	60	60	54	47
				FEED	.0030	.0030	.0030	.0032	.0035	.0045	.0037	.0037	.0035	.0032
			1/2"	SFM	47	43	50	55	54	73	60	60	54	47
				FEED	.0032	.0032	.0033	.0035	.0037	.0045	.0039	.0039	.0037	.0035
			3/4"	SFM	49	45	52	58	56	77	63	63	56	49
				FEED	.0036	.0036	.0037	.0040	.0044	.0055	.0047	.0047	.0044	.0040
			1"	SFM	49	45	52	58	56	77	63	63	56	49
				FEED	.0041	.0041	.0043	.0047	.0051	.0064	.0055	.0055	.0051	.0048
			1¼"	SFM	51	47	55	60	59	80	66	66	59	51
				FEED	.0044	.0044	.0045	.0050	.0058	.0070	.0060	.0060	.0058	.0054
TAPPING		TPI	3-7½	SFM	6	6	6	11	6	11	7	7	6	6
			8-15	SFM	7	7	7	15	7	15	8	8	7	7
			16-24	SFM	11	11	11	19	11	19	12	12	11	11
			OVER 25	SFM	15	15	15	22	15	22	15	15	15	15
THREADING		TPI	3-7½	SFM	8	8	8	15	8	15	10	10	8	8
			8-15	SFM	10	10	10	20	10	20	15	15	10	10
			OVER 16	SFM	15	15	15	25	15	25	20	20	15	15
MILLING				SFM	40-75	35-70	65-80	65-95	35-65	100-130	40-85	40-85	35-65	35-65
				FEED	.002"-.005"	.002"-.005"	.003"-.005"	.003"-.005"	.003"-.005"	.003"-.005"	.003"-.005"	.003"-.005"	.003"-.005"	.003"-.005"
BROACHING				SFM	8-12	8-12	8-15	10-20	8-15	10-20	8-15	8-15	8-15	8-15
				FEED	.001"-.005"	.001"-.005"	.001"-.005"	.001"-.005"	.001"-.005"	.001"-.005"	.001"-.005"	.001"-.005"	.001"-.005"	.001"-.005"
REAMING	UNDER ½"	SIZING		SFM	62	57	67	73	70	90	79	79	70	61
				FEED	.0030	.0030	.0030	.0035	0050	.0060	.0053	.0053	.0050	.0045
		FINISHING		SFM	35	35	35	35	35	40	35	35	35	30
				FEED	.0030	.0030	.0030	.0030	.0030	.0035	.0030	.0030	.0030	.0030
	OVER ½"	SIZING		SFM	62	57	67	73	70	90	79	79	70	61
				FEED	.0045	.0045	.0045	.0050	.0070	.0085	.0073	.0073	.0070	.0065
		FINISHING		SFM	35	35	35	35	35	40	35	35	35	30
				FEED	.0030	.0030	.0030	.0030	.0030	.0035	.0030	.0030	.0030	.0030

Tool steels for press forming

Recent revision of the list of commonly used tool steels focuses attention on the comparative merits of these materials for various uses in pressforming tools

Tool steel requirements for metal stamping operations are influenced by tool design, part requirement, metal thickness and metal condition, plus production requirements.

Cold-work steels predominate. They are produced in a wide range of chemical analyses, resulting in several combinations of properties. Primary requirements are hardness and wear resistance. All cold-work grades are non-deforming in nature. When properly heat treated, they normally have sufficient toughness to resist premature chipping and spalling. Special requirements may involve the use of shock-resisting grades. In situations where heat is applied to improve formability of the work, it may be desirable to use tool-steel grades from the hot-work category.

Hardness and wear resistance are obtained from the high carbon content (0.70% to 2.00%). Normal working hardness range is 61-57 Rockwell C. These properties are boosted with the addition of carbide-forming alloys. Alloy content can range from a low of under 1% to as high as 18%. All grades are furnished in the annealed condition and are generally available in both the hot-rolled and pre-finished (decarb-free) condition.

Tool steels for press-forming applications can be listed in several general classifications. The quenching medium required for heat treating is often implied by the designation given. The categories include water hardening "W," oil hardening "O," air hardening "A," high-carbon high-chromium "D," shock resisting "S" and occasionally hot work "H."

Water-hardening steels

These grades are principally high-carbon compositions. In many instances small amounts of alloying elements are added for additional carbide formation, grain refinement, and hardenability control. These grades develop a hard-case/tough-core condition when treated. This characteristic lends less susceptibility to cracking when welding under less than ideal conditions.

Water-hardening steels have the poorest resistance to distortion in heat treatment of the four categories of cold-work tool steels. Machinability is

Comparison of tool steels for press forming

AISI Type	Tempered Hardness (Rc)	Machinability	Toughness	Wear Resistance	Distortion in Heat Treatment
Water Hardening Tool Steels					
W1	64/50	Highest	High	Low to medium	High
W2	64/50	Highest	High	Low to medium	High
Oil Hardening Cold Work Tool Steels					
O1	62/57	High	Medium	Medium	Very low
O2	62/57	High	Medium	Medium	Very low
O6	63/58	Very high†	Medium	Medium	Very low
O7	64/58	High	Medium	Medium	Very low*
Medium Alloy Air Hardening Cold Work Tool Steels					
A2	62/57	Medium	Medium	High	Lowest
A3	65/57	Medium	Medium	Very high	Lowest
A4	62/54	Low to medium	Medium	Medium to high	Lowest
A6	60/54	Low to medium	Medium	Medium to high	Lowest
A7	67/57	Low	Low	Highest	Lowest
A8	60/50	Medium	High	Medium to high	Lowest
A9	56/35	Medium	High	Medium to high	Lowest
A10	62/55	Medium to high†	Medium	High	Lowest
High Carbon High Chromium Cold Work Tool Steels					
D2	61/54	Low	Low	High to very high	Lowest
D3	61/54	Low	Low	Very high	Very low
D4	61/54	Low	Low	Very high	Lowest
D5	61/54	Low	Low	High to very high	Lowest
D7	65/58	Low	Low	Highest	Lowest
Shock Resisting Tool Steels					
S1	58/40	Medium	Very high	Low to medium	Medium
S2	60/50	Medium to high	Highest	Low to medium	High
S5	60/50	Medium to high	Highest	Low to medium	Medium
S7	57/45	Medium	Very high	Low to medium	Lowest**
Hot Work Tool Steels					
H12	55/38	Medium to high	Very high	Medium	Very low
H21	54/36	Medium	High	Medium to high	Low**

†Contains free graphite in the microstructure to improve machinability

*Oil quench

**Air quench

Source: 'Tool Steel Trends' published by the American Iron & Steel Institute, New York

the best. Water-quenching grades are susceptible to cracking when sharp corners and large (mass) section changes are present. Special attention should be given to tool design to insure safety in heat treatment.

Tools for all press-forming operations can be built of water-hardening tool steels. The Type AISI W2 predominates. But, for maximum life involving severe wear, it is recommended that grades in the "O," "A" or "D" series be considered.

Oil-hardening tool steels

Hardenability is increased by alloy additions so that satisfactory hardening response may be achieved when oil quenching. Through-hardening is generally not obtained in sizes over 2 to 3 in. in cross-section. Manganese is an alloying element that improves hardenability. Additions of chromium and tungsten increase hardenability and also promote the formation of hard carbides for extra wear resistance. The chrome-tungsten, oil-hardening type (O1) is the most widely used, general-purpose tool steel. Distortion (growth) resulting from heat treatment is less than that for water-hardening grades. Machinability is about the same as the "W" grades.

Oil-hardening steels will perform all press-forming operations well under most conditions.

Air-hardening tool steels

The air-hardening steels provide additional wear resistance with certain desirable heat-treatment characteristics. They offer minimum distortion (growth), through-hardening in sections up to 6 in. thick, plus added safety in heat treatment.

Two steels in this category are the so-called low-temperature, air-hardening grades, A4 and A6. They provide minimum distortion in heat treatment with the "A" category, and they are often produced with free-machining additives that make them the easiest to machine in the air-hardening category.

Air-hardening tool steels are used widely in all press-forming tools. The 5% chromium type (AISI A2) is the most popular workhorse.

High-carbon high-chrome steels

These steels provide the ultimate in abrasion resistance for metal stamping operations. Heavily fortified with high percentages of alloys, they offer deep hardenability plus minimum distortion in heat treatment.

The most widely used grade is Type AISI D2, which is hardened by air quenching. Most others require liquid quenching for maximum development of properties.

The high-carbon, high-chromium grades are known primarily for their excellent wear resistance and high compressive strength. They are not recommended in applications involving shock.

The machinability characteristics of D2 are the poorest of all cold-work tool steels. Their sensitivity to stresses induced by a grinding wheel dictates that extreme care be used grinding.

Shock-resisting tool steels

Under certain conditions of press forming, hardness and wear resistance are sacrificed and greater emphasis is placed on shock resistance. In these instances, it is possible to use grades classified in the shock-resisting or "S" category. They are medium-carbon grades containing varying additions of silicon, chromium, molybdenum and/or tungsten. Their normal working hardness is Rockwell C 58/54. A carburized case (61/58 Rockwell C) is sometimes applied to increase wear resistance. Shock-resisting tool steels possess medium hot-work characteristics. Under extremely hot press-forming conditions, it may be advisable to use one of the true hot-work analyses.

The commonly available AISI shock-resisting grades used in cold-work applications are S1, S2, S5, and S7. The quenching media for heat treating include water (S2), oil (S1, S5), and air (S7).

Machinability is similar to that for oil- and air-hardening grades. Distortion is governed largely by the quenching medium.

Hot-work tool steels

Hot-work grades are low-carbon, high-alloy compositions. Their primary characteristics are resistance to softening and good toughness, plus resistance to heat checking when exposed to high-temperature conditions.

Though many hot-work grades are available, either H12 or H21 will do well for most press-forming operations involving high heat.

Machinability and distortion in heat treatment are similar to those of steels in the air-hardening category.

The water-hardening grades provide adequate service life for many applications and offer the desirable features of easy machinability and weldability. Increasing wear resistance is obtained from the oil- and air-hardening grades. They also provide increased safety in hardening.

For maximum abrasion resistance under severe long-run applications, such as progressive dies, it is advisable to use high-carbon, high-chromium steels. Maximum toughness, as might be required in coining or heavy shear operations, often requires shock-resisting tool steels. Hot-work steels offer resistance to tempering. They are occasionally used in operations where parts are formed after being heated to elevated temperatures.

Hot-work tool steels, 1

Selection of a hot-work tool steel for a specific use requires careful appraisal of the need for each of seven characteristics versus service conditions

Three important segments of metalworking activity—forging, extrusion, and die-casting—depend for existence upon hot-work tool steels. By this we mean medium and highly alloyed ferrous materials used as dies and support tooling for service at temperatures in excess of 500 F.

The steels most widely used for this purpose are identified by letter symbol 'H,' followed by a suffix number. An example is H-11. In this discussion hot-work tool steels in general will be considered, followed by the die-casting applications in particular.

The characteristics desired in a good hot-work tool steel are:

1. Retention of hardness and strength at elevated temperature
2. Resistance to heat checking
3. Resistance to erosion or 'washing'
4. Resistance to heat treatment deformation
5. Resistance to shock
6. Machinability
7. Low thermal expansion coefficient
8. High thermal conductivity

The first characteristic is obtained by varying additions of tungsten, molybdenum and vanadium. These are strong complex carbide-forming elements. Chromium, silicon and cobalt also contribute to high temperature strength but to a lesser extent than the previous elements.

Heat checking or 'craze cracking' is associated with thermal fatigue. Properties that oppose this tendency are oxidation resistance and good fatigue and hot strength.

Erosion resistance is usually obtained at some sacrifice of toughness by higher hardness and temper resistance at elevated temperatures.

Maximum freedom from distortion in heat treatment is essential for intricate sections used in today's tooling. Air-hardening steels are the answer if this is a problem.

Shock resistance is synonomous with toughness and is aided by keeping the carbon content as low as practical.

Machinability is related to material uniformity and annealing effectiveness.

Table 1 shows the nominal composition of the principal grades of hot-work tool steel in use. In most cases the carbon content falls between 0.25 and 0.65%, and is usually in the 0.30 to 0.40% range to assure adequate ductility.

The manganese content of the various steels does not vary widely, and silicon is usually either 0.30 or 1.00%. All of the steels contain chromium in amounts from 3.25% to 12.0%. The H-11, 12 and 13 types (Group II)
[Continued on page 243]

By J J McCarthy

Asst metallurgical engineer
Staff of V-P Operations
Bethlehem Steel Corp, Bethlehem, Pa

Nominal composition of hot-work tool steels

AISI type	C	Mn	Si	Cr	Mo	W	V
Group I—3 to 4% chromium die steels							
1.00% C	0.95	0.30	0.30	4.00			
S7	0.50	0.70	0.30	3.25	1.40		
Group II—Chromium-molybdenum steels							
H-10	0.40	0.55	1.00	3.25	2.50		0.40
H-11	0.35	0.30	1.00	5.00	1.50		0.40
H-12	0.35	0.30	1.00	5.00	1.50	1.50	0.40
H-13	0.35	0.30	1.00	5.00	1.50		1.00
Group III—Chromium-tungsten steels							
H-14	0.40	0.30	1.00	5.00		5.00	
H-16	0.55	0.60	0.90	7.00	(Co)	7.00	
H-19	0.40	0.30	0.30	4.25	4.25	4.25	2.00
H-23	0.30	0.30	0.50	12.00		12.00	
Group IV—Tungsten steels							
H-20	0.35	0.30	0.30	2.00		9.00	
H-21	0.35	0.30	0.30	3.50		9.00	
H-22	0.35	0.30	0.30	2.00		11.00	
H-24	0.45	0.30	0.30	3.00		15.00	
H-25	0.25	0.30	0.30	4.00		15.00	
H-26	0.50	0.30	0.30	4.00		18.00	1.00
Group V—Molybdenum steels							
H-15	0.35	0.30	0.40	3.75	6.00	1.00	.75
H-43	0.55	0.30	0.30	4.00	8.00		2.00
H-41	0.65	0.30	0.30	4.00	8.00	1.50	1.00
H-42	0.60	0.30	0.30	4.00	5.00	6.00	2.00
Steels for die-casting dies							
P-20	0.35	0.75	0.45	1.25	0.40		
S-7	0.50	0.70	0.30	3.25	1.40		
H-11	0.35	0.30	1.00	5.00	1.50		0.40
H-13	0.35	0.30	1.00	5.00	1.50		1.00
H-14	0.40	0.30	1.00	5.00		5.00	
H-21	0.35	0.30	0.30	3.50		9.00	
H-23	0.30	0.30	0.50	12.00		12.00	

Applications of steels for die-casting dies

P-20: Zinc, tin, lead
S-7: Zinc
H-11: Aluminum, magnesium
H-13: Aluminum, magnesium
H-14: Aluminum, magnesium
H-21: Copper-base alloys
H-23: Aluminum, copper-base

Hot-work tool steels [Continued]

[Continued from page 242]
differ only in tungsten and vanadium contents. Group III steels contain equal amounts of chromium and tungsten, with one of the four containing cobalt and vanadium also. The primary alloy of Group IV is tungsten, varying in amount from 9 to 18%. Actually the H-26 type in this group is a lower carbon (0.50%C) variety of the regular 18-4-1, or T-1 type of high speed steel.

The last group, V, is designated the molybdenum group, containing from 5 to 8%. Three of these steels, H-41, H-42 and H-43 are lower-carbon versions of M-1, M-2 and M-10 high speed steels. From a tonnage standpoint, approximately 92.5% of the H steels used is confined to seven grades—the four of Group II and the first three of Group III.

Compression applications

Group I steels in the higher-carbon types are useful in compressive type applications where temperatures do not exceed 850 F such as hot-rivet headers, gripper dies, hot bending, etc. The medium carbon S-7 type is used advantageously where service temperatures do not exceed 1000 F, as in hot headers, rivet sets, die-casting dies and shear blades.

The Group II steels are characterized by high hardenability and toughness and excellent fatigue strength, and are employed as forging dies, die casting dies, punches, piercers, mandrels, shear blades, and hot extrusion tooling where the parts may be subjected to temperatures to 1000 F.

Group III steels have somewhat greater hot hardness and wear resistance than the Group II type, but they do not possess similar toughness. The H-14 type is the toughest steel in the group and is used as extrusion tooling—dies, liners and dummy blocks for brass and aluminum, hot punches, aluminum die-casting dies and gripper dies. The H-16 variety is used as hot heading dies and punches, shear blades, gripper dies and such non-tool steel application as balls and ball races, valve parts, etc. Applications for H-19 are similar to those of H-14. The H-23 type, with its higher tungsten content, is the most expensive and is used as extrusion and die-casting dies for brass and for brass and bronze permanent molds.

Hot hardness

The highest hot hardness is found in Group IV tungsten steels. They are not as resistant to thermal shock as Groups I and II, and for that reason they should not be rapidly cooled with water in operation. These steels are used where the maximum hot strength and resistance to tempering are vital, such as in extrusion dies for brass, bronze and steel, hot press, punch and swaging dies and dummy blocks. The last group, Group V molybdenum steels, have hot hardness properties only slightly lower than the tungsten-type steels and their wear resistance and toughness are comparable. Thus their usage is basically similar to that of Group IV materials:

The melting points of die casting alloys influence the choice of die materials. Zinc alloys have good fluidity at 750 to 800 F; aluminum and magnesium alloys both in the vicinity of 1200 F, while copper alloys melt at 1565 F and higher. The last range multiplies die casting problems tremendously.

Of seven commonly used steels for die-casting dies, P-20 is a low-alloy steel normally furnished in the heat-treated condition to approximately 300 Brinell. In addition to its extreme popularity as a zinc die-casting die steel, it enjoys wide usage for injection-type plastic molds. The other steels must be heat treated after finishing the die impression, the usual range being Rockwell C 44 to 48 except that grades used for copper-base alloys are treated to Rc 38/45. Since the steels are air-hardening, deformation as a result of heat treatment is at a minimum.

Vacuum degassing of several of the hot work grades results in steel with a lower content of hydrogen and oxygen plus improved cleanliness. The ultimate effect in some cases has been astonishing. To mention one case history—a vacuum degassed low carbon, multiple-alloy grade used for ram heads in tube extrusion has a service life that has averaged 3 to 1 better than the air-melt equivalent.

Proper heat treatment of the die cannot be overemphasized. Seemingly minor matters such as careful preheat to minimize distortion, proper austenitizing temperatures and times for carbide solution, plus an adequate cooling rate will assure uniform transformation to martensite. The value of multiple tempering has been established. Prior to treatment all mill decarburization should be removed, and during heat treatment the surface chemistry must be maintained by copper plating and packing in carbonaceous material, salt bath heating and the use of various protective atmospheres. More recently vacuum heat treating has expanded, and several excellent furnaces are available.

Hot forming

Ausforming involves the mechanical deformation of austenite at about 1000 F by rolling, forging, extruding or other means, followed by conventional quenching and tempering. The effect is to develop higher strength and wear resistance while maintaining normal ductility and toughness levels. The process appears most applicable to simple shapes, such as pins and punches.

Longer tool life is reported from maraging steels in various hot work applications, but the cost is considerably higher than the 5% chromium steels of the H-13 type, and will have to be justified on an economic basis.■

Mold steels, 1

Method of fabrication, heat-treatment, wear resistance and other essential properties are given for fifteen recommended mold steels, classified into four groups

Plastic-molding die steels are an important and specialized group of tool steels. Rapid growth of the plastics industry has produced an increasing number of abrasive and corrosive molding compounds. In addition, technological advancements in molding methods and equipment clearly dictate the need for greater attention to the selection of plastic mold steels. Characteristics and properties of the plastic mold steel must be matched against mold design, construction factors, operating conditions, and production requirements before selection of the mold steel is made.

Four classes of steels

Steels available for plastic molds fall into one of four general classes. See table on following page.

Cold-hobbing steels are used for both injection and compression molds when the cavity is of such a size and depth that a master hob can be sunk into the soft die steel under high compressive force. This method offers a quick and inexpensive way of producing multiple cavities having identical impressions. Selection among several grades depends on hobbing requirements, minimum core strength, and the amount of distortion permissible during heat treatment. The higher alloyed grades offer both higher core strength and less distortion. Surface hardness is developed by either carburizing or nitriding.

Prehardened steels are the workhorses for injection molds and are used occasionally in compression molds. Although available in a wide range of sizes and chemical-analysis modifications, they have several common characteristics. The first is their prehardened condition of Rc 30/36 (Brinell 285-341), which is the upper limit of machinability if conventional cutting tools are used.

Strength in this hardness range is adequate to meet mechanical loads imposed during the molding operation. Use of prehardened mold steels eliminates the problem of distortion in heat treatment. Welding for repairs or design changes is relatively simple under most conditions, provided proper preheat and postheat temperatures are used. Most steel producers enhance cleanliness by vacuum degassing the prehardened grades of mold steel. Effective hot working from ingot to bar is performed to assure uniform structure, hardness, and freedom from internal defects.

When using conventional cutting tools, a stress relief of 900-1000 F after machining and prior to finish grinding is possibly the cheapest insurance procedure to maintain mold stability and strength.

Service life and resistance to peening of the prehardened steel cavity can often be increased by chrome or nickel plating. Chrome plating may also be used to provide resistance to atmospheric corrosion or chemical attack from resins such as polyvinyl chloride.

Deep-hardening steels are generally chosen for ability to develop high hardness. This feature would be necessary for abrasive wear resistance, as in the molding of nylon or the phenolics. All things being equal, the polishability of steel is in direct proportion to its hardness. For extreme high luster or optical finish, a hardness of at least Rockwell C 54 is regarded as essential and is obtainable with most through-hardening grades. An additional advantage of these steels is the development of high strength to resist the high forces of compression molding. The air-hardening grades exhibit minimal distortion in heat treatment. Machinability in the annealed condition is superior to that of grades in the prehardened condition.

Special-purpose steels are the fourth category of mold steels. This category includes two groups, each of which offers qualities to meet varying specific requirements. Corrosion resistance is the outstanding feature of the Type 420 stainless mold steel. This requirement is desirable for certain plastics such as PVC. In addition, the 420 is deep hardening, has high strength, and has satisfactory hardness for developing an ultra-luster polish. The second group is the precipitation hardening or maraging steels, which are used in place of the through-hardening grades. They offer the advantage of virtually no distortion in heat treatment. They are, however, more difficult to machine and are more expensive.

Making the mold

Mold cavities are usually formed by hobbing or machining. Master hobs do the work in a cold or hot hobbing job. High compressive strength and resistance to abrasion (hardness) are needed for cold hobbing. Steels for hot hobbing must have excellent red hardness characteristics. All master hob steels should have minimum distortion and size change in heat treatment, plus minimum scaling and ability to be polished to a high luster.

Electrical discharge machining (EDM) is a method frequently used in cutting mold cavities in both the prehardened and annealed condition. Precautionary measures aimed at reducing the buildup of harmful stresses suggest the use of a high-frequency, low-energy spark for light finishing. A light grind or lapping will eliminate and minimize the untempered martensitic layer formed during EDM. The last step should be a retempering of the untempered martensitic surface layer by heating the prehardened mold section about 50 F below original tempering temperature.

Source: 'Tool Steel Trends,' published by the Committee of Tool Steel Producers, American Iron & Steel Institute, New York

Mold steels [Continued]

Characteristics of recommended mold steels

AISI Type	Applications and Recommendations	Method of Fabrication	Annealed Hardness (BHN)	Heat Treatment	Usual Working Hardness Rc	Typical Chemical Analysis C	Mn	Ni	Cr	Mo	Other
Cold Hobbing Steels											
P-1	Injection and compression molds	Hobbed	90	Water	60/64*	0.10	—	—	—	—	—
P-2	" "	Hobbed	103	Oil	60/64*	0.07	—	0.50	2.00	0.20	—
P-3	" "	Hobbed	116	Oil	60/64*	0.10	—	1.25	0.60	—	—
P-4	" "	Hobbed	121	Air	60/64*	0.07	—	—	5.00	0.75	—
P-6	" "	Machined	207	Oil	59/61*	0.10	—	3.50	1.50	—	—
Prehardened Steels											
P-20	Injection and compression molds when carburized	Machined	—	—	30/36	0.35	—	—	1.25	0.40	—
P-21	" "	Machined	—	—	36/39	0.20	—	4.00	—	—	Al 1.20
Deep Hardening Steels											
O-1	Injection, compression & transfer molds	Machined	202	Oil	59/61	0.90	1.00	—	0.50	—	W 0.50
A-2	" "	Machined	212	Air	59/62	1.00	—	—	5.00	1.00	—
A-4	" "	Machined	229	Air	59/62	1.00	2.00	—	1.00	1.00	Pb may be added
A-6	" "	Machined	225	Air	58/60	0.70	2.00	—	1.00	1.00	S may be added
D-2	" "	Machined	223	Air	58/61	1.50	—	—	12.00	1.00	—
S-7	" "	Machined	197	Air	56/58	0.50	—	—	3.25	1.50	—
H-13	" "	Machined	223	Air	50/54	0.35	—	—	5.00	1.50	V 1.00
Special Purpose Steels, Corrosion Resistant											
420	Injection, compression & transfer molds	Machined	183	Oil-Air	50/54	0.30	0.25	—	13.00	—	—

AISI Type	Applications and Recommendations	Method of Fabrication	Annealed Hardness (BHN)	Heat Treatment	Usual Working Hardness Rc	C	Mn	Ni	Cr	Mo	Other
Master Hob Steels, Cold and Hot Hobbing											
O-1	Cold hobbing	Machined	202	Oil	59/61	0.90	1.00	—	0.50	—	W 0.50
A-2	Cold hobbing	Machined	212	Air	59/62	1.00	—	—	5.00	1.00	—
S-1	Cold hobbing	Machined	212	Oil	58/60	0.50	—	—	1.50	—	W 2.50
S-7	Hot or cold hobbing	Machined	197	Air	58/60	0.50	—	—	3.25	1.50	—
D-2	Cold hobbing	Machined	223	Air	59/62	1.50	—	—	12.00	1.00	—
H-13	Hot hobbing	Machined	223	Air	50/52	0.35	—	—	5.00	1.50	V 1.00
H-23	Hot hobbing	Machined	241	Salt	40/42	0.30	—	—	12.00	—	W 12.00

*Carburized case

Machining of malleable and ductile irons, 1

Here is updated information, complete with feeds and speeds, on how to machine three malleable irons and two ductile irons with 'steel-cutting' carbide tools

To show where the malleable and ductile irons fit into the general scheme of machining characteristics, we have plotted tool-life curves for these materials and for a variety of other alloys in Fig. 1. The cutting conditions for these tool-life tests are shown directly below.

From the tool-life curves, we find that the cutting speed for turning Inconel 718 nickel-base alloy with carbide tools is less than 100 ft/min. Cutting speeds for a titanium alloy and for an alloy steel of 500 Brinell hardness are 150 ft/min or less. Both the air-quenched and the oil-quenched-and-tempered malleable irons should be machined within the same speed range (425 to 575 ft/min) as an alloy steel that has been spheroidized to 200 Bhn.

The ferritic ductile iron of 143 Bhn should be turned at the same cutting speeds as a 1020 steel of 115 Bhn, somewhere in the range from 725 to 775 ft/min. For the ferritic malleable iron, the speeds would be from 850 to 950 ft/min. Obviously, both the malleable and ductile irons are readily machinable. (Their mechanical properties and chemical analyses are listed in Table 1.)

By **Norman Zlatin**
Vice-president
Metcut Research Associates, Inc
Cincinnati
and **Hans J Heine,** technical director
Malleable Founders Society, Cleveland

Cutting conditions for tests

Tool material: Stated with illustrations

Tool angles:
- Back rake, 5° neg.
- Side rake, 5° neg.
- Relief, 5°
- End-cutting-edge angle, 15°
- Side-cutting-edge angle, 15°
- Nose radius, 0.030 in.

Feed: 0.010 ipr unless otherwise noted

Depth of cut: 0.062 in.

Cutting fluid: Soluble oil (1:20)

Tool-life end point: 0.015-in. uniform wear; 0.030-in. localized wear

Choosing the right carbide

From among the many carbide grades available, the user must select the optimum grade for (1) roughing, (2) semi-finishing, and (3) finishing. Sometimes the same grade can be used for two or even all three operations. Sometimes, but not always. Each case must be checked out separately.

Harder grades of carbides, such as C-7 and C-8 (the latter is a titanium carbide), can be used in the semi-finishing and finishing operations at higher cutting speeds. Some newer carbides have improved transverse rupture strength for interrupted or heavy roughing cuts on malleable and ductile irons. Because of this increased strength, positive rake angles can be used satisfactorily with this tool material. Various shapes and sizes of this micro-grain tungsten-carbide tool material have been made available to industry as duPont's Baxtron DBW Micro-grain Solid and V R Wesson's Ramet 1.

Advances are also being made in ceramic tools. Because of the very high hardness of this type of tool material, wear resistance is exceptionally good. However, because the strength of the average ceramic is appreciably lower than that of tungsten carbide, the ceramic tool is brittle and chips or cracks readily. Higher strength ceramic tools now being developed may have an application in finish-machining of low-strength castings.

Ferritic malleable iron

Machining characteristics of the iron must be known before you can attempt to establish optimum machining conditions.

Consider first a ferritic malleable iron having a fully ferritic microstructure. Should a cutting fluid be used when using carbide tools in turning? Tool life curves, Fig. 2, show that for a tool life of 30 min, cutting dry with C-6 grade carbide and a feed of 0.010 in./rev, you must use a cutting speed

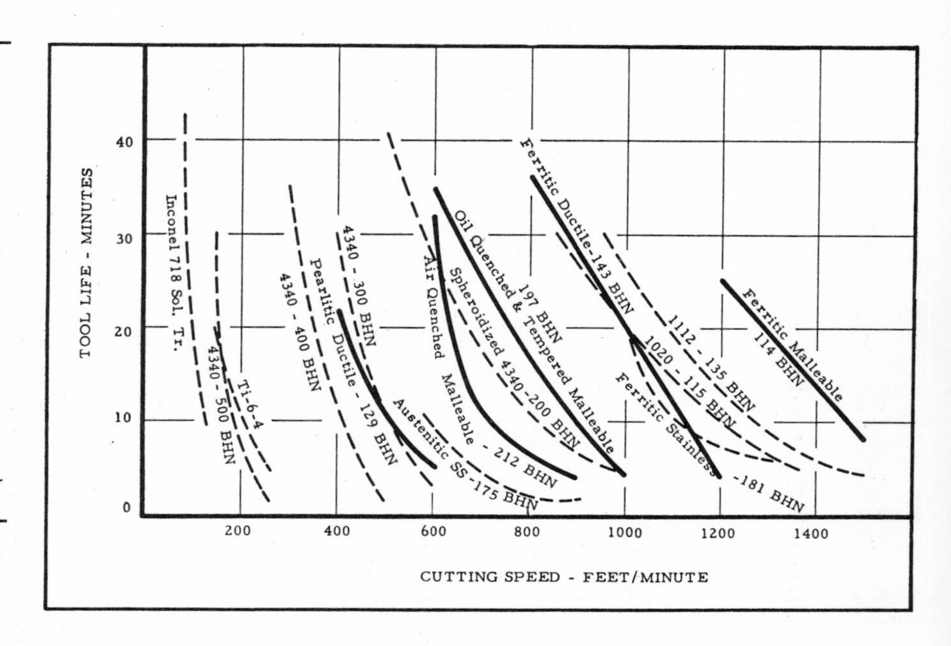

1. **Machinability** of the malleable and ductile irons covers a broad range. All turning was done at 0.010-in. feed

Machining of malleable and ductile irons

[Continued]

Table 1. Properties and analyses of the irons

Mechanical properties

	Yield strength 1000 psi	Ultimate strength 1000 psi	Elongation %
Ferritic malleable	35.8	51.7	17.0
Air-quench malleable	45.0	70.3	4.0
Oil-quenched and tempered malleable	64.7	80.0	5.9
Ferritic ductile	——	——	——
Pearlitic ductile	63.3	100.9	4.4

Chemical analyses, percent

	C	Si	Mn	Cr	Mg	S
Ferritic malleable	2.44	1.44	0.33	0.033	——	——
Air-quench malleable	2.44	1.44	0.33	0.033	——	——
Oil-quenched and tempered malleable	2.44	1.44	0.33	0.033	——	——
Ferritic ductile	3.7	2.47	0.32	0.013	0.050	0.008
Pearlitic ductile	3.77	2.35	0.63	0.035	0.045	0.008

of 600 ft/min. However, by using a soluble oil under the same conditions, you can boost cutting speed 50 percent (reducing machining time 50 percent) without decreasing cutter life.

If you use the harder C-7 grade such as the K7H or the even-harder, more-brittle, titanium carbide (at a light feed of 0.010 in./rev) still higher cutting speeds can be used. For example, as shown in Fig. 3, the cutting speed for a 30-min tool life with C-6 carbide is 900 ft/min; with C-7 (K7H) grade it is approximately 1100 ft/min, and with the titanium carbides more than 1400 ft/min.

Results of machining tests indicate that titanium carbide should only be used for finish cuts at feeds not exceeding 0.010 to 0.015 in./rev. Moreover, this grade of carbide should not be used for machining the cast surface. Development of a higher-strength titanium carbide has been announced, and it may be suitable for use at heavier feeds, but no machining data are yet available.

The C-7 tungsten carbide, being stronger, is recommended for the heavier feeds on ferritic malleable iron. This carbide is less likely to chip or break prematurely.

Very often the major machining that is done on a casting involves removal of the cast surface. A comparison of the tool life curves obtained on the cast surface and under the cast surface respectively, is shown in Fig. 4. Note that the cutting speeds used on the cast surface were about 30 to 50 percent lower than those used under the cast surface.

The actual speed that can be used depends to a great extent on the condition of the cast surface. Porosity, blow holes, burnt-in sand, or oxides on the surface will result in very rapid tool wear or chipping of the tool. Hence, the cutting speed must be reduced appreciably. Again, since the machining of the cast surface is so important in the production of the part, it behooves the manufacturer of the castings to produce good cast surfaces, so that the machine shop will have minimum difficulties in machining them.

Heavier feeds can often be applied to increase the metal removal rate. A slight decrease in cutting speed will frequently allow a large increase in feed, and give a commensurate gain in cubic inches of metal removed.

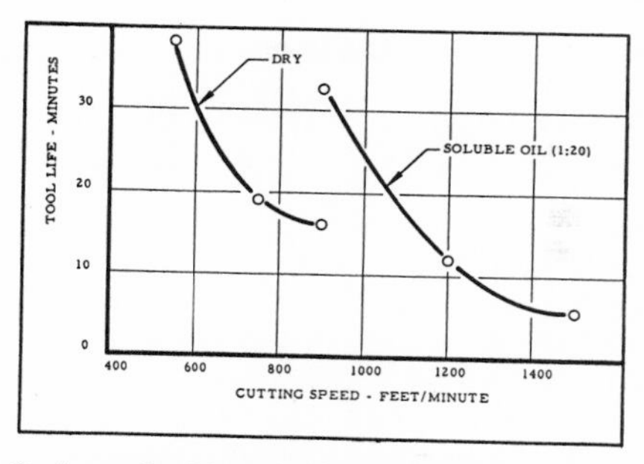

2. A marked increase in tool life is obtained by turning ferritic malleable (114 Bhn) with soluble oil. Tool: C-6

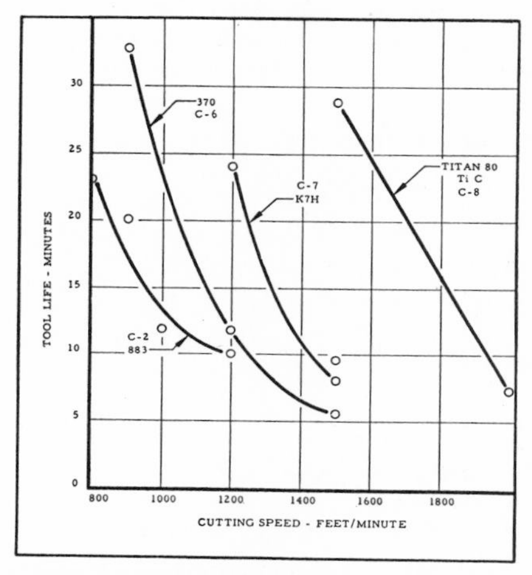

3. When ferritic malleable is cut with a light feed (0.010 ipr) cutting speeds depend on the carbide used

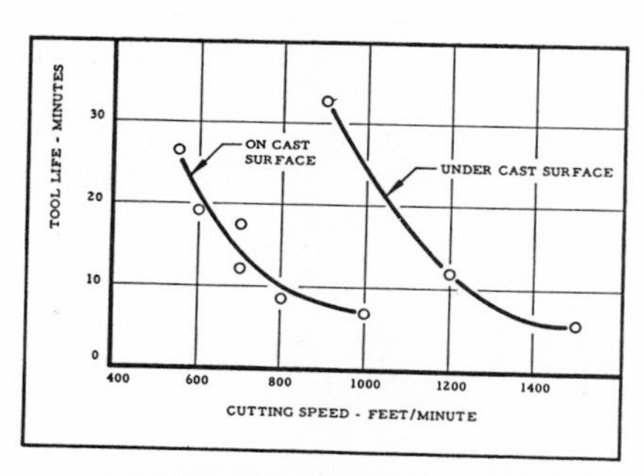

4. Much different cutting speeds are used when cutting on the cast surface or under it. Tool: C-6; soluble oil

Machining of malleable and ductile irons, 2

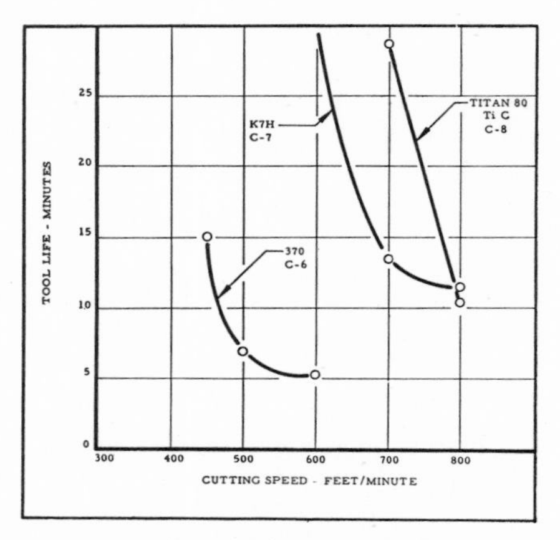

5. Output can be boosted by turning air-quenched malleable iron (197 Bhn) with the C-7 and C-8 carbides

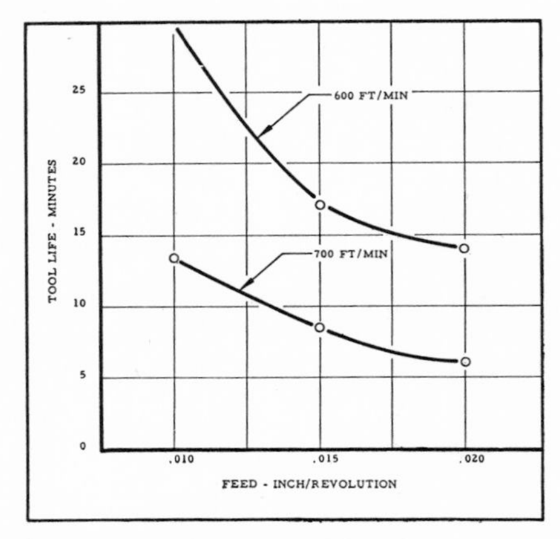

6. Feed rate has a large effect on tool life when air-quenched malleable is turned with C-7 carbide tools

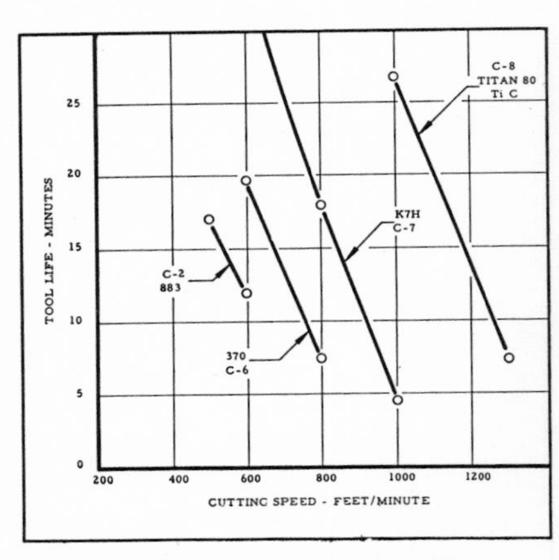

7. Cutting speed and tool material affect tool-life values when turning oil-quenched-and-tempered malleable

Air-quenched malleable iron

The microstructure of pearlitic malleable iron consists of about 75% pearlite and 25% ferrite. This iron has higher strength than the ferritic malleable. Hence, it is questionable whether a titanium carbide with its relatively low strength is suitable for machining this grade. A comparison of the tool-life results obtained with three different grades of carbide, as shown in Fig. 5, indicates again that the titanium carbide has an advantage over both the C-6 and C-7 grades of tungsten carbide at the feed of 0.010 in./rev. The cutting speeds with the titanium carbide were 15% higher than those used with the next best (C-7) grade of carbide.

However, titanium carbide tends to chip more readily at the heavier feeds (above 0.010 in./rev). It is desirable, therefore, to use a C-7 grade of tungsten carbide, even though the cutting speed has to be reduced. Characteristic tool life curves for the C-7 grade (Fig. 6) show that at a cutting speed of 600 ft/min, the tool life drops from 30 minutes at a feed of 0.010 in./rev to slightly under 15 minutes at a feed of 0.020 in./rev.

By **Norman Zlatin**
Vice-president
Metcut Research Associates, Inc
Cincinnati
and **Hans J Heine,** technical director
Malleable Founders Society, Cleveland

Oil-quenched, tempered malleable

Tool-life curves obtained with several different grades of carbide are presented in Fig. 7. Once again the titanium carbide proves superior at a feed of 0.010 in./rev; the cutting speed was 30% higher than with C-7. However, the titanium carbide does not have the strength required for feeds much above 0.010 in./rev on this malleable iron. Here, the C-7 tool life is slightly better and more consistent. The C-6 grade is recommended for rough machining of the cast surface at the heavier feeds.

There is a critical relationship between microstructures and machinability of any cast iron. Tool life curves for a C-6 carbide tool on each of two heats of malleable irons are shown in Fig. 8. Although heats of oil-quenched-and-tempered malleable irons are produced under identical conditions, the attainable speeds may vary by 30%.

Fig. 9 shows the tool-life curves for the three previously discussed malleable irons machined with a C-7 grade of carbide. For a tool life of 25 minutes, the cutting speeds were about 1175 ft/min for ferritic malleable, 700 ft/min for the oil-quenched-and-tempered malleable, and 625 ft/min for the air-quenched-malleable iron.

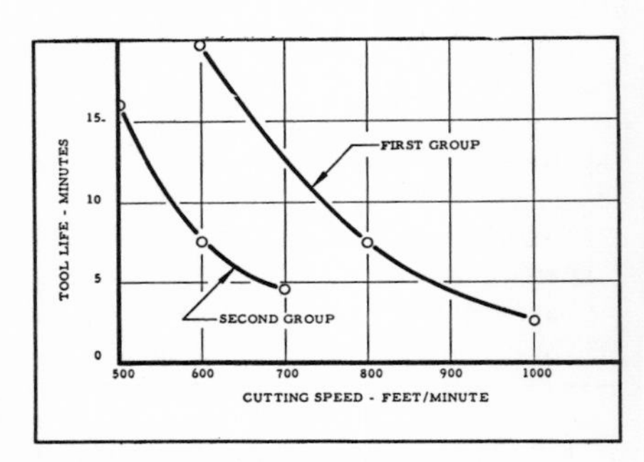

8. Cutting speeds will vary sharply with different heats. Oil-quenched, tempered iron turned with C-6

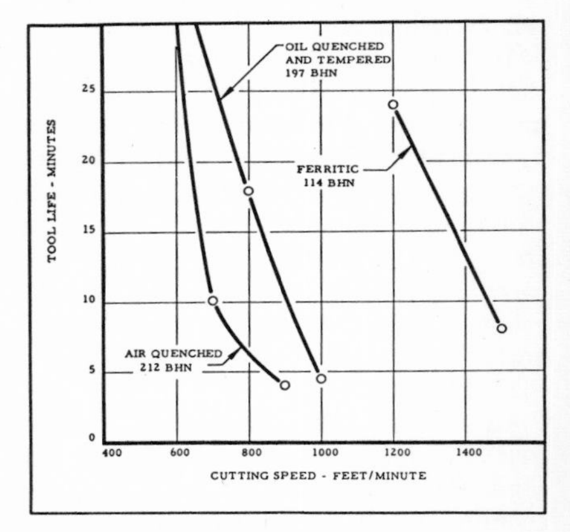

9. Grade of material has a major effect on tool life. Air-quenched grades were cut much slower

Machining of malleable and ductile irons

[Continued]

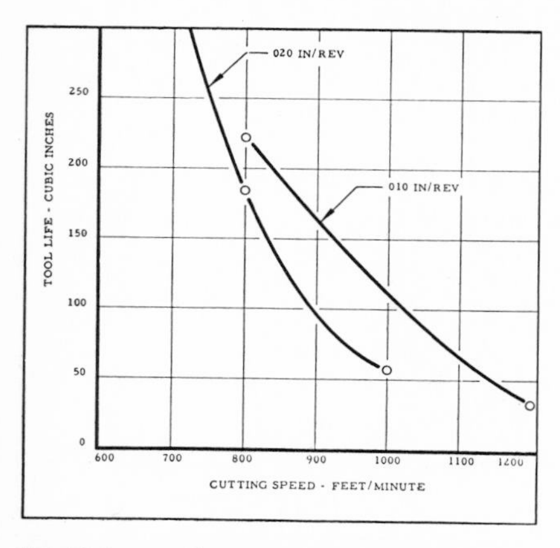

10. Balance the metal-removal rate against the tool life when setting the feed rate. Choose 0.020 ipr

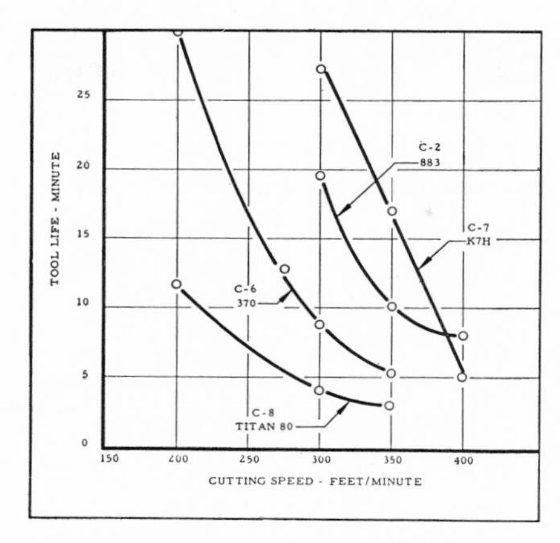

11. Titanium carbide performed poorly when turning pearlitic ductile iron (229 Bhn) at 0.020 ipr

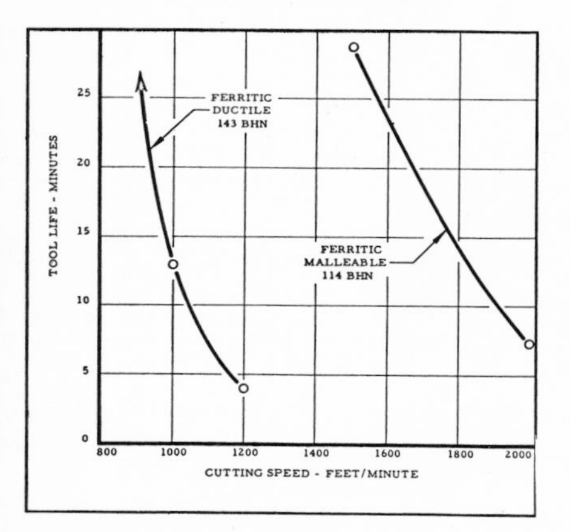

12. Ferritic malleable machines 70% faster than ferritic ductile iron with titanium carbide tools

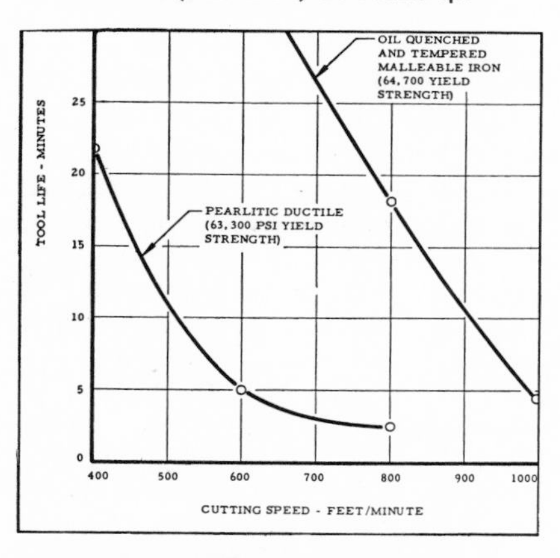

13. Yield strength of the two irons is similar, but the pearlitic ductile is harder by 32 points Brinell

Table 2. Recommended machining conditions

Material	Type of cut	Cutting Speed fpm	Feed in./rev.	Depth of cut in.	Carbide grade
		Malleable irons			
Ferritic 114 Bhn	Roughing (surface)	250-400	0.015	0.100	C-6
	Roughing (below surface)	500-700	0.020	0.062	C-7
	Finishing	800-1000	0.010	0.062	Ti C
Air-quenched 197 Bhn	Roughing (surface)	150-200	0.015	0.100	C-6
	Roughing (below surface)	275-350	0.020	0.062	C-7
	Finishing	500-600	0.010	0.062	Ti C
Oil-quenched & tempered 217 Bhn	Roughing (surface)	175-225	0.015	0.100	C-6
	Roughing (below surface)	350-400	0.020	0.062	C-7
	Finishing	600-800	0.010	0.062	Ti C
		Ductile irons			
Ferritic 143 Bhn	Roughing (surface)	225-350	0.015	0.100	C-7
	Roughing (below surface)	450-600	0.020	0.062	C-7
	Finishing	600-800	0.010	0.062	Ti C
Pearlitic 229 Bhn	Roughing (surface)	125-150	0.015	0.100	C-7
	Roughing (below surface)	200-250	0.020	0.062	C-7
	Finishing	250-300	0.010	0.062	C-7

Notes: Cutting fluid, soluble oil.
Tool Geometry: Back rake, 5° neg. SCEA, 15° Relief, 5°
Side rake, 5° neg. ECEA, 15° Nose radius, 0.030 in.

Ferritic ductile iron

The C-7 tungsten carbide proved to be best for turning the ferritic ductile iron at feeds of 0.010 and 0.020 in./rev. Because of its brittleness, titanium carbide was somewhat more erratic and, hence, is not recommended as the best grade to use.

When selecting a feed rate, examine tool life as well as metal-removal rate before making the final determination. For example, as shown in Fig. 10, at a cutting speed of 800 ft/min, the difference in tool life, in terms of cubic inches of metal removed for the two feed rates, 0.010 and 0.020 in./rev, was not great. However, at a feed of 0.020 in./rev, the machining time was one-half of that required at a feed of 0.010 in./rev.

Pearlitic ductile iron

When machining a pearlitic ductile iron with hardness of 229 Bhn, titanium carbide, being very hard but also very brittle, performed the poorest. The C-7 grade (K7H) was the best at feeds of 0.010 and 0.020 in./rev. See Fig. 11. The difference in the various grades of carbide is quite pronounced. For example, at a feed of 0.020 in./rev and a cutting speed of 300 ft/min, the tool-life values in turning would be 4 minutes for the titanium carbide, 8 minutes for the C-6 grade, 19 minutes for the C-2 grade of carbide, and 27 minutes for the C-7.

Malleable vs ductile iron

With a C-7 carbide tool and feed of 0.010 in./rev, the ferritic malleable iron can be machined 20% faster than the ferritic ductile iron. With a titanium-carbide grade, at the same feed, the cutting speed for the ferritic malleable iron was 70% higher than for the ferritic ductile iron, Fig. 12.

There is a comparison in Fig. 13 of an oil-quenched-and-tempered malleable iron with a pearlitic ductile iron, both of which have yield strengths in the range of 63,000 psi. The hardness of the malleable iron was 197 Bhn, 229 Bhn for the ductile iron. With a C-7 carbide at a feed of 0.010 in./rev, the cutting speed on the oil-quenched-and-tempered malleable iron was 750 ft/min compared with 400 ft/min on the pearlitic ductile iron at a tool life of 22 minutes. (The difference in hardness is only partially responsible for the discrepancy.)■

Machining malleable iron castings

Grade 43010 malleable iron machines like many gray and nodular iron castings

These data represent practice of the Ford Motor Co., and were presented at a seminar on malleable iron castings sponsored by the Malleable Founders Society, Cleveland. Ford's Transmission and Chassis Division selected Grade 43010 malleable as having a hardness and machining response more like general iron castings, nodular iron castings and some forgings that are machined in its plants. This material has a Brinell hardness of 160 to 200. Six other grades of malleable are also used by Ford, but their machining characteristics differ because different hardness ranges and microstructure are involved. Actually some 43 malleable parts are machined in six of the nine plants operated by the Transmission and Chassis Division.

Strength and ductility

Ford designers select grades of iron suitable to safely withstand the forces of motion or pressures to which the part would be exposed. Ferritic malleable iron is chosen as a material for its tensile strength of 50,000 to 60,000 psi, and toughness as the parts will bend or deform if subjected to severe impact or overloads, rather than fracture.

Pearlitic malleable iron is selected for parts designed to withstand rapidly alternating stresses that occur during high-speed operation, because its tensile range approaches 100,000 psi.

Ferritic malleable iron, with a Brinell hardness range of 110-156 may be machined easier than sulphurized mild steel AISI B-1112 used as the standard for machinability ratings.

Under the same conditions, a mild cast iron might be scored only 65% on the machinability scale compared with 120% for ferritic malleable iron.■

Production-proved data for face milling

			High-speed steel			Cast-alloy		Carbides		
Material	BHN	Cut	Speed sfpm	Feed ipt	Grade	Speed sfpm	Feed ipt	Speed sfpm	Feed ipt	Grade
Gray cast iron SAE 110, 111	163 to 207	Rough	85	.014	M-2	110	.014	350	.018	C-2
		Finish	140	.016	M-2	190	.016	450	.020	C-3
Nodular cast iron SAE 60-40-10	140 to 190	Rough	120	.014	M-2	170	.014	440	.018	C-2
		Finish	170	.016	M-2	240	.016	500	.020	C-3
Malleable cast iron grade 43010	160 to 200	Rough	125	.012	M-2	190	.012	375	.014	C-2
		Finish	155	.014	M-2	220	.014	500	.016	C-3
SAE 5120 steel forging	150 to 210	Rough	125	.010	M-2	230	.010	450	.012	C-50, C-6
		Finish	175	.012	M-2	300	.012	525	.014	C-7

Reaming (feeds based on ½-in. reamer, 6 flutes)

Material	BHN	Speed, sfpm	Feed, ipr	Tool material
Gray cast iron SAE 110, 111	163 to 207	65	.010	T-1, M-2
		80	.010	C-2
Nodular cast iron SAE 60-40-10	140 to 190	70	.006	T-1, M-2
		90	.006	C-2
Malleable cast iron grade 43010	160 to 200	80	.015	T-1, M-2
		110	.015	C-2
SAE 5120 steel forging	150 to 210	50	.010	T-1, M-2
		120	.010	C-2

Turning, facing, boring with single-point tools

			High-speed steel			Cast-alloy		Carbides		
Material	BHN	Cut	Speed sfpm	Feed ipr	Grade	Speed sfpm	Feed ipr	Speed sfpm	Feed ipr	Grade
Gray cast iron SAE 110, 111	163 to 207	Rough	75	.015	M-2	140	.015	350	.015	C-2
		Finish	125	.008	M-2	190	.008	500	.008	C-3, C-4
Nodular cast iron SAE 60-40-10	140 to 190	Rough	100	.015	M-2	140	.015	300	.015	C-2
		Finish	140	.008	M-2	200	.008	500	.008	C-3, C-4
Malleable cast iron grade 43010	160 to 200	Rough	100	.015	M-2	150	.015	400	.015	C-6
		Finish	125	.008	M-2	175	.008	550	.008	C-7, C-8
SAE 5120 steel forging	150 to 210	Rough	105	.015	M-2	145	.015	475	.015	C-6
		Finish	140	.008	M-2 T-15	190	.008	600	.008	C-7, C-8

Drilling (feeds based on ½-in drill)

Material	BHN	Speed, sfpm	Feed, ipr	Drill material
Gray cast iron SAE 110, 111	163 to 207	90	.006	M-2, M-10
Nodular cast iron SAE 60-40-10	140 to 190	80	.006	M-2
Malleable cast iron grade 43010	160 to 200	100	.006	M-2, M-10
SAE 5120 steel forging	150 to 210	70	.006	M-2, M-10

Chapter 18

Non-ferrous Metals

Properties of aluminum alloys

Courtesy of the Aluminum Association, New York

Alloy	Nominal Composition, %	Commercial Forms①	Characteristics②					Typical Applications
			Corrosion Resistance	Machinability	Weldability③	Maximum Strength④	Annealed Strength⑤	
EC	99.45 Al min	STEPBW	A-A	D-C	A-A	27	10	Electrical conductors
1100	99.00 Al min	STEBWFO	A-A	D-C	A-A	24	13	Sheet metal work; spun hollow ware
1130	99.30 Al min	S	A-A	D-C	A-A	26	12	Reflector sheet
1145	99.45 Al min	SO	A-A	D-C	A-A	28	12	Foil; light sheet metal work
1060	99.60 Al min	STW	—	—	—	20	10	Chemical equipment; railroad tank cars
2011	5.5 Cu, 0.5 Bi, 0.5 Pb	BW	C-C	A-A	D-D	60	—	Screw machine products
2014	0.8 Si, 4.4 Cu, 0.8 Mn, 0.4 Mg	STEBF	C-C	B-B	B-C	70	27	Truck frames; aircraft structures
2017	4.0 Cu, 0.5 Mn, 0.5 Mg	BW	C	B	B-C	62	26	Screw machine products, fittings
2117	2.5 Cu, 0.3 Mg	WB	C	C	B-C	43	—	Rivets and redraw rod
2018	4.0 Cu, 0.6 Mg, 2.0 Ni	F	C	B	B-C	61	—	Aircraft engine cylinder heads and pistons
2218	4.0 Cu, 1.5 Mg, 2.0 Ni	F	C	B	B-C	48	—	Jet engine impellers and rings
2618	2.3 Cu, 1.6 Mg, 1.0 Ni, 1.1 Fe	F	C	B	B-C	64	—	Aircraft engines (temperatures to 450 F)
2219	6.3 Cu, 0.3 Mn, 0.10 V, 0.15 Zr	SEF	B	B	A	70	25	Structural use at high temperatures (to 600 F); high-strength weldments
2024	4.5 Cu, 0.6 Mn, 1.5 Mg	STPEBW	C-C	B-B	B-B	75	27	Truck wheels; screw machine products; aircraft structures
2025	0.8 Si, 4.5 Cu, 0.8 Mn	F	C-D	B-B	B-B	58	25	Forgings; aircraft propellers
3003	1.2 Mn	All forms	A-A	D-C	A-A	30	16	Cooking utensils; chemical equipment; pressure vessels; sheet metal work; builders' hardware
3004	1.2 Mn, 1.0 Mg	S	A-A	D-C	A-A	41	26	Sheet metal work; storage tanks
4032	12.2 Si, 0.9 Cu, 1.1 Mg, 0.9 Ni	F	C-D	D-C	B-C	55	—	Pistons
4043	5.0 Si	W	—	—	—	—	—	Welding wire
4343	7.5 Si	SW	—	—	—	—	—	Brazing sheet and wire
5005	0.8 Mg	SWO	A-A	D-C	A-A	30	18	Appliances and utensils; architectural; electrical conductor
5050	1.4 Mg	STPO	A-A	D-C	A-A	32	21	Builders' hardware, refrigerator trim and coiled tubes
5052	2.5 Mg, 0.25 Cr	STBWPO	A-A	D-C	A-A	42	28	Sheet metal work; hydraulic tube; appliances
5252	2.5 Mg, 0.25 Cr	S	A-A	D-C	A-A	39	28	Automobile trim
5154	3.5 Mg, 0.25 Cr	STPEBW	A-A	D-C	A-A	48	35	Welded structures; storage tanks; pressure vessels; salt-water service
5454	0.8 Mn, 2.7 Mg, 0.10 Cr	STPEB	A-A	D-C	A-A	44	36	Welded structures and pressure vessels; marine service
5056	0.1 Mn, 5.2 Mg, 0.10 Cr	BW	A-C	D-C	A-A	63	42	Cable sheathing; rivets for magnesium; screen wire; zippers
5356	0.1 Mn, 5.0 Mg, 0.10 Cr	W	—	—	—	—	—	Welding wire
5456	0.8 Mn, 5.1 Mg, 0.10 Cr	SEP	A-B	D-C	A	56	45	High-strength welded structures; storage tanks; pressure vessels; marine applications
5657	0.8 Mg	S	A-A	D-C	A-A	32	19	Anodized auto and appliance trim
5083	0.7 Mn, 4.5 Mg, 0.15 Cr	SEBF	A-C	D-C	A-B	52	42	Unfired, welded pressure vessels; marine, auto and aircraft;
5086	0.5 Mn, 4.0 Mg, 0.15 Cr	SETP	A-C	D-C	A-B	50	38	cryogenics; TV towers; drilling rigs; transportation equipment; missile components
6101	0.5 Si, 0.6 Mg	TPEB	A-B	B-C	A-B	32	14	High-strength bus conductors
6151	1.0 Si, 0.7 Mg, 0.25 Cr	F	A-B	C	A-B	48	—	Moderate-strength intricate forgings for machine and auto parts
6053	0.7 Si, 1.3 Mg, 0.25 Cr	BW	A-B	C	B-C	42	16	Wire and rod for rivets
6061	0.6 Si, 0.25 Cu, 1.0 Mg, 0.20 Cr	STPEBWF	A-A	B-C	A-A	45	18	Heavy-duty structures where corrosion resistance needed; truck and marine; railroad cars, furniture, pipelines
6262	0.6 Si, 0.25 Cu, 1.0 Mg, 0.09 Cr, 0.6 Pb, 0.6 Bi	WB	A-A	A-A	B-B	58	—	Screw machine products
6063	0.4 Si, 0.7 Mg	TPE	A-A	D-C	A-A	42	13	Pipe railing; furniture, architectural extrusions
6463	0.4 Si, 0.7 Mg	E	A-A	D-C	A-A	35	22	Architectural and trim extrusions
6066	1.3 Si, 1.0 Cu, 0.9 Mn, 1.1 Mg	TPEBF	B-C	D-B	A-A	57	22	Forgings and extrusions for welded structures
7001	2.1 Cu, 3.0 Mg, 0.30 Cr, 7.4 Zn	TEB	C	B	D	98	32	High-strength structures
7039	0.2 Mn, 2.7 Mg, 0.20 Cr, 4.0 Zn	S	A-C	B	A	60	32	Welded cryogenic and missile applications
7072	1.0 Zn	S	A-A	D-C	A-A	74	13	Fin stock
7075	1.6 Cu, 2.5 Mg, 0.30 Cr, 5.6 Zn	STEBWF	C	B	D	83	33	Aircraft and other structures; keys
7178	2.0 Cu, 2.7 Mg, 0.30 Cr, 6.8 Zn	STEP	C	B	D	88	33	Aircraft and other structural uses
7079	0.6 Cu, 0.2 Mn, 3.3 Mg, 0.20 Cr, 4.3 Zn	EFS	C	B	D	78	32	Structural parts for aircraft

①B—bar or rod; E—extrusions; F—forgings or forging stock; O—foil; P—pipe; S—sheet or plate; T—tube; W—wire.
②Relative ratings in decreasing order of merit—A, B, C, D. Where applicable, ratings for both annealed and hardest tempers are given (for example, A-C).
③Weldability: A—generally weldable; B—weldable with special techniques for specific applications; C—limited weldability; D—not weldable. Ratings are given for arc welding. Gas welding and brazeability ratings are the same or differ by only one; exceptions are most of the 2000 and 7000 series alloys.
④Typical maximum tensile strength in kips per square inch, for fully work-hardened condition or heat-treated to highest strength level. (Multiply by 0.703 to convert to kilograms per square millimeter.)
⑤Typical annealed tensile strength in kips per square inch. (Multiply by 0.703 to convert to kilograms per square millimeter.)

Fabricating aluminum alloys

Courtesy of the Aluminum Association, New York

Alloy and Temper	Resistance to Corrosion	Workability (Cold)	Machinability	Brazeability	Weldability			Forgeability
					Gas	Arc	Resistance Spot & Seam	
EC-0	A	A	D	A	A	A	B	—
-H12	A	A	D	A	A	A	A	—
-H14	A	A	C	A	A	A	A	—
-H16	A	B	C	A	A	A	A	—
-H19	A	C	C	A	A	A	A	—
1100-0	A	A	D	A	A	A	B	A
-H12	A	A	D	A	A	A	A	A
-H14	A	A	C	A	A	A	A	A
-H16	A	B	C	A	A	A	A	A
-H18	A	C	C	A	A	A	A	A
2011-T3	C	C	A	D	D	D	B	—
-T8	C	D	A	D	D	D	B	—
2014-T4	C	C	B	D	D	B	B	C
-T6	C	D	B	D	D	B	B	C
2017-T4	C	C	B	D	D	B	B	—
2117-T4	C	B	C	D	D	B	B	—
2018-T61	C	—	B	D	D	B	B	C
2218-T72	C	—	B	D	—	B	—	D
2024-T3	C	C	B	D	D	B	B	—
-T4	C	C	B	D	D	B	B	—
-T36	C	D	B	D	D	B	B	—
3003-0	A	A	D	A	A	A	B	A
-H12	A	A	D	A	A	A	A	A
-H14	A	B	C	A	A	A	A	A
-H16	A	C	C	A	A	A	A	A
-H18	A	C	C	A	A	A	A	A
3004-0	A	A	D	B	B	A	B	
-H32	A	B	D	B	B	A	A	—
-H34	A	B	C	B	B	A	A	—
-H36	A	C	C	B	B	A	A	—
-H38	A	C	C	B	B	A	A	—
5005-0	A	A	D	B	A	A	B	—
-H12	A	A	D	B	A	A	A	—
-H14	A	B	C	B	A	A	A	—
-H16	A	C	C	B	A	A	A	—
-H18	A	C	C	B	A	A	A	—
-H32	A	A	D	B	A	A	A	—
-H34	A	B	C	B	A	A	A	—
-H36	A	C	C	B	A	A	A	—
-H38	A	C	C	B	A	A	A	—
5050-0	A	A	D	B	A	A	B	—
-H32	A	B	D	B	A	A	A	—
-H34	A	B	C	B	A	A	A	—
-H36	A	C	C	B	A	A	A	—
-H38	A	C	C	B	A	A	A	—
5052-0	A	A	D	C	A	A	B	—
-H32	A	B	C	D	A	A	A	—
-H34	A	B	C	C	A	A	A	—
-H36	A	C	C	C	A	A	A	—
-H38	A	C	C	C	A	A	A	—
5154-0	A	A	D	D	C	A	B	—
-H32	A	B	D	D	C	A	A	—
-H34	A	B	C	D	C	A	A	—
-H36	A	C	C	D	C	A	A	—
-H38	A	C	C	D	C	A	A	—
-H112	A	B	D	D	C	A	A	—
5056-0	A	A	D	D	C	A	B	—
-H38	C	C	C	D	C	A	A	—
5086-0	A	A	D	D	C	A	B	—
-H32	A	B	D	D	C	A	A	—
-H34	B	B	C	D	C	A	A	—
-H36	B	C	C	D	C	A	A	—
-H38	B	C	C	D	C	A	A	—
6061-0	A	A	D	A	A	A	B	—
-T4	A	C	C	A	A	A	A	—
-T6	A	C	C	A	A	A	A	
6063-0	A	A	D	A	A	A	B	—
-T1†	A	B	C	A	A	A	A	—
-T4	A	B	C	A	A	A	A	—
-T5	A	B	C	A	A	A	A	—
-T6	A	C	C	A	A	A	A	—
-T83	A	C	C	A	A	A	A	—
-T831	A	C	C	A	A	A	A	—
-T832	A	C	C	A	A	A	A	—
6066-0	B	B	D	A	A	A	B	—
-T4	B	C	B	A	A	A	A	—
-T6	B	C	B	A	A	A	A	—
7001-T6	C	D	B	D	D	D	B	—
7075-T6	C	D	B	D	D	D	B	D

Resistance to Corrosion, Workability (Cold), Machinability and Forgeability ratings A, B, C and D are relative ratings in decreasing order of merit. Weldability and Brazeability ratings A, B, C and D are relative ratings defined as follows:

A. Generally weldable by all commercial procedures and methods.

B. Weldable with special technique or on specific applications which justify preliminary trials or testing to develop welding procedure and weld performance.

C. Limited weldability because of crack sensitivity or loss in resistance to corrosion, and all mechanical properties.

D. No commonly used welding methods have so far been developed.

† Formerly designated -T42.

Machinability data for aluminum

Wrought alloys have been classified in five groups in respect to types of chip produced and the quality of finish. These ratings establish the speed-feed ranges

General-purpose twist drills with a bright finish perform satisfactorily on aluminum. For deep holes and with the softer alloys, however, better results may be obtained with drills having a high helix angle. Because of the excellent machining characteristics of aluminum, high rates of penetration can be used. This requires large flute areas to dispose of chips quickly. With soft alloys, the flutes should also be highly polished to minimize chip packing and buildup. The common 118° point angle will perform well on most jobs. In the case of deep holes, however, an angle of 130°–140° will give better performance because it produces a narrower chip, which more readily clears the hole.

Source: 'Alcoa Machinability Handbook,' published by the Aluminum Co of America, Pittsburgh

Power to drill aluminum is:

$$Hp = TN/63{,}025$$

where T = torque, see table, and N = rpm.

Cutting forces are lower for aluminum than for most metals. When taking full advantage of the machinability of aluminum at high cutting speeds, high horsepower is required. Horsepower depends upon both the cutting force, F_c, and

Drilling thrust and torque for aluminum alloys

Feed, ipr	Depth to diam. ratio	Thrust, lb. ¼ in.	Thrust, lb. ⅜ in.	Thrust, lb. ½ in.	Torque, in.lb. ¼ in.	Torque, in.lb. ⅜ in.	Torque, in.lb. ½ in.
			Alloy 2011-T3				
0.002	1	25	34	35	2	4	7
	2	25	34	35	2	4	7
	4	28	35	35	3	5	9
	6	30	35	35	4	8	13
0.004	1	52	69	75	3	7	15
	2	52	69	75	3	8	15
	4	58	71	75	6	10	18
	6	62	71	75	8	15	26
0.008	1	105	140	155	5	13	23
	2	105	140	155	5	13	23
	4	115	148	155	10	16	25
	6	123	148	155	17	30	51
			Alloy 2014-T6				
0.002	1	65	92	112	3	8	15
	2	65	93	140	3	8	16
	4	82	125	173	4	10	22
	6	88	150	208	8	34	38
0.004	1	133	180	224	6	14	30
	2	135	184	290	7	16	30
	4	145	224	352	9	19	44
	6	179	268	370	16	45	71
0.008	1	209	306	405	10	24	47
	2	210	310	405	12	27	50
	4	229	379	455	14	39	67
	6	326	405	578	24	52	117
			Alloy 2024-T351				
0.002	1	60	80	95	3	6	12
	2	60	80	95	3	7	13
	4	72	85	95	5	9	19
	6	88	127	134	17	23	32
0.004	1	120	159	190	6	13	25
	2	120	159	190	6	13	25
	4	130	175	198	10	19	38
	6	180	248	278	34	46	64
0.008	1	194	281	378	9	21	40
	2	194	287	380	10	23	42
	4	281	360	432	15	37	65
	6	360	445	640	51	81	116

Feed, ipr	Depth to diam. ratio	Thrust, lb. ¼ in.	Thrust, lb. ⅜ in.	Thrust, lb. ½ in.	Torque, in.lb. ¼ in.	Torque, in.lb. ⅜ in.	Torque, in.lb. ½ in.
			Alloy 6061-T651				
0.002	1	50	73	100	3	6	12
	2	50	75	100	3	7	13
	4	50	92	115	4	8	19
	6	94	106	132	4	9	27
0.004	1	100	147	204	6	13	24
	2	105	153	204	6	13	24
	4	105	187	232	7	18	37
	6	157	197	240	9	18	55
0.008	1	159	237	312	8	20	37
	2	162	248	318	9	23	41
	4	170	280	385	10	30	75
	6	205	315	390	18	37	112
			Alloy 7075-T651				
0.002	1	75	110	147	4	8	16
	2	75	112	149	4	9	16
	4	82	122	154	5	15	19
	6	107	135	225	16	26	31
0.004	1	150	222	300	7	15	30
	2	150	225	300	8	16	31
	4	164	250	310	10	25	39
	6	208	264	386	27	39	53
0.008	1	224	316	372	10	25	49
	2	225	324	385	11	27	55
	4	254	352	444	17	35	77
	6	383	460	540	36	47	77
			Alloy F132-T5				
0.002	1	32	50	63	2	4	8
	2	32	52	67	2	5	9
	4	38	59	75	3	7	12
	6	44	73	104	6	8	14
0.004	1	66	104	126	4	9	16
	2	66	110	134	4	10	19
	4	75	120	154	6	13	24
	6	90	150	214	12	16	26
0.008	1	135	205	255	6	16	28
	2	136	222	270	7	17	31
	4	154	240	306	10	22	35
	6	183	272	328	20	28	43

Machinability data for aluminum [Continued]

Machinability ratings of wrought alloys

Nonheat-treated alloys

Alloy	Temper	Rating
EC	–O, –H12	E
EC	–H14 to –H19	D
1100	–O, –H12, –H14	E
1100	–H16 to –H18	D
3003(2)	–O, –H12, –H14	E
3003(2)	–H16 to –H18	D
3004(2)	–O, –H32, –H34	D
3004(2)	–H36 to –H38	C
5005	–O, –H12, –H14 –H32, –H34	E
5005	–H16 to –H18	D
5005	–H36 to –H38	D
5050	–O, –H32, –H34	D
5050	–H36 to –H38	C
5052	–O, –H32, –H34	D
5052	–H36 to –H38	C
5056	–O	D
5056	–H18, –H38	C
5083	–O	D
5083	–H112, –H113	C
5086	–O, –H112, –H32	D
5086	–H34, –H36	C
5154	–O, –H112, –H32	D
5154	–H34, –H36, –H38	C
5357	–O, –H32	D
5357	–H34, –H36, –H38	C
5454	–O, –H112	D
5454	–H311 to –H343	C
5456	–O, –H112	D
5456	–H311 to –H343	C

Heat-treated alloys

Alloy	Temper	Rating
2011	–T3, –T8	A
2014(2)	–O	C
2014(2)	–T3, –T4, –T6	B
2017	–O	C
2017	–T4	B
2018	–T61	B
2024(2)	–T3, –T4, –T6, –T8	B
2219	–T3, –T6, –T8	B
2618	–T6	B
4032	–T6	C
6061(2)	–O	D
6061(2)	–T4, –T6	C
6062	–O	D
6062	–T4, –T6	C
6063	–O, –T4	D
6063	–T5, –T6, –T8	C
6262	–T4, –T9	B
6463	–O	D
6463	–T4, –T5, –T6	C
7075(2)	–T6	B
7079	–6	B
7178(2)	–T6	B

The ratings are:

A—Free-cutting, very small broken chips and excellent finish.

B—Curled or easily broken chips and good to excellent finish.

C—Continuous chips and good finish.

D—Continuous chips and satisfactory finish.

E—Satisfactory control of chip and finish with optimum tool design and machine settings.

the speed, S, as shown by the equation: $Hp = F_cS/33{,}000$.

Power at the cutter

	Horsepower/cu. in./min. 0° rake angle	Horsepower/cu. in./min. 20° rake angle
F132-T5	0.20-0.50	0.20-0.40
356-T51	0.30-1.00	0.25-0.45
2011-T3	0.20-0.30	0.15-0.25
2024-T351	0.30-0.50	0.20-0.40
6061-T651	0.30-0.50	0.25-0.35

The horsepower at the toolbit can be found by first calculating the cutting force, F_c, and then using the formula above. The cutting force and feed force, F_f, can be estimated for single-point tools with the formulas in table below, where d and f are depth of cut and feed per revolution or stroke in in., S is speed in fpm, and the values of k are taken from table.

These formulas apply for speeds in the range of 200 to 1,000 fpm.

Drilling speeds and feeds for heat-treated aluminum

Drill size, in.	Feed, ipr.	Speed, sfm
1/16	0.003	50–1,000
3/32	0.007	75–1,000
1/8	0.010	100–1,000
5/32	0.011	125–1,000
3/16	0.012	150–1,000
7/32	0.013	175–1,000
1/4	0.014	200–1,000
9/32	0.015	225–1,000
5/16	0.016	250–1,000
3/8	0.017	300–2,000
7/16	0.017	350–2,000
1/2	0.017	400–3,000
9/16	0.017	450–3,000
5/8	0.017	500–4,000
3/4	0.017	600–4,000

For all nonheat-treated aluminum products, the feed should be reduced about 50 per cent.

Suggested practices for turning aluminum

Variable	Condition	Suggested practice
Back rake angle	–––	5–15 deg.
Side rake angle	Machinability rating A	0–20 deg.
	Machinability rating B	20 deg.
	Machinability rating C	20–30 deg.
	Machinability rating D, E	40 deg.
End relief angle	HSS tool	8–15 deg.
	Carbide tool	6–8 deg.
Side relief angle	HSS tool	8–15 deg.
	Carbide tool	6–8 deg.
End cutting edge angle	–––	5 deg.
Side cutting edge angle	–––	5–15 deg.
Nose radius	Rough cutting	1/32–1/8 in.
	Finish cutting	0.010–0.030 in.
Depth of cut	Small work	1/64–1/4 in.
	Large work	1/2–1 1/2 in.
Feed	Rough cutting	0.006–0.080 ipr
	Finish cutting	0.002–0.006 ipr
Speed	HSS tool	Up to 1000 fpm
	Carbide tool	Up to 15,000 fpm

Cutting force constants

(for equations $F_c = k_1 + k_2df + k_3dfS$ and $F_f = k_4 + k_5df + k_6dfS$

Alloy and temper	k_1	k_2	k_3	k_4	k_5	k_6
		With side rake = 0 deg				
F132-T5	6.2	119,000	−24	5.6	65,000	−23
356-T51	12.3	123,000	−21	8.0	97,000	−32
2011-T3	−0.3	88,000	16	−0.5	26,000	28
2024-T351	6.3	145,000	−32	6.1	66,000	−27
6061-T651	7.7	166,000	−55	7.9	130,000	−71
		With side rake = 20 deg				
F132-T5	4.0	93,000	−13	3.1	32,000	−11
356-T51	6.9	94,000	− 8	2.6	44,000	− 9
2011-T3	1.4	65,000	0	2.6	−2,000	− 2
2024-T351	3.2	129,000	−28	3.6	35,000	−26
6061-T651	1.0	112,000	−13	−0.1	47,000	− 8

Back rake, 0°; end relief, 7°; side relief, 7°; end cutting edge, 15°; side cutting edge, 15°; nose radius, 0.012 in.

How to work tantalum, 1

From milling to deep drawing, here are the basics on working tantalum, along with information on mechanical properties, and tips on joining methods

Tantalum, a ductile, easy-to-fabricate metal, is today finding increased usage for aerospace parts, and heat exchangers. It melts at 5425 F, has a high density (0.60 lb/cu in.), and is resistant to all acids except fuming sulphuric and hydrofluoric —characteristics that are increasing in importance in modern metalworking. For this reason, more and more production men are becoming involved in fabricating equipment and parts from tantalum.

Arc-cast and electron-beam-melted tantalum alloys are supplied in many forms: ingot, bar, rod, wire, sheet, foil, tubing, expanded mesh, as well as powder and sintered shapes made from powders.

Rod, for example, is made in sizes up to 3 in. diameter, sheet and plate in thicknesses to 1.0 in., tubing up to 2 in. OD by 0.250-in. wall, and butt-welded tubing to 12-in. diameter can be made. (Mechanical properties of arc-cast tantalum are given in Fig. 1.)

Machining

Tantalum can be machined without much difficulty by conventional techniques, but it must be worked cold (unless special equipment is used) because heating in air or any common gas causes embrittlement.

Tantalum does have a tendency to seize, tear, and gall but this can be overcome by the right techniques and by using such cutting fluids as carbon tetrachloride and trichloroethane.

Turning. Excellent results can be obtained by using commercially available tool steels operated at low speeds. Cutting fluids are desirable but not absolutely necessary if the tools are kept sharp and the speeds low. Tool rakes and angles must be such that as little of the tool as possible touches the chip or the part being cut. Tools must be kept sharp.

Milling. Staggered-tool HSS cutters should have generous back and side relief. Depth of cut is determined by the same considerations as turning, and ample cutting fluid should be used. Feed rate may be high although cutting speed must be low for best results. For example, the feed rate may be 0.005 ipt and the cutting speed from 100 to 300 fpm.

Thread cutting. Standard chasers are satisfactory. Apply plenty of coolant to prevent galling and tearing of the thread. Work diameter should be 0.004 to 0.008 in. undersize. Dies and taps must be kept free of chips and cleaned frequently. When cutting screw threads in a lathe, the tool should be fed in at ½ thread angle so it cuts one side only.

Grinding. It is difficult to grind annealed tantalum. The metal tends to smear the wheel. Cold-worked tantalum can be ground successfully with the proper wheel.

Table 1. Mechanical properties of arc-cast tantalum

Tensile strength, psi:
- Annealed sheet, 55-75,000
- Worked sheet, 100-150,000
- Annealed wire, 60-80,000
- Worked wire, 120-135,000

Elongation, % in. 2 in.
- Annealed sheet, 40
- Worked sheet, 1
- Annealed wire, 20
- Worked wire, 1

Young's modulus:
- at room temperature, 27×10^6
- At 930 F, 25×10^6
- at 1830 F, 22×10^6

Poissons's ratio, 0.35
Rockwell hardness:
- Annealed, B 50
- Worked, B 90

Vickers hardness, 100-180
Working temperature, room
Recrystall, tmp, 1922-2732 F
Stress relieving temp, 1652 F

Polishing and etching. Tantalum may be polished electrolytically in a solution consisting of 1.25% hydrofluoric acid, 5% sulfuric acid in methanol. It may be polished mechanically with 240 to 600 grit coated abrasive, and rough polished on a geoscience Politex style PA-K cloth or equivalent using a 7

By J Irwin Peters, manager
Technical marketing services
Kawecki Chemical Co
New York

How to work tantalum, 2

micron diamond paste. The specimen is then etched in a one part hydrofluoric acid, one part nitric, two parts sulfuric acid mixture before final polish on microcloth with 1-micron magnesium oxide powder. The preferred final etch is the 1 HF-1 HNO_3-2 H_2SO_4 mixture.

Blanking and punching. Steel dies are used for blanking. Clearance is usually 6% of the thickness of the metal worked.

Bending and forming. Except for precautions to prevent seizing or tearing, tantalum can be handled like mild steel. Dies may be made of steel, but if there is considerable die wear, aluminum bronze or beryllium copper should be used. Rubber or pneumatic die cushions are sometimes required.

Deep drawing. Annealed tantalum can be deep drawn in an aluminum-bronze die, but a steel punch is also suitable. For deep drawing where there is more than one operation, the first draw should have a depth of not more than 40 to 50% of the diameter.

Spinning. Conventional techniques apply and steel rollers are used. Yellow soap and Johnson No. 150 drawing wax have proved to be suitable lubricants.

Joining of tantalum

Mechanical joining. Tantalum can be riveted, bolted or crimped. Such joints should be made cold, if possible.

Brazing. This technique is now seldom used. A vacuum of not less than 0.1 μ (1×10^{-4} mm Hg) or ultra-high-purity gases, such as helium or argon, with less than 5 ppm impurities, will be necessary for gas brazing in a high frequency furnace.

Welding. Rapid reaction of tantalum above 300 F with impurities, and consequent embrittlement, makes a protective atmosphere essential. Size of the heated zone and the time of heating should be minimized. Within these limitations, tantalum can be welded to itself and certain other metals, by resistance and by electron beam welding, and to itself, by inert gas arc welding.

Table 2. HSS tool design and operating data when turning tantalum

Tool design	
Back rake	10 to 20°
Side rake	20 to 30°
End relief	10 to 15°
Side relief	10 to 20°
End cutting edge	8 to 15°
Side cutting edge	15°
Nose radius	1/16 to 1/8 in.

Roughing	Finishing
0.008-0.012 ipr	0.005 ipr
50-60 fpm	100 to 300 fpm
depth of cut 0.015-0.060 in.	depth of cut, as heavy as possible

Resistance welding of thin sheets of tantalum in air has been successful when the welding time is very short. Seam welds—an overlapping series of spot welds—ordinarily must be made under carbon tetrachloride or water. In this manner, the weld is chilled rapidly and the contamination is reduced.

The lap joint, which is generally used, is not as strong as a butt joint; also, a small amount of tantalum is wasted, and design limitations must be considered.

A higher power input must be used for welding tantalum than for many steels, because its melting point is considerably higher, and its resistivity is substantially lower. Too much force on the electrode causes so little interface resistance that no weld is made. Sheets less than 0.040 in. in thickness can easily be resistance welded.

Gas-arc welding. Tungsten-arc and carbon-arc welding are usually used in joining tantalum up to 0.065 in. thick. With the proper precautions, tungsten-inert-gas arc welding (Tig) in open air is possible. However, inert gas must surround all surfaces heated above 600 F. Otherwise oxygen, nitrogen, or hydrogen will be absorbed.

Weld ductilities that allow a 180° bend can be achieved by suitable protection and firm alignment of the work.

Carbon-arc welding (under carbon tetrachloride) provides protection from air and a desirable quick chill. However, carbon pickup in the welds results in lowered ductility and reduces corrosion resistance.

Inert-gas-shielded, consumable-electrode welding (Mig) is useful where heavy sections are involved, and where filler metal must be added. Tantalum sheet 1/16-inch thick has been welded successfully by the Mig method.

Electron-beam welding results in the best possible weld purity. However, high equipment and operating costs tend to restrict use of this method. ■

Machining titanium alloys

Here is a guide not only to turning and grinding titanium alloys but also to the physical and chemical properties that make these important alloys so tricky to machine

Because the structural titanium alloys are ductile, light, and have good fatigue and corrosion resistance properties, they are finding ever wider application. Thus, more and more people are learning that machining these alloys can be a tricky job.

The problems that come up in turning and grinding titanium can better be understood if we look at the physical and chemical properties. These hold the key to successful machining.

The table, right, shows the general properties of commercially pure titanium and compares them with those of SAE 1020 steel, 18Cr-8Ni stainless, and 7075 aluminum. We can see that the specific weight of titanium is about ⅔ that of steel and about 60% higher than that of aluminum. In tensile and shear stiffness, titanium falls between steel and aluminum. But titanium's strength (80,000 psi for pure titanium, 150,000 psi and above for its alloys) is far greater than that of many alloy steels, offering the highest strength-to-weight ratio of any of today's structural mètals.

Thermal properties are something else, however. Titanium alloys have high melting points, which is usually a sign of excellent temperature stability, but the strength of titanium alloys falls off rapidly at temperatures above 800 F or so. And the coefficient of expansion is even less than that of steel. These unusually poor thermal properties account, to a large extent, for the difficulties in machining titanium.

Next, let's look at the structure of titanium alloys. They have a hexagonal close-packed (HCP) lattice structure similar to that of magnesium alloys, but at about 1625 F titanium undergoes an allotropic transformation in which it changes from HCP to body-centered cubic (BCC) structure. These allotropic forms of titanium are known as α and β, respectively.

Alloying elements favor one or the other structure. For example, aluminum stabilizes the structure, raising the transformation temperature; chromium, iron, molybdenum, manganese, and vanadium lower the transformation temperature, thereby making the β phase stable at a lower temperature. Tin, in titanium alloys, is neutral.

Thus, titanium alloys fall into three classes, depending on the structures present: α, β, and an α-β combination. The last phase includes most of the titanium alloys now in use.

Even though commercially pure titanium, undergoes an allotropic transformation, it cannot be hardened by heat treatment; it can, however, be hardened by cold work. It has been found that titanium alloys have about the same work hardening tendencies as ordinary structural steels; so the difficulties encountered in turning and grinding titanium alloys cannot be attributed to increased work hardening.

Some amounts of nitrogen, oxygen, carbon, and iron go into molten titaium and form solid solutions, tending to make the metal harder. The influence of nitrogen and oxygen is approximately additive, and these elements in combination cause about 40 points increase in Brinell hardness for each 0.1% of N_2 + O_2 addition.

As was mentioned, commercially pure titanium can be work hardened; the effect of this can be removed by annealing at about 1250 F. Other non-heat-treatable titanium alloys, such as Ti 100A, have greater percentages of N_2 and O_2 in solution and have correspondingly higher tensile and yield strengths. Titanium alloys with relatively large amounts of Cr, Mo, Al, Mn, or Fe in solution can be quench- and age-hardened. In these alloys, Cr and Fe tend to stabilize the β (BCC) phase of titanium at room temperature. For example, Ti 150A contains 2.7% Cr and 1.5% Fe; these additions make it possible to harden the alloy by adjusting the cooling rate in the range from 1700 F to 1400 F, where the α-β phase transformation normally occurs. Some titanium alloys containing large amounts of carbon (about 1%) are available, but they are usually difficult to machine.

Before titanium alloys came into wide use, the stainless steels were considered to be the most difficult materials to machine, from the viewpoint of adverse thermal properties. Here is

Properties of pure titanium and other structural metals

Property	Titanium	1020 steel	18-8 stainless	7075 alum
Structure*	HCP	BCC	FCC	FCC
Specific wt, ρ, lb/cu in.	0.16	0.28	0.28	0.10
Ult tensile str, annealed, psi	80,000	50,000	90,000	12,000
Young's modulus, E, psi X 10^6	16	30	30	10
Shear modulus, G, psi X 10^6	6	11.5	11.5	4
Melting point, °F	3200	2600	2600	1220
Coef linear exp, per °F X 10^{-6}	5	6.6	9	13
Thermal conductivity, k, Btu/in./sec/(°F/in.) X 10^{-4}	2.0	7.5	2.2	16
Specific heat, c, Btu/lb/°F	0.13	0.13	0.12	0.21
Volume specific heat, ρc, Btu/cu in./°F	0.021	0.036	0.034	0.021
Thermal diffusivity, $K = k/\rho c$, sq in./sec	0.0095	0.021	0.0065	0.076
$k\rho c$	4.2	27	7.5	34

*HCP = hexagonal close-packed lattice; BCC = body-centered cubic; FCC = face-centered cubic

By Dr. C.T. Yang
Mfg. development specialist
Pomona Div., General Dynamics
Pomona, Calif.

Machining titanium alloys

Speeds and feeds for turning titanium

Machine setting	Rough & interrupted turning Titanium grade or condition Pure	Annealed	Heat-treated	Finish turning Titanium grade or condition Pure	Annealed	Heat-treated
Carbide tools:						
Cutting speed, sfm	100 — 240	75 — 120	50 — 95	200 — 350	100 — 300	75 — 275
Feed, ipr	.008 — .015	.008 — .015	.008 — .015	.003 — .012	.003 — .012	.003 — .012
Depth of cut, in.	.100	.100	.100	.003 — .012	.003 — .012	.003 — .012
HSS tools:						
Cutting speed, sfm	25 — 124	25 — 60	10 — 50	76 — 160	45 — 60	30 — 50
Feed, ipr	.004 — .050	.004 — .015	.004 — .015	.002 — .005	.002 — .005	.002 — .005
Depth of cut, in.	.100	.100	.100	.003 — .030	.003 — .030	.003 — .030

Tool geometry for turning titanium

Tool angles	Rough & interrupted turning Carbide tools	HSS tools	Finish turning Carbide tools	HSS tools
Back rake	+5° to −5°	0° to +5°	0° to +5°	0° to +5°
Side rake	0° to −5°	0° to +15°	0° to +15°	0° to +5°
Side cutting edge	+5° to +25°	+6° to +15°	0° to +20°	+5° to +6°
End cutting edge	+6° to +10°	+5° to +6°	+6° to +10°	+5° to +6°
End relief	+5° to +10°	+5° to +7°	+5° to +10°	+5° to +7°
Side relief	+5° to +10°	+5° to +7°	+5° to +10°	+5° to +7°
Nose radius, in.	.03 to .045	.02 to .03	.03 to .045	.02 to .03

where the value of $k\rho c$ (last line in table of properties) comes into play. This value is a measure of thermal diffusivity and is related to tool-tip temperature. The $k\rho c$ value for stainless steel was the lowest among metals (before titanium), making it necessary to cut stainless at a much slower speed than ordinary steel. But the $k\rho c$ value for titanium is far lower yet.

The cutting speed and the $k\rho c$ value enter into the tool-tip temperature equation to the same power; so to maintain the same tool-tip temperature when machining titanium as when machining stainless, it is necessary to reduce the speed used for stainless by a factor of 4.2/7.2, the ratio of their $k\rho c$ values. In other words, titanium is machined at somewhat over one-half the cutting speed used for stainless.

In grinding steels there is relatively little welding or bonding between the abrasive surfaces and the chips. The oxygen in the air is responsible, for it reacts immediately with the freshly cut metal to form an oxide film that prevents the chips from welding to the abrasive grit. This greatly reduces wheel wear.

When titanium alloys are ground in air, however, wheel wear is unusually high. The G ratio (the ratio of workpiece metal removed to the volume of wheel wear) may be as low as 0.5, and even lower when pure water is used as a cutting fluid. At the temperatures reached at the tips of the abrasive grains, the freshly cut titanium has a strong tendency to weld or bond to almost anything it touches. Under these conditions, titanium is an extremely powerful reducing agent.

For example, when a silicon-carbide abrasive is used to grind titanium, the following reaction takes place:

$$Ti + SiC \rightarrow TiC + Si$$

A similar reaction can occur when an aluminum-oxide wheel is used. In both cases, wheel wear is high.

The types of tool wear encountered in turning titanium alloys are similar to those for steel. For example, the tools are found to crater excessively when the cutting speed is too high (especially for high-strength titanium alloys) and to chip excessively at the cutting edge when the speed is too low (especially for softer, weaker alloys).

Here is a listing that shows the relative machinability of several titanium alloys (based on 1045 steel having a rating of 100%):

Ti 100A	70%
Ti 75A	65%
Ti 140A	35%
RC 130B	30%
1045 steel	100%

This means that the tool will last just as long when machining Ti 140A as when machining 1045 steel if the speed used to machine the titanium is 35% of that used to machine the steel. Recommended speeds, feeds, and cutting tool angles for turning titanium alloys are listed in the tables above.

In general, carbide tools are preferable to HSS tools for machining titanium. Only at speeds below 10 to 20 sfm should HSS tools be considered. To keep the tools, work, and machine from overheating, copious amounts of cutting fluid should be used. Because the heat-removal rate of water-based fluid is much higher than that of oil-based fluid, water-based fluids are recommended for high-speed operations, and a weak solution of rust inhibitor and/or water-soluble oil seems most practical.

The proper grinding speed for titanium is about 2000 sfm, with SiC wheels preferred to Al_2O_3 wheels. The best grinding fluids are $NaNO_2$ or $BaNO_2$ in water solution.

Summing it up, there are six basic rules for cutting titanium:

- Use low cutting speeds
- Use heavy feed rates
- Use large volumes of non-chlorinated, water-based cutting fluids
- Use sharp tools and replace them at the first sign of wear
- Never stop feeding while the work and tool are in moving contact
- Use rigid setups. ■

Corrosion Resistance of Metals—I

Abstracted by permission from Manual 700—A "Heat Exchangers," published by the Patterson-Kelley Co., Inc, East Stroudsburg, Pa. Some corrosive chemicals in the original tables have been omitted

Proper selection of materials for heat exchangers, chemical equipment, piping, valves, etc, depends upon their exposure to corrosion. Many forms of corrosive attack can take place — pitting, intergranular attack, and electrochemical reactions.

The accompanying tables give relative resistance of metals to electrochemical attack; that is, when a metal is surrounded by a polar liquid—one that conducts electricity. The tabulated values are supplied as a guide to selecting proper materials, with the understanding that electrochemical reactions are affected by variables such as temperature, concentration, contaminants, aeration, velocity, stresses, metal-surface conditions, purity of the metal, and presence of corrosion inhibitors.

If dissimilar metals are used in a structure exposed to a polar solution, galvanic attack occurs, because electric current flows between them as it does in a battery. One metal, the anode, will corrode by release of metallic ions to the electrolyte, but the cathode does not corrode. Metals are rated according to their galvanic activity. The galvanic series ranges (as tabulated) from magnesium, the most effective anode, to titanium as the least corrodible, or cathodic in nature. The distance between the metals in the galvanic series is a measure of the galvanic corrosion to be expected, when they are used in a structure in the presence of an electrolyte.

Galvanic Series of Metals (in sea water)

ANODE

- Magnesium
- Zinc
- Aluminum
- Steel, low carbon and alloy
- Cast iron
- Stainless steels (active)
- Ni-resist
- Muntz metal
- Yellow brass
- Admiralty metal
- Red brass
- Copper
- Aluminum bronze
- Cupro-nickel
- Nickel
- Inconel
- Tantalum
- Silver
- Stainless steels (passive)
- Monel
- Hastelloy C
- Zirconium
- Titanium

CATHODE

LEGEND
A – Excellent
B – Good
C – Fair
D – Not Suitable
E – Explosive
I – Ignites
• – Information not available

	Concentration, %	Temperature, F	Carbon Steel	Copper	Red Brass	Muntz	Admiralty	Copper Silicon	90-10 Cupro-Nickel	70-30 Cupro-Nickel	Aluminum	304 Stainless Steel	316 Stainless Steel	Nickel	Monel	Inconel	Hastelloy	Titanium	Zirconium	Tantalum
Acetaldehyde	100	70	A	E	E	E	E	E	E	E	A	A	A	A	A	A	A	B	•	A
Acetic Acid (Aerated)	100	70	D	D	D	D	D	D	C	C	B	A	A	D	A	B	A	A	A	A
Acetic Anhydride	100	70	D	B	C	D	C	B	B	B	A	B	B	B	B	B	A	A	A	B
Acetone	100	70	A	A	A	A	A	A	A	A	A	A	A	B	A	A	B	A	•	A
Aluminum Hydroxide	10	70	B	B	B	B	B	B	B	B	B	B	B	B	B	B	B	•	•	B
Ammonia (Anhydrous)	100	70	A	A	A	A	A	A	A	A	A	A	A	B	A	B	B	A	•	A
Ammonium Chloride	10	70	D	D	D	D	D	D	D	D	C	B	B	B	B	B	B	A	A	A
Ammonium Sulfate	10	70	C	C	C	C	C	C	C	C	D	C	C	B	A	B	B	A	A	A
Aniline	100	70	A	D	D	D	D	D	D	D	D	A	A	B	B	B	B	A	•	A
Aroclor	100	70	B	A	A	A	A	A	A	A	A	B	B	A	A	A	A	A	•	A
Barium Chloride	30	70	B	B	B	D	C	B	B	B	B	B	B	B	B	B	B	A	A	A
Benzaldehyde	100	70	B	B	B	B	B	B	B	B	B	B	B	B	B	B	A	A	•	A
Benzene	100	70	A	A	A	A	A	A	A	A	B	B	B	B	B	B	B	A	•	A
Benzoic Acid	10	70	D	B	B	B	B	B	B	B	B	B	B	B	B	B	B	A	•	A
Boric Acid	10	70	D	B	B	B	B	B	B	B	C	A	A	B	B	B	A	A	•	A

Corrosion Resistance of Metals—II

LEGEND
A – Excellent
B – Good
C – Fair
D – Not Suitable
E – Explosive
I – Ignites
• – Information not available

	Concentration, %	Temperature, F	Carbon Steel	Copper	Red Brass	Muntz	Admiralty	Copper Silicon	90-10 Cupro-Nickel	70-30 Cupro-Nickel	Aluminum	304 Stainless Steel	316 Stainless Steel	Nickel	Monel	Inconel	Hastelloy	Titanium	Zirconium	Tantalum
Butane	100	70	A	A	A	A	A	A	A	A	A	A	A	A	A	A	A	A	•	A
Butanol	100	70	A	A	A	A	A	A	A	A	A	A	A	A	A	A	A	A	•	A
Butyl Acetate	100	70	A	B	B	B	B	B	B	B	A	B	B	A	B	A	B	A	•	A
Butyl Chloride	100	70	A	A	A	A	A	A	A	A	A	A	A	A	A	A	A	A	•	A
Calcium Chloride	20	70	B	B	B	D	C	B	B	B	B	C	B	A	A	A	B	A	A	A
Calcium Hydroxide	10	70	B	B	B	B	B	B	B	B	D	B	B	B	B	B	B	A	•	A
Carbon Dioxide (Wet)	100	70	C	C	C	C	C	C	C	C	B	A	A	A	A	A	A	A	•	A
Carbon Tetrachloride (Dry)	100	70	B	B	B	B	B	B	B	B	B	B	B	A	A	A	B	A	A	A
Chlorine Gas (Dry)	100	70	B	B	B	B	B	B	B	B	C	B	B	B	B	A	B	I	A	A
Chloroform (Dry)	100	70	B	B	B	B	B	B	B	B	B	B	B	A	A	B	B	A	A	A
Citric Acid	20	70	D	C	C	D	C	C	C	C	A	C	B	B	B	A	C	A	A	A
Creosote	100	70	B	B	B	B	B	B	B	B	B	B	B	B	B	B	B	A	•	A
Dibutylphthalate	100	70	A	A	A	A	A	A	A	A	B	B	B	B	B	B	B	A	•	A
Dichlodifluoromethane(F-12)	100	70	A	A	A	A	A	A	A	A	A	A	B	B	B	B	A	A	•	A
Diethanolamine	100	85	A	B	B	B	B	B	B	B	A	A	A	A	A	A	A	A	•	A
Diethyl Ether	100	70	B	B	B	B	B	B	B	B	B	B	B	B	B	B	B	A	•	A
Diethylene Glycol	100	70	A	B	B	B	B	B	B	B	B	A	A	B	B	B	B	A	•	A
Diphenyl	100	160	B	B	B	B	B	B	B	B	A	B	B	B	B	B	B	A	•	A
Diphenyl Oxide	100	85	B	B	B	B	B	B	B	B	B	B	B	B	B	B	B	A	•	A
Ethanolamine	100	70	B	B	B	B	B	B	B	B	B	A	B	B	B	B	B	B	•	A
Ether	100	70	B	B	B	B	B	B	B	B	B	B	B	B	B	B	B	A	•	A
Ethyl Acetate (Dry)	100	70	B	B	B	B	B	B	B	B	B	B	B	B	B	B	B	A	•	A
Ethyl Alcohol	100	70	B	B	B	B	B	B	B	B	B	B	B	B	B	B	A	A	A	A
Ethyl Ether	100	70	B	B	B	B	B	B	B	B	B	B	B	B	B	B	B	A	•	A
Ethylene Glycol	100	70	B	B	B	B	B	B	B	B	B	B	B	B	B	B	B	A	•	A
Fatty Acids	100	400	D	D	D	D	D	D	D	D	A	D	A	B	C	B	A	B	•	A
Ferric Sulfate	10	70	D	D	D	D	D	D	D	D	D	B	B	D	D	D	A	A	•	A
Formaldehyde	50	200	D	B	B	D	B	B	B	B	C	B	B	B	B	B	B	B	•	A
Furfural	100	70	B	B	B	D	B	B	B	B	B	B	B	B	B	B	B	A	•	A
Glycerine	100	70	A	A	A	A	A	A	A	A	A	A	A	A	A	A	A	A	•	A
Hydrochloric Acid (Aerated)	38	70	D	D	D	D	D	D	D	D	D	D	D	D	D	D	B	D	D	A
Hydrofluoric Acid (Aerated)	40	70	D	C	D	D	D	D	D	C	D	D	D	D	C	D	A	D	D	D
Iodine	20	70	D	D	D	D	D	D	D	D	D	D	D	D	D	D	B	D	•	A
Isopropanol	100	70	A	B	B	B	B	B	B	B	B	B	B	B	B	B	B	A	•	A
Lactic Acid	50	70	D	B	B	D	C	B	B	B	D	B	A	B	C	A	A	A	A	A

Corrosion Resistance of Metals—III

LEGEND
A – Excellent
B – Good
C – Fair
D – Not Suitable
E – Explosive
I – Ignites
• – Information not available

	Concentration, %	Temperature, F	Carbon Steel	Copper	Red Brass	Muntz	Admiralty	Copper Silicon	90-10 Cupro-Nickel	70-30 Cupro-Nickel	Aluminum	304 Stainless Steel	316 Stainless Steel	Nickel	Monel	Inconel	Hastelloy	Titanium	Zirconium	Tantalum
Lithium Chloride	30	200	B	B	B	D	B	B	B	B	D	B	A	A	A	A	A	•	•	A
Lithium Hydroxide	10	200	B	B	B	D	B	B	B	B	D	B	B	B	B	B	B	•	•	A
Magnesium Chloride	30	70	B	B	B	D	C	B	B	B	C	B	B	A	B	A	A	A	A	A
Magnesium Hydroxide	10	70	B	B	B	B	B	B	B	B	D	B	B	B	B	B	B	A	•	B
Magnesium Sulfate	30	200	B	B	B	B	B	B	B	B	C	A	A	B	B	B	A	A	A	A
Methallyamine	100	70	C	B	B	B	B	B	B	B	B	B	B	B	C	B	B	B	•	A
Methyl Alcohol	100	70	B	B	B	B	B	B	B	B	B	B	B	B	A	B	A	A	A	A
Methylene Chloride (Dry)	100	70	B	B	B	B	B	B	B	B	B	B	B	B	B	B	B	B	•	A
Monochlorobenzene (Dry)	100	70	B	B	B	B	B	B	B	B	A	B	B	A	A	A	B	B	•	A
Monochlorodifluoro Methane (F-22)	100	70	A	A	A	A	A	A	A	A	A	A	A	A	A	A	A	A	•	A
Monoethanolamine	100	200	B	B	B	B	B	B	B	B	B	B	B	B	B	B	•	•	•	A
Naphthalene	100	70	A	B	B	B	B	B	B	B	B	A	A	A	A	A	B	B	•	A
Nickel Sulfate	10	200	D	B	B	D	B	B	B	B	D	B	B	B	B	B	B	B	A	A
Nitric Acid	50	200	D	D	D	D	D	D	D	D	D	B	B	D	D	D	D	A	B	A
Nitrous Acid	10	70	D	D	D	D	D	D	D	D	D	B	B	D	D	D	•	•	•	A
Oleic Acid	100	70	B	B	B	C	B	B	B	B	B	B	B	A	A	A	B	B	B	B
Oxalic Acid	10	70	D	B	B	C	B	B	B	B	C	B	B	C	B	B	B	D	B	A
Perchloroethylene	100	70	A	B	B	C	B	B	B	B	B	B	B	A	A	A	•	A	•	A
Phenol	10	120	B	B	B	B	B	B	B	B	A	B	B	B	A	B	A	A	•	A
Phosphoric Acid (Aerated)	50	200	D	D	D	D	D	D	D	D	D	B	B	D	D	B	A	C	D	B
Potassium Carbonate	40	200	B	B	B	B	B	B	B	B	D	B	B	B	B	B	B	A	•	A
Propylene Glycol	100	70	B	B	B	B	B	B	B	B	B	B	B	B	B	B	B	A	•	A
Pyridine	100	70	A	B	B	B	B	B	B	B	B	B	B	B	B	B	B	B	•	A
Sodium Acetate	10	70	D	B	B	B	B	B	B	B	C	B	B	B	B	B	B	B	•	A
Sodium Hydroxide	50	300	D	D	D	D	D	D	D	D	D	D	D	A	B	B	B	B	B	D
Sodium Sulfate	10	200	B	B	B	B	B	B	B	B	A	B	A	B	B	B	B	A	•	A
Sulfur Dioxide (Dry)	100	300	B	B	B	C	B	B	B	B	B	B	B	B	B	B	B	A	•	A
Sulfuric Acid (Aerated)	60	200	D	D	D	D	D	D	D	D	D	D	D	D	D	D	B	D	A	A
Trichloroethylene (Dry)	100	150	B	B	B	C	B	B	B	B	B	B	B	A	A	B	A	A	A	A
Turpentine	100	70	B	B	B	B	B	B	B	B	B	B	B	B	B	B	B	B	•	A
Vinyl Chloride (Dry)	100	70	A	B	B	D	C	B	B	B	A	B	A	A	A	A	A	A	•	A
Water (Fresh)	100	70	C	A	A	A	A	A	A	A	B	A	A	A	A	A	A	A	A	A
Water (Sea)	100	70	C	B	B	C	A	B	A	A	B	A	A	B	A	B	B	A	A	A
Xylene	100	200	B	A	A	A	A	A	A	A	B	A	A	A	A	A	A	A	A	A
Zinc Chloride	10	70	D	D	D	D	D	D	D	D	C	B	B	B	A	D	B	A	A	A
Zinc Sulfate	20	70	D	B	B	D	B	B	B	B	D	B	A	B	B	A	B	A	•	A

Gall Resistance of Metals..I

These data were compiled by P J Olmstead, chief engineer, Goulds Pumps, with the assistance of G L Cox of The International Nickel Co.

Courtesy Goulds Pumps, Seneca Falls, NY

The chart on page 264 shows the probable galling characteristics of various combination of materials. Either material of a combination may be the stationary or rotating part with no change in the galling characteristics.

The gall resistance of materials is characterized by the following criteria, and the gall resistance increases in the order named:

1. Hardness—such as nitrided or carburized steels.
2. Hardness plus inherent resistance to seizing by formation of silicides, as in "S" Monel, "S" nickel, "S" Inconel, or the like.
3. Inherent resistance to galling as in cast iron, ductile iron, Ni-Resist or "G" Nickel, due to lubricating qualities of graphitic carbon.

Where, and if, possible the following criteria should be used as a guide in selecting and machining materials for gall resistance:

1. Use austenitic against martensitic alloys.
2. Choose mating materials (except bronzes) with a difference in hardness of 50 Brinell numbers.
3. Grind mating surfaces.

Although a combination of materials may indicate non-galling characteristics, consider the possibility of galvanic corrosion arising from the combination of materials and the electrolyte or liquid present.

EXPLANATION OF NOTES

Note 1: The free-machining grade of "K" Monel, known as "KR" Monel, should have somewhat better gall resistance than "K" Monel, as shown.

Note 2: "S" and "G" nickel have better gall resistance than the regular cast nickel. Both are comparable in gall resistance, but "S" nickel will stand heavier loads.

Note 3: "S" Inconel, a high-silicon variety of Inconel, is superior to "S" nickel and "S" Monel in gall resistance.

Note 4: The Ni-Vee bronzes are 5% nickel—5% tin, cast and heat-treatable, similar in balance of composition to the 88-10-2 Cu Sn Zn type: "A"—No lead; "B"—1% lead; "D"—10% lead.

Note 5: Leaded bronze: 85-5-5-5 or 80-10-10.

Note 6: Nickel aluminum bronze is generally somewhat inferior to Ni-Vee "A" in gall resistance and coefficient of friction, but will stand heavier loads in slower motion.

Note 7: Chromium plate varies greatly in gall resistance. To work best, chromium plate must be backed up by hard material, and the plating must bond well to the backing.

This chart was not prepared from an exhaustive study of each combination of metals, under all conditions of galling. The data were first set down 10 years ago from experiences and occasional bits of information. During this period, the chart has proved helpful to people who knew nothing about the subject, and their choices from the chart were reasonably good ones. The current chart incorporates a number of revisions, based on further experience.

Gall Resistance of Metals . . II

	Cast Iron	3% Ni-Cast Iron	Ni-Resist (Type 1, 2)	Ductile Iron	Ductile Ni-Resist	"S" Monel	"K" Monel[1]	"B" Monel	"H" Monel	Duranickel	"G" Nickel[2]	"S" Nickel[2]	Inconel	"S" Inconel[3]	400 Stainless (Soft)	400 Stainless (Hard)	300 Stainless Steel	SAE 1000 to 6000 (Soft)	SAE 1000 to 6000 (Hard)	Bronze (Leaded)[5]	Ni-Vee Bronze "A"[4]	Ni-Vee Bronze "B"	Ni-Vee Bronze "D"	Ni-Al Bronze[6]	Hastelloy "A"—"B"	Hastelloy "C"	Hastelloy "D"	Nitrided Steel	Chrome Plate[7]	Stellite
Cast Iron	S	S	S	S	S	S	S	S	S	S	S	S	S	S	S	S	S	S	S	S	S	S	S	S	S	S	S	S	S	S
3% Ni-Cast Iron	S	S	S	S	S	S	S	S	S	S	S	S	S	S	S	S	S	S	S	S	S	S	S	S	S	S	S	S	S	S
Ni-Resist (Type 1, 2)	S	S	S	S	S	S	S	S	S	S	S	S	S	S	S	S	S	S	S	S	S	S	S	S	S	S	S	S	S	S
Ductile Iron	S	S	S	S	S	S	S	S	S	S	S	S	S	S	S	S	S	S	S	S	S	S	S	S	S	S	S	S	S	S
Ductile Ni-Resist	S	S	S	S	S	S	S	S	S	S	S	S	S	S	S	S	S	S	S	S	S	S	S	S	S	S	S	S	S	S
"S" Monel	S	S	S	S	S	S	F	F	S	S	S	S	F+	S	F	S	F	F	S	S	F	S	S	F	F+	S	S	S	S	S
"K" Monel[1]	S	S	S	S	S	F	F	N	F	F	S	S	N	S	F	F	N	N	F	S	F	S	S	F−	F	F	S	S	S	S
"B" Monel	S	S	S	S	S	F	N	N	F	N	F	F	N	S	N	F	N	N	F	S	F	S	S	F−	N	F	S	S	F	S
"H" Monel	S	S	S	S	S	S	F	F	F	F	F	F	N	S	N	F	N	N	F	S	F	S	S	F−	N	F	S	S	S	S
Duranickel	S	S	S	S	S	S	F	N	F	F	S	S	N	S	N	F	N	N	F	S	F	S	S	F−	F	F	S	S	S	S
"G" Nickel[2]	S	S	S	S	S	S	S	F	F	S	S	F	F	S	F	S	F	F	S	S	S	S	S	S	S	S	S	S	S	S
"S" Nickel[2]	S	S	S	S	S	S	F	F	F	S	F	S	F	S	F	S	F	F	S	S	S	S	S	S	S	S	S	S	S	S
Inconel	S	S	S	S	S	F+	N	N	N	N	F	F	N	F	N	F	N	N	F	S	F	S	S	F−	N	F	S	S	S	S
"S" Inconel[3]	S	S	S	S	S	S	S	S	S	S	S	S	F	S	F	S	F	F	S	S	S	S	S	S	S	S	S	S	S	S
400 Series Stainless Steel (Soft)	S	S	S	S	S	F	F	N	N	N	F	F	N	F	N	F	F	N	F	S	F	S	S	F−	N	F	S	F	F	S
400 Series Stainless Steel (Hard)	S	S	S	S	S	S	F	F	F	F	S	S	F	S	F	S	F	S	S	S	S	S	S	F+	F	S	S	S	S	S
300 Series Stainless Steel	S	S	S	S	S	F	N	N	N	N	F	F	N	F	F	F	N	N	F	S	F	S	S	F−	N	F	S	S	S	S
SAE 1000 to 6000 Steel (Soft)	S	S	S	S	S	F	N	N	N	N	F	F	N	F	N	S	N	N	S	S	S	S	S	F+	N	F	S	S	S	S
SAE 1000 to 6000 Steel (Hard)	S	S	S	S	S	S	F	F	F	F	S	S	F	S	F	S	F	S	S	S	S	S	S	S	F	S	S	S	S	S
Bronze (Leaded)[5]	S	S	S	S	S	S	S	S	S	S	S	S	S	S	S	S	S	S	S	S	S	S	S	S	S	S	S	S	S	S
Ni-Vee Bronze "A"[4]	S	S	S	S	S	F	F	F	F	F	S	S	F	S	F	S	F	S	S	S	F	S	S	F	F	S	S	F	S	S
Ni-Vee Bronze "B"	S	S	S	S	S	S	S	S	S	S	S	S	S	S	S	S	S	S	S	S	F	S	S	F	S	S	S	S	S	S
Ni-Vee Bronze "D"	S	S	S	S	S	S	S	S	S	S	S	S	S	S	S	S	S	S	S	S	S	S	S	S	S	S	S	S	S	S
Ni-Al Bronze[6]	S	S	S	S	S	F	F−	F−	F−	F−	S	S	F−	S	F−	F+	F−	F+	S	S	F	F+	S	F	F−	F+	S	F	S	S
Hastelloy "A"—"B"	S	S	S	S	S	F+	F	N	N	F	S	S	N	S	N	F	N	N	F	S	F	S	S	F−	N	F	S	S	S	S
Hastelloy "C"	S	S	S	S	S	S	F	F	F	F	S	S	F	S	F	S	F	F	S	S	S	S	S	F+	F	F	S	S	S	S
Hastelloy "D"	S	S	S	S	S	S	S	S	S	S	S	S	S	S	S	S	S	S	S	S	S	S	S	S	S	S	S	S	S	S
Nitrided Steel	S	S	S	S	S	S	S	S	S	S	S	S	S	S	F	S	S	S	S	S	F	S	S	F	S	S	S	S	S	S
Chrome Plate[7]	S	S	S	S	S	S	S	S	S	S	S	S	S	S	S	S	S	S	S	S	S	S	S	S	S	S	S	S	?[7]	S
Stellite	S	S	S	S	S	S	S	S	S	S	S	S	S	S	S	S	S	S	S	S	S	S	S	S	S	S	S	S	S	S

Degree of Resistance: **S** *— Satisfactory;* **F** *— Fair;* **N** *— Little or None*

Chapter 19

Heat Treating

Heat treating of aluminum, 1

Heating treating processes for aluminum are precision processes that must have proper thermal conditions and controlled temperature-time cycles

The general types of production heat treatments applied to aluminum and its alloys are:

1. Preheating or homogenizing to reduce chemical segregation of cast structures and to improve their workability.
2. Annealing to soften strain-hardened (work-hardened) and heat treated alloy structures, to relieve stresses, and to stabilize properties and dimensions.
3. Solution heat treatments to effect solid solution of alloying constituents and to improve mechanical properties.
4. Precipitation heat treatments, to provide hardening by precipitation of constituents from solid solution.

Most treatments are performed in controlled-atmosphere or air furnaces. A high rate of forced recirculation commonly is employed to obtain temperature uniformity and efficient heating rates. This is fundamental to all heating equipment, except for induction, salt bath or fluidized-bed facilities.

Table 1. Temper designations for aluminum

F	As fabricated	
O	Annealed, recrystallized (wrought products only)	
H	Strain hardened (wrought products only)	
	H1	Strain hardened only
	H2	Strain hardened and then partially annealed
	H3	Strain hardened and then stabilized
W	Solution heat treated (unstable temper)	
T	Solution heat treated (stable temper)	
	T1	Solution heat treated (modified) and then naturally aged
	T2	Annealed (cast products only)
	T3	Solution heat treated and then cold worked
	T4	Solution heat treated and then naturally aged
	T5	Artificially aged only
	T6	Solution heat treated and then artificially aged
	T7	Solution heat treated and then stabilized
	T8	Solution heat treated, cold worked, and then artificially aged
	T9	Solution heat treated, artificially aged, and then cold worked
	T10	Artificially aged and then cold worked
	T*x*51(a)	Stress relieved by stretching following solution heat treatment
	T*x*52	Stress relieved by compressing following solution heat treatment
	T*x*53	Stress relieved by special thermal treatment
	T*x*510	Stress relieved by stretching following solution heat treatment; no subsequent straightening permitted
	T*x*511	Stress relieved by stretching following solution heat treatment; minor subsequent straightening permitted

(a) "*x*" is replaced by various Arabic numerals in specific designations.

Furnace types

Most standard types of furnaces can be utilized for production heat treatment of aluminum alloys. These include car-bottom and truck-type, vertical-pit and vertical-tower or elevator, horizontal-conveyor, and strip-processing furnaces, as well as induction-heated salt-bath, and fluidized-bed units. When used for heat treating aluminum, the various furnace configurations usually require certain special features.

Car-bottom furnaces are usually indirectly heated, either electrically or by fuel-fired radiant tubes.

Truck-type furnaces resemble the car-bottom type but have an integral insulated bottom structure and a separate truck to move the charge into the furnace.

Vertical pit-type furnaces for aluminum heating always employ a high rate of forced atmosphere recirculation and are heated indirectly by either electrical elements or fuel-fired radiant tubes. They employ removable top closures, involving handling equipment and overhead cranes. Pit furnaces are used primarily for preheating and homogenizing aluminum ingot prior to fabrication.

Vertical-tower furnaces are particularly useful for heat treating long, narrow products, especially extrusions. Hoist facilities raise and lower a rack from which the charge is suspended. A water-quench pit directly beneath the furnace is common, to permit rapid immersion of the charge.

Among conveyor furnaces employed in the aluminum industry there are slat, mesh-belt, walking-beam, pusher, wicket, and pin-type conveyors, as well as parallel chains with attachments to carry rows of parts or cable or stainless steel strand supports. These furnaces are largely direct fired, although some are heated electrically or by radiant tubes. They are built for service temperatures of 200 to 1175 F and generally are used for continuous operation, requiring constant loading and removal.

Furnaces for continuous processing of strip can be of three types: The strip may be supported in the furnace on rolls, in a straight line, an arc, or sinuously; it may pass through the heating chamber in a catenary supported at the extremities by rolls; or it may be moved on a cushion of atmosphere from high-velocity jets to prevent marking of the sheet by hot rolls.

Wrought alloys frequently are treated in molten salt baths. Generally the salt used is sodium nitrate, but a bath employed both for annealing and solution heat-treating requires a lower-melting

---

Extracted from material in a 3-vol. series on **Aluminum**, edited by Dr K R Van Horn, vice president, research and development, Aluminum Co of America, Pittsburgh, and written by over 90 associated experts.
The volumes have now been published by the American Society for Metals, Cleveland

Heat treating of aluminum (Continued)

point mixture of equal parts sodium nitrate and potassium nitrate. To inhibit corrosive effects as these salts become basic, keep approximately 0.5 oz of sodium or potassium dichromate per 100 lb of nitrate in the bath.

Fluidized beds of fine, inert oxide particles such as silica sand or alumina are effective in transferring heat to and from properly oriented surfaces. However, transporting such heated mediums to the product presents some problems. Because fluidization is effected by a vertical upward flow of the fluidizing gas and bed material, flat top surfaces on any part of the charge are in the 'shadow' of this flow. These surfaces collect a stagnant, insulating buildup of bed material, which interferes with uniform heating.

Table 2. Effect of heat-treating atmospheres on tensile properties of 0.064-in 2024-T4 aluminum alloy sheets

Atmosphere	Tensile strength, % decrease	Elongation, % decrease
Helium from cylinder	0	0
Hydrogen from cylinder	0	0
Natural gas	0	0
Dry oxygen	0	0
Dry nitrogen	0	0
Moist nitrogen	0	0
Dry air	0	0
Moist carbon dioxide	0	0
Moist air containing 26% carbon dioxide	0	0
Moist air containing fluoride	0	0
Air containing 50% combustion products from natural gas(b)	0	0
Dry air containing 0.0002% sulfur dioxide	3	27
Air containing 25% combustion products from natural gas(b)	5	75
Dry air containing 0.0012% sulfur trioxide(c)	7	35
Air containing 0.8% water vapor (40 F dew point)	8	40
Dry air containing 0.007% sulfur dioxide	15	68
Sulfur dioxide	20	64
Air containing 3.4% water vapor (80 F dew point)	25	77
Moist oxygen	29	82
Ammonia	29	82
Water vapor (100% saturated steam)	60	95

(a) Heated at 920 to 930 F for 20 hr (unless otherwise noted). (b) Heated for 2 hr. (c) Heated for 30 min.

Temperature control

Many heat treatments must be conducted at a temperature near the eutectic melting point, making it mandatory to control temperature closely throughout a charge during heating, and to maintain the precise temperature during the terminal part of the treatment cycle. Most practices require metal temperature control and furnace uniformity within ± 10 F; in many instances, a distinct advantage accrues if temperatures can be held within ± 5 F.

The instrumentation generally selected for aluminum heat treating is of the thermoelectric type, incorporating a null-balance potentiometer, in which external resistance does not influence sensing accuracy. Both chromel-alumel and iron-constantan calibrations are used in combination with the corresponding thermocouple and lead-wire materials.

Batch-type furnaces lend themselves to control by a load couple. A thermocouple inserted in a piece of aluminum with heating characteristics similar to the product being processed is placed in the furnace with the charge. Because load couples of this type provide the actual temperature of the load, they can be used to establish the high-temperature and low-temperature areas of a charge and permit adjustment of the cycle.

Conveyor and continuous furnaces do not readily permit product temperature monitoring during processing. Performance of this type of equipment therefore requires carefully sustained maintenance and frequent checking of temperature-control instruments.

Atmosphere control

The presence of water vapor, accelerated by certain contaminants such as ammonia and sulfur compounds at elevated heat treating temperatures, can result in destruction of the normal oxide film. Moisture then reacts with the aluminum surface to produce atomic hydrogen, which readily diffuses into the solid metal, precipitating as molecular hydrogen. This may cause subsequent blistering of the product, high-temperature oxidation, porosity that does not heal on subsequent working, or undesirable discoloration.

Dry air is the atmosphere most generally used. Under proper conditions, combustion products of certain gaseous fuels, obtained by direct firing, may be used. The presence of sufficient carbon dioxide in an atmosphere containing 50% or more combustion products inhibits the action of moisture developed during combustion.

The undesirable effects of furnace atmospheres with high moisture content or contaminated with minor amounts of sulfur compounds generally can be alleviated by introducing volatile fluoride-containing compounds—Ammonium, sodium, and potassium fluoborate and boron trifluoride.

To avoid either residual oil stain or discolored oxide coating, sheet and foil products commonly are annealed in a controlled atmosphere, such as the product of the complete combustion of a fuel such as natural gas, subsequently chilled to reduce its moisture content. Frequently an atmosphere chilled approximately 80 F and having a moisture content of slightly over 3% is adequate. ■

Heat treating of aluminum, 2

Homogenizing can control gain growth in ingots and billets; annealing can make the alloys more workable

Preheating and homogenizing

Preheating provides a desirable homogenizing effect, which improves workability and helps control recrystallization and grain growth. A desirable diffusion of the alloying elements takes place. Brittleness of the cast structure is reduced substantially and the characteristics are improved for subsequent fabrication.

The time required for preheating treatments depends on the degree of refinement of the cast structure: the finer the structure, the more rapid the diffusion of soluble alloying elements. Thus, ingots cast by the direct-chill process, having fine dendrite cell size, can be homogenized in less time than ingots produced by the cold mold process, which have a courser structure. An exception is alloy 3003.

The cast microstructure also is affected by ingot size; thicker ingots require longer preheating time than thinner ingots.

Typical preheating conditions for extrusion and forging ingots are given in Table 3. These are based on the use of direct-chill-cast ingots of square or round cross section up to approximately 250 sq in.

After preheating, ingots may be hot worked immediately at the temperature required for the specific operation, or they may be reheated later to the hot working temperature. Ingots of commercially pure aluminum and of alloys with low constituent content, need only be heated to the desired hot working temperature prior to working.

Extracted from material in a 3-vol. series on Aluminum, edited by Dr K R Van Horn, vice president, research and development, Aluminum Co of America, Pittsburgh, and written by over 90 associated experts. The volumes have now been published by the American Society for Metals, Cleveland

Microstructure can affect finishing characteristics of many wrought products. Grain size and orientation, brightness and color of etched and anodized products can be affected both by rate of heating and cooling during preheating and by the maximum temperature and time at that temperature. The maintenance of finish uniformity requires practices that insure uniform heating and cooling conditions for all ingots in a furnace charge. Spacing ingots in suitable racks so as to permit uniform atmosphere circulation during heating and cooling is desirable.

Preheating at high temperatures and for long times requires critical control of furnace atmosphere.

Annealing

Annealing places the metal in a softened, more workable condition for subsequent forming or cold working operations. See Table 4.

Strain-hardened, non-heat-treatable wrought alloys must be heated to the annealing temperature as rapidly as possible to insure a desirable fine recrystallized grain size. The usual commercial practice with batch-type annealing furnaces is to control the furnace at 25 to 50 F higher than the specified temperature until the metal reaches the desired temperature, as indicated by one or more thermocouples. The furnace control is then reset to the specified temperature and the load held for 1 to 2 hr to assure that all parts attain the required temperature.

On completion of the heating cycle, the metal may be withdrawn from the furnace without regard to rate of cooling, if mechanical properties are the only controlling factor. However, if an oxygen-free atmosphere is used to prevent staining and discoloration, some cooling in the furnace may be required before the load is removed. This is important if the furnace charge consists of oily metal in intimate contact. Cooling in the furnace should be continued to a temperature low enough to prevent staining by oxidation of residual oil when the load is removed.

It is permissible to employ higher metal temperatures for annealing in conveyor or continuous furnaces, where individual pieces can be heated quickly to temperature and immediately withdrawn. Practices in general use involve furnace-atmosphere temperatures of 900 to 1000 F, with the metal reaching 800 to 850 F for 30 to 60 sec before being discharged from the heating chamber and cooled. Specific practices must be established by trial.

The practices recommended for obtaining annealed products having normal mechanical properties and grain size presuppose metal sufficiently strain hardened to react properly to the annealing treatment. Best results are obtained when the metal has been strain hardened to the equivalent of a 50% cold reduction or more. With lesser amounts of strain hardening, the probability of excessive grain growth increases. For this reason, discretion must be used when annealing products in the intermediate tempers, approximately H14 or H34 and softer. Where parts partially formed from annealed metal receive intermediate annealing, variable or excessive grain growth is frequent in locations where the degree of strain hardening is inadequate.

Partial annealing of strain-hardened alloys is carried out at temperatures below those normally used to achieve complete recrystallization (full anneal.)

Mill products in H2 temper may be strain hardened substantially

Heat treating of aluminum, [Continued]

more than needed. They are then brought to the required lower level of strength by partial annealing. Partial annealing sometimes is useful where the metal is slightly too hard to permit forming, and a full anneal is not desired. Because final characteristics are influenced by the previous thermal treatments and the degree of cold work, it is impossible to specify precise conditions for partial annealing.

Stabilizing heat treatments are applied to aluminum alloys containing magnesium as a major alloying element to control 'age softening' after strain hardening. These H-3 type treatments provide stable mechanical properties and improve forming characteristics. They consist of short-time treatments in the range of 250 to 350 F. Normally, the strength of the product after stabilization is lower than results from prolonged age softening at room temperature. Products of these alloys provided in the H2-type tempers do not require a final stabilizing treatment as the partial annealing produces stabilization simultaneously.

Heat treatable wrought alloys. When annealed at 650 F, the heat treatable alloys generally soften enough to permit less severe forming or working operations. The cooling rate from 650 F is not important. However, where increased workability is required, or where the hardness of the alloy derives from partial heat treatment effects from prior hot working or an actual heat treatment, an annealing treatment at higher temperature is necessary. Usual practice is to heat for 2 to 3 hr at 775 F (Table 4). This is followed by cooling in the furnace at a rate not in excess of 50 F per hr, to 500 F or lower.

Various modifications of this annealing practice involve slightly lower or higher soaking temperatures, cooling more slowly in the earlier portion of the cooling period, and controlled cooling to a lower temperature (450 F).

For the 7xxx series heat treatable alloys, a two-step treatment is equivalent and generally preferred for maximum softening, workability, and stability of mechanical properties: (a) holding for 2 to 3 hr at 775 F, followed by cooling at any convenient rate to 450 F or below; (b) holding at 450 F for 4 hr, immediately after cooling to 450 F or later, by reheating the metal to this temperature if cooled to a lower temperature. The second step frequently is termed a stabilizing treatment, but should not be confused with the previously described stabilizing treatment. ■

Table 3. Typical preheating treatments for aluminum allloy forging or extrusion ingot(a)

Alloy	Metal temperature, F	Soaking time at temperature, hr(b)
EC, 1060, 1100	No soaking treatment required	
2EC	995 to 1025	4 to 8
2011, 2024	890 to 920	4 to 12
2014, 2017, 2218, 2219, 2618	910 to 940	4 to 12
3003	1145 to 1175	4 to 12
4543	880 to 910	12 to 24
5005, 5050, 5052, 5154	No soaking treatment required	
5083, 5086, 5456	995 to 1025	4 to 12
6061, 6063, 6066, 6070, 6101 6151, 6262, 6351, 6463, 6563	995 to 1025	4 to 12
7001, 7038, 7039, 7075, 7079, 7178	850 to 880	8 to 48

(a) For square or round ingot cast by direct-chill method and with cross-section area up to 250 sq in. (b) The shorter time usually is sufficient for extrusion requirements; the longer time is needed to insure the desired workability for forging.

Table 4. Typical annealing conditions for aluminum alloys

Alloy	Temper	Metal temperature, F	Time at temperature, hr
Strain-hardened, non-heat-treatable wrought alloys			
EC all 1xxx series	O	650	(a)
3003	O	775	(a)
3004	O	650	(a)
4043, 4643	O	650	(a)
All 5xxx series	O	650	(a)
7072	O	650	(a)
Heat treatable wrought alloys			
All 2xxx series	O	775(b)	2 to 3
All 6xxx series	O	775(b)	2 to 3
All 7xxx series	O	775(c)	2 to 3
Non-heat-treatable casting alloys			
A140	T2	600	3 to 4
F214	T2	600	2 to 3
364	T2	650	4 to 5
Heat treatable casting alloys			
142	T2	650	3 to 4
319	T2	650	3 to 4
355	T2	600	3 to 4

(a) Time in the furnace need not be longer than is necessary to bring all parts of load to the annealing temperature. Rate of cooling is unimportant.

(b) Treatments intended to remove effects of solution heat treatment include cooling from the annealing temperature to 500 F at rate of about 50 F per hr. The rate of subsequent cooling is unimportant. To remove only effects of cold work or partially remove effects of heat treatment, if a completely annealed material is not required, treatment at a temperature as low as 650 F may be used.

(c) This treatment is intended to remove the effects of solution heat treatment, and includes air cooling to 450 F or less, followed by reheating at 450 F for 4 hr. Treatment at 650 F, followed by uncontrolled cooling, may be used to remove effects of cold work, or partially remove the effects of heat treatment, if completely annealed material is not required.

Heat treating of aluminum, 3

Regardless of the type of furnace used, aluminum shapes must be supported against potential sagging and warping while they are undergoing heating for solution treatment

Extracted from material in a 3-vol. series on Aluminum, edited by Dr K R Van Horn, vice president, research and development, Aluminum Co of America, Pittsburgh, and written by over 90 associated experts. The volumes have now been published by the American Society for Metals, Cleveland

Solution heat treatment improves mechanical properties by developing the maximum practical concentration of the hardening constituents in the solid solution. This process requires heating the product to a temperature close to the eutectic, holding it there long enough to effect the desired solution, and then quenching fast enough to retain the desired solid solution.

Recommended soaking times for wrought alloy products are given in Table 5. In salt bath furnaces, soaking time begins when the load does not cause the bath temperature to drop more than 10 F below the specified nominal temperature.

Packing for solution heat treatment requires careful consideration because of its effect on heating

Table 5. Recommended soaking time for solution heat treatment of wrought aluminum alloys other than forgings

Minimum thickness of heaviest section, in.	Soaking time, min — Salt bath(a) Min	Max(c)	Air furnace(b) Min	Max(c)
0.016 and under	10	15	20	25
0.017 to 0.020	10	20	20	30
0.021 to 0.032	15	25	25	35
0.033 to 0.063	20	30	30	40
0.064 to 0.090	25	35	35	45
0.091 to 0.125	30	40	40	55
0.126 to 0.250	35	45	55	65
0.251 to 0.500	45	55	65	75
0.501 to 1.000	60	70	90	100
Each additional 0.5 in. or fraction	Add 20 min		Add 30 min	

(a) Soaking time measured either from immersion of load or from recovery of bath to within 10 F of recommended solution temperature, if bath temperature drops due to size of charge.
(b) Soaking time measured from time all furnace control couples indicate furnace temperature has recovered to controller setting. Shorter soaking times are possible when temperature is determined accurately from thermocouple attached to load.
(c) Maximum applicable to Alclad products only.

Table 6. Recommended heat treatments for aluminum alloy sheet and plate products

Alloy	Temper	Solution heat treatment(a)(b) Metal temperature, F(c)	Precipitation heat treatment — Metal temperature F(c)	Time at temperature hr(d)
2014	T4, T42, T451(e)	935	...	..
	T6, T62, T651(e)	935	320	18
2219	T31(f), T351(e), T37(f)	995	...	..
	T42	995	...	..
	T62	995	375	36
	T81(f), T851(e)	995	350	18
	T87(f)	995	325	24
2020	T6, T651(e)	960	320	18
2024	T3(f), T351(e), T36(f)	920	...	..
	T4, T42	920	...	..
	T6, T62	920	375	16
	T81(f), T851(e)	920	375	12
	T86(f)	920	375	8
6061	T4, T451(e)	985	...	..
	T6, T651(e)	985	320	18
X7005	T6	900	225(g)	8
			+300	16
7039	T6	900	240(g)	48
	T61	850	320(h)	18
7075	T6, T651(e)	900	250	24
	T73, T7351(e),	900	225	6
			+325	24
7178	T6, T651(e)	875	250	24
7079	T6, T651(e)	830	200	6
			+250	24

(a) The time of heating varies with the thickness of the product, type of furnace, and size of furnace load. Table 5 gives recommended soaking times. (b) Products should be quenched from solution heat treating temperature as rapidly as possible in room-temperature water suitably cooled to remain below 100 F during the quenching cycle. (c) Nominal metal temperature should be attained as rapidly as possible, and maintained with as little variation as possible during the time at temperature. Furnaces capable of maintaining the temperature well within ±10 F are readily available.
(d) Time of heating depends on rate at which load is brought to temperature. Conditions shown are based on rapid heating, with soaking time measured from time load rises to within 10 F of recommended temperature. (e) Stress relieved by cold stretching to 1.5 to 3% permanent set subsequent to solution heat treatment and prior to any elevated-temperature precipitation treatment
(f) Cold working subsequent to the solution heat treatment and prior to any elevated-temperature precipitation treatment is necessary to secure the specified properties for these tempers. (g) Age at room temperature for 72 hr prior to precipitation heat treatment. (h) Age at room temperature for 48 hr prior to heating to precipitation treating temperature at rate of 35 ± 5 F per hr.

Heat treating of aluminum [Continued]

and quenching conditions, distortion, and residual stress. This applies especially to products of irregular shape, such as castings, forgings, and formed or assembled items. Small parts usually are packed in steel racks or baskets designed to permit free circulation of the heating medium during quenching. In addition, space for circulation around each part is required, requiring careful packing and orientation within the rack or basket.

When packing parts for solution treatment, air pockets or other traps that would interfere with entry of the quenching medium should be oriented to avoid entrapment of salt on removal. Carry-over of salt into the quench water is hazardous, because explosive reactions may occur. Residual salt that may follow through the process also can result in corrosion of the parts.

The low strength of aluminum at solution heat treating temperatures requires support of parts during treatment. This limits the extent to which parts can be stacked in baskets without excessive distortion. The same consideration applies when aluminum parts are loaded on racks and conveyor belts for heat treatment in continuous-type furnaces.

Large castings and forgings, as well as large formed and assembled products, may require special fixtures to support overhanging elements during solution heat treatment.

Proper quenching is equally important to careful observance of heating conditions. Once the metal begins to cool, the alloying elements in solid solution are likely to precipitate immediately. If an excessive degree of uncontrolled precipitation occurs, it can have significant adverse effects on mechanical properties and resistance to corrosion.

Undesirable precipitation varies in its effect on the various aluminum alloys. For those that are most 'quench sensitive,' such as 7075, an extended delay in quenching generally results in mechanical properties below established requirements. Some alloys, such as 2024, are less sensitive to the effect of quench rate on mechanical properties, but can be affected adversely by lowered resistance to corrosion as the quench rate is reduced. The resistance to corrosion of aluminum-copper alloys in the artificially aged condition is less dependent on a rapid quench than when in the naturally aged temper.

The most critical temperature range during quenching begins at approximately 750 F; it extends to approximately 500 F, varying with the alloy. In this range, precipitation occurs quickly if the rate of cooling is not rapid enough. All parts of the load must be at a temperature well above 750 F as the load enters the quenching medium; 775 F usually is considered an absolute minimum.

Table 7. Recommended heat treatments for rolled or drawn aluminum alloy wire, rod, bar, shapes, and tube

Alloy	Temper	Solution heat treatment(a)(b) Metal temperature, F(c)	Precipitation heat treatment: Metal temperature F(c)	Precipitation heat treatment: Time at temperature hr(d)
2011	T3(e)	975(h)	. . .	. .
	T8(e)	975(h)	320	14
2014	T4, T42, T451(f)	935	. . .	. .
	T6, T62, T651(f)	935	320	18
2017	T4, T451(f)	945	. . .	. .
2219	T31(e), T351(f)	995	. . .	. .
	T42	995	. . .	. .
	T6, T62	995	375	36
	T81(e), T851(f)	995	375	18
2024	T3(e), T351(f), T36(e)	920	. . .	. .
	T4, T42	920	. . .	. .
	T6, T62	920	375	16
	T81(e), T851(f)	920	375	11
	T86(e)	920	375	8
6053	T4	975	. . .	. .
	T6	975	320	16
	T61	945	360	7
6061	T4, T451(f)	985	. . .	. .
	T6, T651(f)	985	320	18
	T91(g)	1025	320	24
6262	T6, T651(f), T9(g)	1025	350	12
6066	T6	990	350	8
7001	T6, T651(f)	870	250	24
7075	T6, T651(f)	915(h)	250	24
	T73	915(h)	225	6
			+350	8
7079	T6, T651(f)	830	200	6
			+240	48
7178	T6, T651(f)	900(h)	250	24

(a) The time of heating varies with the thickness of the product, type of furnace, and size of furnace load. Table 5 gives recommended soaking times. (b) Products should be quenched from solution heat treating temperature as rapidly as possible in room-temperature water suitably cooled to remain below 100 F during the quenching cycle. (c) Nominal metal temperature should be attained as rapidly as possible, and maintained with as little variation as possible during the time at temperature. Furnaces maintain the temperature well within ±10 F.

(d) Time of heating depends on rate at which load is brought to temperature. Conditions, shown are based on rapid heating, with soaking time measured from time load rises to within 10 F of recommended temperature. (e) Cold working subsequent to the solution heat treatment and prior to any elevated-temperature precipitation treatment is necessary to secure the specified properties for this temper. (f) Stress relieved by cold stretching to 1 to 3% permanent set subsequent to solution heat treatment and prior to any elevated-temperature precipitation treatment. Tempers not applicable to drawn tube. (g) Cold working subsequent to the precipitation heat treatment is necessary to secure the specified properties for this temper. (h) Solution temperature for drawn tube: 960 F for 2011, 870 F for 7075 and 7178.

Heat treatment of alloy steels

[Directly hardenable grades]

SAE Steels[a]	Normalizing temperature, F	Annealing[d] temperature F	Hardening[e] temperature, F	Quenching medium	Temper
1330	1600–1700[b]	1550–1650	1525–1575	War or oil	To desired hardness
1335 1340 1345	1600–1700[b]	1550–1650	1500–1550	Oil	To desired hardness
4037 4042	–	1500–1575	1525–1575	Oil	To desired hardness
4047	–	1450–1550	1500–1575	Oil	To desired hardness
4130	1600–1700[b]	1450–1550	1500–1600	Water or oil	To desired hardness
4135 4137 4140 4142	–	1450–1550	1500–1600	Oil	To desired hardness
4145 4147 4150	–	1450–1550	1500–1550	Oil	To desired hardness
4161	–	1450–1550	1500–1550	Oil	To desired hardness, 700 F, min
4340	1600–1700[b, c]	1450–1550	1500–1550	Oil	To desired hardness
50B40 50B44 5046 50B46	1600–1700[b]	1500–1600	1500–1550	Oil	To desired hardness
50B50 50B60	1600–1700[b]	1500–1600	1475–1550	Oil	To desired hardness
5130 5132	1600–1700[b]	1450–1550	1525–1575	Water, caustic solution, or oil	To desired hardness
5135 5140 5145	1600–1700[b]	1500-1600	1500–1550	Oil	To desired hardness
5147 5150 5155 5160 51B60	1600–1700[b]	1500-1600	1475–1550	Oil	To desired hardness
50100 51100 52100	–	1350–1450	1425–1475 1500–1600	Water Oil	To desired hardness
6150	–	1550–1650	1550–1625	Oil	To desired hardness
81B45	1600–1700[b]	1550–1650	1500–1575	Oil	To desired hardness
8630	1600–1700[b]	1450–1550	1525–1600	Water or oil	To desired hardness
8637 8640	–	1500–1600	1525–1575	Oil	To desired hardness
8642 8645 86B45 8650	–	1500–1600	1500–1575	Oil	To desired hardness
8655 8660	–	1500–1600	1475–1550	Oil	To desired hardness
8740	–	1500–1600	1525–1575	Oil	To desired hardness
9254 9255 9260	–	–	1500–1650	Oil	To desired hardness
94B30	1600–1700[b]	1450–1550	1550–1625	Oil	To desired hardness

[a] These steels are fine grain unless otherwise specified.
[b] These steels should be either normalized or annealed for optimum machinability.
[c] Temper at 1100-1225.
[d] The specific annealing cycle is dependent upon the alloy content of the steel, the type of subsequent machining operations and desired surface finish.
[e] Frequently, these steels, with the exception of 4337, 4340, 50100, 51100, 52100, and 9840, are hardened and tempered to a final machinable hardness without preliminary heat treatment.

Heat treatment of alloy steels [Continued]

[Carburizing grades]

SAE Steels[a]	Pretreatments: Normalize[b]	Pretreatments: Cycle Anneal[d]	Carburizing temperature,[e] F	Cooling method	Reheat temperature, F	Quenching medium	Tempering[f] temperature, F
4012 4023 4024 4027 4028 4032	Yes	—	1650–1700	Quench in oil	—	—	250–350
4118	Yes	—	1650–1700	Quench in oil	—	—	250–350
4320	Yes	Yes	1650–1700 1650–1700	Quench in oil Cool slowly	— 1525–1550[i]	— Oil	250–350 250–350
4419 4422 4427	Yes	Yes	1650–1700	Quench in oil	—	—	250–350
4615 4617 4620 4621 4626 4718	Yes	Yes	1650–1700 1650–1700 1650–1700	Quench in oil Cool slowly Quench in oil	— 1500–1550[i] 1500–1550[h]	— Oil Oil	250–350 250–350 250–350
4720	Yes	Yes	1650–1700	Quench in oil	1500–1550[h]	Oil	250–350
4815 4817 4820	Yes[c]	Yes	1650–1700 1650–1700 1650–1700	Quench in oil Cool slowly Quench in oil	— 1475–1525[i] 1475–1525[h]	— Oil Oil	250–325 250–325 250–325
5015 5115 5120	Yes	—	1650–1700	Quench in oil	—	—	250–350
6118	Yes	—	1650	Quench in oil	—	—	325
8115 8615 8617	Yes	—					
8620 8622 8625 8627 8720 8822	Yes	Yes	1650–1700 1650–1700 1650–1700	Quench in oil Cool slowly Quench in oil	— 1550–1600[i] 1550–1600[h]	— Oil Oil	250–350 250–350 250–350
9310	Yes[c]	—	1600–1700 1600–1700	Quench in oil Cool slowly	1450–1525[h] 1450-1525[i]	Oil Oil	250–325 250–325
94B17 94B15	Yes	—	1650–1700	Quench in oil	—	—	250–350

[a] These steels are fine grain.

[b] Normalizing temperature should be at least as high as the carburizing temperature followed by air cooling.

[c] After normalizing, reheat to temperature of 1100-1200 F and hold at temperature approximately 1 hr per in. of maximum section or 4 hr minimum time.

[d] Where cycle annealing is desired, heat to at least as high as the carburizing temperature, hold for uniformity, cool rapidly to 1000-1250 F, hold 1 to 3 hr, then air cool or furnace cool to obtain a structure suitable for machining and finish.

[e] It is general practice to reduce carburizing temperatures to approximately 1550 F before quenching to minimize distortion and retained austenite. For 4800 series steels, the carburizing temperature is reduced to approximately 1500 F before quenching.

[f] Tempering treatment is optional. Tempering is generally employed for partial stress relief and improved resistance to cracking from grinding operations.

[h] This treatment is used where the maximum grain refinement is required and/or where parts are subsequently ground on critical dimensions. A combination of good case and core properties is secured with somewhat greater distortion than is obtained by a single quench.

[i] In this treatment the parts are slowly cooled, preferably under a protective atmosphere. They are then reheated and oil quenched. A tempering operation follows as required. This treatment is used when machining must be done between carburizing and hardening or when facilities for quenching from the carburizing cycle are not available.

Heat Treatment of Tool Steels—I

Courtesy of
Latrobe Steel Co, Latrobe, Pa

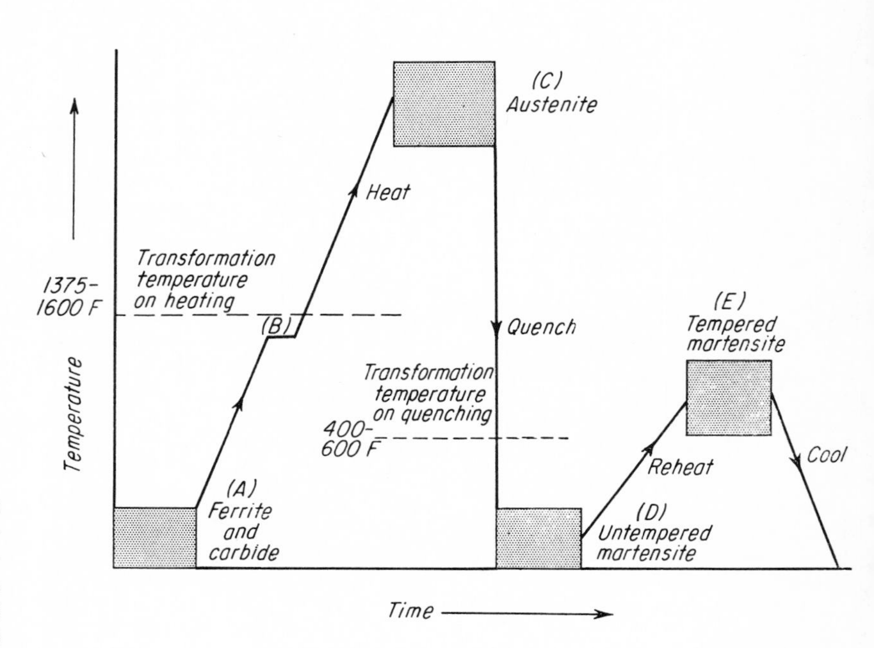

Schematic Diagram of What Happens in Heat Treatment

Preheating

Metallurgically, preheating takes no part in the hardening reaction. However, preheating is usually applied to all tool steels in order to minimize the normal hazards encountered in the heat treatment of these higher alloyed materials.

There are two primary functions of preheating. The first is to reduce thermal shock when a cold tool is placed directly into a hot furnace. The second is to start the austenitizing reaction so that the carbides will be in a more favorable condition for dissolving at high temperature. This latter function is particularly important in the heat treatment of high-speed steels and those die steels requiring a hardening temperature in excess of 2000 F.

Because preheating reduces thermal shock, it is beneficial in minimizing the danger of excessive distortion, warping and cracking. It serves as a stress relief of the internal strains set up during machining. If preheating is carried out just below the critical temperature, these machining stresses are relieved prior to transformation within the steel.

One or more preheating steps also enable the tools to pass through the transformation range as uniformly as possible throughout the entire cross section. Be-

Point *A* represents the annealed steel ready for heat treatment. In the annealed condition the steel is made up of a ferrite and carbide structure, ferrite being incapable of dissolving carbide in solid solution. Preheating of the steel occurs at some stage between *A* and *C*, usually near the transformation temperature *B*.

When heated above the transformation temperature range, the steel undergoes a phase change to a new crystallographic form called austenite, C. Austenite possesses the ability to dissolve carbides in solid solution. The austenitizing treatment is adjusted by temperature and time to dissolve the optimum amount of carbide. These adjustments vary appreciably for different types of steel. The austenite phase represents a more dense structure, which means that the steel occupies a lesser volume when at the hardening temperature.

From C the steel is quenched relatively rapidly by air, oil, or water, depending upon the alloy composition. Below the transformation temperature, austenite is unstable and attempts to regain its room-temperature structure of ferrite and carbide, *A*. The speed of the quench, however, suppresses this reversion, and a new crystallographic form is created called martensite, *D*. Untempered martensite *D* is also unstable and represents a strained condition in the steel from which the hardness is derived. This martensite phase is of greater volume than both the prior austenite phase and the ferrite plus carbide annealed phase, and therefore it is completely normal for steel to increase in dimension upon hardening.

Because untempered martensite is unstable and highly strained, the steel is reheated to produce tempered martensite *E*. This serves to stabilize the martensite and relieve the hardening stresses. Some loss in hardness may result from tempering. A volumetric contraction occurs so that the steel more nearly approaches its original size.

Heat Treatment of Tool Steels—II

cause heating annealed steel to a temperature above the critical results in a decrease in volume, it is important for the tools to go through this phase change slowly and uniformly to minimize distortion.

Preheating of carbon- and low-alloy tool steels normally is carried out in the range of 1200-1300 F, high carbon-high chromium and other highly alloyed die steels at 1200-1450 F, and high-speed and modified high-speed steels at 1500-1600 F. If no protective atmosphere is available, the temperature should be adjusted to the lower side of the range to reduce the danger of excessive scale and decarburization. Double preheating at both 1000-1200 F, and 1500-1600 F is often recommended for treatment of intricate tools and dies, and particularly for high-speed steel when treated in liquid baths.

Austenitizing

Austenitizing is the high-heat phase of the heat-treating operation. Basically, the austenitizing treatment controls the exact hardness which will be obtained, determines the eventual size change, develops the red hardness, and brings out many of the other desirable metallurgical characteristics which are designed into the steel by the chemical composition.

Adequate protection of the steel surface against carburization, decarburization, and excessive scaling is required.

Austenitizing is a time-temperature reaction, in which the more important factor is the temperature. In developing heat-treating data for a new steel, the manufacturer determines the proper temperature range to employ and uses a relatively broad latitude for the time factor. As a result, it is difficult to overheat a piece of steel during austenitizing by using too long a time at temperature, provided the temperature of the steel is within the recommended hardening range. Obviously, this points up the need for accurate pyrometer control of the heating medium with a minimum of temperature fluctuation in order to assure optimum heat-treating results.

When austenitizing temperatures are higher than normal, either intentionally or inadvertently, the time factor becomes increasingly important. The final heat-treating results tend to be erratic and difficult to predict with accuracy.

Carbon- and low-alloy tool steels, which are austenitized at relatively low temperatures, depend upon the solution of iron carbide in austenite for their hardening characteristics. Austenite readily dissolves iron carbide. Therefore, considerable latitude is permissible with respect to both time and temperature in heat treating of these types of steel. The danger of undersoaking is greatly minimized, because of the ease of carbide solution. Usually, it is necessary only to insure that the steel is uniform in temperature and within the proper hardening temperature range to obtain fully satisfactory results. Soaking times for these steels are approximately 5 min per in. of thickness, although longer times are not detrimental.

Higher alloyed die steels, and particularly the high carbon—high chromium types, are much more sluggish in dissolving carbides. The soaking time should be correspondingly increased over the low-alloy steels to avoid the danger of undersoaking. With furnace hardening, a soaking time at temperature of 30 min for the first inch of thickness and 10-15 min for each additional inch is a good rule of thumb. Regardless of how small the size may be, the minimum soaking time should not be less than 20-30 min.

Pack hardening requires additional soaking. The long-established rule of one hour per inch of thickness normally is followed.

High-speed and modified high-speed steels depend upon the solution of various complex alloy carbides during austenitizing to develop their outstanding heat-resisting qualities. These alloy carbides do not dissolve to an appreciable extent unless the steel is heated extremely close to its melting point. Accurate temperature control is mandatory. The factor of time is of greater importance than it is with lower alloy steels hardened at temperatures below 2000 F.

Considerable skill and experience are required to austenitize high-speed steel. In general, heavy-duty tools are hardened from the high side of the recommended temperature range, while fine edge tools are austenitized on the lower side. The soaking time for austenitizing also will vary somewhat, but normally is in the range of 2-5 min at temperature.

Quenching

Quenching may be defined as a controlled cooling operation which allows the steel to harden. While the austenitizing treatment determines the exact hardness that will be obtained, the quenching cycle is the step that permits the steel to attain that hardness.

Quenching is also a time-temperature dependent reaction, with equal emphasis placed on both of these factors. In cooling from the austenitizing temperature, the steel must pass fairly rapidly through the critical range (approximately 1300-1100 F) in order to insure that the hardening reaction will take place. The minimum rate at which the steel

Heat Treatment of Tool Steels—III

must be cooled to develop full hardness can be termed the critical quenching velocity, and is controlled basically by the analysis of the steel and the austenitizing treatment. This critical quenching velocity is quite rapid for low-alloy steels but is much slower for steels containing appreciable amounts of hardenability agents, such as molybdenum, nickel, chromium, and manganese. Steels therefore are commonly classified according to their required quenching velocity, that is, water-, oil-, or air-hardening varieties.

Getting the steel successfully through this critical range of 1300-1100 F is one of the major hurdles in the quenching cycle, but at this point the steel has only promised to harden. Now the reaction becomes more dependent on temperature rather than time, and actual hardening does not begin until a much lower temperature is reached.

The temperature at which hardening begins is below 600 F for most tool steels. The length of time that the steel can be held in the twilight range of 1100 to 600 F depends upon its composition. Existence of this twilight temperature range permits the use of special quenching techniques such as slack quenching, martempering, and salt- and lead-bath quenching. These techniques employ an equalization of temperature before initiating the actual hardening reaction. The final stage of quenching is then carried out at a less drastic rate, resulting in more uniform cooling and lower quenching stresses.

The hardening which occurs in the latter stages of the quench is a volumetric reaction with the relatively soft austenite changing to a harder and more voluminous martensitic structure. This size increase sets up internal transformation stresses in the steel. These stresses, when combined with the thermal stresses introduced by the cooling operation itself, can under adverse circumstances cause warping or even cracking of the steel. An air-hardening steel is less prone to exhibit these quenching problems than is a steel requiring a more drastic quench, such as oil or water.

One common quenching fault is not permitting the steel to cool sufficiently. This results in incomplete hardening and highly stressed finished parts. Tool steels should always be cooled until the parts can be handled comfortably with the bare hands.

During the quenching cycle, the steel is in its most highly stressed condition. Therefore, this step in the heat-treating cycle generally introduces some of the greatest problems associated with the hardening process. Each tool or die must be considered individually, and each presents its own particular challenge.

Tempering

The fourth and final phase of heat treatment, tempering has been defined as reheating of steel, after hardening, to an intermediate temperature range for the primary purpose of toughening the steel.

Upon completion of hardening, the steel is in a highly stressed condition. Tool and die steels generally are not usable in this state, because most applications require some degree of toughness in order to withstand the shock of normal operation.

By reheating the steel to an intermediate temperature, the hard martensitic structure is broken down and some carbon is precipitated from the martensite in the form of fine carbides. This results in most cases in a slight decrease in hardness. While a loss in hardness is not always desirable, the high internal stresses which cause the steel to be brittle are eliminated.

Tempering accomplishes its purpose through a combination of temperature and time. The usual range of tempering temperatures extends from 300 to 1200 F, depending on the grade of steel and the hardness level desired. It is not sufficient merely to heat a hardened part to a certain temperature; it must be soaked at that temperature for a definite length of time. In simple alloys, after the part has reached the tempering temperature throughout the entire section, a one-hour soak is sufficient. However, in more complex alloys where diffusion is slower, as for example in the 5% chromium die-casting steels, a number of hours at temperature is desired. The response to tempering is slow and requires a full measure of time for completeness.

Higher alloyed tool and die steels, which are hardened from temperatures in excess of 2000 F, should always be double tempered in order to achieve maximum benefit. The reason for this is that on cooling from the first temper, additional fresh martensite is formed from retained austenite, and the second temper toughens this newly formed martensite.

If cooling from the quench is interrupted too soon, full hardness will not be obtained. But, on the other hand, if quenched steel is allowed to remain at room temperature for an extended period, the stresses from hardening continue to build up until a point is reached where cracking occurs. Two good rules to follow are:

1. Allow the steel to cool after hardening to a point where it can be comfortably handled with the bare hands.

2. Then, begin the tempering operation immediately.

Hardness Conversion Tables—I

The new American Standard Z 76.4—1961 embodies the collection of standard hardness conversion tables found in Standard E 140—58 issued by the American Society for Testing and Materials, 1916 Race St, Philadelphia 3, Pa. Through the courtesy of ASTM, two tables from E140-58 are reprinted here. The complete standard can be obtained from ASTM.

Both tables present data on the relationships of Brinell hardness, diamond pyramid hardness, Rockwell hardness and Rockwell Superficial hardness for carbon, alloy and tool steels in the as-forged, annealed, normallized and quenched and tempered conditions, provided these materials are homogeneous. The table on this page correlates hardness numbers in relation to the Rockwell C reading, whereas the table on page 278 presents similar data in relation to the Brinell indentation diameter.

Rockwell C Hardness Number	Diamond Pyramid Hardness Number	Brinell Hardness Number*			Rockwell Hardness Number		Rockwell Superficial Hardness Number			Rockwell C Hardness Number
		10-mm Standard Ball, 3000-kg Load	10-mm Hultgren Ball, 3000-kg Load	10-mm Carbide Ball, 3000-kg Load	A Scale, 60-kg Load, Diamond Cone Penetrator	D Scale, 100-kg Load, Diamond Cone Penetrator	15-N Scale, 15-kg Load, Superficial Diamond Cone Penetrator	30-N Scale, 30-kg Load Superficial Diamond Cone Penetrator	45-N Scale, 45-kg Load Superficial Diamond Cone Penetrator	
68	940	..	..	..	85.6	76.9	93.2	84.4	75.4	68
67	900	..	..	..	85.0	76.1	92.9	83.6	74.2	67
66	865	..	..	..	84.5	75.4	92.5	82.8	73.3	66
65	832	..	..	**739**	83.9	74.5	92.2	81.9	72.0	65
64	800	..	..	**722**	83.4	73.8	91.8	81.1	71.0	64
63	772	..	..	**705**	82.8	73.0	91.4	80.1	69.9	63
62	746	..	..	**688**	82.3	72.2	91.1	79.3	68.8	62
61	720	..	..	**670**	81.8	71.5	90.7	78.4	67.7	61
60	697	..	**613**	**654**	81.2	70.7	90.2	77.5	66.6	60
59	674	..	**599**	**634**	80.7	69.9	89.8	76.6	65.5	59
58	653	..	**587**	615	80.1	69.2	89.3	75.7	64.3	58
57	633	..	**575**	595	79.6	68.5	88.9	74.8	63.2	57
56	613	..	**561**	577	79.0	67.7	88.3	73.9	62.0	56
55	595	..	**546**	560	78.5	66.9	87.9	73.0	60.9	55
54	577	..	**534**	543	78.0	66.1	87.4	72.0	59.8	54
53	560	..	**519**	525	77.4	65.4	86.9	71.2	58.6	53
52	544	**500**	**508**	512	76.8	64.6	86.4	70.2	57.4	52
51	528	**487**	494	496	76.3	63.8	85.9	69.4	56.1	51
50	513	**475**	481	481	75.9	63.1	85.5	68.5	55.0	50
49	498	**464**	469	469	75.2	62.1	85.0	67.6	53.8	49
48	484	451	455	455	74.7	61.4	84.5	66.7	52.5	48
47	471	442	443	443	74.1	60.8	83.9	65.8	51.4	47
46	458	432	432	432	73.6	60.0	83.5	64.8	50.3	46
45	446	421	421	421	73.1	59.2	83.0	64.0	49.0	45
44	434	409	409	409	72.5	58.5	82.5	63.1	47.8	44
43	423	400	400	400	72.0	57.7	82.0	62.2	46.7	43
42	412	390	390	390	71.5	56.9	81.5	61.3	45.5	42
41	402	381	381	381	70.9	56.2	80.9	60.4	44.3	41
40	392	371	371	371	70.4	55.4	80.4	59.5	43.1	40
39	382	362	362	362	69.9	54.6	79.9	58.6	41.9	39
38	372	353	353	353	69.4	53.8	79.4	57.7	40.8	38
37	363	344	344	344	68.9	53.1	78.8	56.8	39.6	37
36	354	336	336	336	68.4	52.3	78.3	55.9	38.4	36
35	345	327	327	327	67.9	51.5	77.7	55.0	37.2	35
34	336	319	319	319	67.4	50.8	77.2	54.2	36.1	34
33	327	311	311	311	66.8	50.0	76.6	53.3	34.9	33
32	318	301	301	301	66.3	49.2	76.1	52.1	33.7	32
31	310	294	294	294	65.8	48.4	75.6	51.3	32.5	31
30	302	286	286	286	65.3	47.7	75.0	50.4	31.3	30
29	294	279	279	279	64.7	47.0	74.5	49.5	30.1	29
28	286	271	271	271	64.3	46.1	73.9	48.6	28.9	28
27	279	264	264	264	63.8	45.2	73.3	47.7	27.8	27
26	272	258	258	258	63.3	44.6	72.8	46.8	26.7	26
25	266	253	253	253	62.8	43.8	72.2	45.9	25.5	25
24	260	247	247	247	62.4	43.1	71.6	45.0	24.3	24
23	254	243	243	243	62.0	42.1	71.0	44.0	23.1	23
22	248	237	237	237	61.5	41.6	70.5	43.2	22.0	22
21	243	231	231	231	61.0	40.9	69.9	42.3	20.7	21
20	238	226	226	226	60.5	40.1	69.4	41.5	19.6	20

* The Brinell hardness numbers in boldface type are outside the range recommended for Brinell hardness testing in Section 5(*c*) of the Method of Test for Brinell Hardness of Metallic Materials (ASTM Designation: E 10)

Hardness Conversion Tables—II

	Brinell Hardness Number*				Rockwell Hardness Number			Rockwell Superfical Hardness Number			
Brinell Indentation Diameter, mm	10-mm Standard Ball, 3000-kg Load	10-mm Hultgren Ball, 3000-kg Load[a]	10-mm Carbide Ball, 3000-kg Load	Diamond Pyramid Hardness Number	C Scale, 150-kg Load, Diamond Cone Penetrator	A Scale, 60-kg Load, Diamond Cone Penetrator	D Scale, 100-kg Load, Diamond Cone Penetrator	15-N Scale, 15-kg Load, Superficial Diamond Cone Penetrator	30-N Scale, 30-kg Load, Superficial Diamond Cone Penetrator	45-N Scale, 45-kg Load Superficial Diamond Cone Penetrator	Brinell Indentation Diameter, mm
2.35	..	..	**682**	737	61.7	82.2	72.0	91.0	79.0	68.5	2.35
2.40	..	..	**653**	697	60.0	81.2	70.7	90.2	77.5	66.5	2.40
2.45	..	..	627	667	58.7	80.5	69.7	89.6	76.3	65.1	2.45
2.50	..	..	..	..	..	..	..	..	..	..	2.50
	..	**601**	..	677	59.1	80.7	70.0	89.8	76.8	65.7	
	..	..	601	640	57.3	79.8	68.7	89.0	75.1	63.5	
2.55	..	..	..	..	..	..	..	..	..	..	2.55
	..	**578**	..	640	57.3	79.8	68.7	89.0	75.1	63.5	
	..	..	578	615	56.0	79.1	67.7	88.4	73.9	62.1	
2.60	..	..	..	..	..	..	..	..	..	..	2.60
	..	**555**	..	607	55.6	78.8	67.4	88.1	73.5	61.6	
	..	..	555	591	54.7	78.4	66.7	87.8	72.7	60.6	
2.65	..	..	..	..	..	..	..	..	..	..	2.65
	..	**534**	..	579	54.0	78.0	66.1	87.5	72.0	59.8	
	..	..	534	569	53.5	77.8	65.8	87.2	71.6	59.2	
2.70	..	..	..	..	..	..	..	..	..	..	2.70
	..	**514**	..	553	52.5	77.1	65.0	86.7	70.7	58.0	
	..	..	514	547	52.1	76.9	64.7	86.5	70.3	57.6	
2.75	**495**	..	..	539	51.6	76.7	64.3	86.3	69.9	56.9	2.75
	..	495	..	530	51.1	76.4	63.9	86.0	69.5	56.2	
	..	..	495	528	51.0	76.3	63.8	85.9	69.4	56.1	
2.80	**477**	..	..	516	50.3	75.9	63.2	85.6	68.7	55.2	2.80
	..	477	..	508	49.6	75.6	62.7	85.3	68.2	54.5	
	..	..	477	508	49.6	75.6	62.7	85.3	68.2	54.5	
2.85	**461**	..	..	495	48.8	75.1	61.9	84.9	67.4	53.5	2.85
	..	461	..	491	48.5	74.9	61.7	84.7	67.2	53.2	
	..	..	461	491	48.5	74.9	61.7	84.7	67.2	53.2	
2.90	**444**	..	..	474	47.2	74.3	61.0	84.1	66.0	51.7	2.90
	..	444	..	472	47.1	74.2	60.8	84.0	65.8	51.5	
	..	..	444	472	47.1	74.2	60.8	84.0	65.8	51.5	
2.95	429	429	429	455	45.7	73.4	59.7	83.4	64.6	4[illegible].9	2.95
3.00	415	415	415	440	44.5	72.8	58.8	82.8	63.5	48.4	3.00
3.05	401	401	401	425	43.1	72.0	57.8	82.0	62.3	46.9	3.05
3.10	388	388	388	410	41.8	71.4	56.8	81.4	61.1	45.3	3.10
3.15	375	375	375	396	40.4	70.6	55.7	80.6	59.9	43.6	3.15
3.20	363	363	363	383	39.1	70.0	54.6	80.0	58.7	42.0	3.20
3.25	352	352	352	372	37.9	69.3	53.8	79.3	57.6	40.5	3.25
3.30	341	341	341	360	36.6	68.7	52.8	78.6	56.4	39.1	3.30
3.35	331	331	331	350	35.5	68.1	51.9	78.0	55.4	37.8	3.35
3.40	321	321	321	339	34.3	67.5	51.0	77.3	54.3	36.4	3.40
3.45	311	311	311	328	33.1	66.9	50.0	76.7	53.3	34.4	3.45
3.50	302	302	302	319	32.1	66.3	49.3	76.1	52.2	33.8	3.50
3.55	293	293	293	309	30.9	65.7	48.3	75.5	51.2	32.4	3.55
3.60	285	285	285	301	29.9	65.3	47.6	75.0	50.3	31.2	3.60
3.65	277	277	277	292	28.8	64.6	46.7	74.4	49.3	29.9	3.65
3.70	269	269	269	284	27.6	64.1	45.9	73.7	48.3	28.5	3.70
3.75	262	262	262	276	26.6	63.6	45.0	73.1	47.3	27.3	3.75
3.80	255	255	255	269	25.4	63.0	44.2	72.5	46.2	26.0	3.80
3.85	248	248	248	261	24.2	62.5	43.2	71.7	45.1	24.5	3.85
3.90	241	241	241	253	22.8	61.8	42.0	70.9	43.9	22.8	3.90
3.95	235	235	235	247	21.7	61.4	41.4	70.3	42.9	21.5	3.95
4.00	229	229	229	241	20.5	60.8	40.5	69.7	41.9	20.1	4.00

* The Brinell hardness numbers in boldface type are outside the range recommended for Brinell hardness testing in Section 5(*c*) of the Method of Test for Brinell Hardness of Metallic Materials (ASTM Designation: E 10).

[a] Brinell hardness values are given for the Hultgren ball to 601 Brinell hardness; however, flattening of the ball at the higher hardnesses gives lower Brinell hardness values than obtained with the carbide ball. For example, material having a DPH value of 667 has a carbide ball Brinell hardness of 627, while material having a DPH value of 677 has a Hultgren ball Brinell hardness of only 601.

Chapter 20

Metal Finishing

Barrel-Finishing Guide—I

By William P Nunn, field engineer, Minnesota Mining & Manufacturing Co, St Paul, Minn

New equipment, media and techniques make it possible to barrel finish parts that could not be considered a few years ago. Parts previously finished by rotation alone are now finished in shorter cycles and in larger loads by newly developed machines. Some of these operate on the vibrating-rotating principle or vibration alone. In certain cases, some of these installations have been automated.

Media development is likewise keeping pace with new finishing requirements. For example, ceramic-bonded aluminum oxide is lighter than fused aluminum-oxide chips, to reduce surface impingement and lodging problems. Preformed abrasives may increase cutting rates over 200%, as compared with random-shaped chips. Other new media will barrel finish chrome-plated zinc die castings to resemble a hand-buffed piece.

Performance in barrel finishing is not a happenstance. Carefully planned and recorded test runs will set a pattern for barrel finishing. See rules for making a test run, and the equipment chart.

How to Test Run a Barrel-Finishing Job

1. Select the chip that will not lodge in any part opening, and that can be separated by screening. The chip should be either larger or smaller than the part, depending on the finish required: large for faster cut and coarser finish; small for slower cut and finer finish.

2. Load chips into cylinder to a 40 to 50% level.

3. Add enough work load (parts) to raise the total level to between 50 and 60% of the barrel's capacity.

4. Add abrasive or burnishing compound.

5. Put enough water into the load: a low level for maximum cutting—a high level for burnishing or cleaning.

6. A good trial run speed is 15 rpm in a barrel 30 in. in diameter. As the diameter of the barrel decreases, speeds may be increased.

7. Check the work after 10 turns, after 20 turns, then at 5- and 15-min. intervals. If no undesirable results appear after the 15-min. interval, continue the run with periodic inspection until the finish has been achieved.

If the operation has been a cutting or grinding run, and polish is required, a second stage follows:

8. Flush the barrel thoroughly with water, because traces of abrasive or dirt will retard polishing action. This is done through a screen door.

9. To polish, cover the load with water to 3 or 4 in. above the mass.

10. Add burnishing compound, preferably an all-purpose material for use on all metals.

11. Seal the barrel and run 45 to 60 min.

12. After attaining the desired finish, flush the barrel with clear water and empty the load into a screen coarse enough to allow chips to drop through, but fine enough to retain the parts.

The above procedure is normally used in conventional rotating barrels. In the vibrating-rotating barrel, chips and work load should reach a level of 75 to 80%. Rolling room is not needed as vibratory action replaces sliding action. Barrel speeds at 4 to 8 rpm give best results. Speeds above this point retard the vibratory effect. In tub-type machines, chip levels should be maintained at 80 to 90% capacity.

Equipment Chart for Barrel Finishing

Type of Parts	Conventional Rotating Barrels	Vibrating Rotating Barrels	Vibrating Tub Type Machines
Metal Stampings			
Media-to-parts ratio	2 to 1	1 to 1	1 to 1
Media level	55%	80%	80%
Media type	Natural stone Fused abrasives Ceramic shapes	Same	Same
Water—cut	½ gal./100 lb media	Same	Same
—polish	2 in. over mass	½ gal./100 lb media	¼ gal./100 lb media
Speed of rotation	18 to 20 rpm	6 rpm	
Frequency vibration		2700	2700
Amplitude		maximum	maximum
Machined Parts			
Media-to-parts ratio	4 to 1	3 to 1	3 to 1
Media level	60%	80%	80%
Media type	Fused abrasives Ceramic shapes	Same	Same
Water—cut	½ gal./100 lb media	Same	Same
—polish	2 in. over mass	½ gal./100 lb media	¼ gal./100 lb media
Speed of rotation	12 to 14 rpm	4 rpm	
Frequency vibration		2700	2700
Amplitude		medium	medium

Barrel-Finishing Guide—II

Equipment Chart for Barrel Finishing *(Continued)*

Type of Parts	Conventional Rotating Barrels	Vibrating Rotating Barrels	Vibrating Tub Type Machines
Zinc Die Castings—Preplating finish			
Media-to-parts ratio	5 or 6 to 1	3 to 1	
Media level	65 to 70%	80%	
Media type	Resin-bonded shapes Natural stone	Same	
Water—cut	½ gal./100 lb media	Same	
—polish	1 gal./100 lb media	Same	
Speed of rotation	12 to 14 rpm	4 rpm	
Frequency vibration		2700	
Amplitude		minimum	
Removal flash and parting lines			
Media-to-parts ratio	3 to 1	2 to 1	2 to 1
Media level	50 to 55%	80%	80%
Media type	Natural stone Resin-bonded shapes	Same	Same
Water—cut	½ gal./100 lb media	Same	Same
—polish	2 in. over mass	½ gal./100 lb media	½ gal./100 lb media
Speed of rotation	16 to 18 rpm	4 to 6 rpm	
Frequency vibration		maximum	maximum
Amplitude		medium	medium
Hardware (wrenches, pliers, tools, etc.)			
Media-to-parts ratio	2 to 3 to 1	2 to 1	2 to 1
Media level	50%	80%	80%
Media type	Fused abrasives Ceramic shapes		Same
Water—cut	½ gal./100 lb media	Same	¼ gal./100 lb media
—polish	2 in. over mass	½ gal./100 lb media	½ gal./100 lb media
Speed of rotation	16 to 18 rpm	6 rpm	
Frequency vibration		2700	2700
Amplitude		maximum	maximum
Sand Castings (up to ¼ lb each)			
Media-to-parts ratio	3 to 1	2 to 1	2 to 1
Media level	50%	80%	80%
Media type	Fused abrasives Ceramic shapes	Same	Same
Water—cut	½ gal./100 lb media	½ gal./100 lb media	¼ gal./100 lb media
—polish	2 in. over mass	½ gal./100 lb media	½ gal./100 lb media
Speed of rotation	18 rpm	6 rpm	
Frequency vibration		2700	2700
Amplitude		maximum	maximum
Delicate Wire-like Parts—Thin Fragile Steel Stampings			
Media-to-parts ratio		6 to 8 to 1	
Media level		80%	
Media type		Fused abrasives Ceramic shapes	
Media size		7/32 x 1/8 in. (No. 5)	
Water—cut		½ gal./100 lb media	
—polish		1 gal./100 lb media	
Speed of rotation		4 rpm	
Amplitude		minimum	
Frequency vibration		maximum 2700	
Compound—cut		Cushioning abrasive	
—polish		Burnishing soap 2 oz/gal. water	
Heavy cut			
Media-to-parts ratio			2 to 1
Media level			80%
Media type			Large ceramic-bonded alum. oxide triangles
Water			¼ gal/100 lb media
Amplitude			maximum
Frequency vibration			maximum

Vibratory finishing of zinc die castings

Vibratory finishing removes flash and burrs from zinc die cast parts, and refines surfaces to any desired degree—quickly, cleanly, and economically

Refinement of surface finish, deburring, and deflashing of zinc-based die castings by vibratory finishing is a growing activity. Five elements in the vibratory finishing process include: the equipment, the media or chips which do the work, the compounds used, the parts themselves, and control.

Vibratory equipment

There are several types of vibratory-finishing equipment on the market:

Tub-type. This is the original type of vibratory machine with a U-shaped cross-section, and is a popular configuration. An eccentric weight below the tub moves media and parts in a circular path so that finishing is continuous. Modifications include the inverted keyhole tub design, where both sides are baffled to improve parts and media flow. The single-baffle type eliminates parts back-roll. The tub has a door in the end for discharge of parts and media.

This equipment does a lot of work quickly. It is available in capacities from 1 to more than 40 cu ft. Cycle timers, variable speed, variable amplitude and compound spray systems are available as standard or optional accessories.

Toroidal-type. This popular vibrator is characterized by a bowl of toroidal shape (the bottom half of a doughnut). While it is generally not as aggressive as a tub, it will generate smoother surfaces, using a given media. With a properly adjusted machine, impingement from part-on-part contact is very slight. This type of equipment is available in capacities from 1 to more than 70 cu ft.

Continuous-type. Actually an elongated tub vibrator, this equipment is currently being made in sizes up to 15 ft long, or longer.

Media and parts are fed into one end, take a long spiral path and are discharged at the other end. Media is automatically recirculated. This equipment minimizes impingement due to the spiral path taken by the parts.

Spindle-type. This is a machine designed to hold a part or a cluster of parts on a rotating spindle similar in arrangement to a drill-press. As the spindle is moved down into a revolving basket containing fine media, the parts are plowed through the media mass. The spindle is then raised and the parts are removed. One or more spindles can be used. Cycle times are very short (10 sec to 5 min). This equipment is unusually fast and eliminates impingement.

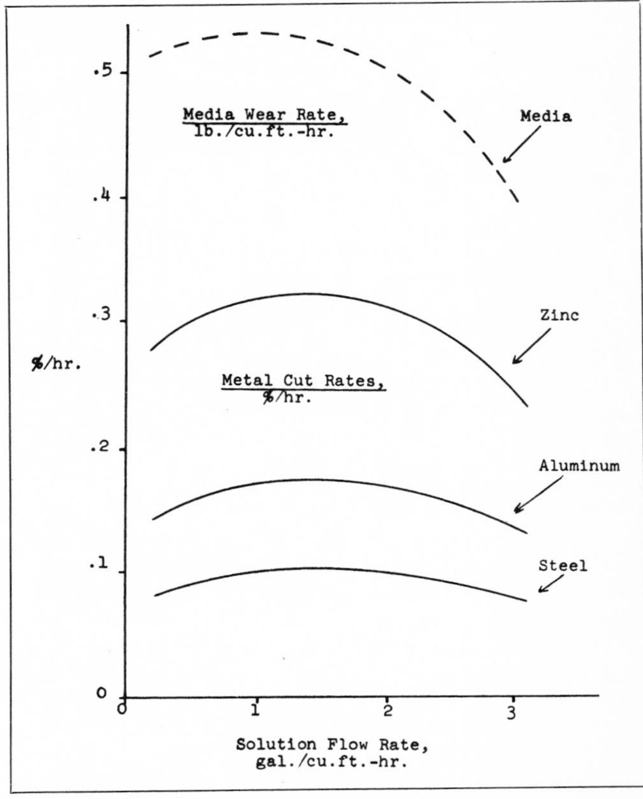

Metal removal rates and media wear are plotted as functions of flow rate in a toroidal bowl vibrator

Newer types. The industry is now getting its second wind in the equipment area with new designs worthy of note. One unit removes metal four to ten times as fast as the tub-type. Another does a good job in restricted intricate interior passages. Equipment of still another type can handle enormous loads such as steel for ball burnishing. A final type, with over 100-cu-ft capacity, is a continuous vibratory unit that should have good retention times.

Media in use

The second element in modern vibratory finishing is the media or chips used to smooth the as-cast surface, to radius edges, to deburr, and to separate parts, while avoiding damage to the parts during processing.

There are three groups of media:

Ceramics include pre-formed synthetics with and without abrasives, natural and synthetic stones, fused aluminum oxide, and related product types. Bulk density of these products will vary between 90 and

By J B Kittredge, supervisor
Industrial Mineral Products Div
3 M Co, St Paul, Minn

Source: Extracted from paper presented before the 5th National Die Casting Congress, Detroit, Nov 4-7, 1968, organized by the Society of Die Casting Engineers

Vibratory finishing of zinc die castings

150 lb/cu ft. Products in this category are low in cost and economical to use. They can deburr parts rapidly, remove flash, and some can 'color' or bring luster to the surface. However, they can also impinge or dent relatively-soft metal surfaces and, therefore, must be avoided when such damage cannot be tolerated. Frequently, ceramic products will also roll burrs and flash over, making their subsequent complete removal an extremely long and arduous task.

Resin-bonded media were developed originally to eliminate the shortcomings of ceramics. Modernized to suit the needs of modern vibratory equipment, these products have low bulk densities (50-80 lb/cu ft), do not impinge soft metals, and will cut off, rather than roll over, burrs and flash. Resin-bonded media products are preformed, are somewhat more expensive than ceramics in operating cost, and give a matte surface to parts, with predictable, low microinch surface finishes. These products are preferred for preplate finishing of zinc die castings.

Other media include steel for ball burnishing (over 300 lb/cu ft), ground corn cob or maize used in drying and for giving a luster to surfaces, carpet tacks, ground walnut shell, and a few others.

Compounds

A preferred compound system is the flow-through or constant spray and drain technique. This system uses a dilute compound solution (0.5 to 1 lb of compound to 55 gallons of water) which is metered into the vibrator continuously and allowed to progress through the media-part mass to drain. This technique has a low operating cost and requires little operator attention. Parts and media are kept clean; parts do not need elaborate cleaning after finishing and clean media will cut as it was designed to cut. Batch-to-batch uniformity is excellent. Solution flow rate must be controlled, preferably by a flow meter. This significant vibratory-finishing variable can be used to real advantage with proper control.

Soluble alkaline cleaners and soaps are used as compounds in the flow-through system. Use of a burnishing-type compound with any media is recommended to improve metal removal rate, thereby reducing time cycles and producing lower microinch surface finishes.

Other compounds include abrasive-containing products that are added directly to the load to increase metal removal rates. They often leave coarser surfaces and can clog drains not designed for their use. Burnishing compounds are also added directly to the load. Descalers and corrosion inhibitors are employed as the metal type and condition require.

Process variables

Control of the following variables will lead to highly improved process control.

Compound flow-through. Compound type, its concentration in the solution, and the solution flow rate are variables which require control. The graph shows the effect of metal removal rate and media wear as a function of flow rate in a toroidal bowl vibrator. Maximum metal removal occurs at a flow rate of about 1.5 gal/cu ft/hour according to the above graph.

A rule of thumb that works well in the absence of flow meters is to use just enough solution to keep media and parts clean. If there is foam in the vibrator, the compound concentration should be reduced or increased.

Amplitude. Amplitude is the total displacement of the vibrator bowl during processing. Easy field measurement of this variable may be performed with an 'Ampli-Check' unit fastened to the vibrator.

In general, higher amplitude produces faster cutting, higher media wear, higher (rougher) microinch surfaces, and more part-on-part impingement. However, amplitude cannot be related to equipment types. Differences in these units will not correlate with amplitude because of their completely different action.

Speed. An obvious variable on some machines, higher speed causes faster metal removal rates, higher media wear, and poorer surface finishes. It also causes more part-to-part impingement.

Materials handling. Impingement due to one part banging into another, causing dents in each, has been mentioned as a problem which can occur during processing. This problem also occurs before and after the vibratory process. How are parts handled after casting? Do they cascade down a chute into larger tote pans? Are these tote pans handled gently or as if they contained forging billets? How is the vibrator loaded?

Procedures to follow

Functional part. Such a part has no special requirements for surface finish. If high volumes are required a continuous vibrator can be used. A ceramic media is economical and fast. Self-tumbling might also be used if large amounts of metal particles are not broken off to interfere with cleaning. A flow-through compound system with cycles of 15 to 45 minutes would be typical.

Plated part—not critical. Ceramic media will usually suffice. One-hour cycles in batch equipment, often with abrasives added during the early stages, will give a satisfactory finish where excellent plating quality is not required.

Plated part—good appearance. A resin-bonded media is needed to produce 4 to 8 microinch finish. Equipment which will not cause part-on-part impingement must be used. Flow-through systems for compound should always be used. Cycles will typically run 1 to 3 hours.

Plated part—good appearance—defects. When casting defects are very deep, more surface metal must be removed. A two-step resin-bonded media system should be used for 2 to 8 hours, followed by a finer, finish media for about one hour.

Pre-color buff. Vibratory finishing prior to a light color buff is accepted practice. This procedure permits good flash removal, blending of edges, removal of most surface defects, and provides surfaces that are actually superior to those resulting from buffing alone. Comets do not appear at holes or ridges, as they do with heavy buffing. Mirror-like platings are produced with excellent economy.■

How to clean copper parts

Removal of grease, oil, dirt and other contamination from copper parts must be done with the correct method, or else the surface will be pitted or tarnished

During fabrication, the metals may become coated with lubricating oils, drawing compounds, greases, oxides, dirt, metallic particles or abrasives. These may be removed thus:

Solvent cleaning—This process is used for removing the greater portion of oils from coated pieces, by immersing in a solvent halfway between kerosene and gasoline. Steel tanks are used. Centrifuging removes excess solvent and pieces are practically free from oil. Solvent cleaning is often used for screw-machine work.

Degreasing—By vapor degreasing quick removal of oil is secured. Condensate washes away the oily coating, leaving the pieces clean and dry. The degreasing unit is self-purifying—oils and waste accumulate at bottom. Solvents used are generally carbon tetrachloride and trichlorethylene.

Alkaline scouring — Hot alkaline cleaning baths are used primarily for the removal of oils, greases, solid particles of dirt, and metal particles. The chemicals saponify or make soap of vegetable and animal oils and fats, emulsifying mineral oils and greases and suspending the solid material. The soil is either removed from the metal surface or its tendency to cling is greatly reduced.

Caustic soda, soda ash and causticized soda form the cheapest and most direct method of producing alkalinity in the bath. However, sodium metasilicate, trisodium phosphate, and similar salts are often used.

Emulsion scrubbing — Proprietary preparations comprise an emulsification agent that disperses organic solvents in water solutions. Emulsifiable cleaners are miscible with oils and can be washed off with water, although a slight film of oil may remain on the work and necessitate a subsequent alkali cleaning treatment. Dragout costs are high.

Electrolytic cleaning — Alkaline materials are used in electrolytic cleaning. The bath is maintained as near boiling as permissible without excessive tarnishing. Work to be cleaned is usually made the cathode.

Tarnish removal—Brass and copper articles often become discolored or tarnished on standing or during the course of alkaline cleaning. This tarnish can be removed when work is free from oil or grease by immersion in a solution containing 4 to 8 oz of sodium cyanide per gal. After discoloration has been removed (usually a matter of seconds), pieces should be thoroughly rinsed. Sodium cyanide is poisonous.

Pickling—This operation removes oxides developed from annealing. Sulphuric acid and hydrochloric acid are commonly used. Either sodium, bichromate or ferric sulphate are used in combination with an acid to remove red stains on brass or to produce a special surface effect. After pickling, it is important that metals be well rinsed and, if permissible, immersed in a neutralizer.

Sulphuric acid pickle

Sulphuric Acid	¼ to 1 gal.
Water	4 gal.
Temperature	100 to 160 F

This is used often after an annealing operation where scale and tarnish have been developed. The heavier the scale, usually the more concentrated and the hotter the pickle.

Sodium bicromate pickle

Sulphuric Acid	¼ gal.
Water	4½ gal.
Sodium Bichromate	2 to 5 oz per gal.
Temperature	80 to 120 F

If pieces remain too long in this solution, or solution is too concentrated, some pitting or etching may result. When properly operated, the solution will give a clear nonglossy yellow finish. Sodium bichromate pickle is used for removing red stains which remain after sulphuric acid pickling.

Bright dip solution

Sulphuric Acid	2 gal.
Nitric Acid	1 gal.
Water	1 qt
Hydrochloric Acid	1 oz. to 5 gal.
Operate	Room Temperature

To obtain a gloss, a bright-dip solution is required. Increasing the nitric acid makes the solution more active. Increasing the sulphuric acid slows down the action.

Scale dip—Preliminary preparation of the surface is sometimes necessary before immersing the pieces in the bright dip, especially when a very bright smooth surface is desired.

(a) When very heavy cutting of the surface is desired, use concentrated nitric acid.

(b) For milder treatment, use the formula:

Nitric Acid	1 gal.
Sulphuric Acid	1 to 2 gal.
Water	1 to 5 gal.

The above solutions are quite active and must be used with care to prevent attacking the brass too deeply. After the bright dipping, either preceded by scale or bichromate dipping, the work should be thoroughly rinsed in water, immersed in a 2 to 4 oz solution of sodium cyanide followed by thorough rinsing. Then immerse in 1 oz solution of neutral soap and thoroughly rinse. Omit the soap solution if pieces are to be soldered. Ferric sulphate can be used instead of sodium bichromate, and is preferred where attack of the base metal must be minimized.

Courtesy Bridgeport Brass Co

Chapter 21

Plating

Facts About Nickel Plating—I

Courtesy International Nickel Co, New York

Plating specialists at International Nickel Co receive each year many questions from practical engineers and shop men about nickel plating. Here are the facts in reply to most of these questions. A review of the answers is timely, because you are now free to employ nickel plating or nickel in any other form as much as you wish. There is an over-supply of the metal now, and according to the experts there should be enough in the future for all conceivable applications.

Practical Questions and Answers About Nickel Plating

What materials can be coated with nickel?
Any material that can be plated can be plated with nickel. Nickel plating combines the properties of nickel with the special properties of other metals—the formability of brass, the impact extrusion of aluminum, the strength of steel, or the design flexibility of die-castings. Nickel can be electroplated on plastics, too, if they are made conductive by special treatments.

Is an undercoating of some other metal required?
Nickel can be applied directly to most metals, but copper is sometimes used as an undercoating. This is usually for economic reasons, such as to provide a surface that is more easily buffed. In coating zinc and aluminum, it is usually necessary to apply a strike coating before applying the main nickel coating. The strike may be copper or it may be nickel applied from a special bath.

What are some advantages of plating other materials with nickel?
These are varied and numerous:
To provide beauty and distinction . . . to increase resistance to corrosion . . . to decrease contact resistance and friction . . . to improve solderability and brazing . . . to increase resistance to galling and wear . . . to salvage worn or mismachined parts. Nickel is a good conductor of electricity . . . conducts heat somewhat better than low-carbon steel . . . has a low coefficient of thermal expansion . . . is magnetic . . . can be electroplated almost as soft as the softest electroplated copper and almost as hard as chromium, and to almost any hardness in between.

Can nickel coatings be applied to rough surfaces to make them smoother?
Yes. Certain bright and semi-bright nickel baths have been specially developed to give them smoothing action. Platers usually refer to this as "leveling." When these baths are employed, the base metal does not have to be completely smooth. This means lower costs for polishing and buffing.

What about the appearance of nickel coatings?
There is a nickel finish for every need. Nickel coatings can be plated bright or they can be plated dull. And the dull ones can be very easily polished or buffed to any desired finish. Sometimes it is better to plate dull coatings and polish them to full brightness. For most applications it is more economical to plate the coating bright.

Why is chromium applied over nickel for decorative applications?
Bright nickel coatings tend to lose luster with time. To avoid this, very thin coatings of bright chromium are applied over the nickel.

Are other coatings applied over nickel?
In the manufacture of printed circuits, rhodium is frequently plated over nickel. In other applications an overlay of gold is used. In still others, tin may be electroplated over nickel to provide certain properties that cannot be achieved in any other practical way.

Can nickel-coated materials be formed?
Many nickel-coated metals—if ductile—can be formed, drawn, or stamped into a variety of shapes.
Soft dull coatings are usually more ductile than hard bright ones. Therefore, when objects are to be plated with bright nickel, they should be stamped or formed before plating.

How strong are nickel coatings?
Their strength depends on the composition and operating conditions of the plating bath. The ultimate tensile strength runs between 50,000 and 220,000 psi, with the coatings that are plated bright having the highest value.

Can nickel coatings be made more workable by heat treatment?
Heat-treating for about 15 min. at 1400 F will increase the ductility of dull nickel as much as 50%. Ductility reaches its peak under these conditions, but hardness and tensile strength are decreased.

Can electroplated nickel be welded?
With special precautions. However, tiny cracks will often develop in the region of the weld. And when they do, cracks are usually so small that they can be ignored for many applications. Consult experts if in doubt.

Can nickel coatings be machined?
Practically all nickel coatings can be machined, but sometimes it is better to grind the harder ones.

Facts About Nickel Plating—II

Do nickel coatings provide good protection outdoors?
Nickel and chromium are a combination that weathers well in practically all atmospheres.

Can nickel coatings be exposed to chemicals?
Electroplated nickel is at home in alkaline mediums: soda ash, soaps, caustic soda and hot caustic liquors.
Nickel coatings also do an economical and satisfactory job in handling a variety of other chemicals, food and acids.

How much surface will a pound of nickel coat?
This will depend on the coating thickness. If a thickness of one mil is desired, one pound of nickel will cover about 22 sq ft.

What are some types of plating baths?
Most dull baths contain no special chemicals and involve no patent rights. There are at least 12 baths that yield dull deposits. By selecting the proper bath and operating conditions you can obtain nickel deposits for a wide variety of applications.
Bright baths are proprietary and contain one or more chemical compounds to make the coatings bright and impart other desirable characteristics.

Who distributes bright baths?
Here are some distributors:
Apothecaries Hall Co, Division of the Hubbard-Hall Chemical Co, Waterbury, Conn
Hanson-Van Winkle-Munning Co, Matawan, N J
Harshaw Chemical Co, Cleveland, Ohio
McGean Chemical Co, Cleveland, Ohio
Seymour Manufacturing Co, Seymour, Conn
Udylite Corp, Detroit, Mich

Can nickel be electroplated on the inside of pipes and tubing?
Yes. Steel pipe and tubing with nickel-plated internal walls are available from the Bart Manufacturing Co, Belleville, N J, under the tradename "Lectro-Clad."

Are nickel-plated sheets also available?
Nickel-plated sheets and plates are also manufactured by Bart.

What is electroforming?
Electroforming is a method of fabricating objects by electroplating nickel (or another metal) as a shell on a removable mold or pattern. No fabricating method gives the designer more latitude.

Questions on Techniques of Nickel Plating

Should bright chromium be applied to nickel before or after forming?
Bright chromium does not usually have good forming characteristics. Hence, it is better practice to apply bright chromium after the article has been formed.

About how thick should a coating be to give good performance?
This is a difficult question to answer, because the thickness required will depend on where and how the plated article will be used. For industrial applications fairly heavy coatings are used: 0.005 to 0.020 in. But for decorative applications much thinner coatings are used. The specifications tabulated are typical of those adopted by some automobile manufacturers for minimum baseplate thicknesses. Such specifications are also applicable to many other fields of decorative plating.
It should be noted, though, that the thicknesses in these specifications are minimums, and better performance can be obtained with heavier coatings.
In the table, letters D, F, K, and Q were arbitrarily adopted to designated grades of plating. Letter S refers to steel as the basis metal; C, to copper and copper alloys containing at least 50% copper; Z, to zinc and zinc alloys containing 80% zinc.
Exposure conditions and use of plated steel are too varied to permit any predictions of the average life of articles plated to the specifications of D.S., F.S., K.S., or Q.S.

Is there any way to measure and control thickness?
One measuring device is well suited to measuring thickness up to about 0.003 in. on copper, brass, and steel. This is non-destructive and utilizes nickel's magnetic properties. There are other non-destructive tests, also used primarily in this thickness range. Destructive tests include measuring the coating with a microscope.
There are meters to measure the electric current going to a particular object. By an accurate check on this and the plating time, a plater can be sure that every object is coated with the same thickness.

Are special precautions required for heat treating nickel coatings?
Yes. In ordinary atmospheres nickel will begin to tarnish at around 750 F. This film will become darker and thicker at higher temperatures. If such films are undesirable, an inert or reducing atmosphere should be used. Whatever the atmosphere, it should be free from sulfur.
Unlike wrought nickel, the electroplated metal contains no sulfur-fixing elements. Hence, even a trace of sulfur in a furnace atmosphere will damage electroplated nickel by precipitating in the grain boundaries and causing serious embrittlement. For the same reason, sulfur in the as-plated coating should also be avoided if the coating is to be heat treated.

Is there a limit to the thickness of nickel that can be electroplated?
For some industrial equipment, coatings of 0.020 to 0.030 in. are common. For some electroformed objects, nickel deposits may be as thick as several inches. Of course, such thick deposits will require al-

Facts About Nickel Plating—III

ternate plating and machining to keep the deposits smooth.

How thin can electroplated nickel be?
Very thin coatings should only be used for special applications, because they will be porous.

Are there other ways besides electroplating for applying nickel coatings?
Commercial methods are: (1) electroless plating, sometimes called chemical reduction, (2) heat reduction, (3) vapor deposition, (4) simple immersion. Not all methods are suited to all applications. For complicated shapes, electroless plating is often the best way. For others, heat reduction may be the answer.

Just what is electroless plating? Is this the same as the Kanigen process?
Electroless plating is a process for plating nickel without the use of electric current. It is sometimes called plating by "chemical reduction" because a chemical is used to plate out the nickel metal.

The method was developed by Dr Abner Brenner and his associates at The National Bureau of Standards and is covered by patents assigned to the U.S. Department of Commerce. This means that they are royalty-free.

Brenner's method is best suited for plating small objects. However, it has been adapted and engineered for medium-scale production by the Metal Processing Co, who market equipment, solutions, and operating instructions under the name "Lustralloy."

Kanigen is also a type of electroless plating. The name stands for "catalytic nickel generation." This process was developed and patented by the General American Transportation Corp, 135 S LaSalle St, Chicago, Illinois, and represents modifications and improvements in the Brenner method. These include improvements in the quality of the coating, the rate of plating, and other changes that make it possible to plate a variety of shapes and sizes from a tube to a tank car.

What are some of the properties of electroless nickel?
Electroless nickel can be applied over many metals—and non-metals, too. It is also suited to applying uniform nickel coatings on a variety of complicated shapes—something that cannot always be easily done by electroplating.

Electroless nickel is extremely fine-grained and hard. Hardness is around 500 Vickers, or 48 Rockwell C, and can be doubled by a mild heat treatment for about one hour.

What about heat-reduction? Is this the same as "Niphos"?
The heat-reduction process is the same as Niphos, which is a patented method for plating nickel without electric current, developed by the New York Testing Laboratories. In this process, a water slurry of nickel oxide and other chemicals is painted on the object to be coated. This slurry is dried and heated at high temperature in a reducing-gas atmosphere. During firing, some diffusion takes place with the base metal, so that the resulting coating is an alloy containing some iron—as well as nickel and phosphorous. Phosphorus content ranges between 1 and 3%.

The coating has useful properties. It is ductile and hard, with a hardness around 400-500 Vickers. As it wets well and can be readily brazed, it can serve a two-fold function of providing protection and joining two, or more, coated parts.

Niphos can be applied to a number of metals, but its chief use is as a coating for steel. Examples: pipe and fittings—especially for cargo heating coils in ocean-going tankers.

What is meant by vapor plating?
The metal to be deposited is evaporated at elevated temperature in a vacuum. Coatings applied are usually quite thin, and best suited to special applications. For example, the thermal decomposition of nickel carbonyl, as employed by the Commonwealth Engineering Co, is said to be ideal for electroforming particular objects. Vacuum deposition is often used to coat plastics and is well suited to the jewelry trade.

What is immersion plating?
Immersion plating is a simple procedure for applying very thin coatings to steel. The part to be plated is immersed in a special water solution of nickel salts, and nickel metal deposits on the surface of the steel. No electric current is required. Coatings applied in this way are primarily used as base coatings for enamels.

Minimum Thicknesses of Electrodeposited Coatings

Nickel and Chromium on Steel

	Type D.S.	Type F.S.	Type K.S.	Type Q.S.
Copper plus nickel, min.	0.0020*	0.0012	0.00075	0.0004
Final nickel, min.	0.0010	0.0006	0.0004	0.0002
Chromium (if required) min.	0.00001	0.00001	0.00001	0.00001

* When copper is used in excess of a thickness of 0.0001 in., its minimum thickness shall be 0.00075 in.

Nickel and Chromium on Copper

	Type F.C.	Type K.C.	Type Q.C.
Nickel, min.	0.0005	0.0003	0.0001
Chromium (if required) min.	0.00001*	0.00001	0.00001

* Chromium coatings 0.00005 in. or more in thickness are likely to cause cracking of nickel deposits on brass. An effort should therefore be made to obtain the required minimum thickness of chromium with as low a maximum as practicable.

Nickel and Chromium on Zinc and Zinc Alloys

	Type F.Z.	Type K.Z.	Type Q.Z.
Copper plus nickel, min.	0.0012	0.00075	0.0005*
Copper, min.	0.0002	0.0002	0.0002
Final nickel, min.	0.0005	0.0003	0.0003
Chromium (if required) min.	0.00001	0.00001	0.00001

* Total of 0.0003 in. if alternative of nickel is used.

Plating of Precision Parts—I

By Robert D'Agostino,
staff mechanical engineer,
Central Development Engr. Dept,
Loral Electronics Corp,
Bronx 72, N Y

The control of plating deposits on precision-fitted parts has assumed special importance, because of the increasingly corrosive environments encountered in new military and industrial applications. As an aid to improved control, a method is presented for estimating the amount and location of plating deposits to assure proper assembly and performance of precision parts.

To determine machining dimensions and tolerances of a part before plating, parts which require a plating deposit may be grouped into four classes (Table 1). Details of these four classes are:

Class 1—Non-Critical parts have fractional and decimal dimensions and tolerances, in which no tolerance is ever less than ±0.005. In this case, dimensions and tolerances as they appear on the drawing may be used for machining the part. No allowances are necessary for plating. Consequently, the inspection department may accept parts having dimensions which exceed stipulated drawing tolerances by the amount of plating deposit.

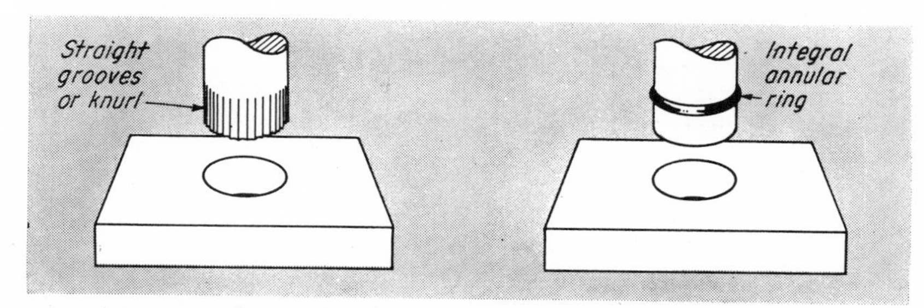

Fig. 1—Alternate solutions for obtaining interference fits on plated parts

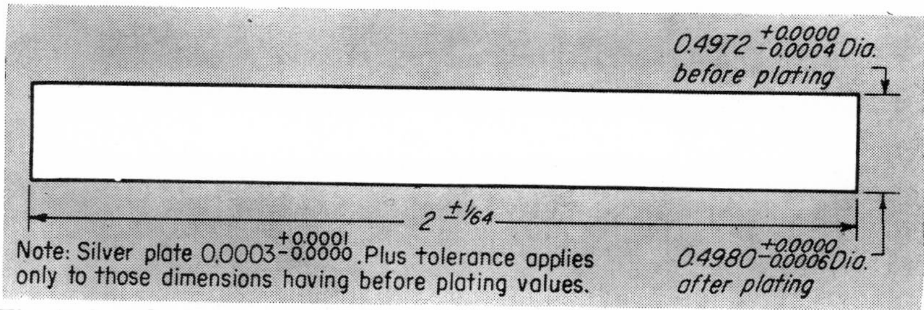

Fig. 2—Sample drawing for precision part of Class 3

Table 1—Tolerances for Four Classes of Parts

Precision Class No.	Part Type	Dimensional Tolerance
1	Non-Critical	±0.005 min.
2	Marginal	±0.0005 to ±0.005
3	Critical	±0.0002 to ±0.0005
4	Super-Critical	±0.0002 max.

Class 2—Marginal parts have decimal dimensions and tolerances in which one or more of the tolerances is less than ±0.005 in., but larger than ±0.0005 in. In this case a conventional tolerance analysis must be made if the surfaces having these tolerances are required to clear a mating part (i.e. a shaft loosely fitted in a hole). If no interference is indicated, the parts may be considered non-critical and treated like Class 1 parts. If an interference is indicated, "Before Plating" and "After Plating" dimensions and tolerances may be determined in accordance with the formulas listed in Table 2.

Class 3—Critical parts have decimal dimensions and tolerances in which one or more of the tolerances is less than ±0.0005 in. but larger than ±0.0002 in. In this case "Before Plating" and "After Plating" dimensions and tolerances may be determined in accordance with the formulas listed in Table 2. If any surface on the part is required to accomplish an interference fit with another part, it may be possible to use a knurl or annular ring as an alternate solution, as shown in Fig. 1. This solution will eliminate the need for exacting control during the machining and plating operations.

Class 4—Super critical parts have decimal dimensions and tolerances in which one or more of the tolerances is less than ±0.0002 in. This tolerance does not permit adequate allowances for both manufacturing and plating. Parts having such surfaces must, therefore, be manufactured and plated in one of four ways:

1. Those surfaces on the part that are super-critical may be masked to prevent plating. This practice of course increases costs.
2. The parts may be selectively chosen for assembly after plating. This of course will insure corrosion resistance but raise unit costs due to the need for individual handling and identification of parts.
3. Those surfaces on the part which are super-critical and circular in shape may be designed with the male portion having a knurl or annular ring, Fig. 1.
4. Those surfaces on the part which are super-critical may be plated and machined to final di-

Plating of Precision Parts—II

mension and tolerance after plating. This approach reduces plating costs because the masking operation or close plating control is eliminated. However, machining costs will rise dependent upon the complexity of the part, its size and material. This technique is most effective where soft plating deposits are required, because hard deposits such as nickel or chrome tend to flake when machined and may peel off other areas.

Every time a Class 2, 3 or 4 part is used, costs rise proportionately. Therefore, every effort should be made to reduce the number of these classes on any part to a minimum. The need for Class 2, 3 or 4 parts should be carefully scrutinized, because these parts require special precautions and procedures by the plater, causes hardship in manufacturing and extra careful handling by the Quality Control Dept. Even when these classes must be used, effective savings are often realized by reducing Class 4 parts to Class 2 or 3 parts. •

Table 2—Formulas for Plating Allowances on Precision Parts

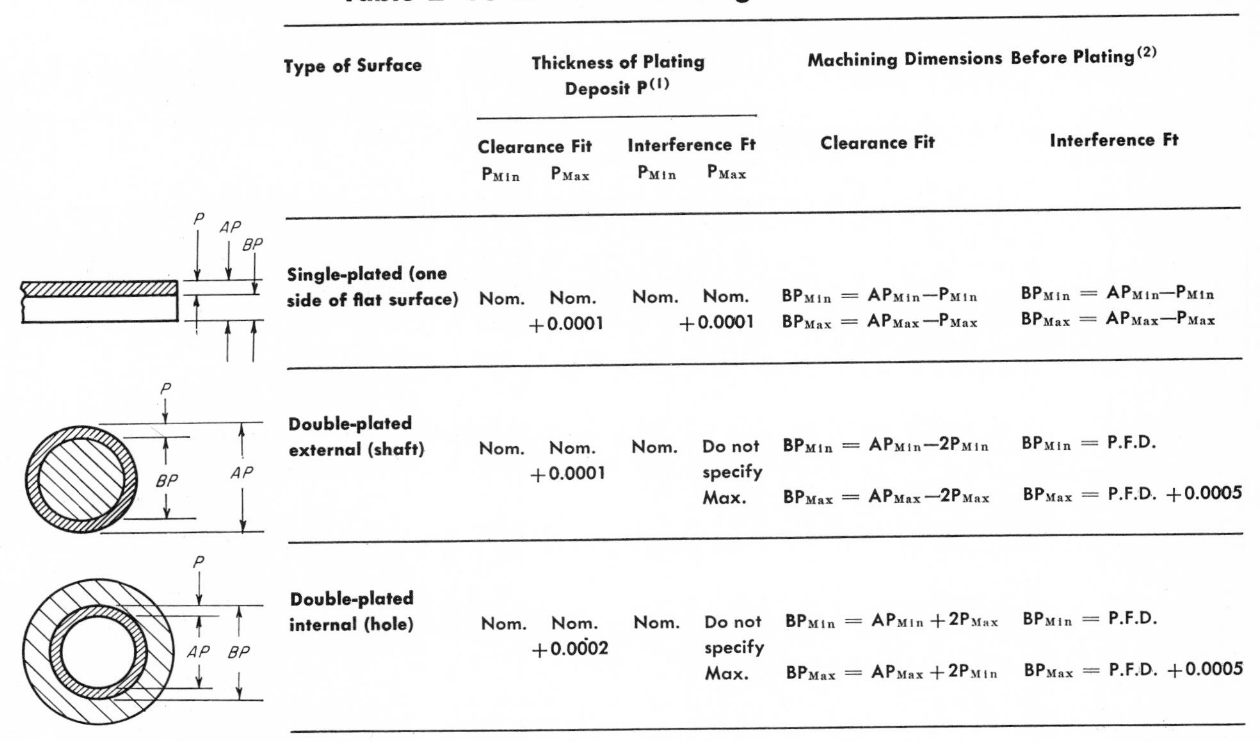

Type of Surface	Thickness of Plating Deposit P[1]				Machining Dimensions Before Plating[2]	
	Clearance Fit		Interference Ft		Clearance Fit	Interference Ft
	P_{Min}	P_{Max}	P_{Min}	P_{Max}		
Single-plated (one side of flat surface)	Nom.	Nom. +0.0001	Nom.	Nom. +0.0001	$BP_{Min} = AP_{Min} - P_{Min}$ $BP_{Max} = AP_{Max} - P_{Max}$	$BP_{Min} = AP_{Min} - P_{Min}$ $BP_{Max} = AP_{Max} - P_{Max}$
Double-plated external (shaft)	Nom.	Nom. +0.0001	Nom.	Do not specify Max.	$BP_{Min} = AP_{Min} - 2P_{Min}$ $BP_{Max} = AP_{Max} - 2P_{Max}$	BP_{Min} = P.F.D. BP_{Max} = P.F.D. +0.0005
Double-plated internal (hole)	Nom.	Nom. +0.0002	Nom.	Do not specify Max.	$BP_{Min} = AP_{Min} + 2P_{Max}$ $BP_{Max} = AP_{Max} + 2P_{Min}$	BP_{Min} = P.F.D. BP_{Max} = P.F.D. +0.0005

Nomenclature

AP_{Min}—After-plating minimum dimension
AP_{Max}—After-plating maximum dimension
BP_{Min}—Before-plating minimum dimension
PFD—Nominal press-fit diameter
P_{Min}—Minimum plating deposit
P_{Max}—Maximum plating deposit

Notes:

1. The minimum and maximum values of plating deposit indicate tolerances that can usually be held by the plater on significant surfaces (i.e. surfaces which can be touched by a ¾ in. spherical ball).

2. The values for "After Plating" are the functional values which have been calculated by the designer, based on initial design requirements. They indicate performance of the part.

EXAMPLE: A shaft must have a final after-plating diameter of 0.4980 in. +0.0000, —0.0006 to fit the mating part with clearance. Find the "Before Plating" (machining) dimension and tolerance if the plating is to be 0.0003 minimum thickness of silver.

Step 1: This part falls into Class 3 (less than ±0.0005 in. but more than ±0.0002 in.).

In Table 2 for a double-plated external surface it can be seen that the maximum plating thickness (P_{Max}) can be controlled within 0.0001 in. Therefore P_{Max} = 0.0004 in.

Step 2: In Table 2 for a double-plated external surface, the formulas for a clearance fit are:

(a) $BP_{Min} = AP_{Min} - 2\,P_{Min}$
(b) $BP_{Max} = AP_{Max} - 2\,P_{Max}$

Substituting in (a):

$BP_{Min} = 0.4974 - 2(0.0003) = 0.4966$

Substituting in (b):

$BP_{Max} = 0.4980 - 2(0.0004) = 0.4972$

Therefore: "Before Plating" dimension of 0.4972 in. +0.000, —0.0004 is written on detail drawing of part as well as the "After Plating" dimension of 0.4980 in., +0.0000, —0.0006. See Fig. 2.

Limits for Plated Threads—I

By Robert D'Agostino, staff engineer, Central Development Engrg Dept, Loral Electronics Corp, New York

To deposit a predetermined amount of plating on a threaded part, it is essential to control the pitch diameter and quality of these parts *prior* to plating. The reason for this is the fact that the plating thickness affects the pitch diameter by a 4:1 ratio.

The tables on this page give the actual limits of the thread at the pitch diameter both before and after plating for two thicknesses of deposit. These limits can be used to order gages, setup the machine, and inspect the product.

Plating Deposit 0.0003-0.0005 In.

No. Th'ds Size & Class	External Th'd Pitch Dia Before Plating (specify on dwg)		External Th'd Pitch Dia After Plating (inspect to)		Internal Th'd Pitch Dia Before Plating (specify on dwg)		Internal Th'd Pitch Dia After Plating (inspect to)	
	Max	Min	Max	Min	Min	Max	Min	Max
	Class 2A				Class 2B			
0-80-UNF-2A	0.0499	0.0489	0.0519	0.0501	0.0539	0.0554	0.0519	0.0542
1-64-UNC-2A	0.0609	0.0597	0.0629	0.0609	0.0649	0.0667	0.0629	0.0655
1-72-UNF-2A	0.0620	0.0609	0.0640	0.0621	0.0660	0.0677	0.0640	0.0665
2-56-UNC-2A	0.0724	0.0711	0.0744	0.0723	0.0764	0.0784	0.0744	0.0772
2-64-UNF-2A	0.0739	0.0727	0.0759	0.0739	0.0779	0.0798	0.0759	0.0786
3-48-UNC-2A	0.0835	0.0820	0.0855	0.0832	0.0875	0.0897	0.0855	0.0885
3-56-UNF-2A	0.0854	0.0840	0.0874	0.0852	0.0894	0.0914	0.0874	0.0902
4-40-UNC-2A	0.0938	0.0921	0.0958	0.0933	0.0978	0.1003	0.0958	0.0991
4-48-UNF-2A	0.0965	0.0949	0.0985	0.0961	0.1005	0.1028	0.0985	0.1016
5-40-UNC-2A	0.1068	0.1050	0.1088	0.1062	0.1108	0.1133	0.1088	0.1121
5-44-UNF-2A	0.1082	0.1065	0.1102	0.1077	0.1122	0.1146	0.1102	0.1134
6-32-UNC-2A	0.1157	0.1137	0.1177	0.1149	0.1197	0.1226	0.1177	0.1214
6-40-UNF-2A	0.1198	0.1180	0.1218	0.1192	0.1238	0.1264	0.1218	0.1252
8-32-UNC-2A	0.1417	0.1396	0.1437	0.1408	0.1457	0.1487	0.1437	0.1475
8-36-UNF-2A	0.1440	0.1420	0.1460	0.1432	0.1480	0.1508	0.1460	0.1496
10-24-UNC-2A	0.1609	0.1584	0.1629	0.1596	0.1649	0.1684	0.1629	0.1672
10-32-UNF-2A	0.1677	0.1655	0.1697	0.1667	0.1717	0.1748	0.1697	0.1736
12-24-UNC-2A	0.1869	0.1843	0.1889	0.1855	0.1909	0.1945	0.1889	0.1933
12-28-UNF-2A	0.1908	0.1884	0.1928	0.1896	0.1948	0.1982	0.1928	0.1970
12-32-NEF-2A	0.1937	0.1914	0.1957	0.1926	0.1977	0.2010	0.1957	0.1998
1/4-20-UNC-2A	0.2155	0.2126	0.2175	0.2138	0.2195	0.2235	0.2175	0.2223
1/4-28-UNF-2A	0.2248	0.2223	0.2268	0.2235	0.2288	0.2323	0.2268	0.2311
1/4-32-NEF-2A	0.2277	0.2253	0.2297	0.2265	0.2317	0.2351	0.2297	0.2339
5/16-18-UNC-2A	0.2744	0.2712	0.2764	0.2724	0.2784	0.2829	0.2764	0.2817
5/16-24-UNF-2A	0.2834	0.2805	0.2854	0.2817	0.2874	0.2914	0.2854	0.2902
5/16-32-NEF-2A	0.2902	0.2878	0.2922	0.2890	0.2942	0.2976	0.2922	0.2964
3/8-16-UNC-2A	0.3324	0.2288	0.3344	0.3300	0.3364	0.3413	0.3344	0.3401
3/8-24-UNF-2A	0.3459	0.3429	0.3479	0.3441	0.3499	0.3540	0.3479	0.3528
3/8-32-NEF-2A	0.3527	0.3501	0.3547	0.3513	0.3567	0.3603	0.3547	0.3591
7/16-14-UNC-2A	0.3891	0.3852	0.3911	0.3864	0.3931	0.3984	0.3911	0.3972
7/16-20-UNF-2A	0.4030	0.3996	0.4050	0.4008	0.4070	0.4116	0.4050	0.4104
7/16-28-UNEF-2A	0.4123	0.4095	0.4143	0.4107	0.4163	0.4201	0.4143	0.4189
1/2-12-N-2A	0.4439	0.4393	0.4459	0.4405	0.4479	0.4541	0.4459	0.4529
1/2-13-UNC-2A	0.4480	0.4438	0.4500	0.4450	0.4520	0.4577	0.4500	0.4565
1/2-20-UNF-2A	0.4655	0.4620	0.4675	0.4632	0.4695	0.4743	0.4675	0.4731
1/2-28-UNEF-2A	0.4748	0.4719	0.4768	0.4731	0.4788	0.4828	0.4768	0.4816

No. Th'ds Size & Class	External Th'd Pitch Dia Before Plating (specify on dwg)		External Th'd Pitch Dia After Plating (inspect to)		Internal Th'd Pitch Dia Before Plating (specify on dwg)		Internal Th'd Pitch Dia After Plating (inspect to)	
	Max	Min	Max	Min	Min	Max	Min	Max
	Class 2A				Class 2B			
9/16-12-UNC-2A	0.5064	0.5020	0.5084	0.5032	0.5104	0.5164	0.5084	0.5152
9/16-18-UNF-2A	0.5244	0.5207	0.5264	0.5219	0.5284	0.5335	0.5264	0.5323
9/16-24-NEF-2A	0.5334	0.5303	0.5354	0.5315	0.5374	0.5417	0.5354	0.5405
5/8-11-UNC-2A	0.5640	0.5593	0.5660	0.5605	0.5680	0.5744	0.5660	0.5732
5/8-12-N-2A	0.5689	0.5643	0.5709	0.5655	0.5729	0.5792	0.5709	0.5780
5/8-18-UNF-2A	0.5869	0.5830	0.5889	0.5842	0.5909	0.5961	0.5889	0.5949
5/8-24-NEF-2A	0.5959	0.5927	0.5979	0.5939	0.5999	0.6043	0.5979	0.6031
11/16-12-N-2A	0.6314	0.6268	0.6334	0.6280	0.6354	0.6417	0.6334	0.6405
11/16-24-NEF-2A	0.6584	0.6552	0.6604	0.6564	0.6624	0.6668	0.6604	0.6656
3/4-10-UNC-2A	0.6830	0.6779	0.6850	0.6791	0.6870	0.6939	0.6850	0.6927
3/4-12-N-2A	0.6939	0.6892	0.6959	0.6904	0.6970	0.7043	0.6959	0.7031
3/4-16-UNF-2A	0.7074	0.7032	0.7094	0.7044	0.7114	0.7171	0.7094	0.7159
3/4-20-UNEF-2A	0.7155	0.7119	0.7175	0.7131	0.7195	0.7244	0.7175	0.7232
13/16-N-2A	0.7564	0.7517	0.7584	0.7529	0.7604	0.7668	0.7584	0.7656
13/16-16-UN-2A	0.7699	0.7658	0.7719	0.7670	0.7739	0.7794	0.7719	0.7782
13/16-20-UNEF-2A	0.7780	0.7744	0.7800	0.7756	0.7820	0.7869	0.7800	0.7857
7/8-9-UNC-2A	0.8008	0.7953	0.8028	0.7965	0.8048	0.8122	0.8028	0.8110
7/8-12-N-2A	0.8189	0.8142	0.8209	0.8154	0.8229	0.8293	0.8209	0.8281
7/8-14-UNF-2A	0.8266	0.8220	0.8286	0.8232	0.8306	0.8368	0.8286	0.8356
7/8-16-UN-2A	0.8324	0.8283	0.8344	0.8295	0.8364	0.8419	0.8344	0.8407
7/8-20-UNEF-2A	0.8405	0.8369	0.8425	0.8381	0.8445	0.8494	0.8425	0.8482
15/16-12-UN-2A	0.8814	0.8765	0.8834	0.8777	0.8854	0.8920	0.8834	0.8908
15/16-16-UN-2A	0.8949	0.8907	0.8969	0.8919	0.8989	0.9046	0.8969	0.9034
15-16-20-UNEF-2A	0.9030	0.8993	0.9050	0.9005	0.9070	0.9121	0.9050	0.9109
1-8-UNC-2A	0.9168	0.9108	0.9188	0.9120	0.9208	0.9288	0.9188	0.9276
1-12-UNF-2A	0.9439	0.9388	0.9459	0.9400	0.9479	0.9547	0.9459	0.9535
1-16-UN-2A	0.9574	0.9532	0.9594	0.9544	0.9614	0.9671	0.9594	0.9659
1-20-UNEF-2A	0.9655	0.9618	0.9675	0.9630	0.9695	0.9746	0.9675	0.9734
1-1/16-12-UN-2A	1.0064	1.0015	1.0084	1.0027	1.0104	1.0170	1.0084	1.0158
1-1/16-16-UN-2A	1.0199	1.0157	1.0219	1.0169	1.0239	1.0296	1.0219	1.0284
1-1/16-18-NEF-2A	1.0244	1.0205	1.0264	1.0217	1.0284	1.0338	1.0264	1.0326
1-1/8-7-UNC-2A	1.0302	1.0238	1.0322	1.0250	1.0342	1.0428	1.0322	1.0416
1-1/8-8-N-2A	1.0418	1.0357	1.0438	1.0369	1.0458	1.0540	1.0438	1.0528
1-1/8-12-UNF-2A	1.0689	1.0637	1.0709	1.0649	1.0729	1.0799	1.0709	1.0787
1-1/8-16-UN-2A	1.0824	1.0782	1.0844	1.0794	1.0864	1.0921	1.0844	1.0909
1-1/8-18-NEF-2A	1.0869	1.0830	1.0889	1.0842	1.0909	1.0963	1.0889	1.0951

Limits for Plated Threads—II

Note that maximum pitch diameter of the plated external thread exceeds the standard dimension for uncoated threads by an amount equal to the allowance. This practice is consistent with a recommendation on p 23, Handbook H 28-1957, or Screw Thread Standards for Federal Services." Thus, line-to-line contact with the female thread can occur at the pitch diameter Such practice minimizes corrosion between threaded members, overcomes lead-error problems, and the softer plating may be compressed to provide a snug, gage-like fit.

Plating Deposit = 0.0005-0.0007 In.

No. Th'ds Size & Class	External Th'd Pitch Dia, Class 2A: Before Plating (specify on dwg) Max	Before Plating Min	After Plating (inspect to) Max	After Plating Min	Internal Th'd Pitch Dia, Class 2B: Before Plating (specify on dwg) Min	Before Plating Max	After Plating (inspect to) Min	After Plating Max
0-80-UNF-2A	0.0491	0.0481	0.0519	0.0501	0.0547	0.0562	0.0519	0.0542
1-64-UNC-2A	0.0601	0.0589	0.0629	0.0609	0.0657	0.0675	0.0629	0.0655
1-72-UNF-2A	0.0612	0.0601	0.0640	0.0621	0.0668	0.0685	0.0640	0.0655
2-56-UNC-2A	0.0716	0.0703	0.0744	0.0723	0.0772	0.0792	0.0744	0.0772
2-64-UNF-2A	0.0731	0.0719	0.0759	0.0739	0.0787	0.0806	0.0759	0.0786
3-48-UNC-2A	0.0827	0.0812	0.0855	0.0832	0.0883	0.0905	0.0855	0.0885
3-56-UNF-2A	0.0846	0.0832	0.0874	0.0852	0.0902	0.0922	0.0874	0.0902
4-40-UNC-2A	0.0930	0.0913	0.0958	0.0933	0.0986	0.1011	0.0958	0.0991
4-48-UNF-2A	0.0957	0.0941	0.0985	0.0961	0.1013	0.1036	0.0985	0.1016
5-40-UNC-2A	0.1060	0.1042	0.1088	0.1062	0.1116	0.1141	0.1088	0.1121
5-44-UNF-2A	0.1074	0.1057	0.1102	0.1077	0.1130	0.1154	0.1102	0.1133
6-32-UNC-2A	0.1149	0.1129	0.1177	0.1149	0.1205	0.1234	0.1177	0.1214
6-40-UNF-2A	0.1190	0.1172	0.1218	0.1192	0.1246	0.1272	0.1218	0.1252
8-32-UNC-2A	0.1409	0.1388	0.1437	0.1408	0.1465	0.1495	0.1437	0.1475
8-36-UNF-2A	0.1432	0.1412	0.1460	0.1432	0.1488	0.1516	0.1460	0.1496
10-24-UNC-2A	0.1601	0.1576	0.1629	0.1596	0.1657	0.1692	0.1629	0.1672
10-32-UNF-2A	0.1669	0.1647	0.1697	0.1667	0.1725	0.1756	0.1697	0.1736
12-24-UNC-2A	0.1861	0.1835	0.1889	0.1855	0.1917	0.1953	0.1889	0.1933
12-28-UNF-2A	0.1900	0.1876	0.1928	0.1896	0.1956	0.1990	0.1928	0.1970
12-32-NEF-2A	0.1921	0.1906	0.1957	0.1926	0.1985	0.2018	0.1957	0.1998
1/4-20-UNC	0.2147	0.2118	0.2175	0.2138	0.2203	0.2243	0.2175	0.2223
1/4-28-UNF-2A	0.2240	0.2215	0.2268	0.2235	0.2296	0.2331	0.2268	0.2311
1/4-32-NEF-2A	0.2269	0.2245	0.2297	0.2265	0.2325	0.2359	0.2297	0.2339
5/16-18-UNC-2A	0.2736	0.2704	0.2764	0.2724	0.2792	0.2837	0.2764	0.2817
5/16-24 UNF-2A	0.2826	0.2797	0.2854	0.2817	0.2882	0.2922	0.2854	0.2902
5/16-32-NEF-2A	0.2894	0.2870	0.2922	0.2898	0.2950	0.2984	0.2922	0.2964
3/8-16-UNC-2A	0.3316	0.3280	0.3344	0.3300	0.3372	0.3421	0.3344	0.3401
3/8-24-UNF-2A	0.3451	0.3421	0.3479	0.3441	0.3507	0.3548	0.3479	0.3528
3/8-32-NEF-2A	0.3519	0.3493	0.3547	0.3513	0.3575	0.3611	0.3547	0.3591
7/16-14-UNC-2A	0.3883	0.3844	0.3911	0.3864	0.3939	0.3992	0.3911	0.3972
7/16-20-UNF-2A	0.4022	0.3988	0.4050	0.4008	0.4078	0.4124	0.4050	0.4104
7/16-28-UNEF-2A	0.4115	0.4087	0.4143	0.4107	0.4171	0.4209	0.4143	0.4189
1/2-12-N	0.4431	0.4385	0.4459	0.4405	0.4487	0.4549	0.4459	0.4529
1/2-13-UNC	0.4472	0.4430	0.4500	0.4450	0.4528	0.4585	0.4500	0.4565
1/2-20-UNF-2A	0.4647	0.4612	0.4675	0.4632	0.4703	0.4751	0.4675	0.4731
1/2-28-UNEF-2A	0.4740	0.4711	0.4768	0.4731	0.4769	0.4836	0.4768	0.4816
9/16-12-UNC-2A	0.5056	0.5012	0.5084	0.5032	0.5112	0.5172	0.5084	0.5152
9/16-18-UNF-2A	0.5596	0.5559	0.5624	0.5579	0.5652	0.5343	0.5624	0.5323
5/8-11-UNC-2A	0.5632	0.5585	0.5660	0.5605	0.5688	0.5752	0.5660	0.5732
5/8-12-N-2A	0.5681	0.5635	0.5709	0.5655	0.5737	0.5800	0.5709	0.5780
5/8-18-UNF-2A	0.5861	0.5822	0.5889	0.5842	0.5917	0.5969	0.5889	0.5949
5/8-24-NEF-2A	0.5951	0.5919	0.5979	0.5939	0.6007	0.6051	0.5979	0.6031
11/16-12 N-2A	0.6306	0.6260	0.6334	0.6280	0.6362	0.6425	0.6334	0.6405
11/16-24 NEF-2A	0.6576	0.6544	0.6604	0.6564	0.6632	0.6676	0.6604	0.6656
3/4-10 UNC-2A	0.6822	0.6771	0.6850	0.6791	0.6878	0.6947	0.6850	0.6927
3/4-12 N-2A	0.6931	0.6884	0.6959	0.6904	0.6987	0.7051	0.6959	0.7031
3/4-16 UNF-2A	0.7066	0.7024	0.7094	0.7044	0.7122	0.7179	0.7094	0.7159
3/4-20 UNEF-2A	0.7147	0.7111	0.7175	0.7131	0.7203	0.7252	0.7175	0.7232
13/16-12-N-2A	0.7556	0.7509	0.7584	0.7529	0.7612	0.7676	0.7584	0.7656
13/16-16-UN-2A	0.7691	0.7650	0.7719	0.7670	0.7747	0.7802	0.7719	0.7782
13/16-20 UNEF-2A	0.7772	0.7736	0.7800	0.7756	0.7828	0.7877	0.7800	0.7857
7/8-9 UNC-2A	0.8000	0.7945	0.8028	0.7965	0.8056	0.8130	0.8028	0.8110
7/8-12 N-2A	0.8181	0.8134	0.8209	0.8154	0.8237	0.8301	0.8209	0.8281
7/8-14 UNF-2A	0.8258	0.8212	0.8286	0.8232	0.8314	0.8376	0.8286	0.8356
7/8-16 UN-2A	0.8316	0.8275	0.8344	0.8295	0.8372	0.8427	0.8344	0.8407
7/8-20 UNEF-2A	0.8397	0.8361	0.8425	0.8381	0.8453	0.8502	0.8425	0.8482
15/16-12 UN-2A	0.8806	0.8757	0.8834	0.8777	0.8862	0.8928	0.8834	0.8908
15/16-16 UN-2A	0.8941	0.8899	0.8969	0.8919	0.8997	0.9054	0.8969	0.9034
15/16-20 UNEF-2A	0.9022	0.8985	0.9050	0.9005	0.9078	0.9129	0.9050	0.9109
1-8 UNC-2A	0.9150	0.9100	0.9188	0.9120	0.9216	0.9296	0.9188	0.9276
1-12 UNF-2A	0.9431	0.9380	0.9459	0.9400	0.9487	0.9555	0.9459	0.9535
1-16 UN-2A	0.9566	0.9524	0.9594	0.9544	0.9622	0.9679	0.9594	0.9659
1-20 UNEF-2A	0.9647	0.9610	0.9675	0.9630	0.9703	0.9754	0.9675	0.9734
1-1/16-12 UN	1.0056	1.0007	1.0084	1.0027	1.0112	1.0178	1.0084	1.0158
1-1/16-16 UN-2A	1.0191	1.0149	1.0219	1.0169	1.0247	1.0304	1.0219	1.0284
1-1/16-18 NEF-2A	1.0236	1.0197	1.0264	1.0217	1.0292	1.0346	1.0264	1.0326
1-1/8-7 UNC-2A	1.0294	1.0230	1.0322	1.0250	1.0350	1.0436	1.0322	1.0416
1-1/8-8 N-2A	1.0410	1.0349	1.0438	1.0369	1.0466	1.0548	1.0438	1.0528
1-1/8-12 UNF-2A	1.0681	1.0629	1.0709	1.0649	1.0737	1.0807	1.0709	1.0787
1-1/8-16 UN-2A	1.0816	1.0774	1.0844	1.0794	1.0872	1.0929	1.0844	1.0909
1-1/8-18-NEF-2A	1.0861	1.0822	1.0889	1.0842	1.0917	1.0971	1.0889	1.0951

Chapter 22

Jig and Fixture Details

Locking-type indexing fingers, 1

These fingers are required in jigs and fixtures, dividing heads, chucks and dial-feed tables to hold parts securely but temporarily in precise relationship

In any indexing device, there are at least three basic parts: a plunger (key, gage, male or movable part), a bushing (female or stationary hole) and a guide for the male part. In the majority of cases, there is also a spring and sometimes also a housing (body).

The actual design of an indexing device varies enormously from one application to the next, and is influenced by several factors—accuracy, production rate and function.

Classification

The basic form of the indexing gage depends primarily upon the kind of immobilization required. Thus, there are three main groups:

1. Locking. The indexing finger must be made inactive positively when the movable member must pass to the next position

2. Automatic bi-directional. The index finger is retracted (lifted) by itself when a certain pre-established load is applied

3. Automatic uni-directional. The same arrangement as in (2) but with the limitation that the mechanism-component may be moved only in one direction (it is locked in the opposite direction)

The shape of the indexing finger may be (in increasing order of accuracy):

1. Steel balls
2. Cylindrical plungers with:
   semi-spherical point
   90° degree taper point
   cylindrical point
   tapered point (about 10-20°)

For flat indexing gages, either parallel or tapered points are used, with the tapered design preferred.

By Federico Strasser
Vice-president
Alpha Products, Inc, Chicago

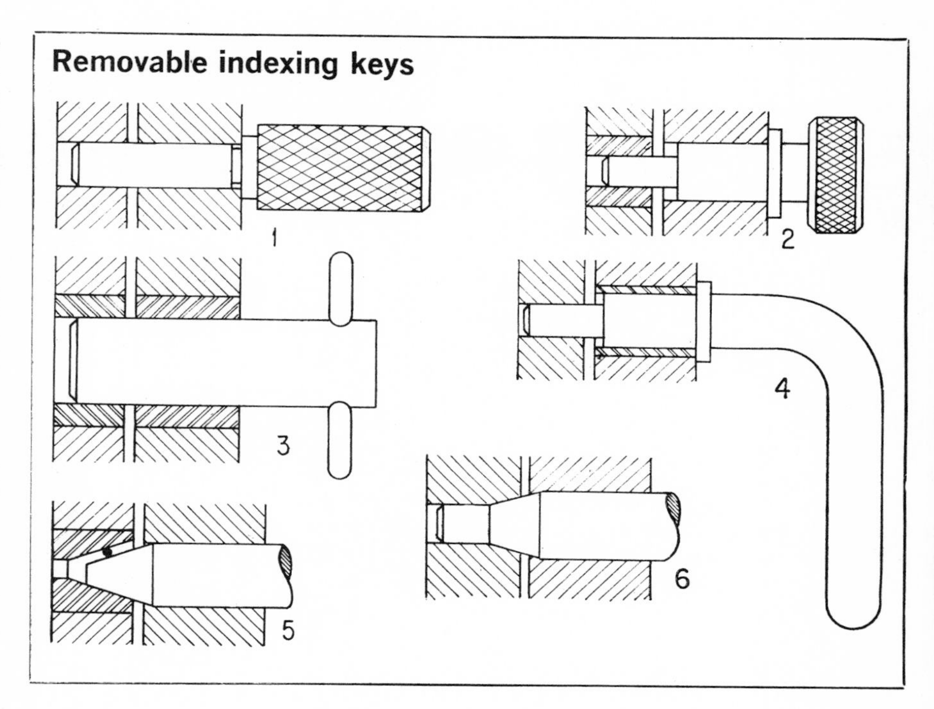

Indexing fingers are usually made from case-hardened steel, and the aligning surfaces must be ground after heat-treatment.

Design principles for numerous indexing devices are given.

Removable indexing keys

These keys offer the advantage of low tooling costs but are slow in use. First, they must be inserted manually before the machining operation. After the operation is completed, the keys must be withdrawn, the work moved to the next station and keys re-inserted.

The simplest design consists of a round plug, provided with a handle (preferably knurled). The cylindrical part is a slip fit in reamed holes in the stationary and the movable member, Fig. 1.

When the two members are not of the same diameter (Fig. 2), the key is stepped

If the mechanism members to be aligned are comparatively soft, it is good practice to use a hardened and ground steel bushing for guiding the indexing gage, increasing accuracy and life. The bushing may be used in either of the toolmembers (Figs. 2 and 4) or in both of them (Fig. 3), according to conditions.

Indexing plungers over 1 in. dia should be hollow, Fig. 3. In this case the handle is a pin put through the head of the gage. Another design consists of bending the protruding part of the gage, (Fig. 4), to 90°, to make a handle. This design is preferred in case of small diameters.

Misalignment

Although cylindrical holes are commonly used for aligning purposes, sometimes toolmakers prefer tapered holes and gages. This

Locking-type indexing fingers [Continued]

design is somewhat objectionable, because any dirt or foreign matter on the gaging surfaces will change the accuracy of alignment, Fig. 5.

The best solution of this problem, but expensive, is a combination of cylindrical and tapered gage ends. In this case, the maximal error (if any) may be only the allowance between the cylindrical hole diameter and the diameter of the cylindrical gage-end, Fig. 6.

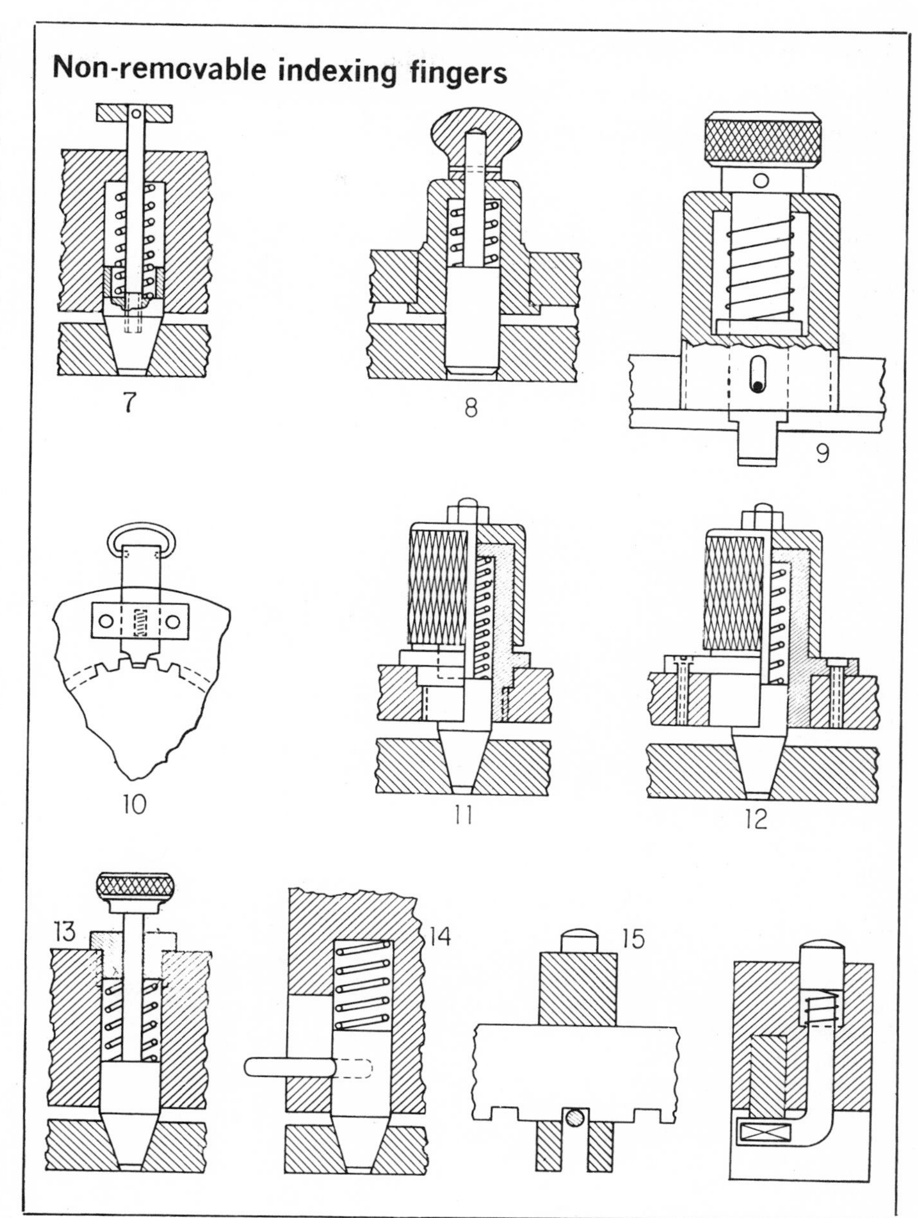

Non-removable indexing fingers

These non-removable indexing fingers are preferred for high production tools. In their simplest form the plungers are actuated by a compression spring. For release the plunger is usually pulled out of its seat by means of a suitable handle, which may have any reasonable form, Fig. 7.

If the tool member is thin or soft, the plunger-housing is made as a separate item and a press fit, Fig. 8.

Fig. 9 shows a round indexing gage with a rectangular indexing point. In this case, the shaft must be guided, (prevented from turning radially) while longitudinal movement is effected freely. Therefore, a retaining pin operates in a slot in the housing.

The same indexing task is solved with a flat index gage, Fig. 10. Here a slide is moved in and out by means of a handle.

In the case of movable fixtures (those mounted on lathe faceplates) it is necessary to avoid accidents. For such cases the handle is made in the form of a sleeve (Fig. 11). This design offers another advantage: If the indexing point is too tight in its seat, the gage must be loosened before it can be pulled out. This separation is obviously easier to effect with a large knurled handle bushing than with a mere knob. Furthermore, mounting of the housing is made by means of a threaded portion; however, alignment is insured by means of the shoulder.

In another case (Fig. 12) the housing has a large flange, which is screwed to the body.

The plunger is not always mounted on the interior of the mechanism. If it is mounted from the outside, a threaded, knurled bushing holds the plunger in place. See Fig. 13. Now, the index finger may be removed without dismantling other tool members.

If the index finger is mounted in a part that is too large for axial actuation, the motion may be effected by means of lateral pins (Fig. 14).

The plunger is not always pulled out of the tool-member for making the indexing inactive; sometimes it is pushed in, Fig. 15 Here the bent gage is guided.

Locking-type indexing fingers, 2

Some indexing fingers must be actuated indirectly; others require a rest position so that the operator need not hold onto them. Here are 14 designs

Removable and non-removable indexing fingers (**AM**—Nov 21, '66, p139) are commonly actuated directly: that is, co-axially with the indexing point. But there are several designs of fingers which are pivoted and indirectly actuated. In such cases the spring may not be co-axial with the indexing plunger. Then, the spring will be located outside of the device, especially when the finger is not round. Here are several examples:

In Fig. 16, the indexing finger consists of a tapered detent on a ball-handled lever. Normally the detent lies in the part to be indexed, being held there by a spring- backed pin. When the lever is pulled up, the mechanism can be indexed.

In the remaining designs, Fig. 17 to 21, the actuating lever is pushed manually or hydraulically to retract the index finger.

To keep out of trouble in such applications, try to foresee potential sources of erratic behavior. For example, in both Figs. 16 and 17, the tool engineer will specify an arc-like pivoting movement of the indexing point to avoid impairing the alignment.

Sometimes the actuating lever is pushed instead of pulled, as in Figs. 18 and 19. Or indirect action may be secured by a handle-operated plunger that is backed by a spring, Fig. 20. When the handle is pushed down, the detent will be lifted out of engagement with the part to be indexed, and when the handle is released the spring will hold the detent in the next notch.

Remote or automatic control can be achieved by hydraulic oil or compressed air. The fluid enters a cylinder and pushes upward a piston connected through a stuffing box to the plunger. When the fluid is released, the spring depresses the plunger, as shown in Fig. 21.

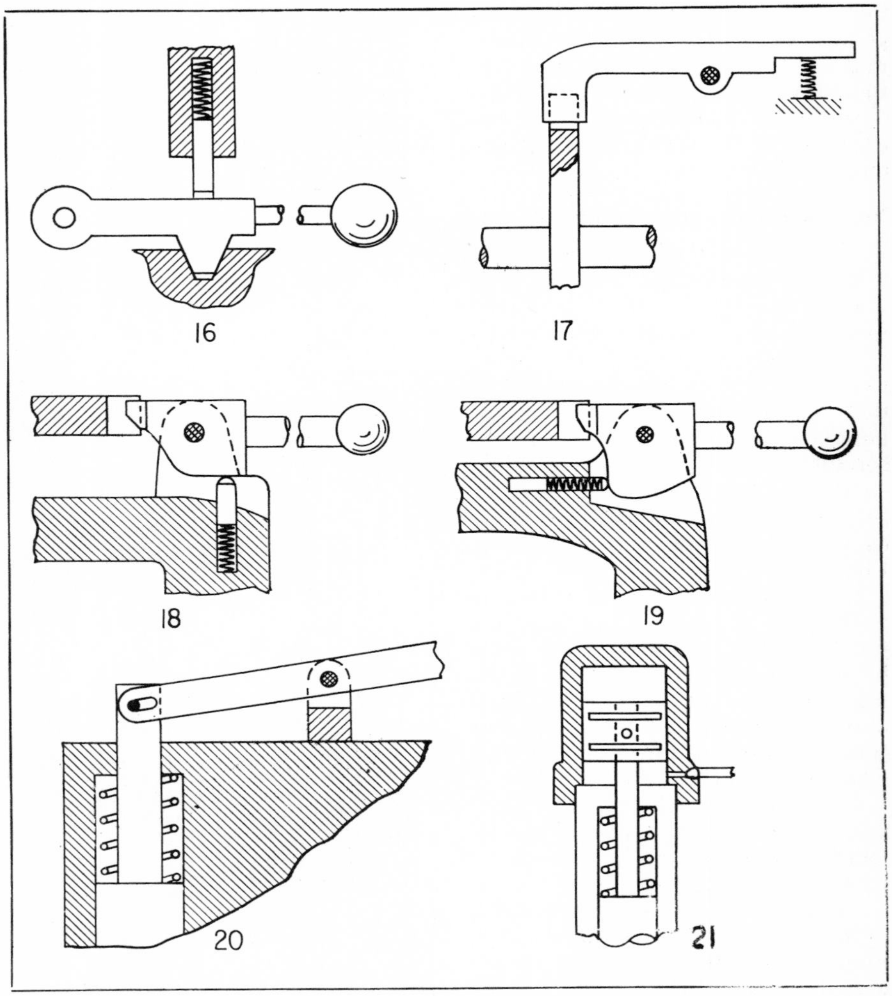

By Federico Strasser
Vice president
Alpha Products, Inc, Chicago

16

17

18

19

20

21

Indexing fingers with rests

So far, the indexing fingers must be held out of engagement by the operator. Sometimes, this is inconvenient or tiring. Therefore a mechanical means of disengaging the finger may be required. There are many ways to accomplish this. Eight examples are shown on page 297.

A very common method is to mill a slot in housing for the finger and drive a pin into a blind hole drilled in the plunger shaft. The

Locking-type indexing fingers [Continued]

pin ordinarily lies in the slot. When the indexing finger is to be made inactive, the knob is pulled out, rotated say 90°, and let go. Now the pin rests on the top of the housing and the plunger can't engage the indexing notch, Fig. 22.

Two other ways of accomplishing the same objective are shown in Fig. 23 and 24. Here the pin engages an appropriate hole drilled to the proper depth

A bayonet lock is always an easy device to use and it's positive. The sleeve handle, Fig. 25, has the bayonet slot cut in it, and a pin is fitted to the housing. This type of finger is engaged or disengaged with a twist of the wrist.

Safe locking of the index finger can be obtained by engaging a spring-backed steel ball into a spherical or notched recess. This design, Fig. 26, requires little effort.

A modification of the wedge idea is the use of a pin in a fast-helix groove. The plunger is prevented from turning, Fig. 27, by a setscrew.

In many cases, a simple eccentric cam (not shown) will lock and unlock the plunger. When a cam can't be placed on the housing head, use a "cam" that acts laterally. In this case a pin in the end of a shaft engages a slot in the plunger and lifts or lowers the latter as the actuating handle is moved.

Racks are a good means for operating heavy plungers. The rack can be cut directly into the plunger body, Fig. 28, while the pinion is operated by the handle.

Our last design, Fig. 29, works without the aid of a spring. In this case the plunger is guided and it is raised or lowered by the knob and screw. The screw is retained in the housing by a pin engaging a circumferential slot.

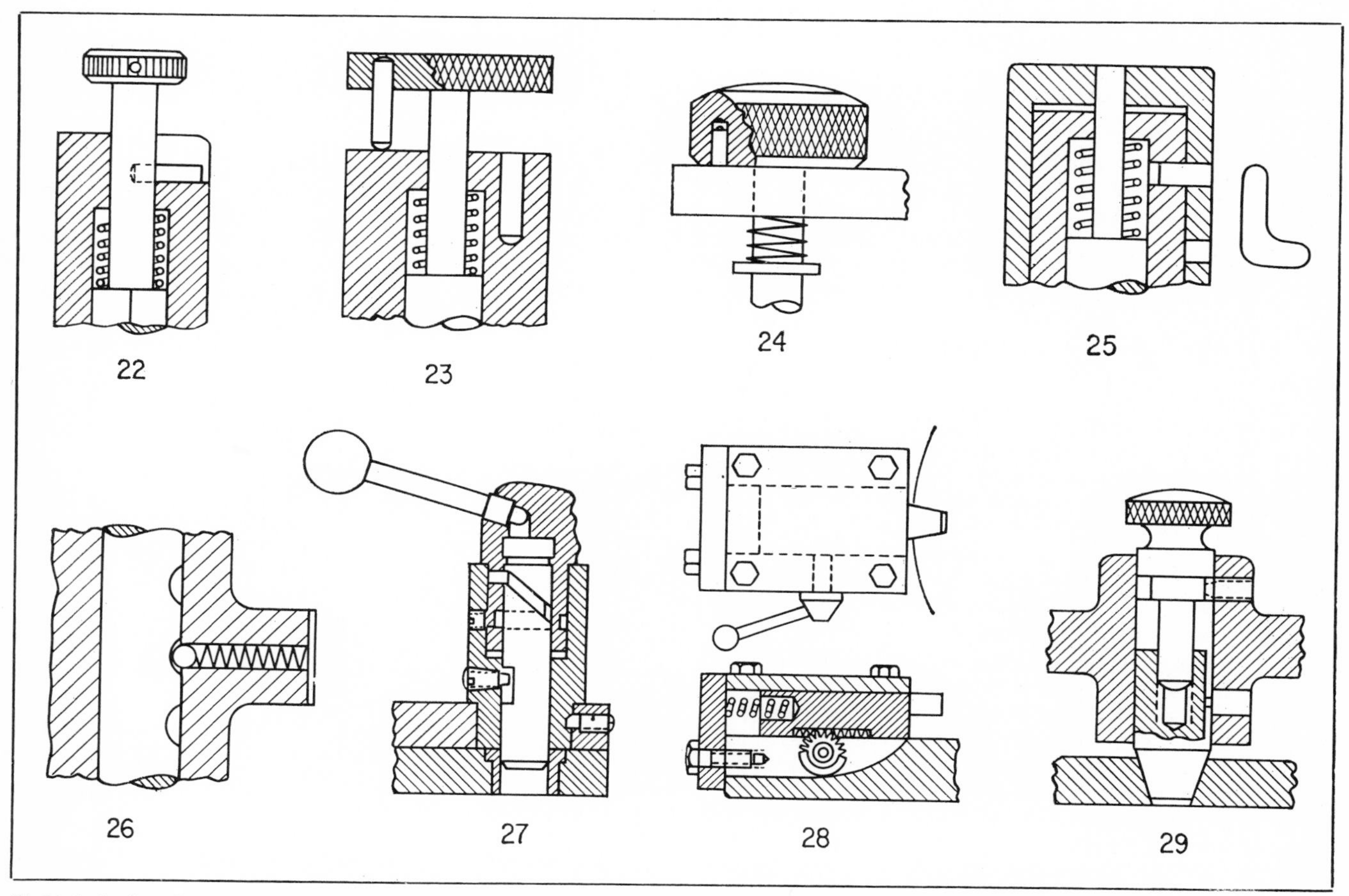

Eight indexing fingers with rest positions: 22—Pin in slot; 23—Pin enters hole in housing; 24—Pin enters knob; 25—Sleeve with bayonet lock; 26—Ball engages recess; 27—Pin in helical groove; 28—Rack-actuated slide and 29—Screw-operated finger

Standard Jig and Fixture Parts

Courtesy: National Institute of Jig & Fixture Component Manufacturers

Dimensions for quarter-turn screws and rest buttons have been standardized by the Institute, as the first step in a program to apply standard dimensions and tolerances to all tooling components produced by member firms. The purpose is to lower the cost of tooling and to achieve interchangeability.

It is the Institute's intention to standardize shoulder screws, T-slot nuts, swing bolts, hand knobs, handwheels and other items in the next five to six years. Firms approving the efforts include: Carr Lane Mfg. Co., St. Louis; Jergens Tool Specialty Co., Cleveland; Vlier Engineering Corp, Los Angeles and Downers Grove, Ill.; Universal Engineering Co., Frankenmuth, Mich.; Morton Machine Works, Millersburg, Pa.; Harco Engineering Co, Detroit; Lodding, Inc., Detroit; Monroe Engineering Products, Detroit; Northwestern Tools, Inc., Dayton, Ohio, and Tomco, Inc., Racine, Wisconsin.

Descriptions of Items

QUARTER-TURN SCREW — This device is extensively used to hold the hinged top plate of a drill jig in position. The device has a threaded shank and a rectangular head. Thickness of the head is equal to the diameter of the threaded shank. This arrangement allows a flat plate containing a slotted hole, the width of which is slightly greater than the diameter of the quarter-turn screw, and the length of which is slightly greater than the width of the head of the quarter turn screw, to slip on and off over the head of the screw. After the plate is positioned over the quarter-turn screw, the screw may be tightened until it clamps against the flat plate.

There are two positions in each revolution of the screw in which the flat plate may be removed. By adjusting the position of the top of the flat plate, or by filing the bottom surface of the ears of the quarter-turn screw, the final clamped position of the plate can be obtained when the ears of the screw are at 90° to the slot in the plate. By turning the screw a quarter of a turn (counter-clockwise), the plate is unclamped and the slot will be in a position such that the plate may be removed.

REST BUTTON—The rest button has many uses. It is a hardened-steel cylindrical button with a shank slightly smaller in diameter than the button. The shank diameter is such that it will press into a standard reamed hole. The height of the button is held to a close tolerance.

Rest buttons may be used as resting points for a part where it is desirable not to have the part lie on a large flat surface. Rest buttons may also be used as jig feet. The fact that they are hardened makes them useful in providing a small wear surface. Another application is for hardened cutter sets. The close tolerance on the height of the button makes it possible to press four or more into a flat plate and have their top surfaces in a flat plane. When the buttons are pressed into a surface which is not flat, the tops may be ground to clean up and this provides a flat surface. •

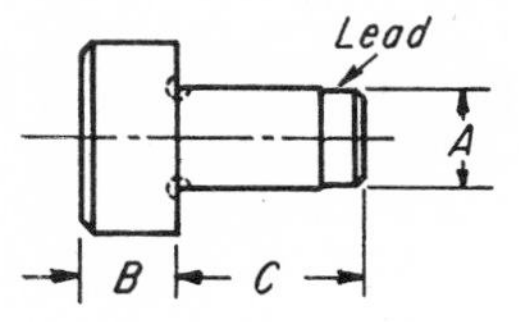

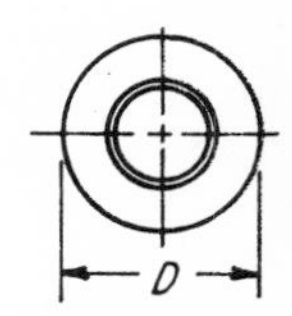

Dimensions of Standard Rest Buttons, Inches

Part No.	A	B	C	D
201	0.1880	0.250	3/8	3/8
	0.1885	0.251		
202	0.1880	0.375	3/8	3/8
	0.1885	0.376		
203	0.1880	0.500	3/8	3/8
	0.1885	0.501		
204	0.2505	0.250	1/2	1/2
	0.2510	0.251		
205	0.2505	0.375	1/2	1/2
	0.2510	0.376		
206	0.2505	0.500	1/2	1/2
	0.2510	0.501		
210	0.3755	0.250	5/8	5/8
	0.3760	0.251		
211	0.3755	0.375	5/8	5/8
	0.3760	0.376		
213	0.3755	0.500	5/8	5/8
	0.3760	0.501		
214	0.3755	0.625	5/8	5/8
	0.3760	0.626		
217	0.5005	0.375	5/8	7/8
	0.5010	0.376		
219	0.5005	0.500	5/8	7/8
	0.5010	0.501		
220	0.5005	0.625	5/8	7/8
	0.5010	0.626		
223	0.6255	0.375	3/4	1
	0.6260	0.376		
225	0.6255	0.500	3/4	1
	0.6260	0.501		
227	0.6255	0.750	3/4	1
	0.6260	0.751		
230	0.7510	0.500	7/8	1 1/4
	0.7515	0.501		
231	0.7510	0.625	7/8	1 1/4
	0.7515	0.626		

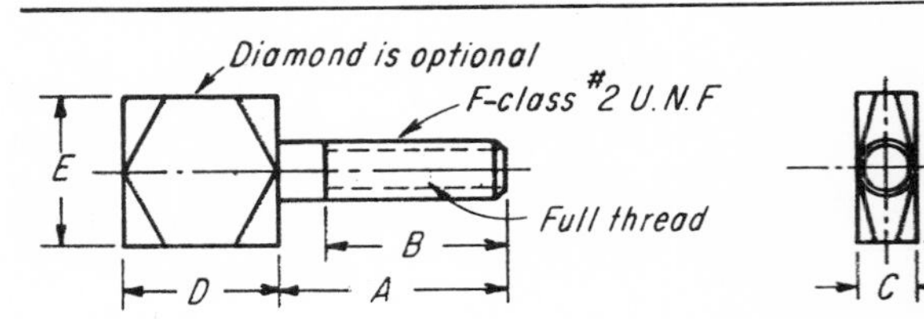

Dimensions of Quarter-Turn Screws, Inches

Part No.	A	B	C	D	E	F
101	1	7/8	3/16	1/2	1/2	10-32
*102	1	7/8	3/16	1/2	1/2	10-32
113	1 1/4	1	1/4	5/8	3/4	1/4-20
*114	1 1/4	1	1/4	5/8	3/4	1/4-20
119	1 1/2	1 1/4	5/16	3/4	1	5/16-18
*120	1 1/2	1 1/4	5/16	3/4	1	5/16-18
125	1 1/2	1 1/4	3/8	1	1	3/8-16
126	1 1/2	1 1/4	3/8	1	1	3/8-16
131	1 3/4	1 1/2	1/2	1	1 1/4	1/2-13
*132	1 3/4	1 1/2	1/2	1	1 1/4	1/2-13

*Hardened

Chapter 23

Fasteners

Grade Markings for Bolts—I

Grade markings on bolt heads identify material, strength and the specifying group or user. These markings serve as a quick visual check of the product upon receipt. They also aid the assembly operator to install the correct grade of fastener

By Frank W Wood, Jr, Advanced Designs, Inc, Vienna, Virginia

Carbon-Steel Fasteners

Markings	Specifications	Used by
C	Low-carbon 60,000 YS*	Army Ord.
60	Commercial steel 48,000 TS† Fine-thrd, hex-hd bolt	Air Force
65	Commercial steel 48,000 TS Coarse-thrd hex bolt	Air Force
70	Commercial steel 48,000 TS Carriage bolt	Air Force
E4	Low-carbon (AMS 5061-cold drawn) 70,000 TS; Rh B80-100	Air Force
	Medium carbon, SAE Grade 3 110,000 TS up to ½″ 100,000 TS over ½″	SAE & Army Ord.
B	Medium carbon 75,000 YS	Army Ord.
	Medium carbon Grade 5—SAE Mil-B-857A, ASTM A325 120,000 TS to ¾″ 111,000 TS over ¾″ 105,000 TS over 1″ 90,000 TS over 1½″	Military, SAE Navy BuShips, ASTM
	Medium carbon SAE Grade 6 140,000 TS to ⅝″ 133,000 TS over ⅝″	SAE & Army Ord.
E	Medium carbon 95,000 YS	Army Ord.

* YS = yield strength

† TS = tensile strength

Alloy-Steel Fasteners

Markings	Specifications	Used by
	Medium-carbon SAE Grade 7 130,000 TS	SAE & Army Ord.
	Medium-carbon: 150,000 TS—Grade 8 SAE or Mil-B857A	SAE, Military Navy BuShips
	Quenched & tempered— ASTM A354, Grade BD 150,000 TS—	ASTM
	Grade 8 125,000 TS over 1½″	Military

Grade Markings for Bolts—II

Alloy-Steel Fasteners—*Continued*

Markings	Specifications	Used by
BB	Quenched & tempered ASTM A354, Grade BB 105,000 TS up to 2″ 100,000 TS over 2″	ASTM
BC	Quenched & tempered ASTM A354, Grade BC 125,000 TS to 2½″ 115,000 TS over 2½″	ASTM
H	Alloy steel 130,000 YS	Army Ord.
E1	Medium carbon AMS 6320 Rh: C 26-32	Air Force
E11	Medium carbon AMS 6322 Hex bolts—Rh: C26-32 Socket bolts—Rh: C32-36	Air Force Army Ord.
E11	Medium carbon, AMS 6322 Rh C26-32	Air Force
X	Medium carbon (Cr, Ni, Mo) Mil-B-6812 125,000 TS	Air Force Army Ord.
X	Medium carbon (Cr, Ni, Mo) Mil-S-7839 125,000 TS	Air Force
B7	Chrome-moly 4140, 4142, 4145 105,000 TS to 2½″	ASTM
B7A	Chrome-high moly Grade B7A (A193) 125,000 TS to 2½″	ASTM
B14	Chrome-moly-vanadium Grade B14 (A193) 125,000 TS to 2½″	ASTM
B16	Chrome-moly-vanadium Grade B16 (A193) 125,000 TS to 2½″	ASTM
L7	Chrome-moly Grade L 7 (A320) 4140, 4142, 4145 125,000 TS to 2½″	ASTM
L9	Nickel steel AISI 2340 Grade L 9 (A320) 125,000 TS to 2½″	ASTM
L10	Nickel steel AISI 2317— Grade L10 (A320) 70,000 TS to 4″	ASTM
L43	Nickel-chrome-moly AISI 4340 or Grade L 43 (A320) 125,000 TS to 4″	ASTM

Maximum Torque for Screws

By John A Stevenson,

The H M Harper Co, Morton Grove, Ill

Source: "Fasteners," Vol. 14, No. 4, published by Industrial Fasteners Institute, 1517 Terminal Tower, Cleveland 13, Ohio

Torquing values for screws made of various materials are offered as a guide. Tests were conducted on dry, or near dry, product. Mating parts were wiped clean of chips. Bolt tension is held at a factor somewhat less than the yield point.

Bolt Size	Low-Carbon Steel	18-8 Stainless	Brass	Silicon Bronze	Aluminum 24ST-4	316 Stainless	Monel
	in.-lb	in.-lb	in.-lb	in.-lb	in.-lb	in.-lb	
2-56	2.2	2.5	2.0	2.3	1.4	2.6	2.5
2-64	2.7	3.0	2.5	2.8	1.7	3.2	3.1
3-48	3.5	3.9	3.2	3.6	2.1	4.0	4.0
3-56	4.0	4.4	3.6	4.1	2.4	4.6	4.5
4-40	4.7	5.2	4.3	4.8	2.9	5.5	5.3
4-48	5.9	6.6	5.4	6.1	3.6	6.9	6.7
5-40	6.9	7.7	6.3	7.1	4.2	8.1	7.8
5-44	8.5	9.4	7.7	8.7	5.1	9.8	9.6
6-32	8.7	9.6	7.9	8.9	5.3	10.1	9.8
6-40	10.9	12.1	9.9	11.2	6.6	12.7	12.3
8-32	17.8	19.8	16.2	18.4	10.8	20.7	20.2
8-36	19.8	22.0	18.0	20.4	12.0	23.0	22.4
10-24	20.8	22.8	18.6	21.2	13.8	23.8	25.9
10-32	29.7	31.7	25.9	29.3	19.2	33.1	34.9
¼-20	65.0	75.2	61.5	68.8	45.6	78.8	85.3
¼-28	90.0	94.0	77.0	87.0	57.0	99.0	106.0
5/16-18	129	132	107	123	80	138	149
5/16-24	139	142	116	131	86	147	160
⅜-16	212	236	192	219	143	247	266
⅜-24	232	259	212	240	157	271	294
7/16-14	338	376	317	349	228	393	427
7/16-20	361	400	327	371	242	418	451
½-13	465	517	422	480	313	542	584
½-20	487	541	443	502	328	565	613
9/16-12	613	682	558	632	413	713	774
9/16-18	668	752	615	697	456	787	855
⅝-11	1000	1110	907	1030	715	1160	1330
⅝-18	1140	1244	1016	1154	798	1301	1482
¾-10	1259	1530	1249	1416	980	1582	1832
¾-16	1230	1490	1220	1382	958	1558	1790
⅞-9	1919	2328	1905	2140	1495	2430	2775
⅞-14	1911	2318	1895	2130	1490	2420	2755
1-8	2832	3440	2815	3185	2205	3595	4130
1-14	2562	3110	2545	2885	1995	3250	3730
	ft-lb	ft-lb	ft-lb	ft-lb	ft-lb	ft-lb	ft-lb
1⅛-7	340	413	337	383	265	432	499
1⅛-12	322	390	318	361	251	408	470
1¼-7	432	523	428	485	336	546	627
1¼-12	396	480	394	447	308	504	575
1½-6	732	888	727	822	570	930	1064
1½-12	579	703	575	651	450	732	840

Torque for Machine Screws—I

(For sizes 0 to 10 inclusive)

By E F Perout, standards planning engineer, Hawthorne Works, Western Electric Co.

Intelligent use of threaded fasteners requires consideration of five types of stresses, namely: tension, torsion, impact, shear and fatigue.

On the assembly line our sole interest is in tensioning threaded fasteners sufficiently to result in secure assemblies. Here, torque is a practical measurable means of developing desired tension.

In establishing suitable driving torques for threaded fasteners, it will be found that numerous variables affect the applied torsional force adversely. Friction is developed between the screw head and the mating surface, as well as between the threads. The amount of friction depends on surface finish, cleanliness, and length of thread engagement.

Consideration must also be given to variations in the materials for the threaded fastener and the engaging member; also, their tolerances. In addition, relaxation of tension occurs sometime after tightening because of creep, thereby resulting in a lower torque.

Prior to our tests little had been published concerning fastening strengths of number-size screws in various metals and non-metals. The demand for improved reliability indicated a need for torque values for tightening and testing in a form suitable for us by operators and inspectors. Hence, numerous tests were conducted on selected sizes of rolled-thread machine screws to ascertain the ultimate torsional strength of each size of fastener.

These tests were made with a commercial torque-testing fixture wherein the screw is held firmly in a split collet having threads of the same diameter and pitch as that of the screw under test, and a bottoming pin to prevent the screw from threading through. The value obtained in these tests are the ultimate average torsional strengths or the torsional forces required to break the screws.

To ascertain the torsional forces required to overcome the friction between head and seating surface and between the threaded surfaces, the above obtained values were compared with average ultimate torsional strength values of electro-plated screws seating against a smooth hardened-steel block without additive finish such as plating.

The results of these tests permitted establishment of "suitable" driving and testing torques for selected screw sizes. Suitable driving torque is a relactive factor, an indirect way of tensioning the threaded fastener sufficiently to result in a secure assembly.

Fastening problems do not end with selection of proper driving torques. Careful consideration must be given to type of drive and the clutch, close regulation of air pressure and proper handling and application of the driver on the assembly line. The driving and testing torque tables, while not complete as to material coverage, have proved satisfactory in our field.

The headings in the accompanying tables are Max (maximum), Min (minimum), Insp (inspection), and Final Check (for quality control purposes). The values under each of these headings are:

Max & Min:

The recommended tightening torque fails anywhere within this range but shall not be below the minimum value nor should it exceed the maximum value.

Insp:

The recommended testing torque to which the gages are set for the inspection groups. In general, these checks are made either in the assembly line or shortly after the product leaves the assembly line. There shall

Average Maximum Torsion to Break Screws

Size	Threads Per Inch	Torque at Failure, lb-in.	
		Steel	Brass
0.060	80	4.8	2.8
0.073	64 72	5.8 7.5	4.4 4.3
0.086	56 64	6.8 9.3	6.9 7.6
0.112	40 48	12.6 13.4	10.3 16.9
0.125	40 44	19.0 27.6	19.0 22.0
0.138	32 40	23.1 22.4	14.3 18.3
0.164	32 36	42.3 50.4	30.8 35.8
0.190	24 32	49.1 78.6	38.6 54.9
0.216	24 28	81.0 100.0	66.0 81.0

All pound inch values denote screw breakage.

Torque for Machine Screws—II

be no perceptible movement when this torsional force is applied.

Final Check

The recommended testing torque to which the gages are set for the quality-control groups when checking the product after a time interval of days has elapsed. The values are less than those for the inspection groups because of the "relaxation" that occurs after a time interval. There shall be no perceptible movement when this torsional force is applied.

The manner in which the tables are used is as follows:

Example:

A 0.190-32 steel screw is driven into 1/4 in. thick steel (or 1/4 in. depth of thread in a blind hole).

Locate screw size and material thickness or depth of tapping in proper table and the recommended tightening and testing torque values, in pound inches, are then found in line with and to the right of same, i.e. Max—25.0, Min—17.5, Insp—17.5 and Final Check—16.0.

In the tables, the recommended values are those that, in general, will result in a secure and satisfactory fastening (assembly) based on numerous tests and trials conducted over an extended period of time. This approach permitted a reduction in the number of driver sizes and fixed setting torque testing gages required as well as a reduction in the number of types of replacement parts for maintenance of the drivers and gages. In addition, the values are within the output range of drivers, the body sizes of which permit their use more readily in multiple-unit setups wherein anywhere from two to five drivers are assembled and operated as a single unit. •

Master Table for Setting Driving Torque

Directions for Use:

With calibrated driver, apply torque to fastener until threads strip or screw breaks. Find average from at least five samples and refer to Col. 1. Opposite figure corresponding to test average, find the safe torques for max, min, inspect and final check values. The driver setting in lb-in. is also given.

Strip or Break Torque, lb-in.	Working Torques, lb-in. Max	Min	Insp	Final Check	Driver Setting, lb-in.
0.18- 0.22, incl	0.125	0.094	0.094	0.078	0.109
0.23- 0.34, incl	0.156	0.109	0.109	0.094	0.141
0.35- 0.52, incl	0.25	0.188	0.188	0.172	0.234
0.53- 0.72, incl	0.375	0.266	0.266	0.234	4.344
0.73- 1.0, incl	0.5	0.344	0.344	0.313	0.453
1.1 - 1.4, incl	0.75	0.531	0.531	0.469	0.683
1.5 - 1.8, incl	1.0	0.75	0.75	0.688	0.906
1.9 - 2.2, incl	1.25	0.875	0.875	0.813	1.125
2.3 - 2.6, incl	1.5	1.063	1.063	1.0	1.375
2.7 - 2.9, incl	1.75	1.25	1.25	1.125	1.563
3.0 - 3.6, incl	2.0	1.5	1.5	1.25	1.75
3.7 - 4.6, incl	2.5	1.75	1.75	1.5	2.25
4.7 - 5.6, incl	3.0	2.0	2.0	1.75	2.5
5.7 - 6.5, incl	3.5	2.5	2.5	2.0	3.0
6.6 - 7.4, incl	4.0	3.0	3.0	2.5	3.5
7.5 - 8.4, incl	5.0	3.5	3.5	3.0	4.5
8.5 - 9.4, incl	5.5	4.0	4.0	3.5	5.0
9.5 -10.3, incl	6.0	4.0	4.0	3.5	5.5
10.4 -11.3, incl	6.5	4.5	4.5	4.0	6.0
11.4 -11.7, incl	7.0	5.0	5.0	4.5	6.0
11.8 -12.6, incl	7.5	5.0	5.0	4.5	6.5
12.7 -13.4, incl	8.0	5.5	5.5	4.5	7.0
13.5 -14.4, incl	8.5	6.0	6.0	5.5	7.5
14.5 -15.0, incl	9.0	6.0	6.0	5.5	8.0
15.1 -15.9, incl	9.5	6.5	6.5	6.0	8.5
16.0 -16.7, incl	10.0	7.0	7.0	6.5	9.0
16.8 -17.5, incl	10.5	7.5	7.5	6.5	9.5
17.6 -19.5, incl	11.0	7.5	7.5	6.5	10.0
19.6 -23.2, incl	12.0	8.0	8.0	7.0	11.0
23.3 -24.4, incl	13.0	9.0	9.0	8.0	11.5
24.5 -25.2, incl	13.5	9.5	9.5	8.5	12.0
25.3 -27.0, incl	14.0	10.0	10.0	9.0	12.5
27.1 -31.5, incl	15.0	10.5	10.5	9.5	13.5
31.6 -35.3, incl	16.0	11.0	11.0	10.0	14.5
35.4 -39.0, incl	18.0	12.5	12.5	11.0	16.0
39.1 -51.0, incl	20.0	14.0	14.0	12.5	18.0
51.1- 80.0, incl	25.0	17.5	17.5	16.0	22.5
80.1-100.0, incl	30.0	21.0	21.0	19.0	27.0
100.1 and over	35.0	24.5	24.5	22.0	31.05

Torque for Machine Screws—III

By E F PEROUT, standards planning engineer, Western Electric Co, Hawthorne Works, Chicago

Screw, Finished or Unfinished			Steel Screws in Aluminum				Steel Screws in Magnesium, Heat-Treated				Brass Screws in Brass			
			Driving and Testing Torque, lb-in.				Driving and Testing Torque, lb-in.				Driving and Testing Torque, lb-in.			
Size	Threads Per Inch	Thread Engagement, in.	Max	Min	Insp	Final Check	Max	Min	Insp	Final Check	Max	Min	Insp	Final Check
0.060	80	1/32 to 3/64	0.50	0.344	0.344	0.313	0.75	0.531	0.531	0.469	0.75	0.531	0.531	0.469
		3/64 and over	3.0	2.0	2.0	1.75	3.0	2.0	2.0	1.75	1.75	1.25	1.25	1.125
0.073	64	1/32 to 3/64	1.75	1.25	1.25	1.125	1.5	1.063	1.063	1.0	1.0	0.75	0.75	0.688
		3/64 and over	3.5	2.5	2.5	2.0	3.5	2.5	2.5	2.0	2.5	1.75	1.75	1.5
	72	1/32 to 3/64	2.0	1.5	1.5	1.25	1.0	0.75	0.75	0.688	0.75	0.531	0.531	0.469
		3/64 to 5/64	3.0	2.0	2.0	1.75	2.0	1.5	1.5	1.25	* 2.5	1.75	1.75	1.5
		5/64 and over	5.0	3.5	3.5	3.0	5.0	3.5	3.5	3.0	—	—	—	—
0.086	56	1/32 to 3/64	1.5	1.063	1.063	1.0	1.75	1.25	1.25	1.125	2.5	1.75	1.75	1.5
		3/64 and over	4.0	3.0	3.0	2.5	4.0	3.0	3.0	2.5	4.0	3.0	3.0	2.5
	64	1/32 to 3/64	1.5	1.063	1.063	1.0	1.25	0.875	0.875	0.813	2.5	1.75	1.75	1.5
		3/64 to 5/64	3.5	2.5	2.5	2.0	3.5	2.5	2.5	2.0	* 5.0	3.5	3.5	3.0
		5/64 to 7/64	3.5	2.5	2.5	2.0	* 5.5	4.0	4.0	3.5	—	—	—	—
		7/64 and over	5.5	4.0	4.0	3.5	—	—	—	—	—	—	—	—
0.112	40	1/16 to 3/32	2.5	1.75	1.75	1.5	3.5	2.5	2.5	2.0	* 6.0	4.0	4.0	3.5
		3/32 to 9/64	6.0	4.0	4.0	3.5	* 7.5	5.0	5.0	4.5	—	—	—	—
		9/64 and over	7.5	5.0	5.0	4.5	—	—	—	—	—	—	—	—
	48	1/16 to 3/32	4.0	3.0	3.0	2.5	2.0	1.5	1.5	1.25	* 10.5	7.5	7.5	6.5
		3/32 and over	8.0	5.5	5.5	4.5	* 8.0	5.5	5.5	4.5	—	—	—	—
0.125	40	1/16 to 3/32	3.5	2.5	2.5	2.0	7.5	5.0	5.0	4.5	* 11.0	7.5	7.5	6.5
		3/32 to 5/32	7.5	5.0	5.0	4.5	* 11.0	7.5	7.5	6.5	—	—	—	—
		5/32 to 7/32	10.5	7.5	7.5	6.5	—	—	—	—	—	—	—	—
		7/32 and over	11.0	7.5	7.5	6.5	—	—	—	—	—	—	—	—
	44	1/16 to 3/32	4.0	3.0	3.0	2.5	7.5	5.0	5.0	4.5	* 12.0	8.0	8.0	7.0
		3/32 to 5/32	11.0	7.5	7.5	6.5	* 15.0	10.5	10.5	9.5	—	—	—	—
		5/32 to 7/32	14.0	10.0	10.0	9.0	—	—	—	—	—	—	—	—
		7/32 and over	15.0	10.5	10.5	9.5	—	—	—	—	—	—	—	—
0.138	32	1/16 to 3/32	3.0	2.0	2.0	1.75	6.5	4.5	4.5	4.0	* 8.5	6.0	6.0	5.5
		3/32 to 5/32	9.5	6.5	6.5	6.0	* 12.0	8.0	8.0	7.0	—	—	—	—
		5/32 and over	12.0	8.0	8.0	7.0	—	—	—	—	—	—	—	—
	40	1/16 to 3/32	5.5	4.0	4.0	3.5	7.0	5.0	5.0	4.5	* 11.0	7.5	7.5	6.5
		3/32 to 5/32	10.0	7.0	7.0	6.5	* 12.0	8.0	8.0	7.0	—	—	—	—
		5/32 and over	12.0	8.0	8.0	7.0	—	—	—	—	—	—	—	—
0.164	32	1/16 to 3/32	5.5	4.0	4.0	3.5	9.0	6.0	6.0	5.5	* 15.0	10.5	10.5	9.5
		3/32 to 5/32	15.0	10.5	10.5	9.5	16.0	11.0	11.0	10.0	—	—	—	—
		5/32 and over	20.0	14.0	14.0	12.5	20.0	14.0	14.0	12.5	—	—	—	—
	36	1/16 to 3/32	5.0	3.5	3.5	3.0	5.5	4.0	4.0	3.5	* 18.0	12.5	12.5	11.0
		3/32 to 5/32	10.0	7.0	7.0	6.5	18.0	12.5	12.5	11.0	—	—	—	—
		5/32 to 7/32	14.0	10.0	10.0	9.0	* 20.0	14.0	14.0	12.5	—	—	—	—
		7/32 and over	20.0	14.0	14.0	12.5	—	—	—	—	—	—	—	—
0.190	24	1/16 to 3/32	12.0	8.0	8.0	7.0	12.0	8.0	8.0	7.0	* 18.0	12.5	12.5	11.0
		3/32 and over	20.0	14.0	14.0	12.5	20.0	14.0	14.0	12.5	—	—	—	—
	32	1/16 to 3/32	5.5	4.0	4.0	3.5	5.5	4.0	4.0	3.5	* 25.0	17.5	17.5	16.0
		3/32 to 5/32	13.0	9.0	9.0	8.0	* 25.0	17.5	17.5	16.0	—	—	—	—
		5/32 to 7/32	18.0	12.5	12.5	11.0	—	—	—	—	—	—	—	—
		7/32 and over	25.0	17.5	17.5	16.0	—	—	—	—	—	—	—	—
0.216	24	1/16 to 3/32	12.0	8.0	8.0	7.0	18.0	12.5	12.5	11.0	* 25.0	17.5	17.5	16.0
		3/32 to 5/32	25.0	17.5	17.5	16.0	* 30.0	21.0	21.0	19.0	—	—	—	—
		5/32 to 7/32	25.0	17.5	17.5	16.0	—	—	—	—	—	—	—	—
		7/32 and over	30.0	21.0	21.0	19.0	—	—	—	—	—	—	—	—
	28	1/16 to 3/32	9.0	6.0	6.0	5.5	16.0	11.0	11.0	10.0	* 30.0	21.0	21.0	19.0
		3/32 to 5/32	18.0	12.5	12.5	11.0	* 30.0	21.0	21.0	19.0	—	—	—	—
		5/32 to 7/32	20.0	14.0	14.0	12.5	—	—	—	—	—	—	—	—
		7/32 to 9/32	25.0	17.5	17.5	16.0	—	—	—	—	—	—	—	—
		9/32 and over	30.0	21.0	21.0	19.0	—	—	—	—	—	—	—	—

* Entries in all four columns are for the thread engagement given at left plus "and over"

Torque for Machine Screws—IV

Steel Screw, Finished or Unfinished			Tapped Hole in Finished or Unfinished Steel*				Tapped Hole in Finished or Unfinished Brass				Tapped Hole in Half-Hard Copper			
			Driving and Testing Torque, lb-in.				Driving and Testing Torque, lb-in.				Driving and Testing Torque. lb-in.			
Size	Threads Per Inch	Thread Engagement, in.	Max	Min	Insp	Final Check	Max	Min	Insp	Final Check	Max	Min	Insp	Final Check
0.060	80	1/32 to 3/64	2.5	1.75	1.75	1.5	1.0	0.75	0.75	0.688	0.375	0.266	0.266	0.234
		3/64 and over	3.0	2.0	2.0	1.75	3.0	2.0	2.0	1.75	3.0	2.0	2.0	1.75
0.073	64	1/32 to 3/64	3.0	2.0	2.0	1.75	1.5	1.063	1.063	1.0	2.0	1.5	1.5	1.25
		3/64 and over	3.5	2.5	2.5	2.0	3.5	2.5	2.5	2.0	3.5	2.5	2.5	2.0
	72	1/32 to 3/64	3.0	2.0	2.0	1.75	2.0	1.5	1.5	1.25	1.25	0.875	0.875	0.813
		3/64 and over	5.0	3.5	3.5	3.0	5.0	3.5	3.5	3.0	5.0	3.5	3.5	3.0
0.086	56	1/32 to 3/64	3.5	2.5	2.5	2.0	3.0	2.0	2.0	1.75	1.75	1.25	1.25	1.125
		3/64 and over	4.0	3.0	3.0	2.5	4.0	3.0	3.0	2.5	4.0	3.0	3.0	2.5
	64	1/32 to 3/64	4.0	3.0	3.0	2.5	3.0	2.0	2.0	1.75	1.75	1.25	1.25	1.125
		3/64 and over	5.5	4.0	4.0	3.5	5.5	4.0	4.0	3.5	5.5	4.0	4.0	3.5
0.112	40	1/32 to 3/64	5.5	4.0	4.0	3.5	4.0	3.0	3.0	2.5	2.5	1.75	1.75	1.5
		3/64 and over	7.5	5.0	5.0	4.5	7.5	5.0	5.0	4.5	7.5	5.0	5.0	4.5
	48	1/32 to 3/64	3.0	2.0	2.0	1.75	3.5	2.5	2.5	2.0	2.5	1.75	1.75	1.5
		3/64 and over	8.0	5.5	5.5	4.5	7.5	5.0	5.0	4.5	5.5	4.0	4.0	3.5
		5/64 and over	—	—	—	—	8.0	5.5	5.5	4.5	8.0	5.5	5.5	4.5
0.125	40	1/32 to 3/64	8.0	5.5	5.5	4.5	8.0	5.5	5.5	4.5	2.5	1.75	1.75	1.5
		3/64 to 5/64	10.5	7.5	7.5	6.5	11.0	7.5	7.5	6.5	8.0	5.5	5.5	4.5
		5/64 and over	11.0	7.5	7.5	6.5	—	—	—	—	11.0	7.5	7.5	6.5
	44	1/32 to 3/64	5.0	3.5	3.5	3.0	7.5	5.0	5.0	4.5	3.0	2.0	2.0	1.75
		3/64 to 5/64	12.0	8.0	8.0	7.0	15.0	10.5	10.5	9.5	6.5	4.5	4.5	4.0
		5/64 to 7/64	15.0	10.5	10.5	9.5	—	—	—	—	13.0	9.0	9.0	8.0
		7/64 and over	—	—	—	—	—	—	—	—	15.0	10.5	10.5	9.5
0.138	32	1/32 to 3/64	8.0	5.5	5.5	4.5	8.0	5.5	5.5	4.5	5.0	3.5	3.5	3.0
		3/64 and over	12.0	8.0	8.0	7.0	12.0	8.0	8.0	7.0	12.0	8.0	8.0	7.0
	40	1/32 to 3/64	6.0	4.0	4.0	3.5	6.5	4.5	4.5	4.0	5.0	3.5	3.5	3.0
		3/64 to 5/64	10.5	7.5	7.5	6.5	7.0	5.0	5.0	4.5	12.0	8.0	8.0	7.0
		5/64 and over	12.0	8.0	8.0	7.0	12.0	8.0	8.0	7.0	—	—	—	—
0.164	32	1/32 to 3/64	9.0	6.0	6.0	5.5	8.5	6.0	6.0	5.5	—	—	—	—
		3/64 to 5/64	15.0	10.5	10.5	9.5	20.0	14.0	14.0	12.5	14.0	10.0	10.0	9.0
		5/64 and over	20.0	14.0	14.0	12.5	—	—	—	—	20.0	14.0	14.0	12.5
	36	1/32 to 3/64	15.0	10.5	10.5	9.5	14.0	10.0	10.0	9.0	—	—	—	—
		3/64 and over	20.0	14.0	14.0	12.5	16.0	11.0	11.0	10.0	16.0	11.0	11.0	10.0
0.190	24	1/32 to 3/64	13.5	9.5	9.5	8.5	11.0	7.5	7.5	6.5	—	—	—	—
		3/64 and over	20.0	14.0	14.0	12.5	20.0	14.0	14.0	12.5	15.0	10.5	10.5	9.5
	32	1/32 to 3/64	18.0	12.5	12.5	11.0	15.0	10.5	10.5	9.5	—	—	—	—
		3/64 and over	25.0	17.5	17.5	16.0	25.0	17.5	17.5	16.0	20.0	14.0	14.0	12.5
0.216	24	1/32 to 3/64	12.0	8.0	8.0	7.0	14.0	10.0	10.0	9.0	—	—	—	—
		3/64 to 5/64	20.0	14.0	14.0	12.5	30.0	21.0	21.0	19.0	—	—	—	—
		5/64 to 7/64	25.0	17.5	17.5	16.0	—	—	—	—	20.0	14.0	14.0	12.5
		7/64 and over	30.0	21.0	21.0	19.0	—	—	—	—	30.0	21.0	21.0	19.0
	28	1/32 to 3/64	12.0	8.0	8.0	7.0	12.0	8.0	8.0	7.0	—	—	—	—
		3/64 to 5/64	25.0	17.5	17.5	16.0	25.0	17.5	17.5	16.0	20.0	14.0	14.0	12.5
		5/64 and over	30.0	21.0	21.0	19.0	30.0	21.0	21.0	19.0	30.0	21.0	21.0	19.0

*For high tensile steel screws in high tensile material add approximately 30 per cent to the above values.

Chapter 24

Mathematics and Tables

COMPOUND ANGLES—I

By D E Sweet

Attempts to derive formulas and tables for accurate computation of compound angles can be discouraging. Each case is usually presented in a manner such that it is often easier to consider each problem separately. There is one case, however, that can be standardized to the extent that formulas and tables can be a big help. An example is shown in Fig. 1. Here an angle is to be relieved down and away: it could be found on a cutting tool (for example, a counterbore blade).

Because the "down" and "away" angles are usually arbitrarily selected, the tables and charts given will aid the computation for the true angle. For convenience in using the table and chart, the "down" angle is designated as the "tipping angle" and the "away" angle as the "turning angle." The case angle A is tipped to angle ϕ and then turned to angle α; the resultant angle is the true angle or T.

In the table, constants are given for combinations of "tip" and "turn" angles from 5° to 45°. There are two constants for each combination, namely, constant 1 and constant 2.

To determine the true angle (using table)

Rule: multiply constant 1 by the tangent of the case angle A and *add* constant 2. The result of this addition is the tangent of the true angle.

If it is necessary to compute constants for tip-turn angles not given in the table, they can be found in these formulas:

Constant 1 $= \cos \alpha \div \cos \phi$
Constant 2 $= \sin \alpha \tan \phi$

Example: Case angle $A = 29°$, tip and turn angles $= 5°$. What is the true angle ***T*?**

Tan 29° $= 0.55431$; constant 1 $=$ 1.0000; constant 2 $= 0.00762$. Then $\tan T = 1 \times 0.55431 + 0.00762 = 0.56193$.

$T = 29° 20'$.

CONSTANT TABLE

TIP ANGLES ϕ	CONSTANT	TURN ANGLES α 5°	10°	15°	20°	25°	30°	35°	40°	45°
5°	1	1.0000	0.9886	0.9696	0.9433	0.9098	0.8693	0.8223	0.7690	0.7098
	2	.00762	.01519	.02264	.02992	.03697	.04374	.05018	.05624	.06186
10°	1	1.0115	1.0000	0.9808	0.9542	0.9203	0.8794	0.8318	0.7779	0.7180
	2	.01537	.03062	.04563	.06031	.07452	.08816	.10114	.11334	.12468
15°	1	1.0313	1.0195	1.0000	0.9728	0.9383	0.8966	0.8480	0.7931	0.7320
	2	.02335	.04653	.06935	.09164	.11324	.13397	.15369	.17223	.18947
20°	1	1.0601	1.0480	1.0279	1.0000	0.9645	0.9216	0.8717	0.8152	0.7525
	2	.03172	.06320	.09420	.12448	.15382	.18198	.20876	.23395	.25736
25°	1	1.0992	1.0866	1.0658	1.0368	1.0000	0.9555	0.9038	0.8452	0.7802
	2	.04064	.08097	.12069	.15948	.19707	.23315	.26746	.29974	.32973
30°	1	1.1503	1.1372	1.1153	1.0851	1.0465	1.0000	0.9459	0.8845	0.8165
	2	.05032	.10025	.14943	.19746	.24400	.28867	.33115	.37111	.40825
35°	1	1.2161	1.2022	1.1792	1.14715	1.1064	1.0572	1.0000	0.9352	0.8632
	2	.06103	.12159	.18123	.23948	.29592	.35010	.40162	.45008	.49512
40°	1	1.3004	1.2856	1.2609	1.2267	1.1831	1.1305	1.0693	1.0000	0.9230
	2	.07313	.14571	.21717	.28699	.35462	.41955	.48129	.53936	.59333
45°	1	1.4088	1.3927	1.3660	1.3289	1.2817	1.2247	1.1584	1.0833	1.0000
	2	.08749	.17365	.25882	.34202	.42262	.5000	.57358	.64279	.70711

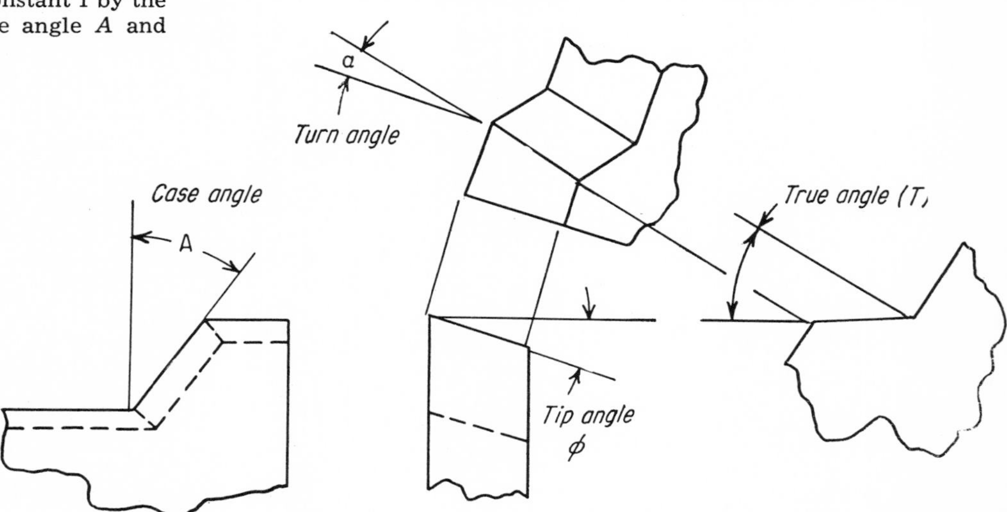

COMPOUND ANGLES—II

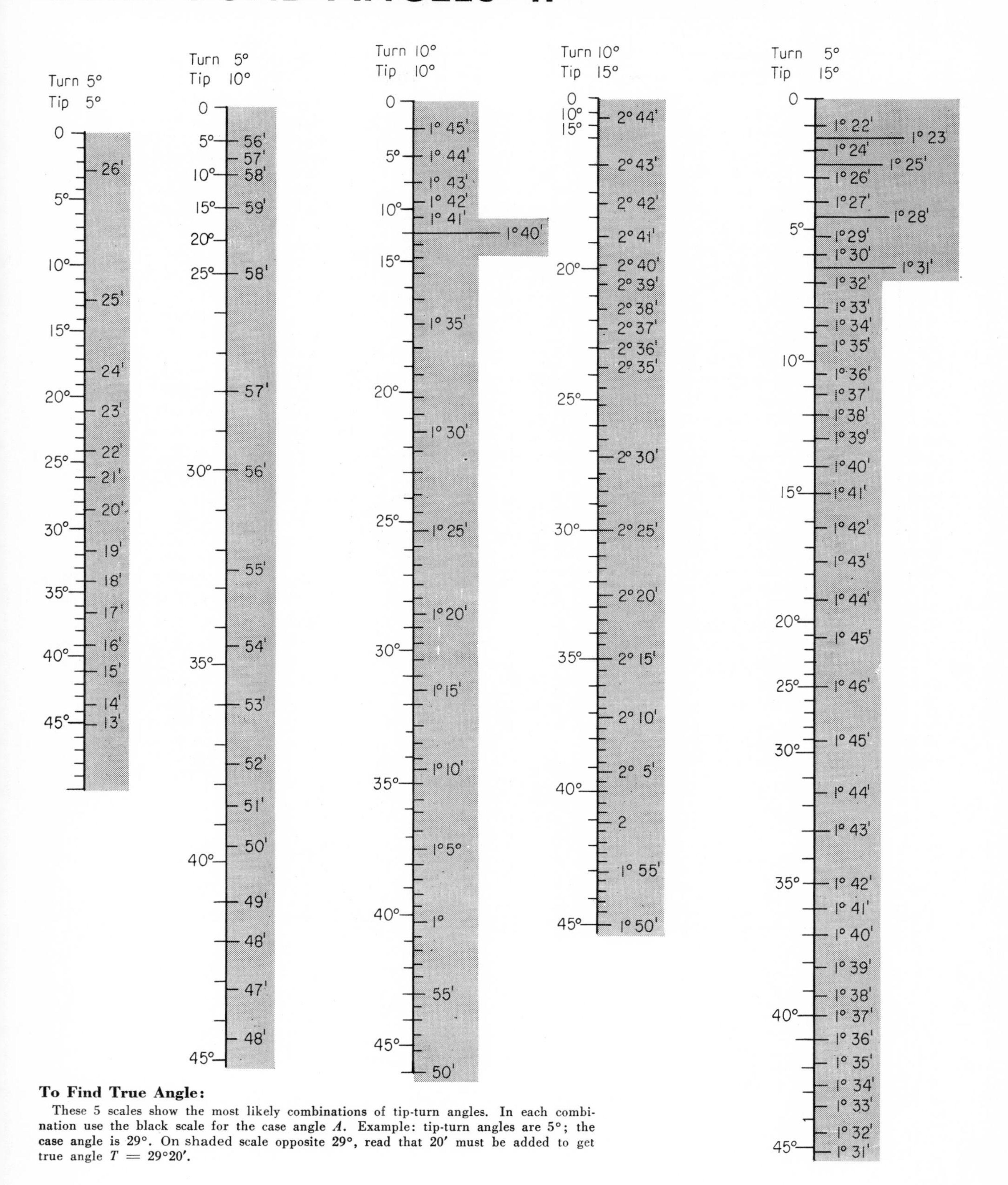

To Find True Angle:

These 5 scales show the most likely combinations of tip-turn angles. In each combination use the black scale for the case angle A. Example: tip-turn angles are 5°; the case angle is 29°. On shaded scale opposite 29°, read that 20′ must be added to get true angle $T = 29°20′$.

COMPOUND ANGLES—III

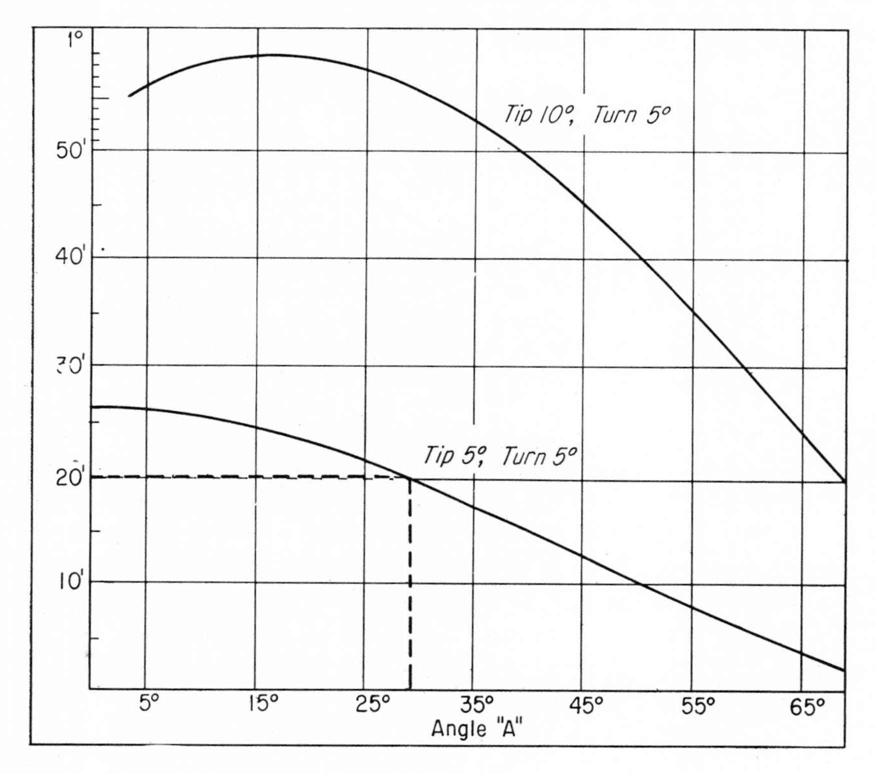

Graphical Solution of True Angle This chart is laid out with two curves for tip-turn angle combinations. Example: case angle $A = 29°$, tip and turn angles are 5°. Follow construction lines to see that 20′ must be added to angle $A = 29°$, to get true angle $T = 29°20'$.

Proof

Given θ, α and ϕ

To Find T

Solution:

$AC = 1 = AB'$

$CB = \tan\theta = B'C$

$B'C' = \sin\phi = CD$

$B'C'' = \cos\alpha\,(B'C + CE)$

$CE = \tan\alpha\; CD$

$$B'C'' = (\tan\theta + \tan\alpha \sin\phi) \times \cos\alpha$$

$$= \tan\theta\cos\alpha + \frac{\sin\alpha}{\cos\alpha} \times \cos\alpha\sin\phi$$

$$= \tan\theta\cos\alpha + \sin\alpha\sin\phi$$

$AC' = \cos\phi = A'C''$

$$\tan T = \frac{B'C''}{A'C''}$$

$$= \frac{\tan\theta\cos\alpha + \sin\alpha\sin\phi}{\cos\phi}$$

$$= \tan\theta\left(\frac{\cos\alpha}{\cos\phi}\right) + \sin\alpha\tan\phi$$

$$\text{Constant 1} = \frac{\cos\alpha}{\cos\phi}$$

$$\text{Constant 2} = \sin\alpha\tan\phi$$

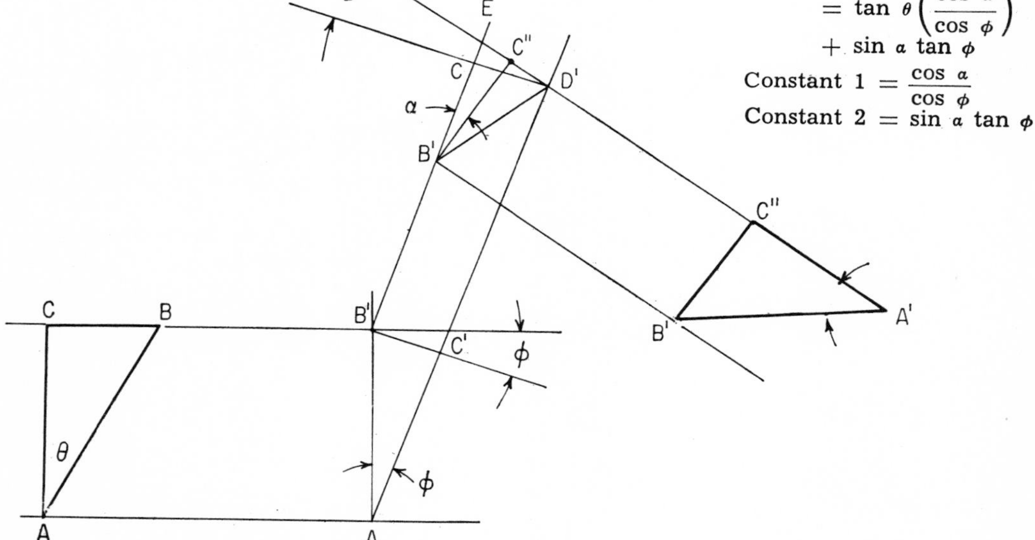

Frustums of Cones—I

By Ronald L Wakelee, Chicago, Illinois

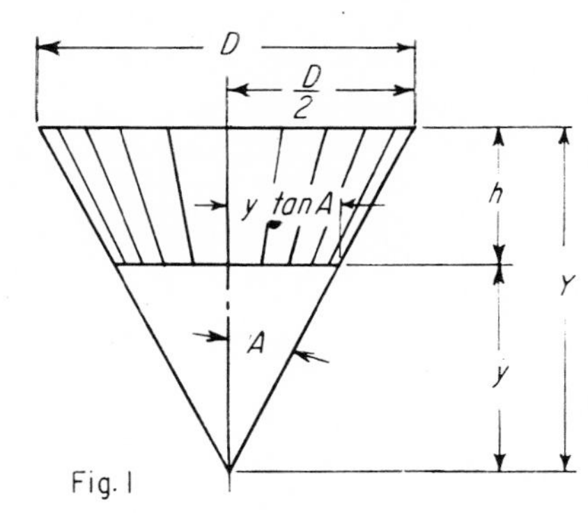

Fig. 1

Frustrum of Right circular cone

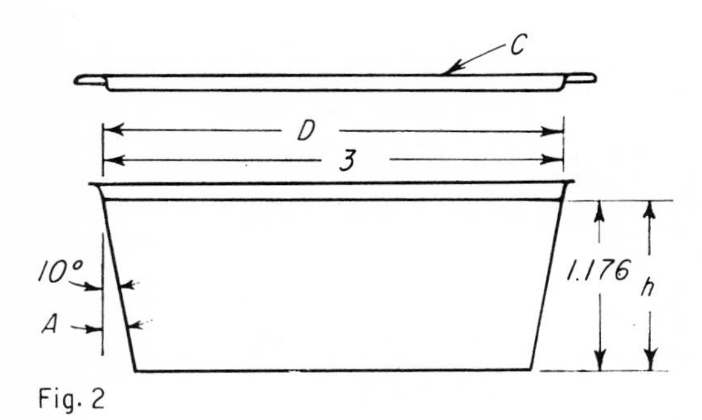

Fig. 2

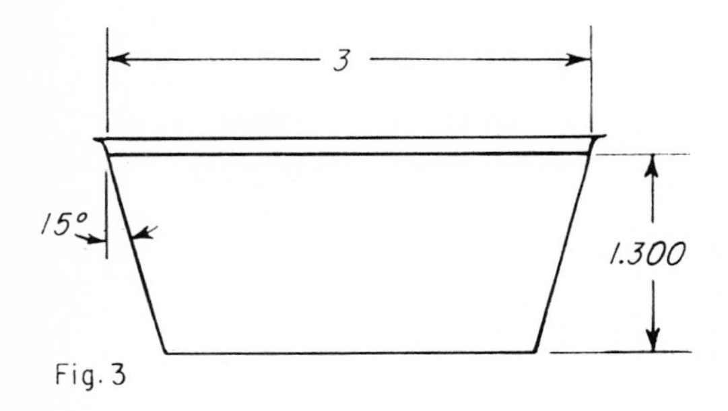

Fig. 3

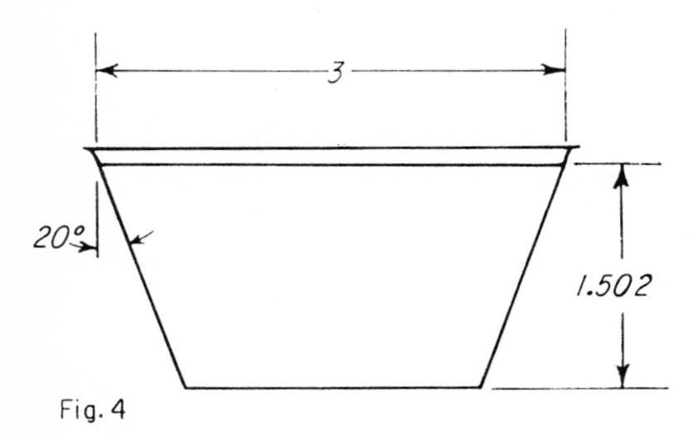

Fig. 4

Efficient solution of problems relating to frustums of cones may be hampered for lack of quantity A, Fig. 1. This quantity does not appear in the conventional formula for the volume of a frustrum:

$$V = 0.2618\, h\, (D^2 + Dd + d^2)$$

Quantity A, Fig. 1, was needed for solution of a problem involving thin-wall containers of frustrum shape. The customer wished to pack a standard volume in a container with a pleasing profile or side wall angularity.

Selection of can profile starts with certain known factors: Volume V to be packed, open-end diameter D, and an approximate ratio of $D/h = 2\frac{1}{2}$, or $h = 0.4\, d$ as suggested by customer. If possible a standard packer's can end is selected. In this case D is taken as 3. The problem is to find the exact value of h.

First step is to draw up tentative can profiles, Figs. 2, 3, 4, of equal volume, which must be 7.218 cu in. Note that side angularity A is 10°, 15°, and 20° for the three profiles. The value of h in each case is calculated from the formula 7, page 312:

$$h = \frac{D - \sqrt[3]{D^3 - 7.64\, V \tan A}}{2 \tan A}$$

The customer-selected profile is shown in Fig. 2. Here $D = 3$, $A = 10°$ and $V = 7.218$ cu in. By substituting these values in equation 7, $h = 1.176$.

Bottom diameter d of the container is easily found from the formula:

$$\begin{aligned} d &= D - 2 \tan A \times h \\ &= 3 - (2 \tan 10° \times 1.176) \\ &= 3 - (2 \times 0.17633 \times 1.176) \\ &= 2.585 \end{aligned}$$

Frustums of Cones—II

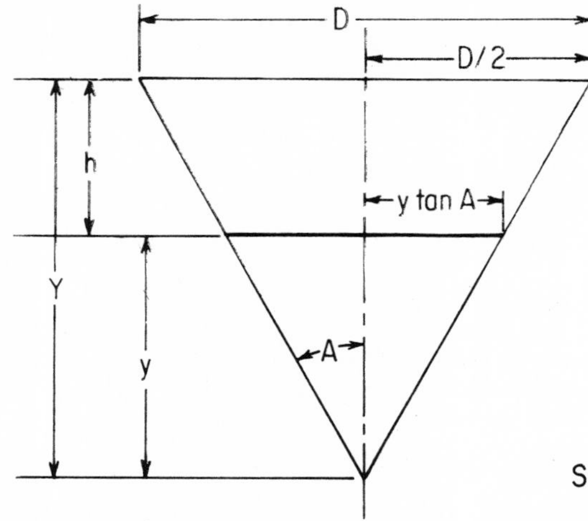

The problem: to find h, as for a container, when the volume V is known and the side-wall angle A is known

Solution:

$$\text{Volume of cone} = \frac{\pi}{3}\left(\frac{D}{2}\right)^2 Y = \frac{\pi D^2 Y}{12} \quad \text{.....①}$$

$$\text{Volume of conical segment} = \frac{\pi (y \tan A)^2 y}{3} = \frac{\pi y^3 \tan^2 A}{3} \quad \text{.....②}$$

$$\text{Volume of frustum} \quad \text{eq. 1 minus eq. 2} = V = \frac{\pi D^2 Y}{12} - \frac{\pi y^3 \tan^2 A}{3} \quad \text{.....③}$$

$$Y = \frac{D}{2 \tan A} \quad \text{.....④}$$

$$\text{Substituting ④ in ③} \quad V = \frac{\pi D^2 D}{24 \tan A} - \frac{\pi y^3 \tan^2 A}{3}$$

$$\text{Transposing terms,} \quad \frac{\pi y^3 \tan^2 A}{3} = \frac{\pi D^3}{24 \tan A} - V$$

$$\text{or,} \quad \pi y^3 \tan^2 A = \frac{\pi D^3}{8 \tan A} - 3V$$

$$\text{Solving for } y^3, \quad y^3 = \frac{D^3}{8 \tan^3 A} - \frac{3V}{\pi \tan^2 A}$$

$$\text{or,} \quad y^3 = \frac{D^3 - \frac{24 V \tan A}{\pi}}{8 \tan^3 A} = \frac{D^3 - 7.64 V \tan A}{8 \tan^3 A}$$

$$\text{Solving for } y, \quad y = \sqrt[3]{\frac{D^3 - 7.64 V \tan A}{8 \tan^3 A}}$$

$$\text{or,} \quad y = \frac{\sqrt[3]{D^3 - 7.64 V \tan A}}{2 \tan A} \quad \text{.....⑤}$$

$$h = Y - y \quad \text{.....⑥}$$

$$\text{Substituting ⑤ in ⑥} \quad h = \frac{D}{2 \tan A} - \frac{\sqrt[3]{D^3 - 7.64 V \tan A}}{2 \tan A}$$

$$\text{or,} \quad h = \frac{D - \sqrt[3]{D^3 - 7.64 V \tan A}}{2 \tan A} \quad \text{.....⑦}$$

Example, Fig. 1: D = 3", A = 10° and V = 7.218 cu in.

$$\text{Substituting values in ⑦} \quad h = \frac{3 - \sqrt[3]{3^3 - 7.64 \times 7.218 \times 0.17633}}{2 \times 0.17633} = 1.176''$$

Radius of a Segment

By CARL A JOHNSON

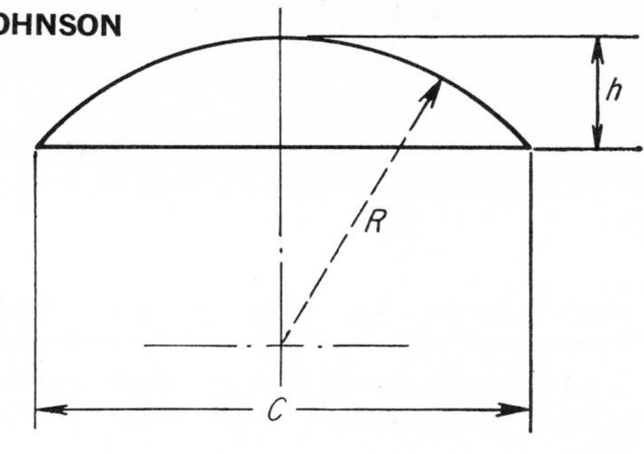

Measure height *h* and width *C* to find ratio *h*/*C* or height/width.

Then find quotient *h*/*C* in column 1 and opposite in column 2 find the radius *R* for segment of chord width *C* = 1.

Multiply the constant in column 2 by the actual width of segment to find the radius *R* of that segment.

Example: If the height *h* of the chord is 0.750 and width *C* is 3.0000, *h*/*C* = 0.250.

In table the constant for *h*/*C* = 0.250 is 0.62500. Multiply chord width 3 by constant 0.62500 and find radius is 1.875.

h/C	CONSTANT	h/C	CONSTANT	h/C	CONSTANT	h/C	CONSTANT	h/C	CONSTANT	h/C	CONSTANT	h/C	CONSTANT	h/C	CONSTANT	h/C	CONSTANT	h/C	CONSTANT
.010	12.50500	.060	2.11333	.110	1.19136	.160	.86125	.210	.70024	.260	.61077	.310	.55823	.360	.52722	.410	.50988	.460	.50174
1	11.36914	1	2.07968	1	1.18163	1	.85690	1	.69792	1	.60943	1	.55743	1	.52676	1	.50964	1	.50165
2	10.42267	2	2.04713	2	1.17207	2	.85260	2	.69562	2	.60810	2	.55664	2	.52630	2	.50940	2	.50156
3	9.62188	3	2.01563	3	1.16269	3	.84837	3	.69335	3	.60679	3	.55586	3	.52585	3	.50916	3	.50148
4	8.93557	4	1.98513	4	1.15349	4	.84420	4	.69111	4	.60548	4	.55509	4	.52541	4	.50893	4	.50140
5	8.34083	5	1.95558	5	1.14446	5	.84008	5	.68890	5	.60420	5	.55433	5	.52497	5	.50870	5	.50132
6	7.82050	6	1.92694	6	1.13559	6	.83601	6	.68670	6	.60292	6	.55357	6	.52453	6	.50848	6	.50124
7	7.36144	7	1.89917	7	1.12688	7	.83200	7	.68454	7	.60166	7	.55282	7	.52410	7	.50826	7	.50117
8	6.95344	8	1.87224	8	1.11832	8	.82805	8	.68239	8	.60042	8	.55208	8	.52367	8	.50804	8	.50109
9	6.58845	9	1.84609	9	1.10992	9	.82415	9	.68028	9	.59918	9	.55135	9	.52325	9	.50783	9	.50102
.020	6.26000	.070	1.82071	.120	1.10167	.170	.82029	.220	.67818	.270	.59796	.320	.55062	.370	.52284	.420	.50762	.470	.50096
1	5.96288	1	1.79606	1	1.09356	1	.81649	1	.67611	1	.59675	1	.54991	1	.52243	1	.50741	1	.50089
2	5.69282	2	1.77211	2	1.08559	2	.81274	2	.67406	2	.59556	2	.54920	2	.52202	2	.50721	2	.50083
3	5.44628	3	1.74883	3	1.07776	3	.80904	3	.67204	3	.59438	3	.54850	3	.52162	3	.50701	3	.50077
4	5.22033	4	1.72619	4	1.07006	4	.80539	4	.67004	4	.59320	4	.54780	4	.52122	4	.50681	4	.50071
5	5.01250	5	1.70417	5	1.06250	5	.80179	5	.66806	5	.59205	5	.54712	5	.52083	5	.50662	5	.50066
6	4.82069	6	1.68274	6	1.05506	6	.79823	6	.66610	6	.59090	6	.54644	6	.52045	6	.50643	6	.50061
7	4.64313	7	1.66188	7	1.04775	7	.79471	7	.66416	7	.58976	7	.54576	7	.52006	7	.50624	7	.50055
8	4.47829	8	1.64156	8	1.04056	8	.79125	8	.66225	8	.58864	8	.54510	8	.51969	8	.50606	8	.50051
9	4.32484	9	1.62178	9	1.03349	9	.78782	9	.66035	9	.58753	9	.54444	9	.51932	9	.50588	9	.50046
.030	4.18167	.080	1.60250	.130	1.02654	.180	.78444	.230	.65848	.280	.58643	.330	.54379	.380	.51895	.430	.50570	.480	.50042
1	4.04776	1	1.58371	1	1.01970	1	.78111	1	.65663	1	.58534	1	.54314	1	.51858	1	.50552	1	.50038
2	3.92225	2	1.56539	2	1.01297	2	.77781	2	.65479	2	.58426	2	.54251	2	.51823	2	.50535	2	.50034
3	3.80438	3	1.54752	3	1.00635	3	.77456	3	.65298	3	.58320	3	.54188	3	.51787	3	.50518	3	.50030
4	3.69347	4	1.53010	4	.99984	4	.77135	4	.65119	4	.58214	4	.54125	4	.51752	4	.50502	4	.50026
5	3.58893	5	1.51309	5	.99343	5	.76818	5	.64941	5	.58110	5	.54063	5	.51718	5	.50486	5	.50023
6	3.49022	6	1.49649	6	.98712	6	.76504	6	.64766	6	.58006	6	.54002	6	.51683	6	.50470	6	.50020
7	3.39688	7	1.48028	7	.98091	7	.76195	7	.64593	7	.57904	7	.53942	7	.51650	7	.50454	7	.50017
8	3.30847	8	1.46445	8	.97480	8	.75889	8	.64421	8	.57803	8	.53882	8	.51616	8	.50439	8	.50015
9	3.22463	9	1.44899	9	.96878	9	.75588	9	.64251	9	.57703	9	.53823	9	.51584	9	.50424	9	.50012
.040	3.14500	.090	1.43389	.140	.96286	.190	.75289	.240	.64083	.290	.57603	.340	.53765	.390	.51551	.440	.50409	.490	.50010
1	3.06928	1	1.41913	1	.95702	1	.74995	1	.63917	1	.57505	1	.53707	1	.51519	1	.50395	1	.50008
2	2.99719	2	1.40470	2	.95128	2	.74704	2	.63753	2	.57408	2	.53650	2	.51488	2	.50381	2	.500065
3	2.92848	3	1.39059	3	.94563	3	.74417	3	.63590	3	.57312	3	.53593	3	.51457	3	.50367	3	.500050
4	2.86291	4	1.37679	4	.94006	4	.74133	4	.63430	4	.57217	4	.53537	4	.51426	4	.50353	4	.500036
5	2.80028	5	1.36329	5	.93457	5	.73853	5	.63270	5	.57123	5	.53482	5	.51396	5	.50340	5	.500025
6	2.74039	6	1.35008	6	.92916	6	.73576	6	.63113	6	.57030	6	.53427	6	.51366	6	.50327	6	.500016
7	2.68307	7	1.33716	7	.92384	7	.73302	7	.62957	7	.56938	7	.53373	7	.51336	7	.50314	7	.500009
8	2.62817	8	1.32451	8	.91859	8	.73031	8	.62803	8	.56846	8	.53320	8	.51307	8	.50302	8	.500004
9	2.57552	9	1.31213	9	.91343	9	.72764	9	.62651	9	.56756	9	.53267	9	.51278	9	.50290	9	.500001
.050	2.52500	.100	1.30000	.150	.90833	.200	.72500	.250	.62500	.300	.56667	.350	.53214	.400	.51250	.450	.50278	.500	.500000
1	2.47648	1	1.28812	1	.90331	1	.72239	1	.62351	1	.56578	1	.53163	1	.51222	1	.50266		
2	2.42985	2	1.27649	2	.89837	2	.71981	2	.62203	2	.56491	2	.53111	2	.51195	2	.50255		
3	2.38499	3	1.26509	3	.89349	3	.71726	3	.62057	3	.56404	3	.53061	3	.51167	3	.50244		
4	2.34181	4	1.25392	4	.88869	4	.71474	4	.61913	4	.56318	4	.53011	4	.51141	4	.50233		
5	2.30023	5	1.24298	5	.88395	5	.71225	5	.61770	5	.56234	5	.52961	5	.51114	5	.50223		
6	2.26014	6	1.23225	6	.87928	6	.70979	6	.61628	6	.56150	6	.52912	6	.51088	6	.50212		
7	2.22148	7	1.22172	7	.87468	7	.70736	7	.61488	7	.56067	7	.52864	7	.51063	7	.50202		
8	2.18417	8	1.21141	8	.87014	8	.70496	8	.61350	8	.55984	8	.52816	8	.51037	8	.50193		
9	2.14814	9	1.20129	9	.86566	9	.70259	9	.61213	9	.55903	9	.52769	9	.51012	9	.50183		

Conversion Table for Ten-Thousandths to MM.

COURTESY NATIONAL JET COMPANY

This table is useful in micro-drilling and instrument work.

INCH	MM	INCH	MM	INCH	MM	INCH	MM	INCH	MM	INCH	MM	INCH	MM	INCH	MM
.0010	.02540	.0060	.15240	.0110	.27940	.0160	.40640	.0210	.53340	.0260	.66040	.0310	.78740	.0360	.91440
.0011	.02794	.0061	.15494	.0111	.28194	.0161	.40894	.0211	.53594	.0261	.66294	.0311	.78994	.0361	.91694
.0012	.03048	.0062	.15748	.0112	.28448	.0162	.41148	.0212	.53848	.0262	.66548	.0312	.79248	.0362	.91948
.0013	.03302	.0063	.16002	.0113	.28702	.0163	.41402	.0213	.54102	.0263	.66802	.0313	.79502	.0363	.92202
.0014	.03556	.0064	.16256	.0114	.28956	.0164	.41656	.0214	.54356	.0264	.67056	.0314	.79756	.0364	.92456
.0015	.03810	.0065	.16510	.0115	.29210	.0165	.41910	.0215	.54610	.0265	.67310	.0315	.80010	.0365	.92710
.0016	.04064	.0066	.16764	.0116	.29464	.0166	.42164	.0216	.54864	.0266	.67564	.0316	.80264	.0366	.92964
.0017	.04318	.0067	.17018	.0117	.29718	.0167	.42418	.0217	.55118	.0267	.67818	.0317	.80518	.0367	.93218
.0018	.04572	.0068	.17272	.0118	.29972	.0168	.42672	.0218	.55372	.0268	.68072	.0318	.80722	.0368	.93472
.0019	.04826	.0069	.17526	.0119	.30226	.0169	.42926	.0219	.55626	.0269	.68326	.0319	.81026	.0369	.93726
.0020	.05080	.0070	.17780	.0120	.30480	.0170	.43180	.0220	.55880	.0270	.68580	.0320	.81280	.0370	.93980
.0021	.05334	.0071	.18034	.0121	.30734	.0171	.43434	.0221	.56134	.0271	.68834	.0321	.81534	.0371	.94234
.0022	.05588	.0072	.18288	.0122	.30988	.0172	.43688	.0222	.56388	.0272	.69088	.0322	.81788	.0372	.94488
.0023	.05842	.0073	.18542	.0123	.31242	.0173	.43942	.0223	.56642	.0273	.69342	.0323	.82042	.0373	.94742
.0024	.06096	.0074	.18796	.0124	.31496	.0174	.44196	.0224	.56896	.0274	.69596	.0324	.82296	.0374	.94996
.0025	.06350	.0075	.19050	.0125	.31750	.0175	.44450	.0225	.57150	.0275	.69850	.0325	.82550	.0375	.95250
.0026	.06604	.0076	.19304	.0126	.32004	.0176	.44704	.0226	.57404	.0276	.70104	.0326	.82804	.0376	.95504
.0027	.06858	.0077	.19558	.0127	.32258	.0177	.44958	.0227	.57658	.0277	.70358	.0327	.83058	.0377	.95758
.0028	.07112	.0078	.19812	.0128	.32512	.0178	.45212	.0228	.57912	.0278	.70612	.0328	.83312	.0378	.96012
.0029	.07366	.0079	.20066	.0129	.32766	.0179	.45466	.0229	.58166	.0279	.70866	.0329	.83566	.0379	.96266
.0030	.07620	.0080	.20320	.0130	.33020	.0180	.45720	.0230	.58420	.0280	.71120	.0330	.83820	.0380	.96520
.0031	.07874	.0081	.20574	.0131	.33274	.0181	.45974	.0231	.58674	.0281	.71374	.0331	.84074	.0381	.96774
.0032	.08128	.0082	.20828	.0132	.33528	.0182	.46228	.0232	.58928	.0282	.71628	.0332	.84328	.0382	.97028
.0033	.08282	.0083	.21082	.0133	.33782	.0183	.46482	.0233	.59182	.0283	.71882	.0333	.84582	.0383	.97282
.0034	.08636	.0084	.21336	.0134	.34036	.0184	.46736	.0234	.59436	.0284	.72136	.0334	.84836	.0384	.97536
.0035	.08890	.0085	.21590	.0135	.34290	.0185	.46990	.0235	.59690	.0285	.72390	.0335	.85090	.0385	.97790
.0036	.09144	.0086	.21844	.0136	.34544	.0186	.47244	.0236	.59944	.0286	.72644	.0336	.85344	.0386	.98044
.0037	.09398	.0087	.22098	.0137	.34798	.0187	.47498	.0237	.60198	.0287	.72898	.0337	.85598	.0387	.98298
.0038	.09652	.0088	.22352	.0138	.35052	.0188	.47752	.0238	.60452	.0288	.73152	.0338	.85852	.0388	.98552
.0039	.09906	.0089	.22606	.0139	.35306	.0189	.48006	.0239	.60706	.0289	.73406	.0339	.86106	.0389	.98806
.0040	.10160	.0090	.22860	.0140	.35560	.0190	.48260	.0240	.60960	.0290	.73660	.0340	.86360	.0390	.99060
.0041	.10414	.0091	.23114	.0141	.35814	.0191	.48514	.0241	.61214	.0291	.73914	.0341	.86614	.0391	.99314
.0042	.10668	.0092	.23368	.0142	.36068	.0192	.48768	.0242	.61468	.0292	.74168	.0342	.86868	.0392	.99568
.0043	.10922	.0093	.23622	.0143	.36322	.0193	.49022	.0243	.61722	.0293	.74422	.0343	.87122	.0393	.99822
.0044	.11176	.0094	.23876	.0144	.36576	.0194	.49276	.0244	.61976	.0294	.74676	.0344	.87376	.0394	1.00076
.0045	.11430	.0095	.24130	.0145	.36830	.0195	.49530	.0245	.62230	.0295	.74930	.0345	.87630	.0395	1.00330
.0046	.11684	.0096	.24384	.0146	.37084	.0196	.49784	.0246	.62484	.0296	.75184	.0346	.87884	.0396	1.00584
.0047	.11938	.0097	.24638	.0147	.37338	.0197	.50038	.0247	.62738	.0297	.75438	.0347	.88138	.0397	1.00838
.0048	.12192	.0098	.24892	.0148	.37592	.0198	.50292	.0248	.62992	.0298	.75692	.0348	.88392	.0398	1.01092
.0049	.12446	.0099	.25146	.0149	.37846	.0199	.50546	.0249	.63246	.0299	.75946	.0349	.88646	.0399	1.01346
.0050	.12700	.0100	.25400	.0150	.38100	.0200	.50800	.0250	.63500	.0300	.76200	.0350	.88900	.0400	1.01600
.0051	.12954	.0101	.25654	.0151	.38354	.0201	.51054	.0251	.63754	.0301	.76454	.0351	.89154	.0401	1.01854
.0052	.13208	.0102	.25908	.0152	.38608	.0202	.51308	.0252	.64008	.0302	.76708	.0352	.89408	.0402	1.02108
.0053	.13462	.0103	.26162	.0153	.38862	.0203	.51562	.0253	.64262	.0303	.76962	.0353	.89662	.0403	1.02362
.0054	.13716	.0104	.26416	.0154	.39116	.0204	.51816	.0254	.64516	.0304	.77216	.0354	.89916	.0404	1.02616
.0055	.13970	.0105	.26670	.0155	.39370	.0205	.52070	.0255	.64770	.0305	.77470	.0355	.90170	.0405	1.02870
.0056	.14224	.0106	.26924	.0156	.39624	.0206	.52324	.0256	.65024	.0306	.77724	.0356	.90424	.0406	1.03124
.0057	.14478	.0107	.27178	.0157	.39878	.0207	.52578	.0257	.65278	.0307	.77978	.0357	.90678	.0407	1.03378
.0058	.14732	.0108	.27432	.0158	.40132	.0208	.52832	.0258	.65532	.0308	.78232	.0358	.90932	.0408	1.03632
.0059	.14986	.0109	.27686	.0159	.40386	.0209	.53086	.0259	.65786	.0309	.78486	.0359	.91186	.0409	1.03886

Index